Mobile Hydraulic Technology

Bulletin 0274-B1

Technology in Motion

Preface

Mobile Hydraulic Technology and its associated material are a foundation for a course in mobile hydraulics, covering the widest range of mobile hydraulic applications found around the world.

Fluid power is evident in transportation as power and control systems of ships, planes and automobiles. The technology of fluid power is found hard at work improving the environment by compacting waste materials and controlling flood gates of huge dams. Construction equipment is another area of fluid power involvement. Applications for fluid power are only limited by the imagination.

This *Mobile Hydraulic Technology* textbook has been prepared by Parker Hannifin Corporation to provide the user of mobile hydraulic components and equipment a detailed reference.

This textbook has been arranged to begin with a discussion of the basic principles of hydraulics and the typical components found in a mobile hydraulic system. Each subsequent chapter covers individual components found in a mobile hydraulic system. The text materials were compiled through the combined efforts of Parker mobile engineers and field application specialists.

No attempt is made to show component selection criteria or detailed operating characteristics of a component of a system. This level of information is beyond the scope of this text. Numeric values have been rounded for ease of calculation; these values are not exact.

The color coding of the drawings in this text correspond to the following: RED - operating or system pressure; YELLOW - reduced flow or pressure; GREEN - intake; and BLUE - return line.

Exercises at the end of each lesson are designed to be a summary of the lesson high points and, at the same time, are intended to be interesting and self-checking in many instances.

Mobile Hydraulic Technology is intended to be presented within a 26 - 33 hour period. Depending upon the instructor, the size, makeup and experience of the course attendees this time frame will vary. To try and cover all of the aspects of mobile hydraulics in one text, intended for a one-semester course (26-33 hours) would be a near impossibility. In fact, such a detailed text would very quickly expand into a multi-volume set.

We would like to thank all the people who contributed to the writing and organization of the material contained in this textbook. We hope the student will find the textbook and information contained herein to be logical, easily understood, valuable and useful. Future editions of this textbook will include additional information and revisions suggested by those who use this text. Therefore, your comments and suggestions are cordially requested.

Please address your comments to:

Parker Hannifin Corporation
Motion & Control Training Department
MS: W3MC01
6035 Parkland Blvd.
Cleveland, OH 44124-4141
Phone: 216/896-2495
Fax: 216/514-6738
E-mail: mctrain@parker.com

Printed in the United States of America

Table of Contents

Chapter 1

Introduction to mobile hydraulics

Today, hydraulics is found almost everywhere and is very common in industrial, marine and military applications, as well as in mobile on- and off-road vehicles. Today, it would be almost impossible to think of an excavator (fig. 1-1) designed with only mechanical or all-electric components.

Fig. 1-1 Crawler excavator

Advantages of hydraulic power transmission

Hydraulic power transmissions are utilized because of features like:

- compact components - powerful, lightweight and with small installation dimensions
- flexible installation - components can be located in the best possible position
- power is transmitted easily through pipes and hoses
- transmission cooling and lubrication easy to arrange
- increased productivity - the operator can control several functions simultaneously with comfortable, remote control lever units and foot pedals
- precise movements - very accurate and smooth positioning
- overload protection easy to obtain - no component breakdown or engine stall

Some considerations

As with everything else in life, there are some considerations with hydraulics:

- noise
- external leakage - a small amount of mineral oil can ruin a very large amount of ground water; today, however, biodegradable fluids are becoming more and more common
- contamination - dirt in the hydraulic fluid causes wear and machine performance deterioration
- temperature - too high or too low temperatures are damaging, affecting both machine performance and component reliability
- air in the hydraulic fluid causes a softer (less stiff) transmission and pump damage (because of cavitation)

Hydraulic motor:
- Displacement: 5 cm^3/r *(0.30 cu in/r)*
- Continuous speed 8500 rpm
- Continuous power 13 kW *(17.5 hp)*
- Length 134 mm *(5.28 in)*
- Weight 5 kg *(11 lb)*

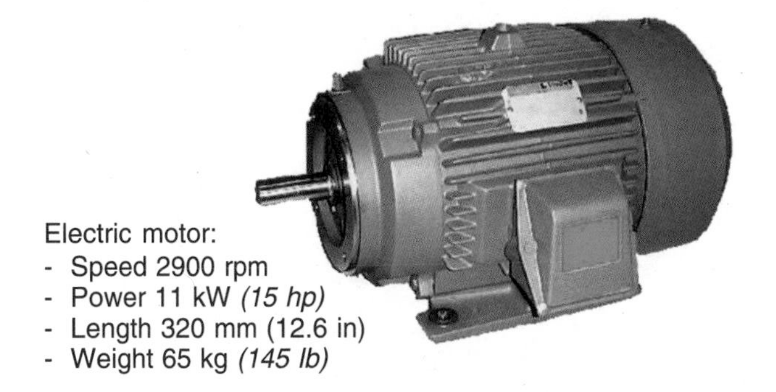

Electric motor:
- Speed 2900 rpm
- Power 11 kW *(15 hp)*
- Length 320 mm (12.6 in)
- Weight 65 kg *(145 lb)*

Fig. 1-2 Hydraulic versus electric motor of approximately the same power

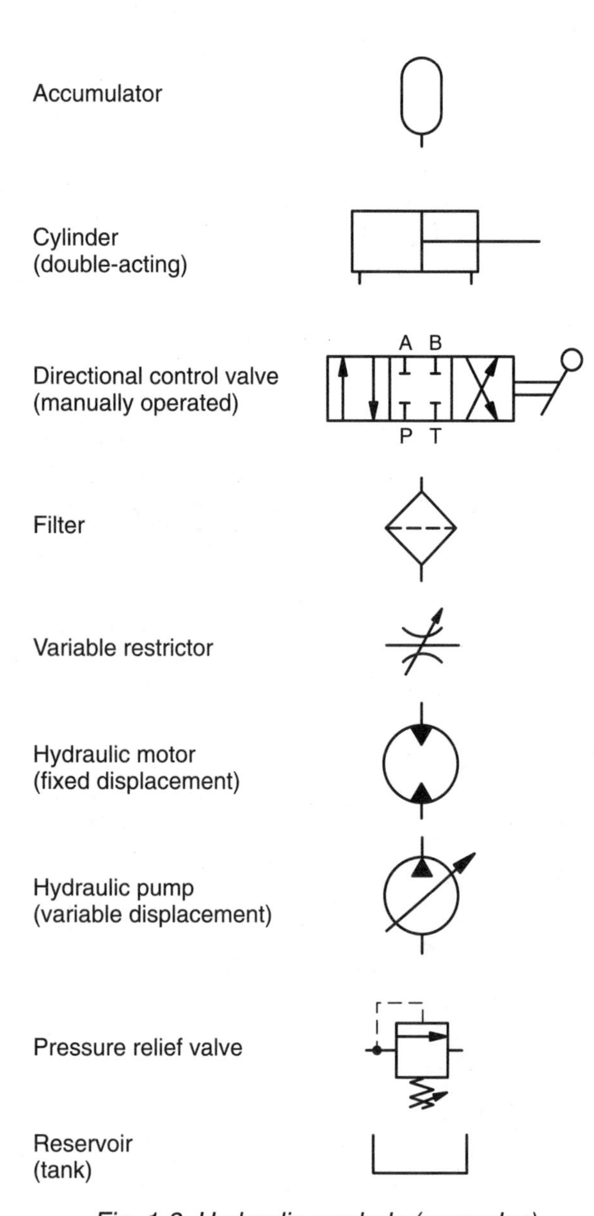

Fig. 1-3 Hydraulic symbols (examples)

In this chapter

Some basic components utilized to build hydraulic systems will be shown in 'Basic components that are used in hydraulic systems' (pages 1-3 through 1-5) including a picture and the corresponding hydraulic symbol.

Some mobile application examples will be presented at the end of this chapter (pages 1-6 through 1-12):

- drill rig
- excavator
- forwarder
- wheel loader
- reach stacker
- crane
- refuse collecting vehicle

In the following chapters

In order to understand hydraulics, some knowledge about 'basic hydraulic principles' is required; this subject is covered in chapter 2.

More detailed descriptions of pumps, accumulators, motors, cylinders, pressure/flow/directional control valves, and remote control systems are found in later chapters.

In addition, 'Common hydrostatic systems', servo valves and hydraulic fluids, as well as fluid conditioners (filters, reservoirs, coolers and heaters) and fluid conductors are covered in subsequent chapters.

Finally, at the end of the book, the following reference information is presented:

- hydraulic symbols (examples in fig. 1-3)
- common hydraulic formulas
- basic hydraulic circuit diagrams
- subject index

Basic components that are used in hydraulic systems

A simple hydraulic system

A hydraulic system can take many different shapes depending upon where it is found, i.e. in aerospace, aircraft, military, industrial or mobile applications. However, all systems operate generally on the same principles, and contain basically the same components.

Fig. 1-4 shows, as an example, a hydraulic system on a machine designed for the simple task of lifting a load. As the load is assumed to be substantial and the max lifting speed high, the operator of the machine is typically located away from the main hydraulic components, in a position to see what is going on.

A main directional valve is therefore chosen which is pilot operated by a small remote control valve.

The main advantage with this arrangement is that the machine designer can select the best possible position for the operator.

An additional benefit is the positioning of the load which can be extremely smooth with a remote control system; this is very important in delicate loading operations.

The following sections describe, briefly, the most common mobile system components; most are shown in fig. 1-4. Individual products as well as various circuits and hydraulic systems will be covered in more detail in succeeding chapters.

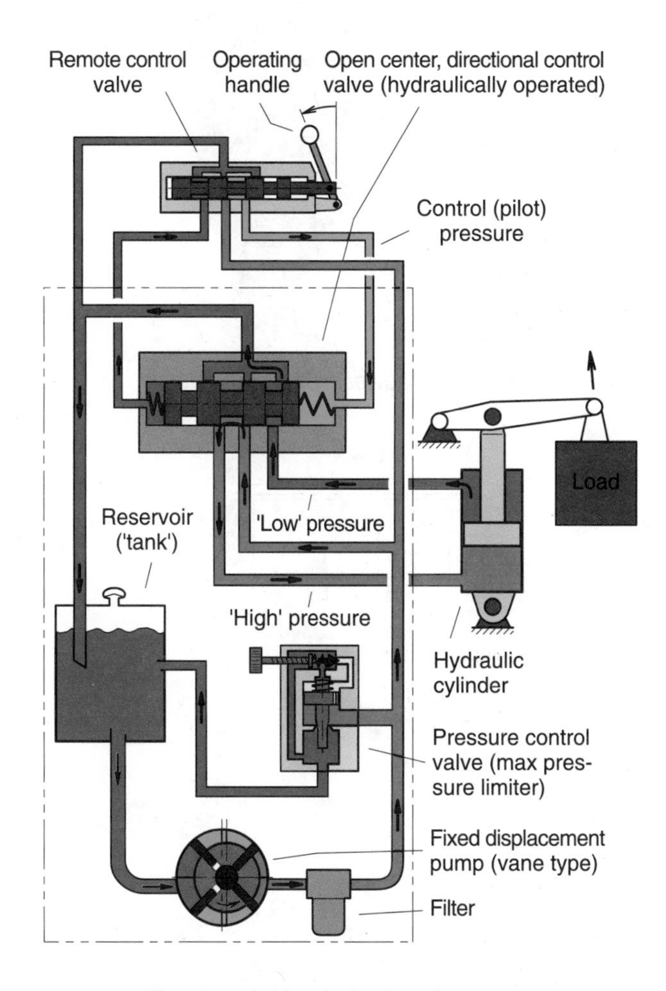

Fig. 1-4 A simple hydraulic system

Pumps

The pump is the heart in the hydraulic system. It supplies a certain flow of fluid under pressure, which is determined by the load, into the system.

The load pressure may be developed in a hydraulic cylinder lifting a log or in a hydraulic motor driving the vehicle.

The most common, rotary pump types in mobile hydraulics are:

- gear
- vane
- axial piston

Fig. 1-5 Variable displacement, axial piston pump

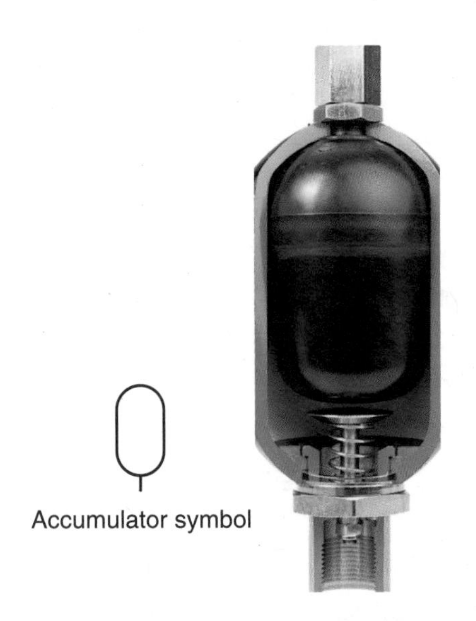

Fig. 1-6 Bladder type accumulator

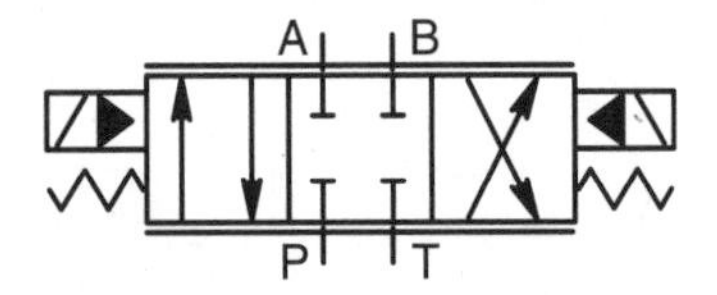

Fig. 1-7 Valve symbol

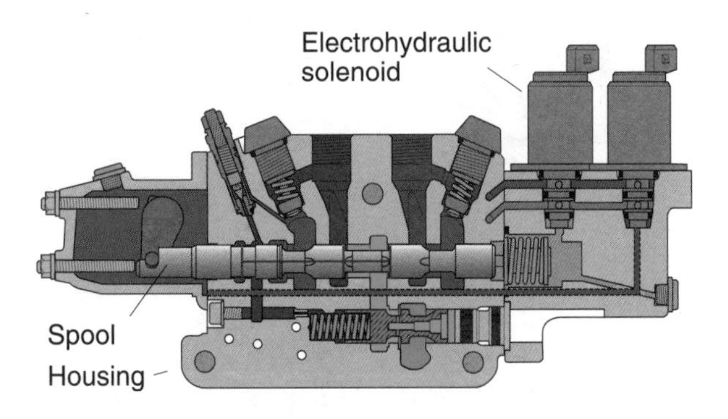

Fig. 1-8 Directional control valve (cross section) with built-in flow and pressure controls; the spool is shifted to the left

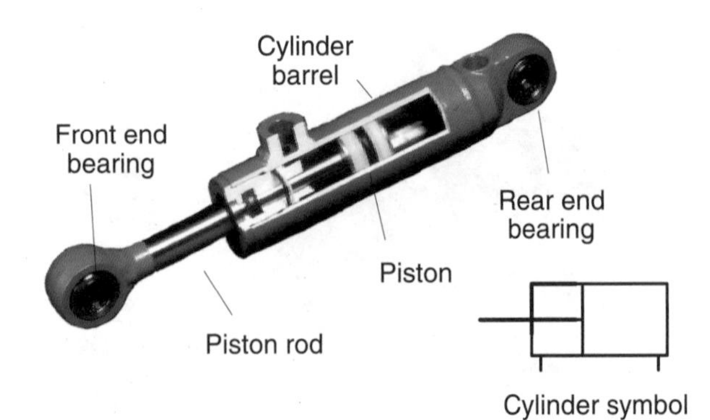

Fig. 1-9 Double-acting hydraulic cylinder (cutaway)

Gear pumps are "fixed displacement", delivering a specific amount of fluid for each pump shaft revolution.

Vane and axial piston pumps can be either fixed or "variable displacement", depending upon the system.

A variable displacement pump delivers fluid flow according to demand (up to its maximum capacity).

Accumulators

As the name implies, an accumulator stores fluid. Normally, the pump supplies pressurized fluid into the accumulator, which in turn, is available to supplement the pump flow when needed.

They are also used to damp pressure shocks and absorb pressure pulses that can otherwise damage the system.

There are three types of accumulators:
- bladder
- diaphragm
- piston

Valves

The directional control valve consists basically of a housing and a moving part such as a spool. The valve connects or disconnects fluid passages within the valve housing (fig. 1-8).

As an example, the port supplied with fluid from the pump can be connected to an actuator such as a hydraulic motor (P-to-B in the valve symbol illustration, fig. 1-7). At the same time the return fluid from the motor goes back to the reservoir (A-to-T).

In addition to the distribution of flow, other valve functions can be included in the valve housing (also available as separate units):
- pressure limiting or reduction
- flow limiting

Fig. 1-8 shows a valve where the position of the spool is determined by a remote electrical signal to one of the two solenoids, which in turn produces a hydraulic pressure pushing on the end of the spool.

There are several other ways of controlling the spool position (e.g. handle or pneumatic pressure).

Hydraulic cylinders

The hydraulic cylinder is called a linear actuator. When subject to hydraulic pressure, it produces a pushing or pulling force. Its main parts are a cylinder barrel and a piston with piston rod (fig. 1-9).

There are basically three cylinder versions:

- double-acting
- single-acting
- telescopic

Single- and double-acting cylinders have numerous mobile applications, whereas the telescopic cylinder is mainly used as a dumping cylinder for the tipper body of a truck or construction dumper.

Motors

The hydraulic motors have numerous applications such as driving winches, wheels and tracks, saw chains and cooling fans, as well as excavator swing functions and the rod rotation function in a drill rig.

The motor is a rotary actuator, very similar in design to the hydraulic pump. The main difference is that the motor in general is bidirectional, i.e. it can operate in both directions, left hand and right hand. The two main ports must consequently be able to withstand max operating pressure, as well as short duration 'peaks' that may develop in the hydraulic system.

Motors in mobile applications are available in the following versions:

- gear (including gerotor types)
- vane
- radial piston
- axial piston

Axial piston motors can be subdivided into swashplate and bent-axis, both in fixed and variable displacement versions.

Components for remote control systems

A remote control system consists of levers, foot pedals and various pushbuttons that are used to operate main hydraulic components such as pumps, directional control valves and motors.

Fig. 1-10 Fixed displacement motor

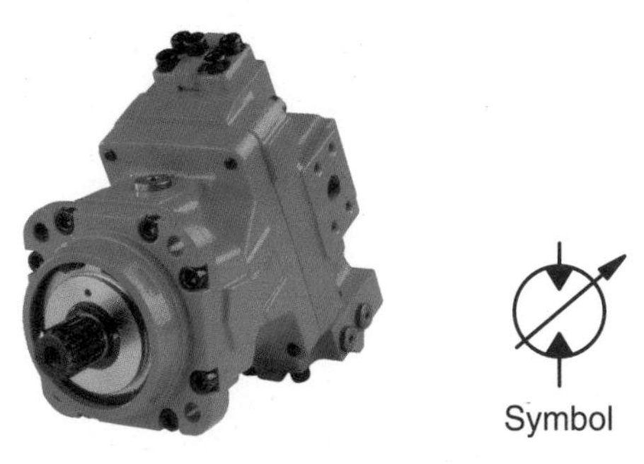

Fig. 1-11 Variable displacement motor

Fig. 1-12 Hydraulic joystick and foot pedal

Fig. 1-13 Electric remote control multi-function joystick and electronic control box.

The controlling or pilot signal from the remotely located component can be hydraulic, electric, pneumatic, or any combination. In mobile machines the all-hydraulic and electrohydraulic combinations are most common. In work platforms for use around high tension power lines, radio or fiber optic control signals are also utilized.

The remote control system, as the name implies, facilitates the control of all machine functions at a distance, e.g. from the operator's seat in the cab of an excavator or front end loader.

In addition, the low-pressure or electric signals minimize heat generation and noise, and increase the operator's level of safety and comfort in the cab.

Even more important may be the low operating forces inherent in the remote control system in addition to other ergonomically favorable measures that can be built into a machine. This reduces the risk of operator fatigue at the same time as it increases the productivity of the machine.

Typical mobile applications
Drill rig - Description

The term 'drill rig' covers several types of machines utilized in such industries as construction and water well drilling, mining and quarrying, tunneling, and for surface as well as underground work.

On modern drill rigs, all functions are hydraulically operated in order to make the machine flexible and user friendly with good efficiency.

The basic technology can be divided into:

- percussive drilling
- rotary percussive drilling
- rotary drilling
- pile drilling

Percussive and rotary percussive drilling are used in hard rock. Rotary drilling is used in hard as well as soft rock drilling, while pile drilling is used in soft rock.

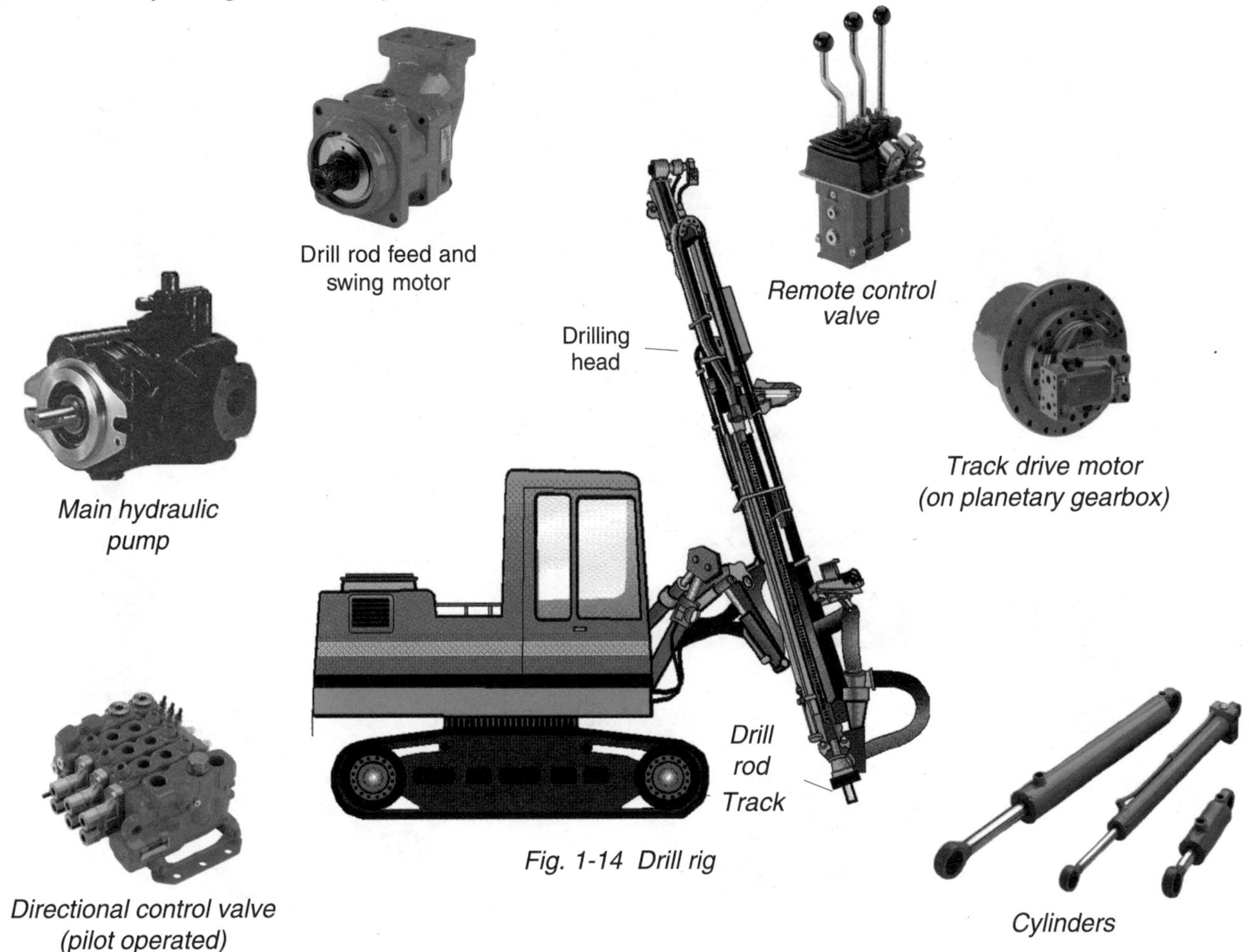

Fig. 1-14 Drill rig

The hydraulic system

The main hydraulic pumps are driven by a diesel engine (or, on underground mining drill rigs, by an electric motor). They deliver fluid to the drilling head which hammers on, and turns, the drill rod.

The pumps also supply all other work functions such as the positioning cylinders and track drive motors, as well as fan drive and drill rod feed motors.

The directional control valves receive pump flow and, on command, direct all or part of the flow to the various work functions.

In the cab, the machine operator has several low pressure and/or electrohydraulic, hand operated remote controls at his disposal, by which he signals the directional control valves in order to activate the chosen work functions.

Excavator - Description

Excavation is one of the most common operations in construction work. Over the past four to five decades, excavators have developed into very efficient machines.

The number of machines produced worldwide is about 150,000 per annum, including small machines of less than one ton to very large machines weighing more than 100 tons, all operated hydraulically.

Excavators can be divided into three groups:

- mini excavators
- wheeled excavators
- crawler (tracked) excavators

The main functions of the machine (boom, stick, bucket, swing and tracks) have different duty cycles. The swing can be operating 75% of the time while the tracks are used maybe only 10%.

All work functions have one thing in common, however, they are subject to very tough operating conditions.

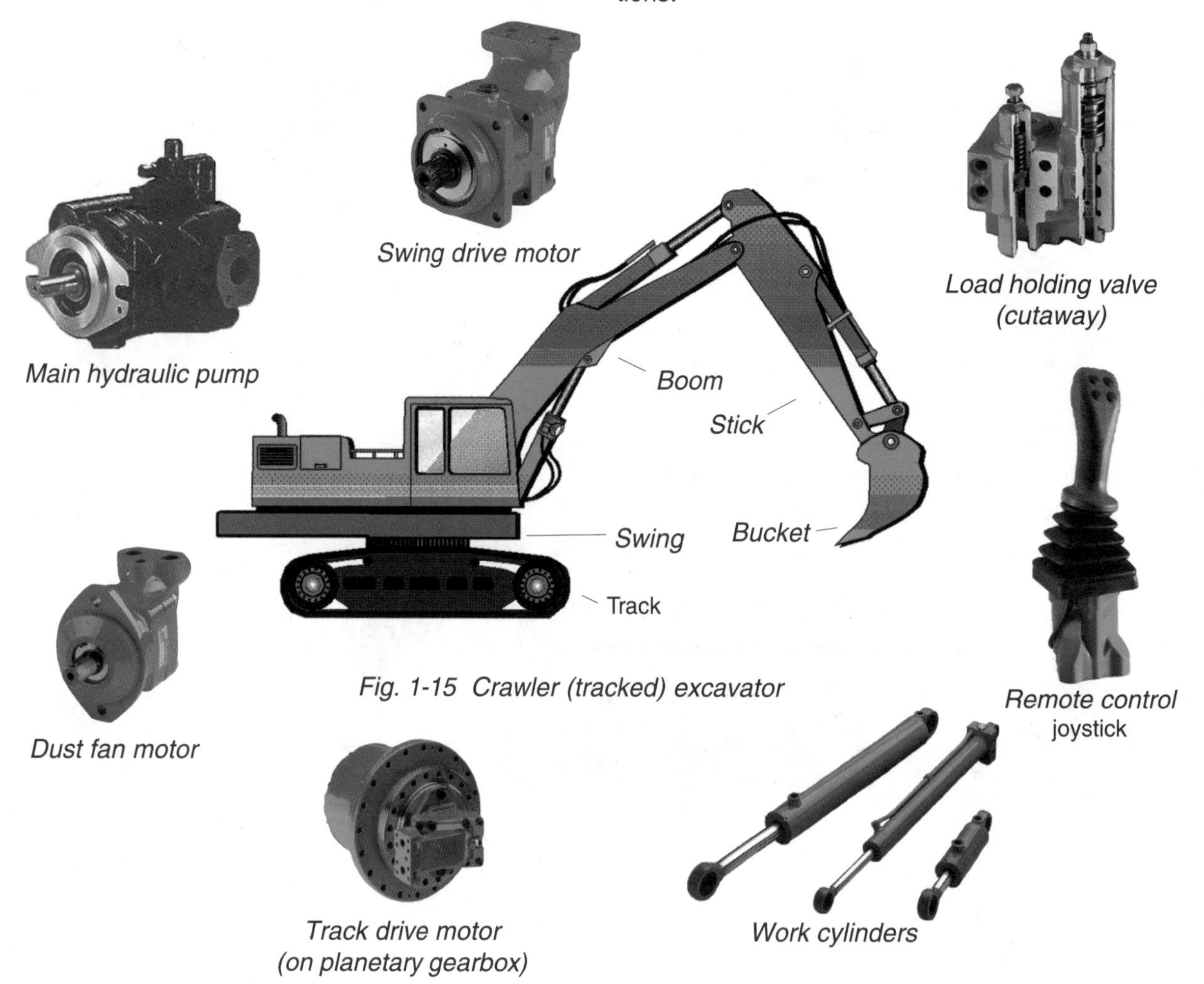

Fig. 1-15 Crawler (tracked) excavator

The hydraulic system

Two (or more) pumps are mounted on a gearbox, driven by the diesel engine, delivering fluid to the main work functions: boom, stick, bucket and swing.

Alternatively, fluid is directed to the two track motors supplied by one or two pumps; it is thus possible to turn the machine “on a dime” as one motor is directed to operate clockwise and the other counter clockwise.

Two ‘hand-operated’ remote control ‘joysticks’ are installed in front of the armrests of the driver's seat, each of them controlling two or more functions. With the joysticks and two foot pedals the operator controls all main work and track functions.

In addition, thanks to the remote control system, both high speed movements and low speed precision maneuvers are possible.

Forwarder (log loader) Description

There are two different methods of wood harvesting. One is the "full tree" method, when the entire tree including branches, after being cut at the root, and is 'skidded' to the roadside.

The other is the "cut-to-length" method where the tree is being processed by a harvester; the tree is then cut at the root, felled and delimbed, and the trunk cut into suitable lengths.

A forwarder (log loader), illustrated below, then transports the logs out of the forest to the roadside, where a truck takes them to a plant for further processing into e.g. planks or paper pulp.

The load carrying capacity ranges from 8 to 20 tons, and engine power from 60 to 150 kW (80 to 200 hp).

Main hydraulic pump (variable displacement)

Electrical remote control joystick

Electronic control box

Directional control valve (pilot operated)

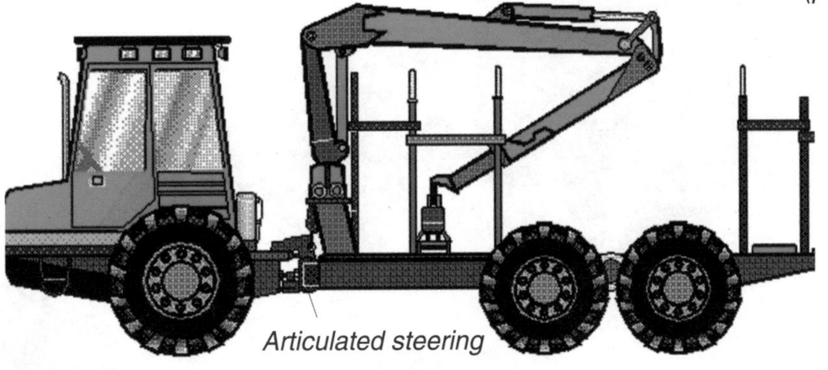

Wheel transmission drive motor (variable displacement)

Manifold for special functions

Work and steering cylinders

Fig. 1-16 Forwarder

The hydraulic system

The diesel driven, main hydraulic pumps supply all work and transport functions on the machine.

The operator handles the six crane or 'loader' functions with two 'joysticks' installed in front of the chair armrests. He can then select to operate only one function at a time, or several functions simultaneously. The utilized remote control system permits both high multifunction speeds and delicate maneuvering.

Articulated steering and forward/reverse speed control is available to the operator in the two driving positions, one when operating the loader, the other (opposite) during transportation.

The hydrostatic wheel transmission with variable displacement pump and motor, automatically limits the drive force, thus preventing engine stall. In addition, it offers very smooth operation in both directions.

Wheel loader Description

The total market for wheel loaders is about 55,000 machines per year worldwide. Wheel loaders can basically be split into small and large machines.

Small front end wheel loaders (fig. 1-17) with engine power up to 50 kW *(65 hp)* have 'articulated frame' steering; almost all are equipped with hydrostatic transmissions (using hydraulic pumps and motors). They are usually utilized as multipurpose machines with multiple attachments like forks, etc. The machine is often used as a carrier of various tools.

Larger front end loaders, above 75 kW *(100 hp)*, are usually designed with hydrodynamic* transmissions and articulated frame steering.

Large loaders are mostly used as production machines in quarries, open pit mining, construction sites and various industrial applications.

* In a hydrodynamic transmission, an engine driven pump fan hurls fluid at high speeds towards a motor fan which drives the vehicle through a gearbox.

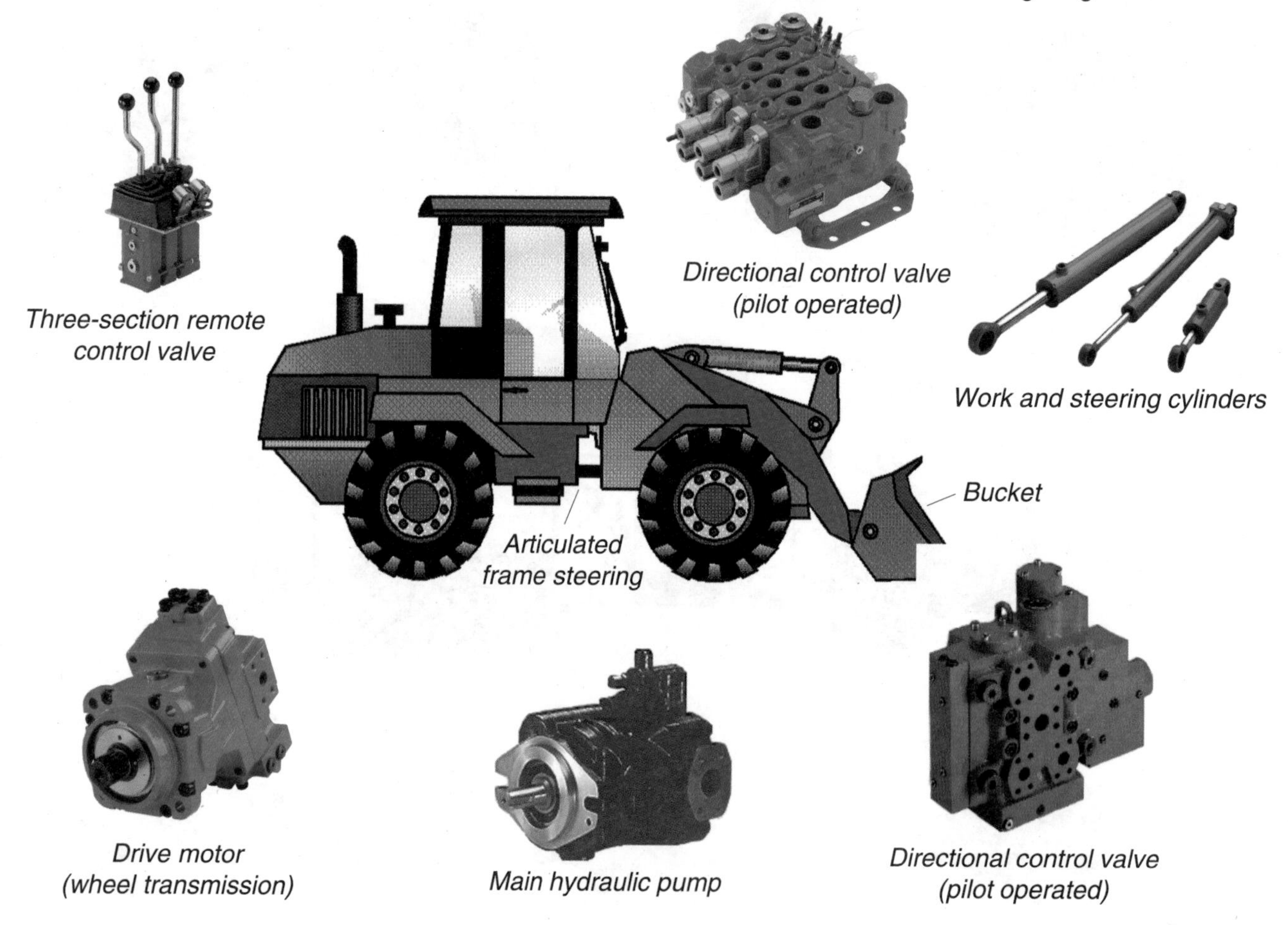

Fig. 1-17 Front end wheel loader

The hydraulic system

The two main functions of the machine are driving (forward and reverse) and bucket excavation. The bucket is controlled from hydraulic remote control levers, or 'joysticks', placed in an ergonomically suitable position.

The articulated frame steering cylinders on larger machines are supplied with fluid from a specific, steering wheel operated metering valve. It outputs an amount of fluid which is proportional to the amount the wheel is turned.

On machines with a hydrostatic four-wheel propel system, the variable pump and motor provide high speed during transport, and high driving force during the digging mode. The lever operated, remote control system also limits the force, automatically, preventing the engine from stalling.

Implement functions such as the bucket are supplied from a separate pump; a fixed displacement version on machines below 45 kW *(60 hp)* and variable versions on some of the larger types.

Reach stacker
Description

The reach stacker can be seen in harbors and truck or railway terminals handling large, heavy containers. It must be capable of high speeds in both the loading and transport modes, with maintained stability.

But most important, the reach stacker must be able to provide precision maneuvers at very low speeds when the container is being positioned.

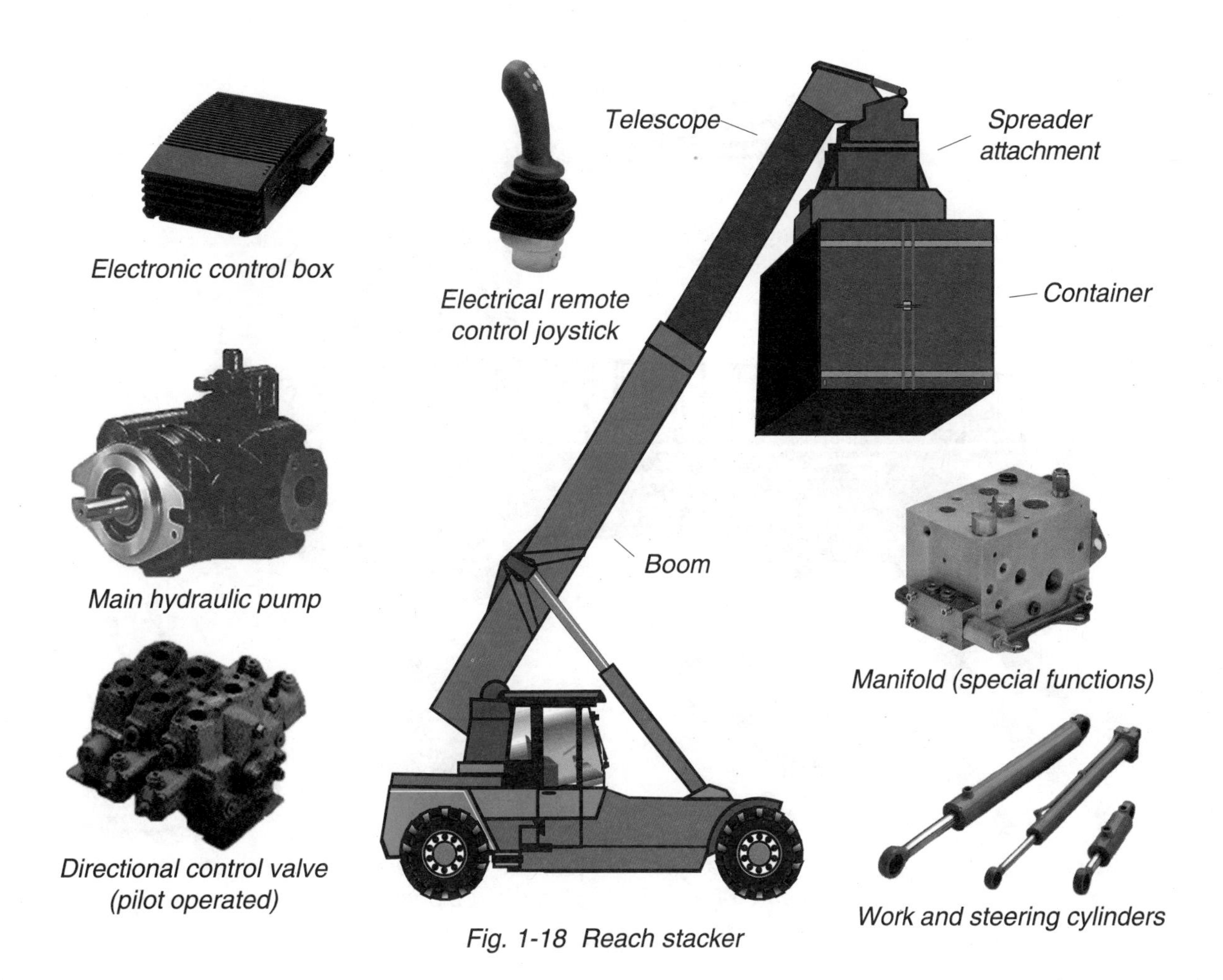

Fig. 1-18 Reach stacker

The hydraulic system

The diesel driven, normally variable displacement pumps provide high fluid flow to the boom and telescope functions, and, with priority, to the rear-wheel steering.

Auxiliary functions such as the spreader attachment is typically supplied from a smaller, fixed displacement pump.

All functions are usually electrohydraulically controlled from the cab with a multifunction joystick.

When both boom and telescope are operated simultaneously, a flow sharing system is employed. Flow sharing means splitting up the pump circuit in two parts, one part for each function.

On machines with a more sophisticated hydraulic system, the splitting is handled by an electronic control system.

Cranes - Description

Self-loading trucks (lorries) are used to load and unload cargo. The crane (hoist) is installed on a commercially produced truck chassis.

Cranes for truck applications come in various shapes such as:

- knuckle boom (shown below)
- stiff boom utilizing a winch
- tool carrier for a wide spectrum of everyday jobs

Smaller cranes are controlled manually from a valve that can be operated from both sides of the truck. Larger cranes are, more and more, operated with a radio controlled, hydraulic system.

According to EC regulations in Europe, all cranes, four metric ton-meters and up, are equipped with overload protection and emergency stop.

Fig. 1-19 Selfloading truck

The hydraulic system

The hydraulic pump is driven off a PTO (power take-off) on the gearbox. The PTO can be engaged and disengaged with a dashboard switch. Depending upon crane type and intended use, the pump can be of a single- or twin-flow, fixed displacement type, or less common, a variable displacement version.

In connection with a suitable directional control valve, a twin-flow pump (providing two independent flows) can be used e.g. on a truck with crane and tipable platform. High flow at relatively low pressure is directed to the telescopic cylinder (tilting the platform), and low flow at much higher pressure to the boom cylinder of the crane.

Today, the directional multi-spool control valve must be set up to offer good maneuverability of the main work functions in all operating conditions (low or high speeds, as well as low or high loads).

The overload protection device is preferably built into the directional control valve for the boom function. It prevents any tendency to exceed the capacity of the crane, which can otherwise occur e.g. when raising the jib.

Refuse collecting vehicle
Description

A refuse collecting vehicle picks up garbage containers from industries, homes and apartment buildings, and unloads the garbage at a refuse site.

The vehicle is basically built on a truck chassis which has been suitably modified to accept a collecting system. Five types can be identified:

- side loader
- rear-end loader
- front-end loader
- skip loader
- hook loader

The side loader, shown below, grabs the bin, lifts it and dumps the contents into the forward hold of the vehicle. When the hold is full, a very powerful ram moves the garbage into the container and compresses it. Additional garbage can now be collected.

Finally, at the refuse site, the entire contents are pushed out through the rear door opening by the ram; the ram will then retract to the forward position, ready for another compacting round.

The productivity of these machines is extremely high. 1,000 to 2,000 bins is the normal daily capacity, which puts high demands on ergonomics and safety.

Fig. 1-20 Refuse collecting vehicle

Main hydraulic pump (variable displacement)

Directional control valve

Joystick with electronic control box

Work cylinders

The hydraulic system

The hydraulic system is typically split into two independent circuits because of differences in pressures and operating modes.

One circuit is handling the packing mechanism and auxiliary functions like locks for the rear door and container. The other circuit is operating the loading arm, which is controlled from the cab with a joystick.

The control system

Refuse collecting vehicles are surrounded by a lot of safety regulations and interlock requirements which makes an electronic control system very suitable.

The introduction of 'refuse weighing' calls for an even more accurate and smooth operation of the bin - with remained, high loading speed.

Chapter 1 exercise
hydraulic transmission of energy

INSTRUCTIONS: For each incomplete sentence, one choice will correctly complete the statement. After reading the sentence and the possible choices, circle the letter next to the most correct answer. After all six statements have been completed, place the letter for each answer in the appropriately labeled box. The letter combination should form a word. Total points possible 35; 5 points per answer, 5 points for correct word.

1. Hydraulic power transmission is utilized because of features such as _____ and _____ .
   B. noise, seals
   R. flow, resistance
   M. speed, load
   E. compactness, precision

2. In mobile hydraulics, the most common rotary type pumps are the gear, vane and _____ .
   L. axial piston
   R. hand pump
   A. gerotor
   E. radial

3. Two of mobile hydraulics biggest considerations are _____ and _____ .
   I. viscosity, friction
   S. contamination, temperature
   N. viscosity, changing direction
   F. velocity, liquid

4. Accumulators are common in mobile hydraulic systems. The type(s) of accumulator(s) is/are _____ .
   U. bladder
   O. diaphragm
   N. piston
   P. all of the above

5. Axial piston motors can be subdivided into _____ and _____ .
   A. swashplate, bent-axis
   R. radial, bent-axis
   I. single-acting, swashplate
   G. single-acting, double-acting

6. One function of a mobile directional control valve is the distribution of flow. However, other valve functions, such as _____ and _____ can be included in the housing.
   N. sequencing, flow limiting
   L. direction of rotation, sequencing
   S. proportional control, anti-cavitation
   M. pressure limiting, flow limiting

Answer 3	Answer 5	Answer 6	Answer 4	Answer 2	Answer 1

Chapter 2

Basic hydraulic principles

Mobile equipment was invented to perform work; an example is shown in fig. 2-1. Physical laws or properties affect the performance of these machines.

To understand how a machine operates, this chapter will attempt to define some of these laws or properties, and then determine how mobile equipment is affected by them.

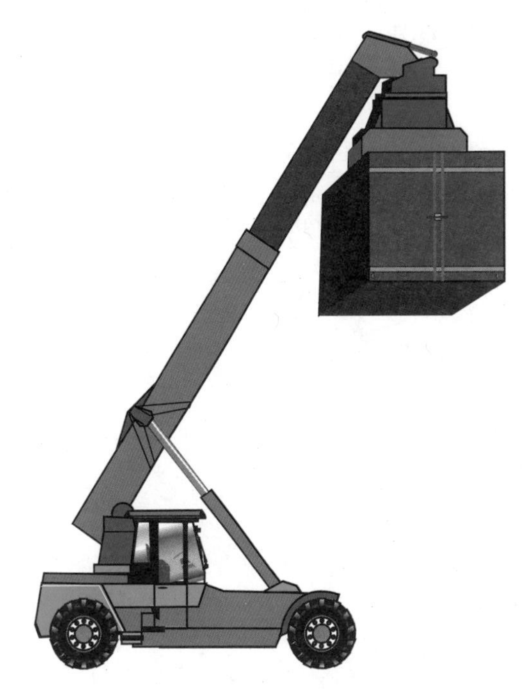

Fig. 2-1 Reach stacker

Force

A force is any influence capable of producing a change in the motion of a body. In the international system, SI, the unit for force is the Newton, N (after Isaac Newton); by definition 1N gives the mass of 1 kg, an acceleration of $1 m/s^2$.

The unit of force in the imperial system is the pound-force *(lbf)*; by definition 1 *lbf* gives the mass of 1 *lb* an acceleration of *1* ft/s^2.

Changes in motion

Force, as dealt with in this text, can change the motion of a body in basically three ways:

- it can cause a body to move
- it can retard or stop a body, which is moving
- it can change the direction of motion

Resistance

Any force which can stop or retard the movement of a body is a resistance. Examples of resistance are friction and inertia (when a body is being accelerated).

Friction as resistance

Frictional resistance is always present between the contacting surfaces of two objects when they are moving past one another (fig. 2-2).

The introduction of hydraulic oil between the two objects lessens the friction as well as reduces the heat that is generated and the amount of force or energy needed to move the object.

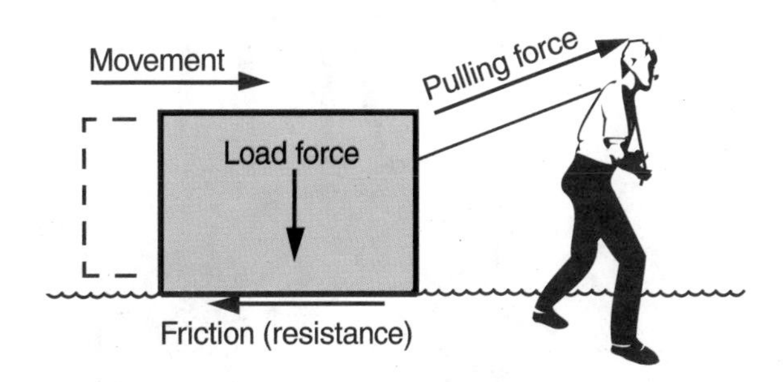

Fig. 2-2 Resistance when pulling a load

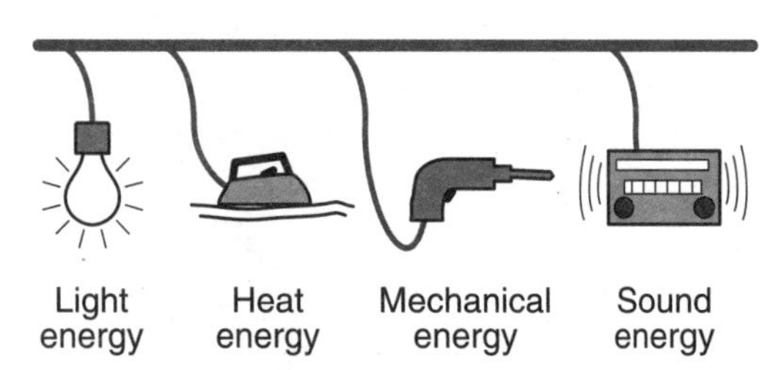

Fig. 2-3 Examples of forms of energy

Energy

A force that causes a body to move is energy. Some forms of energy (as exemplified in fig. 2-3) are:

- mechanical energy
- heat energy
- electrical energy
- light energy
- chemical energy
- sound energy

Law of conservation of energy

The principle of conservation of energy says that energy can neither be created nor destroyed, although it may change from one form to another.

Energy changes form

Energy exists in various forms and has the ability to change from one form to another. For instance, electrical energy may be changed to several other forms. Depending upon what device or appliance is plugged into the electrical outlet, electrical energy can change to light energy, heat energy, mechanical energy, or sound energy.

Another example of energy changing form is a person sliding down a rope. When it becomes time to stop or slow down, the rope is squeezed and some mechanical energy of the falling body is changed into heat energy, as most people are well aware.

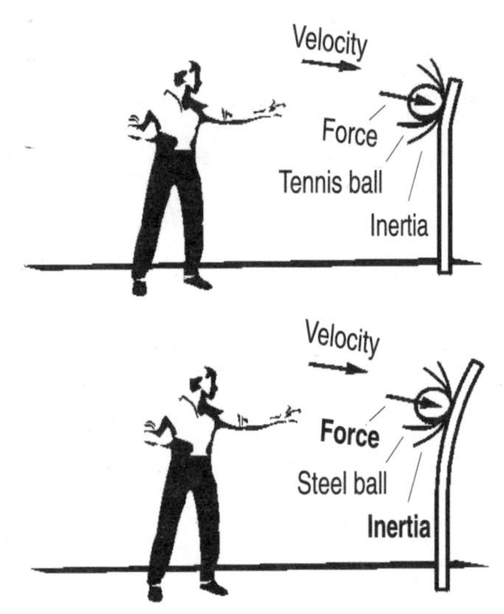

Fig. 2-4 A tennis ball has less inertia than a steel ball of the same size when thrown at the same velocity.

Inertia as energy

Inertia, being the reluctance of a body to change its motion, can also be energy. A moving body exhibits a reluctance to be stopped and can, therefore, strike another body and cause motion. With a tennis ball and a steel ball of the same size moving at the same speed (fig. 2-4), the steel ball exhibits more inertia (energy) since it is more dense, therefore difficult to stop. The steel ball has more *kinetic energy* than the tennis ball.

Two m^3 *(2.6 cubic yards)* of gravel in the bucket of a front end loader (fig. 2-5) represents a tremendous amount of inertia or energy when the bucket is lowered fast. This inertia or energy must be controlled or equipment damage or personnel injury could result.

Fig. 2-5 Front end loader

In hydraulic systems, inertia is encountered whenever a load is accelerated from rest (resistance) and whenever an attempt is made to slow or stop a load that is moving (energy).

States of energy

An important consideration when dealing with energy is the state or condition in which it is found. It can be found in one of two states, *kinetic* or *potential* (fig. 2-6).

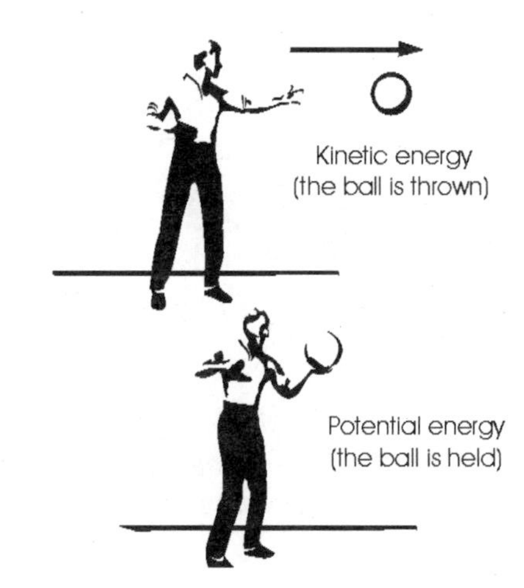

Fig. 2-6 Kinetic versus potential energy

Kinetic state of energy

Energy in the kinetic state is moving energy. It is an indication of the amount of work done on an object, or the amount of work an object can do.

In a hydraulic system, kinetic energy is represented by the system flow (the weight and the velocity of the fluid).

Potential state of energy

When in the potential state, energy can be stored (as in an accumulator; fig. 2-7). It is waiting to spring into action, to change to a kinetic state as soon as an opportunity arises.

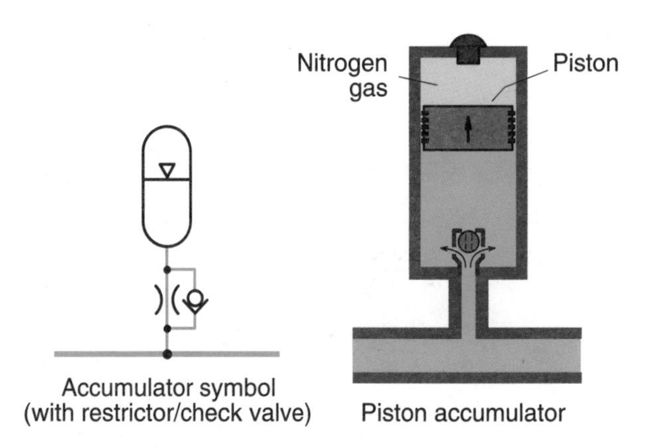

Fig. 2-7 Potential energy stored in an accumulator

The hydraulic accumulator may store hydraulic oil under pressure that is potential energy. This potential energy can turn to working energy or kinetic energy (flow and pressure). In the illustration (fig. 2-7) the pressure is stored in the piston accumulator precharged with nitrogen gas.

Potential energy has the ability to become kinetic because of its physical makeup or its position above a reference point.

Because of its elevation, the water contained in a water tower (fig. 2-8) possesses potential energy. It has the ability to be drawn off as kinetic energy at a household water tap at a lower level. Likewise, a vehicle battery, when not connected in a circuit, is in a state of potential energy.

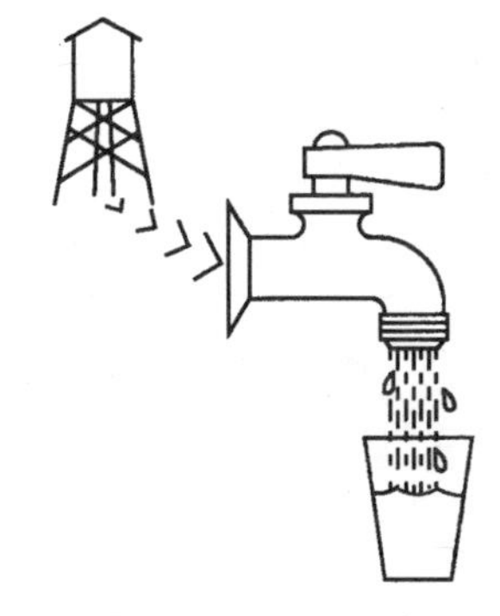

Fig. 2-8 Potential energy stored in a water tower

Because of their physical makeup, the chemicals in a battery have the ability to change to electrical, kinetic energy.

Energy changes state

As has been discussed, potential energy has the ability to change to kinetic. But, kinetic energy can change to potential as well.

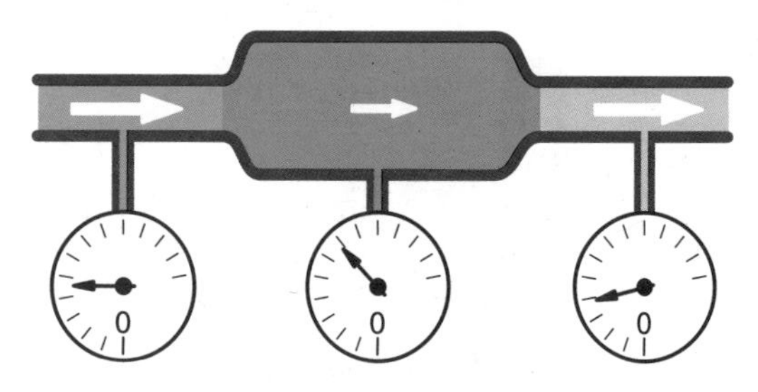

Fig. 2-9 Pressure increases when flow decreases and vice versa

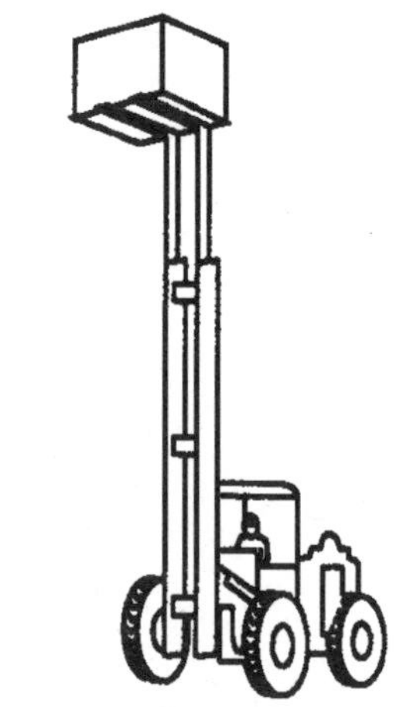

Fig. 2-10 Lifting a pallet is work

Formula 2-1 Work is a function of force and distance

W = F x l (work = force x distance)

where: 'W' is work in Nm *(lbf-ft)*
'F' force exerted in N *(lbf)*
'l' distance in m *(ft)*.

Formula 2-2 Power is work over time

$$P = \frac{W}{t} \quad (\text{power} = \frac{\text{work}}{\text{time}})$$

where: 'P' is power in W (watt)
'W' work in Nm
't' time in s (seconds)

NOTE: In the imperial system, the formula is:

$$P(hp) = \frac{W\ (lbf\text{-}ft)}{t\ (seconds)} \times \frac{1}{550}$$

The water in the water tower (fig. 2-8) is potential energy that changes to hydraulic kinetic energy at a water tap. This kinetic energy changes to a potential state while it fills a glass.

In a hydraulic system, the working energy consists of kinetic energy (flow) and potential energy (pressure). When, for example, a pipe diameter changes size, this will directly affect the relationship between potential and kinetic energy (fig. 2-9). If the diameter increases, the flow speed decreases and pressure increases, all according to 'the law of conservation'.

Work

Work, as it is dealt with in this text, is the application of a force to cause movement of an object through a distance. *Work* is getting things done.

An example of doing work would be a forklift lifting a pallet (fig. 2-10). If the forklift exerts a force of 5 000 N *(^{3}1 100 lbf)* over a vertical distance of 2 m *(6.6 ft)* to load a pallet, then 10 000 Nm (10 kNm) *(7260 lbf ft)* of work would be done.

The SI unit for measuring work is the Joule (J) or Newton-meter (Nm). 1 Nm is approx. *0.74 lbf-ft.* Formula 2-1 shows the relationship between work, force and distance.

Power

Most work is done within a certain time, typically seconds. Power is the speed or rate at which work is done (formula 2-2).

Horsepower

The SI unit for measuring power is the 'watt' (W); the imperial unit is the 'horsepower' *(hp).*

James Watt, the inventor of the steam engine, wanted to compare the amount of power his engine could produce against the power produced by an average work horse. From his experiments, Watt discovered that a horse could lift a *550 lbf* load a vertical distance of *1 ft* in one second, i.e. one horsepower (which corresponds to 746 watts or 0.746 kW).

If, in the forklift example (fig. 2-10), the work (10 kNm) were done in 5 seconds, the rate of doing work would be:

$$P = \frac{10\,000 \text{ Nm}}{5 \text{ seconds}} = 2\,000 \text{ W} = 2 \text{ kW}$$

or:

$$P = \frac{7260 \; lbf}{5 \; seconds} \times \frac{1}{550} = 2.6 \; hp$$

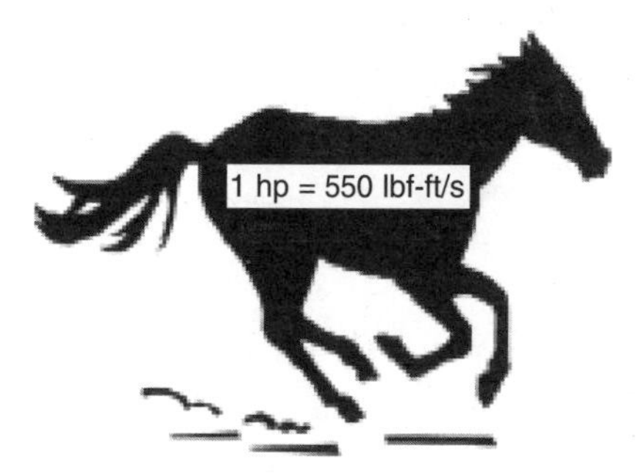

Fig. 2-11 Horsepower

Liquid characteristics

Like all substances, liquids are made up of molecules. Unlike a solid, liquid molecules are not as tightly held together, but on the other hand, not as loosely as gas molecules.

Therefore, the liquid molecules can slip and slide past one another. Because of this slipping and sliding action, a liquid is able to take the shape of any container.

As liquid molecules are in close contact with one another, liquids exhibit, in some respects, the characteristic of a solid. Liquids are considered relatively *non-compressible*. Since liquids are considered non-compressible and can take the shape of any container, they possess certain advantages for transmitting energy.

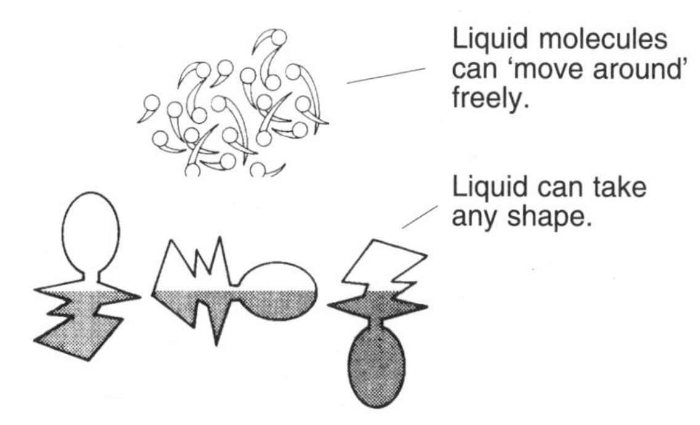

Fig. 2-12 Liquid behavior

Force transmitted through a liquid

A force applied to a confined liquid is transmitted equally throughout the liquid in the form of hydraulic pressure.

If a downward force is applied on a container filled with liquid (fig. 2-13), the confined liquid will transmit pressure equally whether the applied force came from a hammer blow, by hand, weight, fixed or variable spring, compressed air, or any combination of forces.

Since the liquid takes the shape of any container, pressure will be transmitted regardless of the shape of the container.

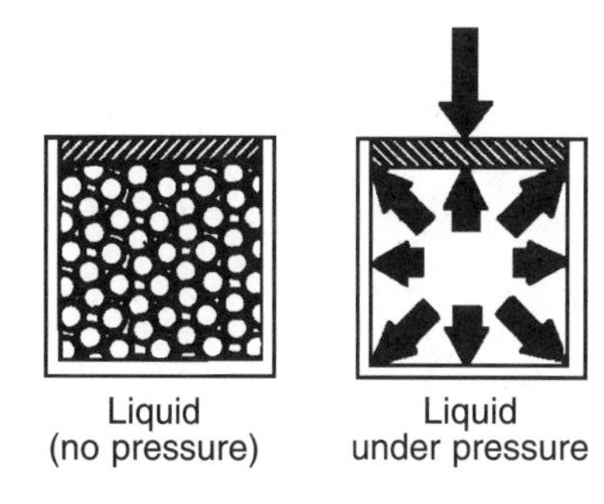

Fig. 2-13 Liquid transmits pressure equally

Pascal's Law

The property of a liquid to transmit pressure equally throughout itself is known as *Pascal's Law*, in honor of Blaise Pascal, who discovered it in the 17th century.

The mathematical expression that describes Pascal's Law is the same as the one describing pressure (formula 2-3).

Pressure

Pressure is a measure of the *intensity of a force.* Many times, the intensity is of more interest, and of greater concern, than the actual force itself.

To determine pressure, the intensity of a force, the total force must be divided by the area on which it is acting. The result is the pressure (or the amount of force per area unit).

A 100 Newton *(20 lbf)* force acting on an area of 100 cm^2 *(20 in^2)* exerts a pressure of 1 N/cm^2 *(1 psi)* on a stationary surface upon which it is acting (fig. 2-14). The same force acting on a square steel rod with a base area of 1 cm^2 *(0.2 in^2)* exerts a pressure of 100 N/cm^2 *(100 psi).* The forces are equal, but the intensities differ greatly.

100 N *(20 lbf)*

100 lbs.
444 N

100 lbs.
444 N

100 cm^2 *(20 in^2)*

1 cm^2 *(0.2 in^2)*

Fig. 2-14 Pressure is force over surface area

An example of this from everyday life would be the difference in pressures generated by heels of various styles of shoes. Anyone who has had their foot stepped on by a woman wearing high-heeled shoes (fig. 2-15) with a tiny heel area knows what it feels like. The same woman wearing low, large heel area shoes would most likely cause much less discomfort.

Fig. 2-15 Pressure intensifies when exerted on a smaller area.

The expression describing pressure is shown in formula 2-3.

Formula 2-3 Pressure depends on force and area

In the metric system:

$$p = \frac{F}{A} \times \frac{1}{10}$$

where: 'p' is pressure in bar
'F' force exerted in N
'A' area in cm^2

In the Imperial system:

$$p = \frac{F}{A}$$

where: 'p' is pressure in *psi*
'F' force exerted in *lbf*
'A' area in *in^2.*

The SI unit for pressure is the pascal, 'Pa'. It is, however, a very small unit, and in hydraulics, the commonly used unit is the 'bar':

- 1 bar = 10^5 N/m^2= 10^5 Pa = 0.1 MPa (megapascal)
- 1 bar = *14.5 psi.*

When calculating in pascal units, larger multiple units are commonly used:

- 1 kPa = 10^3 Pa; 1 MPa = 10^6 Pa.

The weight of the air acting on 1 m^2 surface, called *atmospheric pressure,* is equal to approx. 1 bar.

Atmospheric pressure is normally used as a reference level. A pressure may be given as 200 bar (20 MPa) which means 200 bar above atmospheric pressure *(2900 psi).*

If a pressure level is below atmospheric, a vacuum, it is assigned a minus (-). The highest achievable negative pressure is -1 bar (-0.1

MPa) which would be equivalent to a vacuum of 760 mm Hg *(29.92 inches of mercury, Hg).*

In hydraulic systems, pump cavitation can be experienced because of insufficient pump inlet pressure, which in turn causes noise and pump damage.

A pump manufacturer may state that the 'under-pressure' (vacuum) on the inlet or 'suction' side of a specific pump at sea level must not exceed -0.2 bar, which corresponds to a vacuum of *6 in. Hg.*

To talk about suction is somewhat misleading. The fact is that atmospheric air pressure is pushing on the oil surface in the reservoir, causing the oil to move in a direction of a lower pressure area - the pump inlet (fig. 2-16).

To increase the pump inlet pressure, some hydraulic reservoirs are pressurized internally and the reservoir is then sealed from the outside atmosphere. The reservoir can also be placed higher than the pump in order to increase the potential energy; more about this in chapter 3, Pumps.

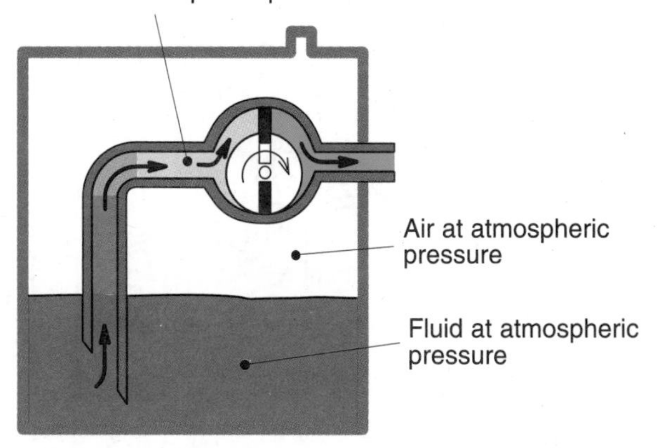

Fig. 2-16 Atmospheric and inlet pressures

Pressure gauge

A pressure gauge (fig. 2-17) is a device, which measures the intensity of a force applied to a liquid. Two types of pressure gauges are commonly used in hydraulic systems:

- Bourdon tube gauge
- plunger gauge

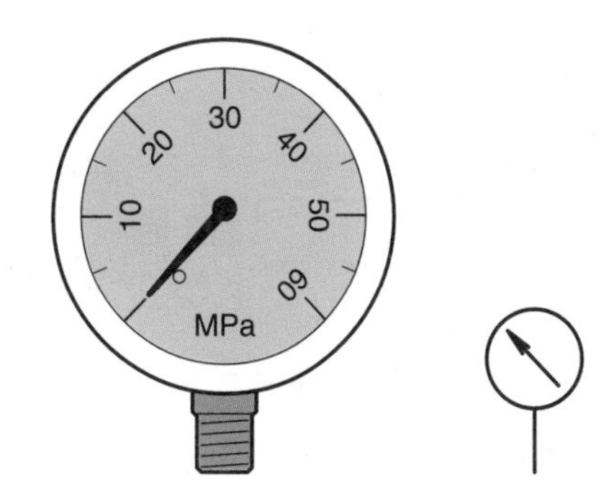

Fig. 2-17 Pressure gauge with symbol

Bourdon tube pressure gauge

A Bourdon tube gauge (fig. 2-18) basically consists of a dial face (calibrated e.g. in bar, MPa, kPa or *psi),* and a needle pointer attached through a linkage to a flexible metal coiled tube, the 'Bourdon tube'.

One end of the Bourdon tube is connected to system pressure. As pressure in the system rises, the Bourdon tube tends to straighten out because of the area difference between its inside and outside diameters. This action causes the pointer to move and indicate the appropriate pressure on the dial face.

Bourdon tube gauges are generally precision instruments with an accuracy ranging from 0.1% to 3.0% of the full scale reading.

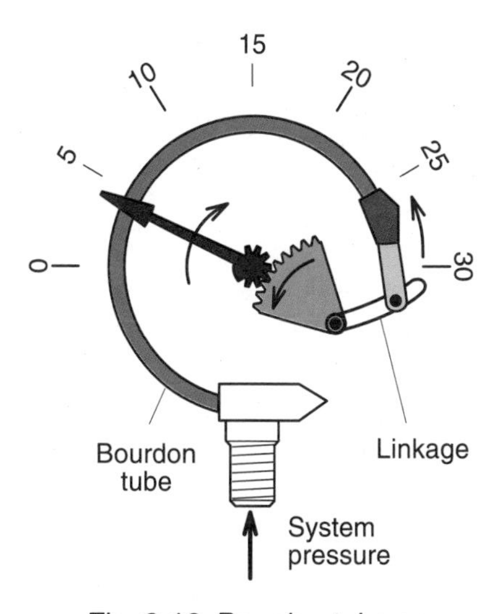

Fig. 2-18 Bourdon tube gauge

Bourdon tube gauges are frequently used for laboratory purposes and on systems where accurate pressure determination is important.

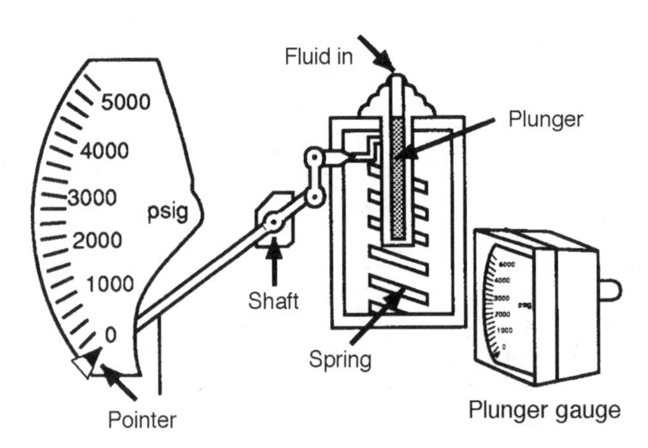

Fig. 2-19 Plunger gauge

Plunger pressure gauge

A plunger gauge consists of a plunger connected to system pressure, a bias spring, a pointer, and a scale calibrated in appropriate pressure units (fig. 2-19).

As pressure in a system rises, the pressure acting against the force of the bias spring moves the plunger. This movement causes the pointer attached to the plunger to indicate the appropriate pressure on the face of the gauge.

Plunger gauges are a durable and economical means of measuring the pressure in a system.

Hydraulic transmission of energy

In the hydraulic transmission of energy, energy in the form of pressurized liquid flow is transmitted and controlled through a conduit system (piping, tubing and hose) to a hydraulic actuator at the point of work. For almost all machines, the energy that does the ultimate work is mechanical energy. For this reason, they require an actuator before the point of work.

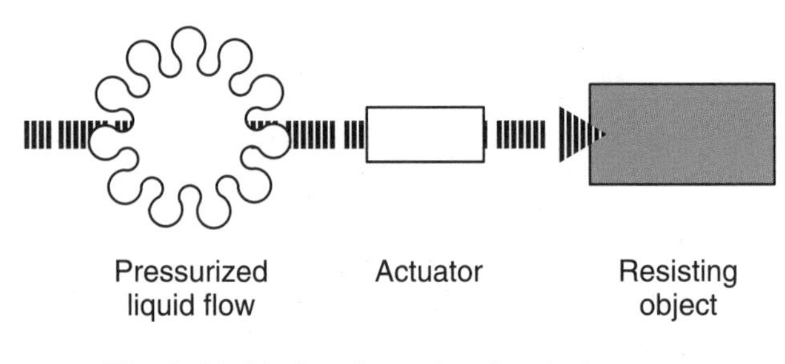

Fig. 2-20 Hydraulic-to-mechanical energy transformation

Actuators transform (convert) electrical, pneumatic, and hydraulic energy into mechanical energy (fig. 2-20).

Each method of energy transmission has its own advantages and disadvantages. Therefore, a machine may be equipped with a combination of mechanical, electrical, pneumatic, and hydraulic systems.

Inefficiency

The objective of the various transmission systems is to perform useful work, that is, to move a resisting object through a distance with as little loss of energy as possible (energy loss is exemplified in fig. 2-21).

Useful work is performed by the application of kinetic energy and pressure to a surface of the resisting object. This is defined as *working energy* (flow and pressure).

Energy transmitted through the various systems is also working energy. The conductors in each system are physical objects with surfaces, which offer resistance to this energy.

The kinetic energy applies pressure to the conductors' surface. This energy performs non-useful work, since no resisting object is moved.

Fig. 2-21 Energy loss (heat is wasted as it goes up the chimney)

Travelling through the system, the pressure of the working energy becomes lower and lower as it gets to the point of work. This pressure is not destroyed, but changes to the form of heat energy because of friction.

A mainstream of fluid generates heat when it crashes into other liquid molecules while forced to change direction because of a bend or elbow (fig. 2-22). Depending upon the conduit size, one 90° elbow could generate as much heat (energy loss) as several meters *(feet)* of straight conduit (pipe, tubing, hose). The faster a liquid travels, the more heat is generated. The degree to which this happens is a measure of a system's inefficiency.

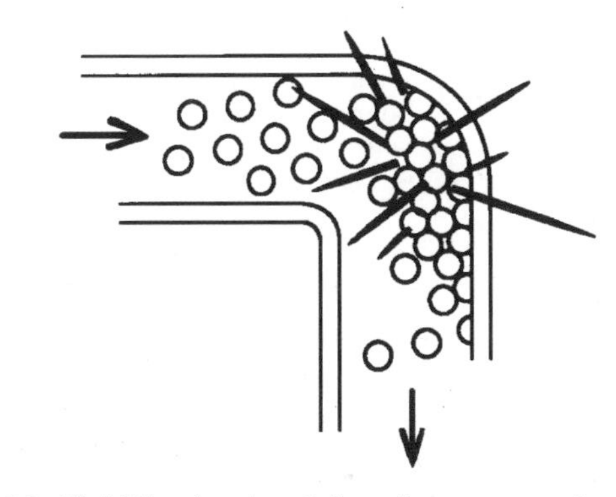

Fig. 2-22 Fluid flowing in a tube elbow generates heat (that is usually wasted).

Pressure differential and pressure drop

Pressure differential, ΔP, is simply the difference in pressure between any two points in a system.

The ΔP is a symptom of what is happening in the system, indicating that working energy in the form of a moving, pressurized liquid is present in the system. In the illustration (fig. 2-23), the pressure differential, or pressure drop, ΔP, between the two gauge points is 0.03 bar *(0.5 psi),* which indicates that:

1. Working energy is moving from gauge 1 towards gauge 2
2. While moving between the two gauge points, 0.03 bar *(0.5 psi)* of the working energy is transformed into heat energy because of liquid resistance, or DP = 0.03 bar *(0.5 psi).*

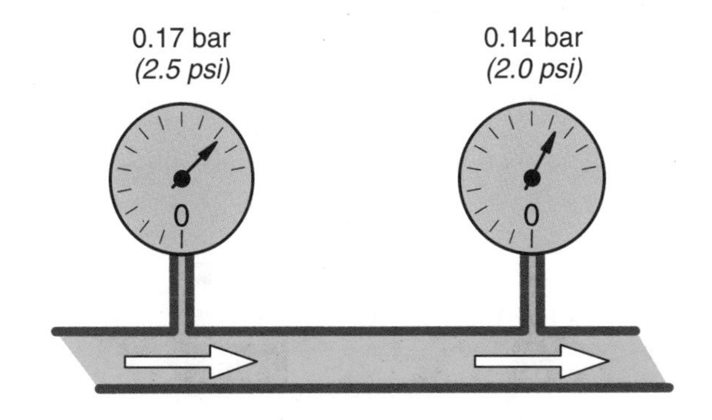

Fig. 2-23 Pressure drop in a conduit

Hydraulic system design to avoid heat generation

The generation of heat, by working energy travelling through a system to a point of work, is system inefficiency.

To make a hydraulic system more efficient, a designer chooses oil with the appropriate viscosity,

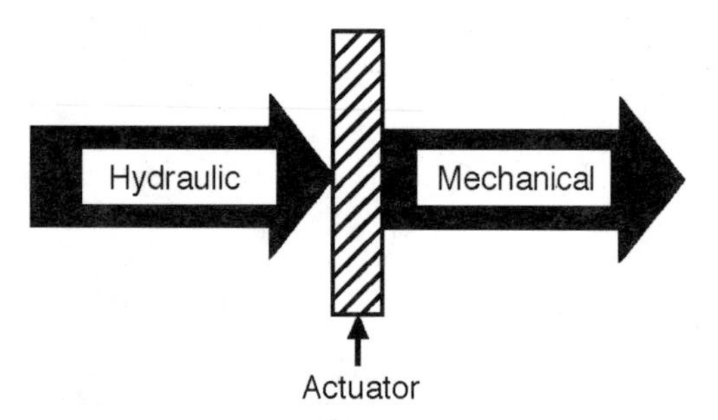

Fig. 2-24 Hydraulic pressure converted to force

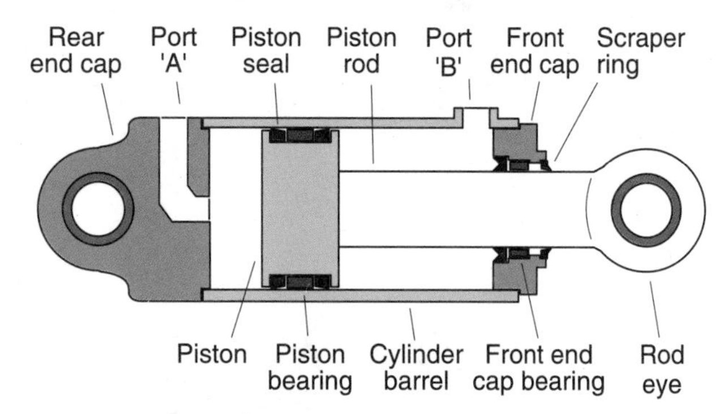

Fig. 2-25 Cylinder nomenclature

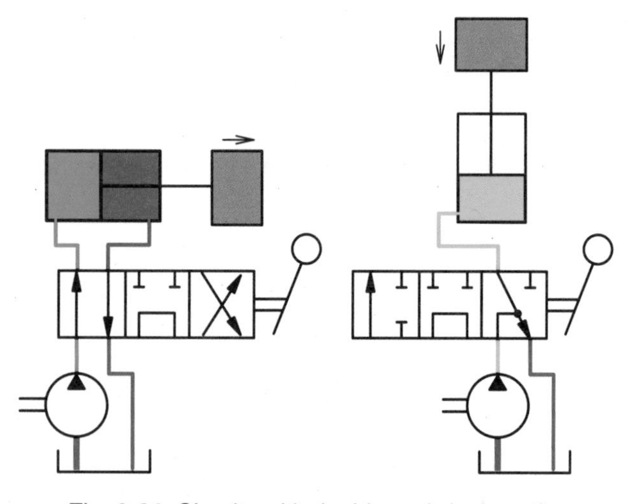
Fig. 2-26 Circuits with double and single acting cylinders

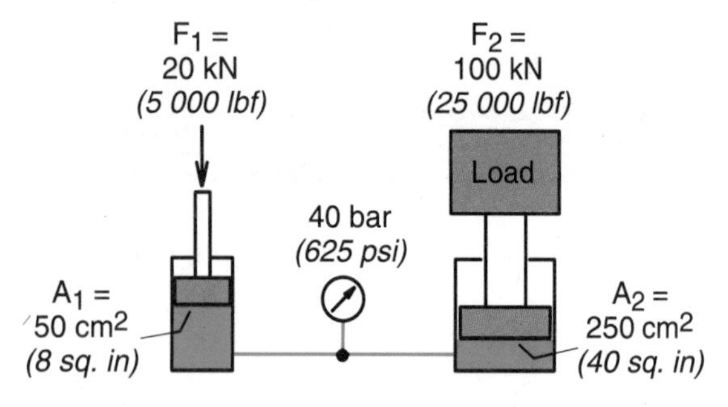

Fig. 2-27 Force multiplication using cylinders

using piping (tubing or hose) of a proper size, and keeping the number of bends to a minimum.

Converting hydraulic pressure to mechanical force

The function of a hydraulic actuator is to convert hydraulic pressure into mechanical force (fig. 2-24). The actuator could be a motor, a rotary actuator, or a cylinder, the last being the most common actuator.

A hydraulic cylinder (fig. 2-25) basically consists of rear end cap, cylinder barrel, piston and rod, front end cap, front end cap bearing, and seals.

In each end cap is a port. The rear end cap port is sometimes called the inlet port, and the front end cap port is called the vent or outlet port.

These same ports may also be referred to as 'A' and 'B', or sometimes 'C1' and 'C2'. Cylinders that consist of an inlet port, on the cap or base end, and a vent port, on the rod end, are known as 'single acting' cylinders.

Fig. 2-26 shows that if pressurized fluid enters the inlet port of the cylinder, it forces the piston and the attached load to move. In order to lower the load, the fluid is allowed to exhaust from the cylinder and the load will move down.

It should be pointed out that this type of cylinder, in order to operate properly, must be mounted in such a way that an external force can be used to return the cylinder. A near vertical to vertical position (refer to fig. 2-26) would use gravity as the external force.

A spring could also be used to supply the required force. Cylinders that consist of an inlet and an outlet port (working ports at both ends) are known as double-acting cylinders. When pressurized fluid enters an inlet port, any fluid in the opposite end of cylinder will exhaust, or return, to tank via outlet. It forces the piston and load to move (extend).

By directing pressurized fluid into the opposite, outlet port and exhausting the fluid from the inlet port or cap end, the piston and load are forced to move in opposite direction (retract).

In the two examples of cylinders and their operation, when the pressure acted upon the piston, the result was a mechanical force created to move the load.

Mechanical force multiplication

Mechanical forces can be multiplied using hydrostatic power. One determining factor for force multiplication is the area on which the pressure is applied (Formula 2-4).

Since the piston area in the left cylinder (A_1) is smaller than that of the right cylinder (A_2), the output force (F2) lifting the load will be greater than input force (F1).

As seen in Formula 2-4, force output is inversely proportional to the area relationship.

Formula 2-4 When pressure is the same, force is proportional to area

As pressure is the same in the two cylinders, then:

$$p = \frac{F_1}{A_1} = \frac{F_2}{A_2} \quad \text{and:} \quad F_2 = F_1 \times \frac{A_2}{A_1}$$

where: 'p' is pressure
'F' force
'A' area

(Compare with Formula 2-3 and fig. 2-27.)

Positive displacement pumps

The supplier of hydraulic energy is the pump, driven by the vehicle's engine. With constant cycling of its movable member(s), a positive displacement pump will deliver a constant flow of liquid and will apply, within limits of its prime mover, whatever pressure is required to overcome any resistance.

NOTE: A pump only generates a liquid flow; the pressure developed is a result of resistance to that flow. If there is no resistance there is no pressure. In other words, a positive displacement pump develops working energy, i.e. kinetic energy (flow) and pressure.

A simple hydraulic system

The very simple pump shown in fig. 2-28 has a plunger for a movable member. The shaft of the plunger is connected to the prime mover of the machine (an engine) which produces a constant reciprocating (back-and-forth) motion of the plunger.

The inlet port is connected to tank, and a check valve in the line allows fluid to flow in only one direction. The outlet is connected to an actuator such as a cylinder, and a check valve in the line prevents a reverse flow from the actuator.

The illustration shows how fluid is drawn into the pump and delivered to the actuator. All pumps draw fluid into their housing by creating an increasing volumetric space at the inlet. In other words, as the plunger is pulled out by the prime mover, a less than atmospheric pressure or

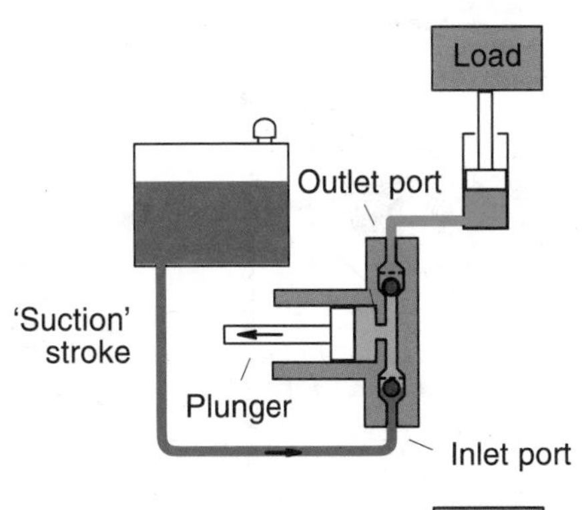

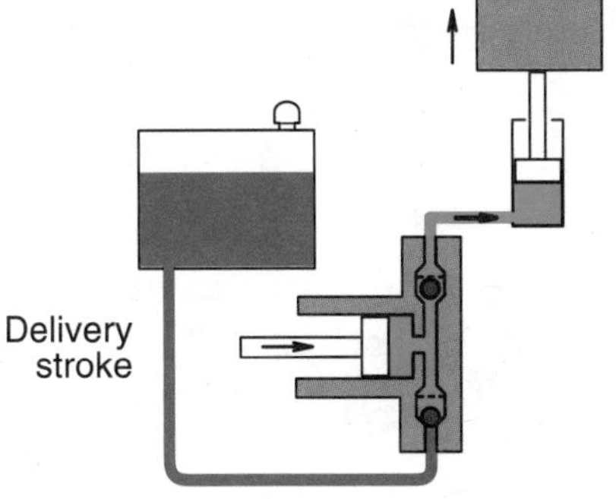

Fig. 2-28 Pump strokes

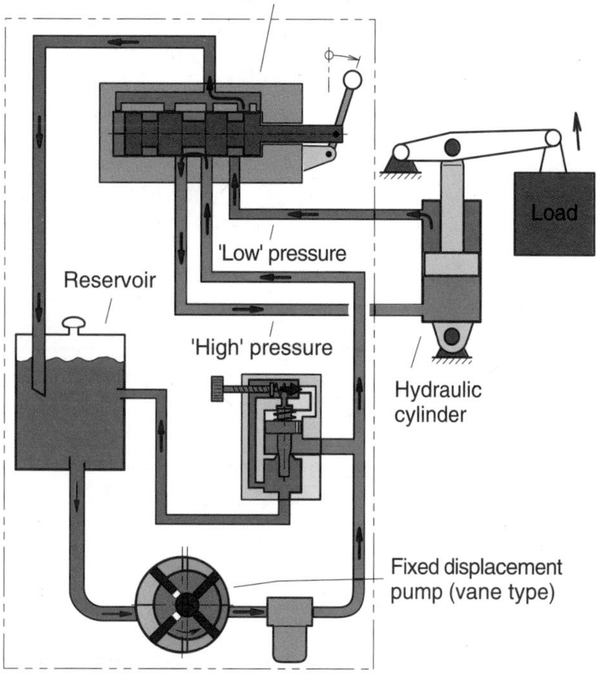

Open center, directional control valve (manually operated)

Reservoir

'Low' pressure

'High' pressure

Hydraulic cylinder

Load

Fixed displacement pump (vane type)

Fig. 2-29 Hydraulic circuit (load being lifted)

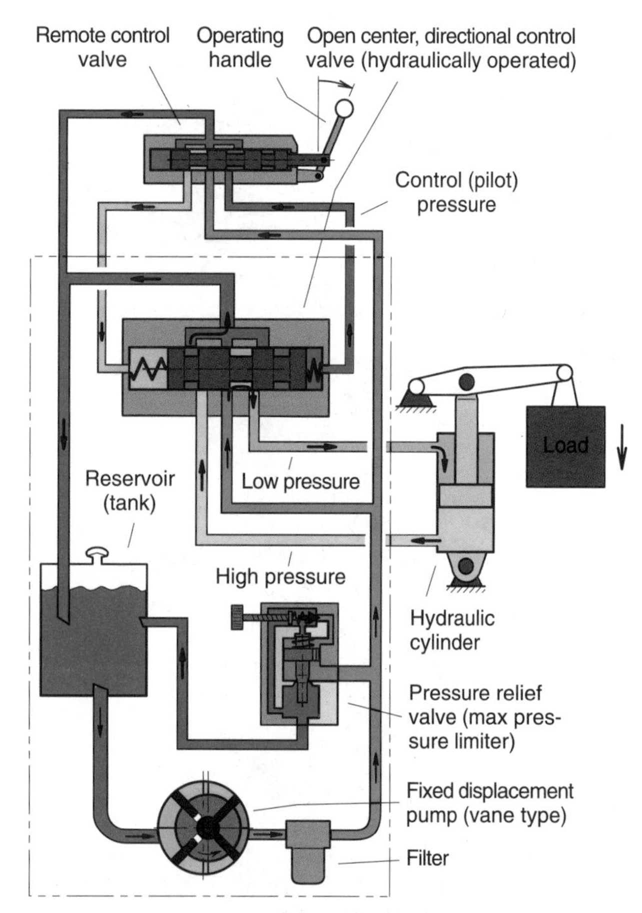

Fig. 2-30 Hydraulic circuit (load being lowered slowly by 'throttling' the control valve)

vacuum is created and fluid is forced into the pump, from the reservoir.

When the plunger is pushed in by the prime mover, a decreasing volumetric space is generated. If the outlet is open to the atmosphere, the pressure developed by the pump will be equal to the resistance of the check valve. By placing a resistive load at the outlet, the pressure will increase to overcome the load and extend the cylinder.

The constant cycling of the plunger results in a pulsating rate of flow at whatever pressure is required. Also, this system does not allow for the actuator to lower the load. To get a continuous operation of this system, some modifications are required.

In the hydraulic circuit (fig. 2-29), the plunger type pump has been replaced by a rotary type pump.

The outlet check valve is replaced by a directional valve, the single acting cylinder by a double acting cylinder, and a pressure relief valve is added to the system.

Operating principles of a simple hydraulic system

The rotary pump is representative of a fixed positive displacement vane pump. This means that for every revolution of the pump, the amount of fluid moved, i.e. the pump's displacement, is the same regardless of any resistance downstream of the pump. As the pump is driven by the prime mover, an increasing volume at its inlet and a decreasing volume at its outlet is created.

Fluid is then directed, by piping, tubing, hose, etc., to a relief valve and to a directional control valve.

The relief valve in any hydraulic system is used to limit the maximum pressure desired within the system. Two different versions of relief valves, direct acting and pilot operated, could be used in the system; these valves will be discussed in a later chapter.

To direct the fluid flow into and out of the actuator, a directional control valve is added to the system.

The directional control valve used in the hydraulic example (figs. 2-29 and 2-30) is an open center valve. It allows the pump output flow to go through the valve back to the reservoir (tank) while the spool is centered in the valve body. The valve spool blocks the actuator ports while the spool is centered trapping fluid on both sides of the cylinder.

If the valve spool is moved to the right (fig. 2-29), pump flow is directed out to the cap or base end inlet port of the cylinder. The flow and pressure from the pump causes the cylinder to lift the load. At the same time, the fluid in the rod end of the cylinder is pushed out the outlet, or rod end port to the directional valve, where it is directed back to the reservoir.

When the spool of the directional valve is moved to the right (fig. 2-30), the fluid in the piston end of the cylinder is pushed out through the valve because of the load. The flow from the pump is directed to fill the increasing rod end volume. In order to prevent an uncontrolled lowering of the load, the flow must be limited by “throttling” the valve spool, i.e. by moving the spool only partially.

The relief valve limits the maximum pressure attainable in the hydraulic system. If the cylinder is allowed to fully stroke and the directional valve stays in either shifted condition, the pump meets a much higher resistance to flow. This higher resistance causes the pressure within the system to rise.

When the pressure in the system reaches the pressure setting of the relief valve, it will open. The pump flow is then returned to the reservoir, thus limiting the pressure (fig. 2-31).

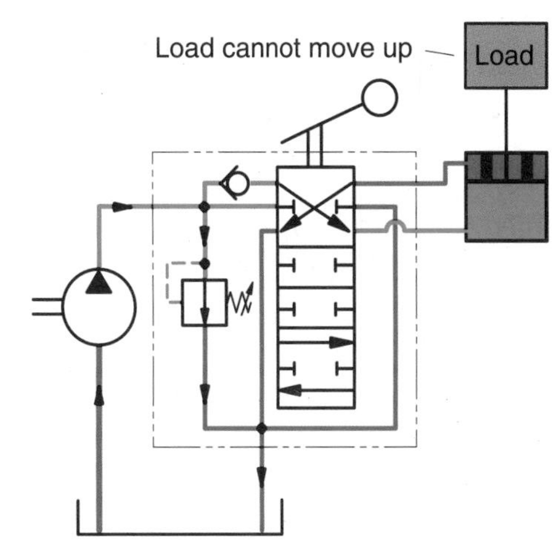

Fig. 2-31 Hydraulic circuit (entire pump flow goes over the relief valve)

Hydrodynamics versus hydrostatics

The field of hydraulics can be divided into two areas:

- hydrodynamics
- hydrostatics

Hydrodynamics is the study and use of liquids in motion, such as in the torque converter of an automobile or tractor transmission. The illustration (fig. 2-32) shows how hydrodynamics is applied.

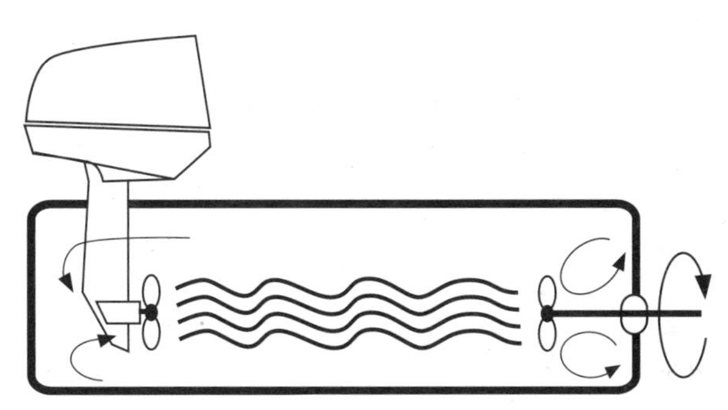
Fig. 2-32 Principle of a hydrodynamic torque converter
Adapted from The Off-Road Vehicle Volume 1, courtesy of CPPA

The propeller of a running outboard motor on a boat hurls a stream of water rearward, forcing the boat forward. If another propeller would be placed behind the outboard motor as shown in the illustration, the propeller would start to turn. This indicates, that torque is being transferred,

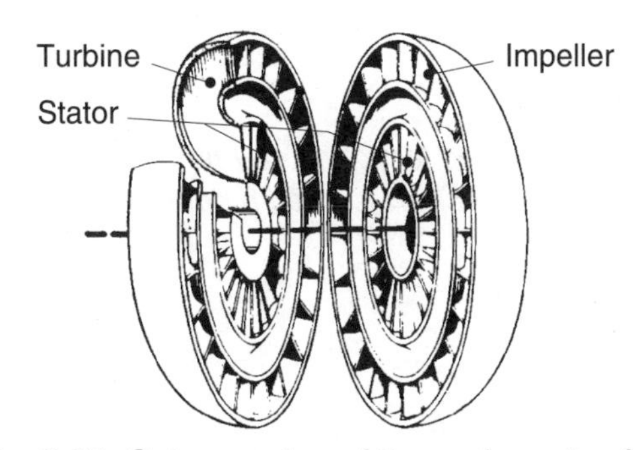

Fig. 2-33 Cutaway view of the main parts of a torque converter
Reprinted from The Off-Road Vehicle Volume 1, courtesy of CPPA

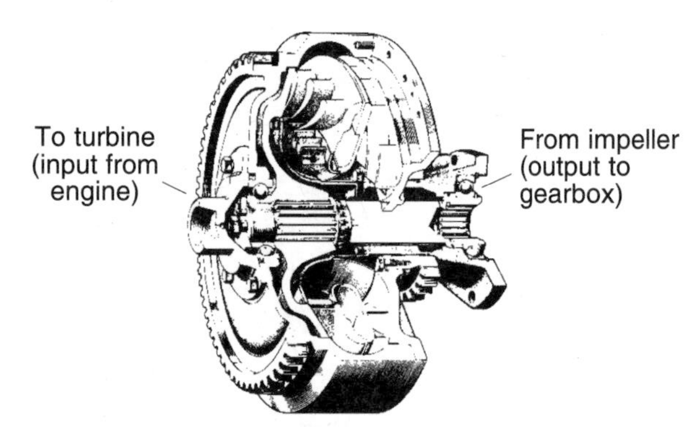

Fig. 2-34 Torque converter cutaway
Reprinted from The Off-Road Vehicle Volume 1, courtesy of CPPA

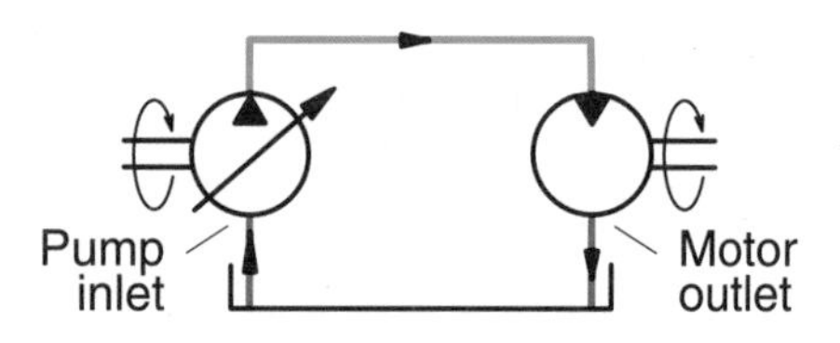

Fig. 2-35 Open loop system (simplified); without a directional valve, the motor can operate in only one direction.

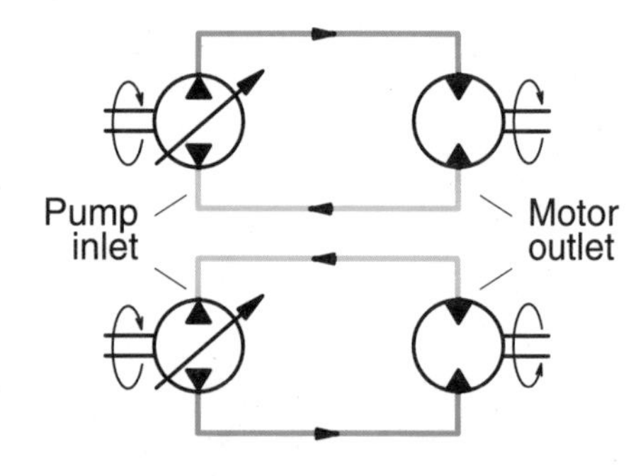

Fig. 2-36 Closed loop system (simplified); the motor can operate in both directions

caused by the impact of water against the propeller.

A torque converter consists of an impeller, a stator and a turbine (fig. 2-33). As the engine of the vehicle drives the turbine, fluid is directed towards the impeller. When the fluid strikes the impeller, energy is transmitted by the impact of this moving fluid against the blades of the impeller. In other words the kinetic energy of the fluid, or the energy of the fluid-in-motion, is used to drive the impeller.

In this example a stator is introduced. The stator's function within the torque converter is to redirect the fluid back into the turbine in an efficient manner in order to reduce drag on the turbine. Fig. 2-34 shows a cutaway of an automotive type torque converter.

Hydraulics, as it will be discussed in this text, will look at the area known as ***hydrostatics***. In a hydrostatic device, energy is transmitted through a static force, ***pressure,*** as shown earlier in this chapter.

Hydrostatics can further be divided into:

- open loop systems
- closed loop systems

Open loop systems

In an open loop system (fig. 2-35) the main system flow circulates from the reservoir, through the system and back to the reservoir. The pump always has an inlet and an outlet, representing the low and high pressure sides respectively.

Open loop systems, in turn, can be divided into three groups:

- constant flow, 'CF'
- constant pressure, 'CP'
- load sensing, 'LS'

These systems will be examined in more detail in chapter 8, Directional control valves.

Closed loop systems

In a closed loop system (fig. 2-36), the main flow goes from the pump outlet, through the system, and then back to the pump inlet. The pump inlet and outlet ports interchange as soon as the working direction is changed. This is the case in the illustrated hydrostatic transmission, which can be utilized to propel a vehicle.

Hydrostatic transmissions are covered in more detail in chapter 4, Motors.

Summary

As has been shown, through the development of a simple hydraulic system, the mission of a hydraulic system is to transmit energy from the prime mover to a location where this energy can do work.

The pump of any hydraulic system provides the flow, and the resistance to that flow generates the pressure. The pump converts the mechanical power of the prime mover into hydraulic power.

A hydraulic cylinder or motor converts the hydraulic power of the system back into mechanical power, the work force of the system. The hydraulic valves direct flow and control pressure within the system.

The reservoir (tank) is a storage and conditioning unit for the system fluid where a filter, a cooler and a heater can be installed if required.

Chapter 2 exercises
Exercise 1 - Basic hydraulic principles

Instructions: Match the word with the phrase that best describes that word.

____ 1. Force
____ 2. Friction
____ 3. Inertia
____ 4. Potential energy
____ 5. Work
____ 6. Horsepower
____ 7. Pressure
____ 8. Inefficiency
____ 9. Hydrostatic
____ 10. Closed-loop system

A. A form of resistance
B. Can be energy or resistance
C. Force moved through a distance
D. Measure of a force's intensity
E. Heat is an indication of this
F. Any influence capable of producing a change in the motion of a body
G. Stored energy
H. The SI unit is the "watt"
I. The pump inlet and outlet ports interchange depending upon working direction
J. Energy transmitted through a static force
K. Main system flow circulates from the reservoir back to the reservior
L. A simple hydraulic system
M. Connecting pressure to mechanical force

Chapter 2 exercises (cont'd.)
Exercise 2 - Basic hydraulic principles

1. What does the gage read?

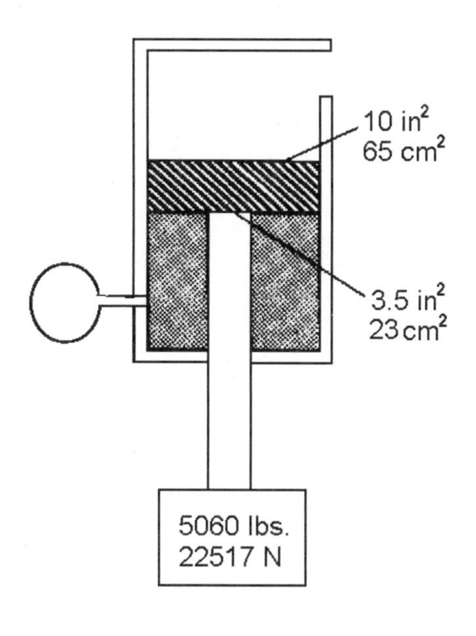

2. What does the gage read?

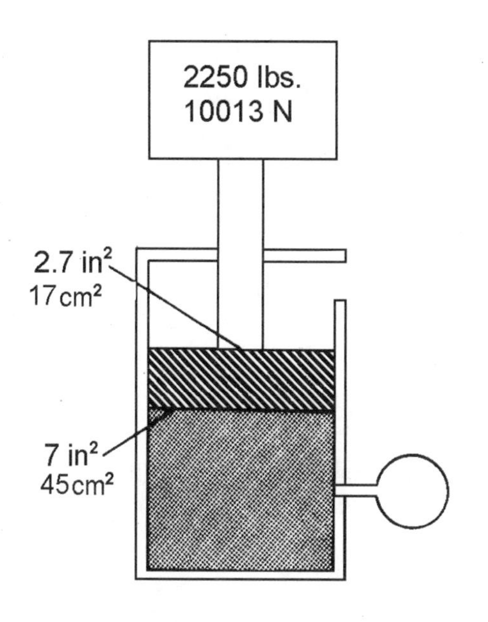

3. What pressure is needed to suspend the 8811 N *(1980 lb.)* load?

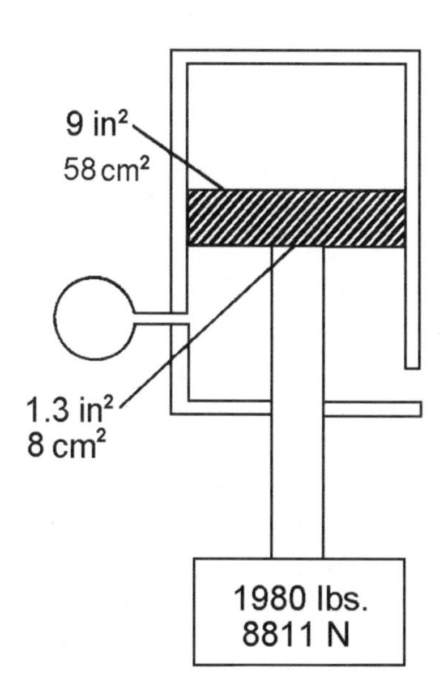

4. What pressure is needed to raise the 13 929 N *(3130 lb.)* load?

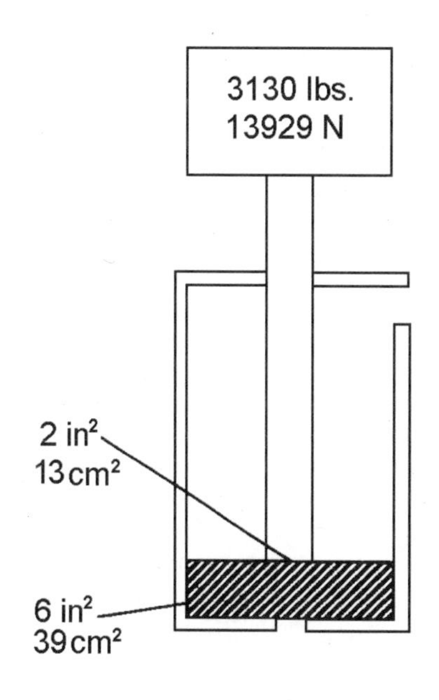

Chapter 2 exercises (cont'd.)
Exercise 2 - Basic hydraulic principles

5. What pressure is needed to move out the load?

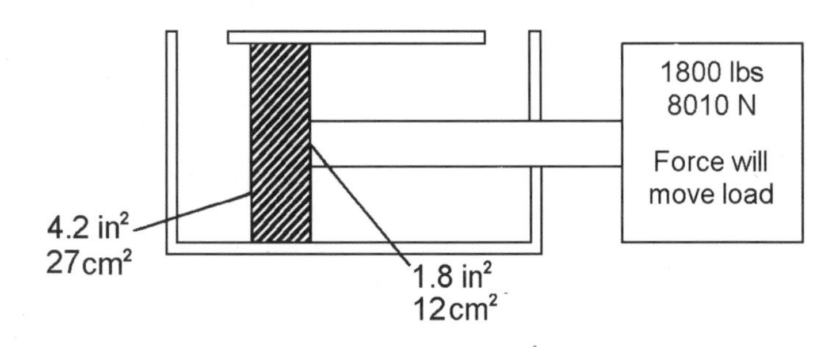

6. What pressure at the intensifier inlet will result in 897 bar *(13,000 psi)* at the outlet?

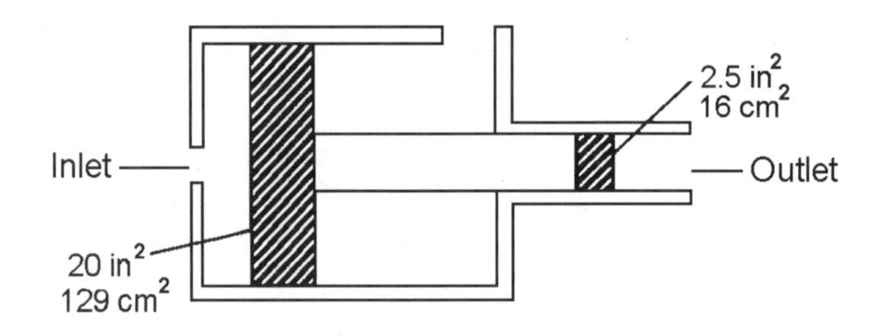

CHAPTER 3

Mobile hydraulic pumps

Hydraulic pumps convert the mechanical energy (rpm and torque) transmitted by a prime mover (electric motor, internal combustion engine, etc.) into hydraulic working energy (flow and pressure). Pumping action is the same for every pump. All pumps generate an increasing volume at the suction side and a decreasing volume at the pressure side. However, the elements which perform the pumping action are not the same in all pumps. The type of pump used in a mobile hydraulic system is a positive displacement pump. There are many types of positive displacement pumps. For this reason, we must be selective and concentrate on the most popular. These are gear and piston pumps. There is a third style of positive displacement pump, the vane, though encountered in mobile applications it is not common. But because it is encountered, it will be included in this discussion on pumps. The first group of pumps discussed will be fixed displacement pumps. The symbol for this style pump is shown in fig. 3-1.

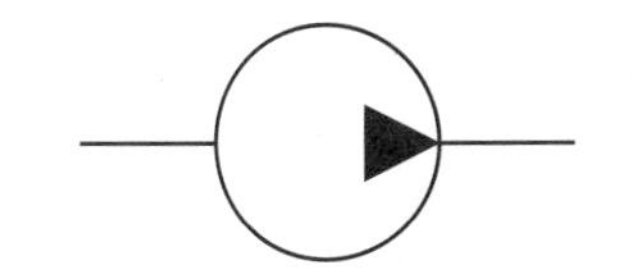

Fig. 3-1 Hydraulic pump symbol

Vane pumps

Vane pumps generate a pumping action by causing vanes to track along a ring.

What a vane pump consists of

A pumping mechanism of a vane pump basically consists of rotor, vanes, ring and a port plate with kidney-shaped inlet and outlet ports (fig. 3-2).

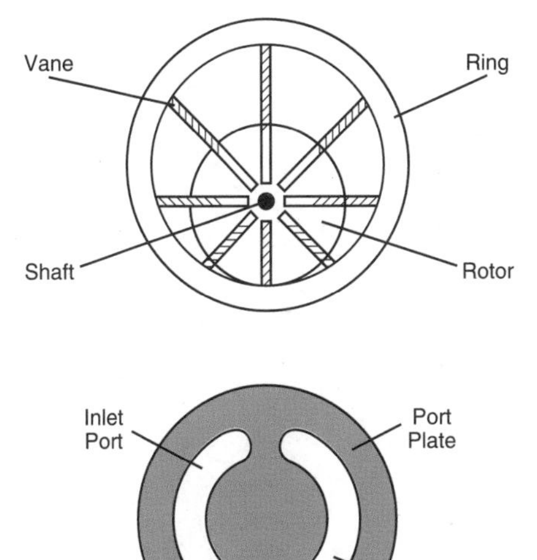

Fig. 3-2 Vane pump components

How a vane pump works

The rotor of a vane pump houses the vanes and it is driven by a shaft which is connected to a prime mover. As the rotor is turned, vanes are thrown out by centrifugal force and track along a ring. (The ring does not rotate.) As the vanes make contact, a positive seal is formed between vane tip and ring.

The rotor is positioned off-center to the ring. As the rotor is turned, an increasing and decreasing volume is formed within the ring (fig. 3-3).

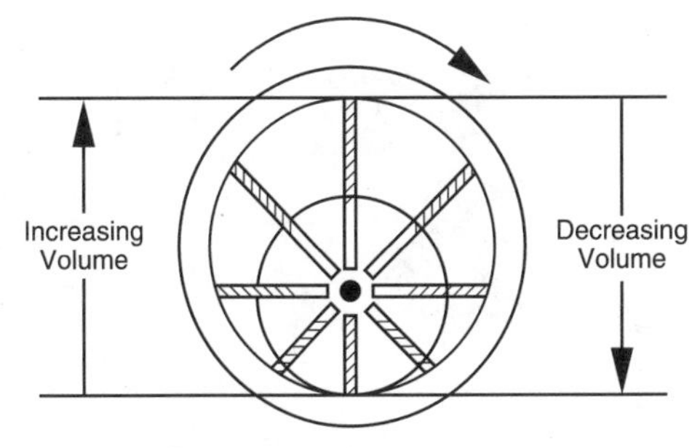

Note: Ring Does Not Rotate

Fig. 3-3 The rotor of a vane pump increases and decreases volume.

Since there are no ports in the ring, a port plate is used to separate incoming fluid from outgoing fluid. The port plate fits over the ring, rotor, and

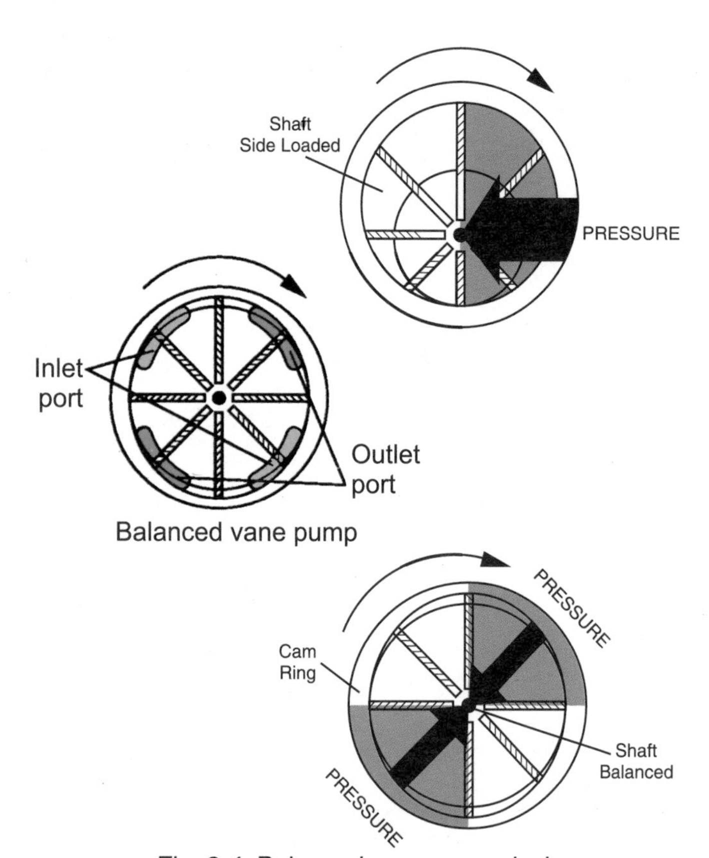

Fig. 3-4 Balanced vane pump design

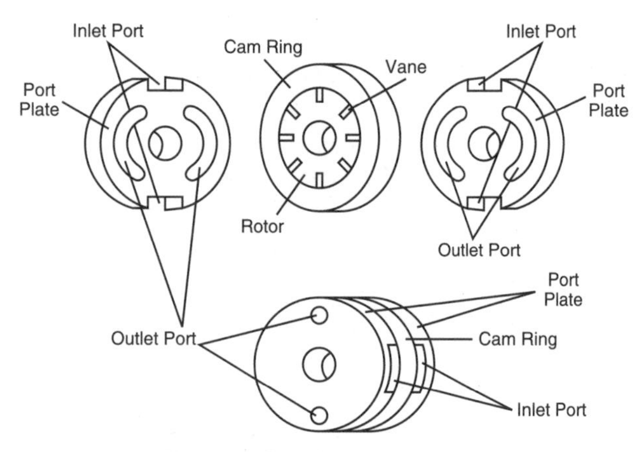

Fig. 3-5 Cartridge assembly

Fig. 3-6 Easy pump service

vanes. The inlet port of the port plate is located where the increasing volume is formed. The port plate's outlet port is located where the decreasing volume is generated. All fluid enters and exits the pumping mechanism through the port plate. (The inlet and outlet ports in the port plate are, of course, connected respectively to the inlet and outlet ports in the pump housing.)

Balanced vane pump design

In a pump, two very different pressures are involved — working pressure of a system and less-than-atmospheric pressure. In the vane pump which has been described, one half of the pumping mechanism is at less-than-atmospheric pressure. The other half is subjected to full system pressure. This results in side loading the shaft which could be severe when high system pressures are encountered. To compensate for this condition, this ring is changed from circular to cam-shaped. With this arrangement, the two pressure quadrants oppose each other and the forces acting on the shaft are balanced (fig. 3-4). Shaft side loading is eliminated.

Therefore, a balanced vane pump consists of a cam ring, rotor, vanes, and a port plate with inlet and outlet ports opposing each other. (Both inlet ports are connected together, as are the outlet ports, so that each can be served by one inlet or one outlet port in the pump housing).

Constant volume, positive displacement vane pumps are generally of the balanced design.

Cartridge assembly

The pumping mechanism of vane pumps is often an integral unit called a cartridge assembly (fig. 3-5). A cartridge assembly consists of vanes, rotor, and a cam ring sandwiched between two port plates. (Note that the port plates of the cartridge assembly are somewhat different in design than the port plates previously illustrated).

An advantage of using a cartridge assembly is easy pump servicing. After a period of time when pump parts naturally wear, the pumping mechanism can be easily removed and replaced with a new cartridge assembly (fig. 3-6). Also, if for some reason the pump's volume must be increased or decreased, a cartridge assembly with the same outside dimension, but with the appropriate volume, can be quickly substituted for the original pumping mechanism.

Vane loading

Before a vane pump can operate properly, a positive seal must exist between vane tip and cam ring. When a vane pump is started, centrifugal force is relied on to throw-out the vanes and achieve a seal. (This is the reason that the minimum operating speed for most vane pumps is 600 rpm). This and other design considerations are the reasons for the **1200-1800 rpm maximum** operating speed limitation for most valve pumps. However, some designs can operate at speeds up to 2700 rpm.

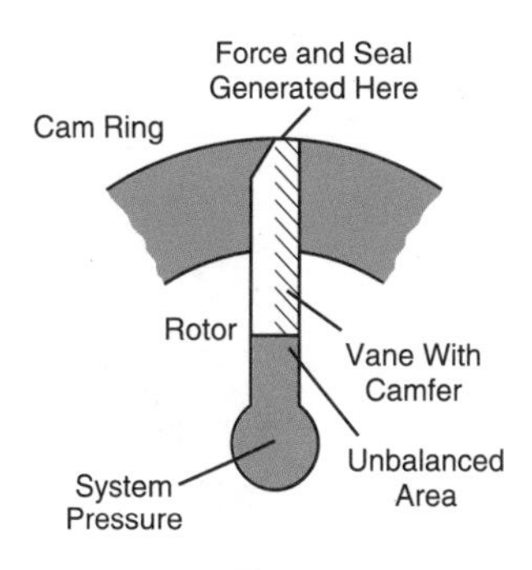

Fig. 3-7

Once the pump is primed and system pressure begins to rise, a tighter seal must exist at the vane so that leakage does not increase across the vane tip. To generate a better seal at high pressures, industrial vane pumps direct system pressure to the underside of the vane. With this arrangement, the higher system pressure becomes, the more force is developed to push the vane out against the cam ring.

Hydraulically loading a vane in this manner develops a very tight seal at the vane tip. But, if the force loading the vane is too great, vanes and cam ring would wear excessively and the vanes would be a source of drag.

As a compromise between achieving the best seal and causing the least drag and wear, manufacturers design their pumps so that the vanes are only partially loaded.

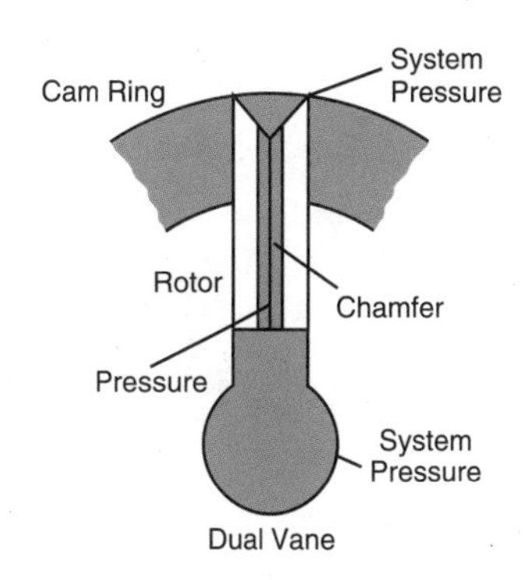

Fig. 3-8 Dual vane construction

The use of vanes with a chamfer or beveled edge (fig. 3-7) is one way in which high vane loading is eliminated. With these vanes, the complete underside vane area is exposed to system pressure as well as a large portion of the area at the top of the vane. This results in a balance of most of the vane. The pressure which acts on the unbalanced area is the force which loads the vane.

In high pressure systems, the use of a vane with a beveled edge still results in too much wear and too much drag. The use of this type vane in a high pressure pump is not satisfactory. Another arrangement is used.

Common vane construction of high pressure vane pumps consists of dual vanes (fig. 3-8), intra-vanes, spring-loaded vanes, pin-vanes, and angled vanes. The dual vane construction consists of two vanes in each vane slot. Each vane

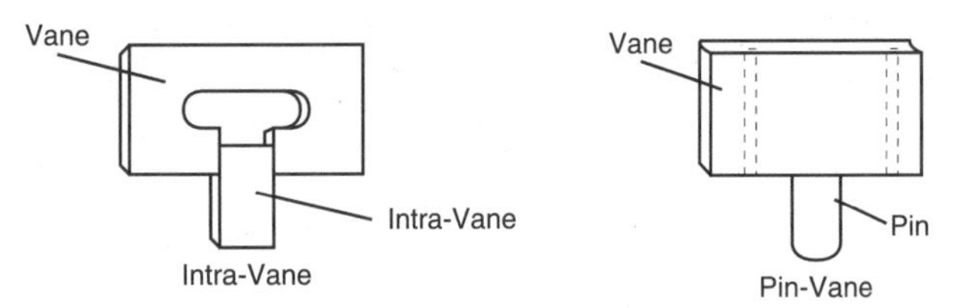

Fig. 3-9 Intra-vane construction

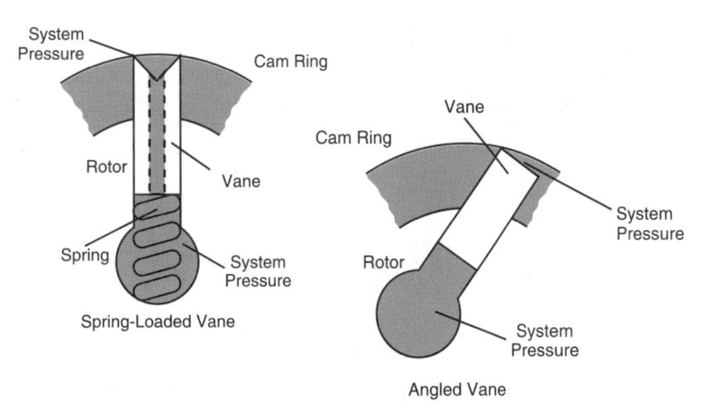

Fig. 3-10 Spring loaded vane (left) and vane loading

Fig. 3-11a Tandem Vane pump

Fig. 3-11b Tandem gear pump

Fig. 3-11c Thru-Drive vane and piston pumps

is almost completely balanced. And, a good seal is achieved because two vanes are used.

The intra-vane (fig. 3-9) is another type of vane construction which consists of a small vane within a large vane with a beveled edge. System pressure is directed to the area above the small vane. This again results in less vane loading.

Very similar to the intra-vane construction is the pin-vane construction. In pin-vane construction pressure is directed to the underside of a pin. The pin then forces the vane out against the cam ring.

With a spring-loaded vane (fig. 3-10), spring pressure at the bottom of the vane is primarily what loads the vane.

Another means of reducing vane loading is positioning the vanes in the rotor at an angle (fig. 3-10). This results in a slight loading of the vane without using any other mechanical device.

Tandem and thru-drive pumps

The vane pump which has been described is referred to as a single pump; that is, it consists of one inlet, one outlet and a single cartridge assembly. Vane and gear pumps are available as "tandem" pumps (fig. 3-11a).

A tandem vane pump consists of a housing with two cartridge assemblies, one or two inlets, and two separate outlets. In other words, a double pump consists of two pumps in one housing. A tandem pump can discharge two different flow rates from each outlet. Since both pump cartridges are connected to a common shaft, one prime mover is used to drive the whole unit.

A tandem gear pump is similar to the tandem vane pump. The tandem gear pump has two pumping units in one housing. However, unlike the vane design, each pumping unit has it's own inlet and outlet ports (fig. 3-11b).

Thru-drive pumps, often refered to as "piggy back" pumps, are many times used in hi-lo circuits and where two different flow rates are supplied from the same power unit. Thru-drive pumps are two or more pumps bolted together, in effect, a common shaft running through all of the pumps (fig. 3-11c). One pump could be used to supply flow to the machine steering system, while the other pump supplies flow to the main hydraulic system.

Gear pumps

Gear pumps (fig. 3-12) generate a pumping action by causing gears to mesh and unmesh.

Fig. 3-12 Gear pumps

What a gear pump consists of

A gear pump basically consists of a housing with inlet and outlet ports, and a pumping mechanism made up of two gears (fig. 3-13). One gear, the drive gear, is attached to a shaft which is connected to a prime mover. The other gear is the driven gear.

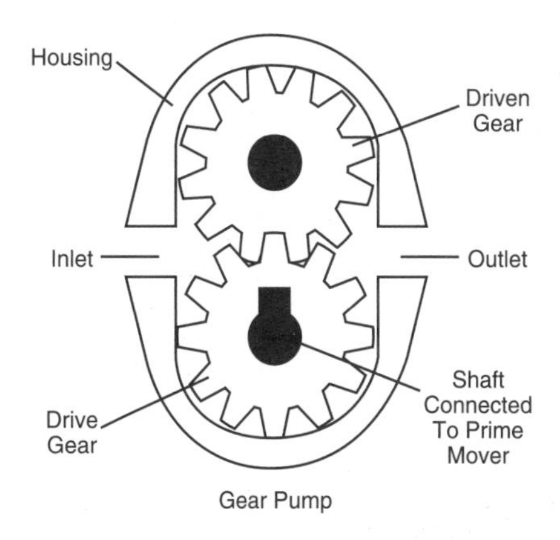

Fig. 3-13 Gear pump components

How a gear pump works

As the drive gear is turned by a prime mover, it meshes with and rotates the drive gear. The action of teeth meshing and unmeshing generates an increasing and decreasing volume. At the inlet where gear teeth unmesh (increasing volume), fluid enters the housing. The fluid is then trapped between the gear teeth and housing, and carried to the other side of the gear (fig. 3-14). At this point, the gear teeth mesh (decreasing volume) and force the fluid out into the system.

A positive seal in this type pump is achieved between the teeth and the housing, and between the meshing teeth themselves.

Gear pumps are generally an unbalanced design.

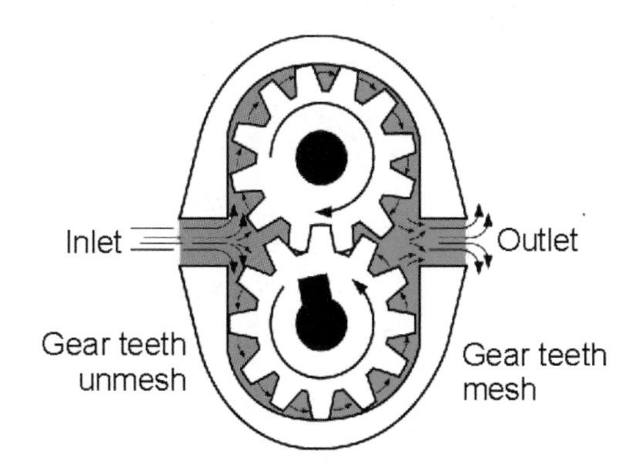

Fig. 3-14 Meshing and unmeshing of gear teeth carry fluid through the pump.

External gear pumps

The gear pump that has been described above is an external gear pump; that is, both meshing gears have teeth on their outer circumferences. These pumps are sometimes referred to as gear-on-gear pumps.

There are basically three types of gears used in external gear pumps — spur, helical, and herringbone (fig. 3-15). Since the spur gear is the easiest to manufacture, this type pump is the most common and the least expensive of the three.

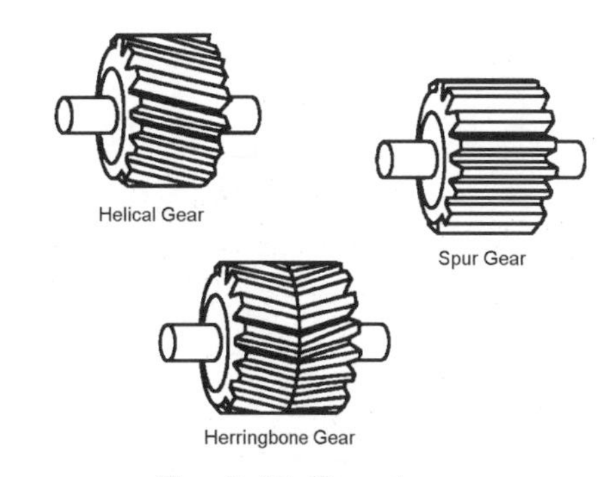

Fig. 3-15 Gear types

internal gear pump

An internal gear pump consists of one external gear which meshes with the teeth on the inside circumference of a larger gear. This type pump is sometimes referred to as gear-within-gear pump. The most common type of internal gear

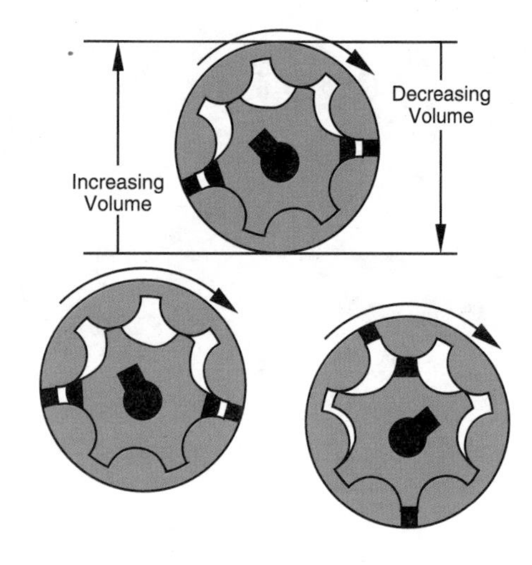

Fig. 3-16 Gerotor pumps use internal gears.

pump in industrial systems is the gerotor pump (fig. 3-16).

Gerotor pump

A gerotor pump is an internal gear pump with an inner drive gear and an outer driven gear. The inner gear has one less tooth than the outer gear.

As the inner gear is turned by a prime mover, it rotates the larger outer gear. On one side of the pumping mechanism, an increasing volume is formed as gear teeth unmesh. On the other half of the pump, a decreasing volume is formed. A gerotor pump has an unbalanced design.

Fluid entering the pumping mechanism is separated from the discharge fluid by means of a port plate as in a vane pump.

While fluid is carried from inlet to outlet, a positive seal is maintained as the inner gear teeth follow the contour of crests and valleys of the outer gear.

Piston pumps

Piston pumps generate a pumping action by causing pistons to reciprocate within a piston bore.

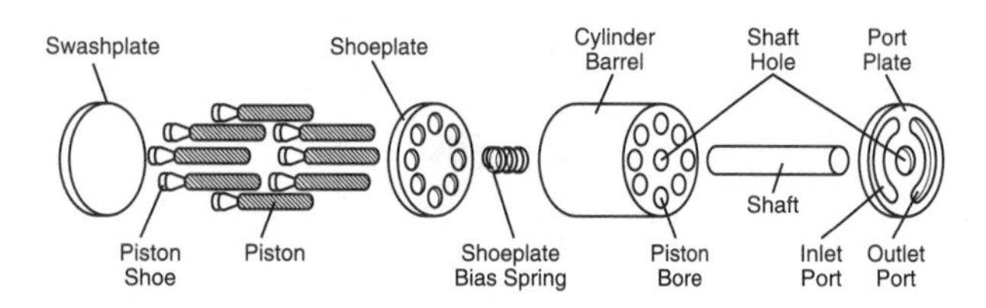

Fig. 3-17 Piston pump components

What a piston pump consists of

The pumping mechanism of a piston pump basically consists of a cylinder barrel, pistons with shoes, swashplate, shoeplate, shoeplate bias spring and port plate (fig. 3-17).

How a piston pump works

Earlier, we have seen one example of a piston pump. This pump generates an increasing and decreasing volume by means of a plunger being pulled and pushed in and out of a cylinder body. It was pointed out that the disadvantages of this type of pump were that the pump developed a pulsating flow and that it could not be easily operated by an electric motor or internal combustion engine.

However, a piston can be made to reciprocate easily by the turning motion of a prime mover as well as developing a smooth flow.

In the example illustrated (fig. 3-18), a cylinder barrel with one piston bore is fitted with one piston. A swashplate is positioned at an angle. The shoe of the piston rides on the surface of the swashplate.

As the cylinder barrel is rotated, the piston shoe follows the surface of the swashplate. (The swashplate does not rotate.) Since the swashplate is at an angle, this results in the piston reciprocating within the bore. In one half of the circle of rotation, the piston moves out of the cylinder barrel and generates an increasing volume.

In the other half of the circle of rotation, this piston moves into the cylinder barrel and generates a decreasing volume (fig. 3-19). In actual practice, the cylinder barrel is fitted with many pistons. The shoes of the pistons are forced against the swashplate surface by a shoeplate and bias spring. To separate the incoming fluid from the discharge fluid, a port plate is positioned at the end of the cylinder barrel opposite the swashplate.

A shaft is attached to the cylinder barrel which connects it with the prime mover. This shaft can be located at the end of the barrel where the porting is taking place. Or, more commonly, it can be positioned at the swashplate end. In this case, the swashplate and shoeplate have a hole in their centers to accept the shaft. If the shaft is positioned at the other end, just the port plate has a shaft hole.

The piston pump which has been described above is known as an axial or in-line piston pump; that is, the pistons are rotated about the same axis as the pump shaft.

Axial piston pumps are the most popular piston pumps in industrial applications. Other types of piston pumps include the bent-axis and radial piston pumps.

Bent-axis piston pump

The pumping mechanism of a bent-axis piston pump consists of a cylinder barrel, pistons, port plate, flange, piston linkages, and a drive shaft (fig. 3-20). In this pump the cylinder barrel is not in-line with the drive shaft, but is at an angle to the shaft (fig. 3-21). Because of this arrangement, as the shaft is turned the pistons are pulled out of the barrel during one half of the barrel rotation. This generates an increasing volume. On the other half of the barrel rotation, the pistons are pushed in and a decreasing volume is formed. In this pump, incoming fluid is sepa-

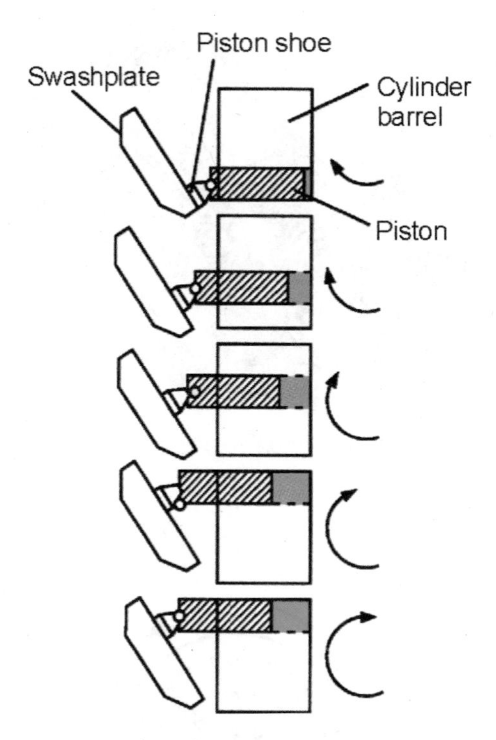

Fig. 3-18

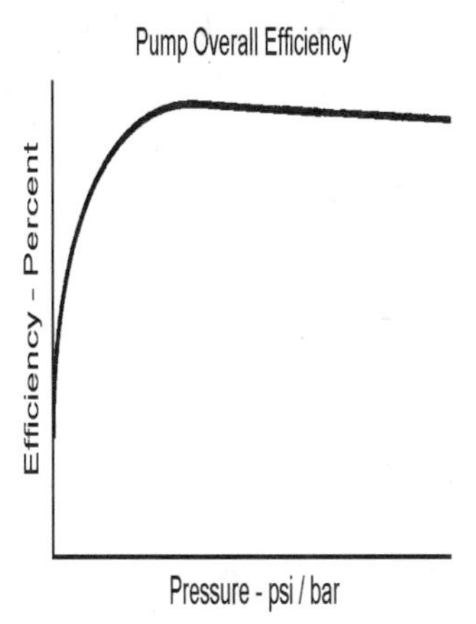

Fig. 3-19

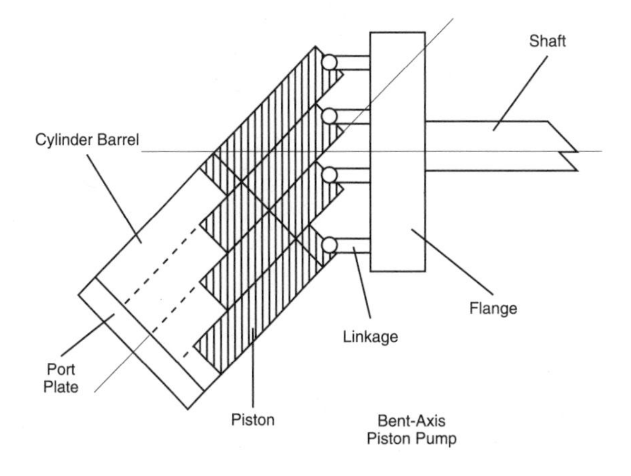

Fig. 3-20 Bent-axis piston pump components

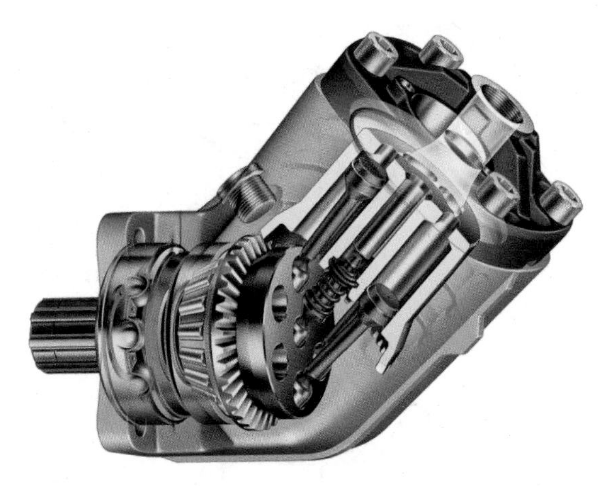

Fig. 3-21 Cylinder barrel is angled to drive shaft

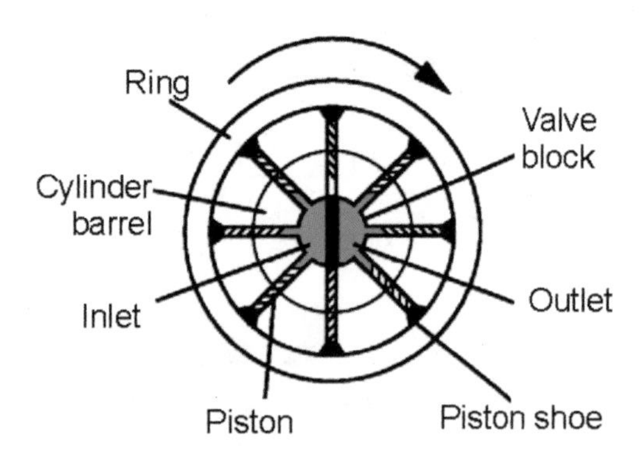

Fig. 3-22 Radial piston pump components

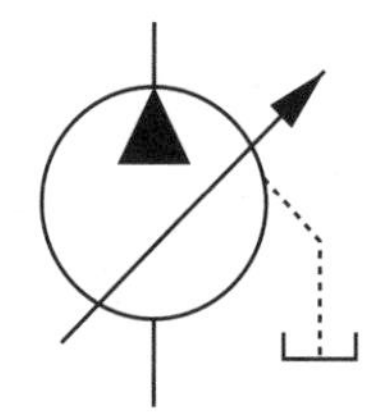

Fig. 3-23 Variable volume pump symbol

rated from discharge fluid by means of a port plate.

Radial piston pump

The pumping mechanism of a radial piston pump basically consists of a cylinder barrel, pistons with shoes, a ring, and a valve block (fig. 3-22).

The action of a radial piston pump is quite similar to a vane pump. But, instead of using vanes to track along a ring, the pump uses pistons.

The cylinder barrel, which houses the pistons, is positioned off-center to the ring. As the cylinder barrel is rotated, an increasing volume is formed within the cylinder barrel during one half of the barrel rotation. During the other half, a decreasing volume is formed.

Fluid enters and is discharged from the pump through the valve block in the center of the pump.

Variable volume pumps

A positive displacement pump delivers the same volume of fluid for each revolution. Industrial pumps are generally operated at constant 1200 or 1800 rpm. On the other hand, mobile pumps operate at speeds up to 2500 to 3000 rpm. This infers that the pump flow rate remains constant.

In some cases, it is desirable that a pump's flow rate be variable. One way of accomplishing this is by varying the speed of the prime mover. This is usually economically impractical. The only other way then, to vary the output of a pump is to change its displacement. The symbol for such a pump is shown in fig. 3-23.

The amount of fluid such a pump displaces can be changed by moving the cam ring in the case of a vane pump, or the swashplate in a piston pump. This is known as a variable volume or displacement pump.

Variable volume from a gear pump

The output volume of a gear pump is determined by the volume of fluid each gear displaces and by the rpm at which the gears are turning. Consequently, the output volume of gear pumps can be altered by replacing the original gears with gears of different dimensions or by varying the rpm.

Gear pumps, whether of the internal or external variety, do not lend themselves to a change in displacement while they are operating. There is nothing that can be done to vary the physical dimensions of a gear while it is turning.

One practical way then to vary the output flow from a gear pump is to vary the speed of its prime mover. This can be easily done when the pump is being driven by a variable speed internal combustion engine. It can also be accomplished electrically by using a variable speed electric motor.

Regulating hydraulic power with variable volume pumps

Since hydraulic power consists of l/min *(gpm)* and bar *(psi)*, power generation by pump/motor or engine can be controlled by reducing the generation of flow as well as limiting pressure. This is accomplished with the use of a variable volume vane or piston pump which is pressure compensated. These pumps have the ability to reduce their displacement once system pressure reaches a preset level.

The intent of this lesson is to describe how pressure compensated, variable volume pumps can more easily match hydraulic power generated with mechanical power output and how wear affects their operation.

Fig. 3-24 shows the symbol for a pressure compensated variable volume pump.

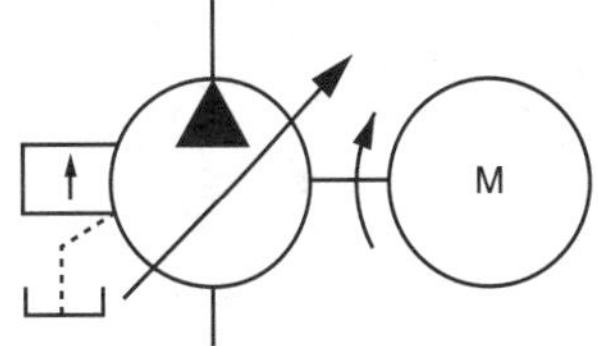

Fig. 3-24 Pressure compensated variable volume pump symbol

What a variable volume vane pump consists of

The pumping mechanism of a variable volume vane pump basically consists of a rotor, vanes, a cam ring which is free to move from side to side, a port plate, a thrust bearing to guide the cam ring, and something to vary the position of the cam ring. In fig. 3-25, a screw adjustment is used.

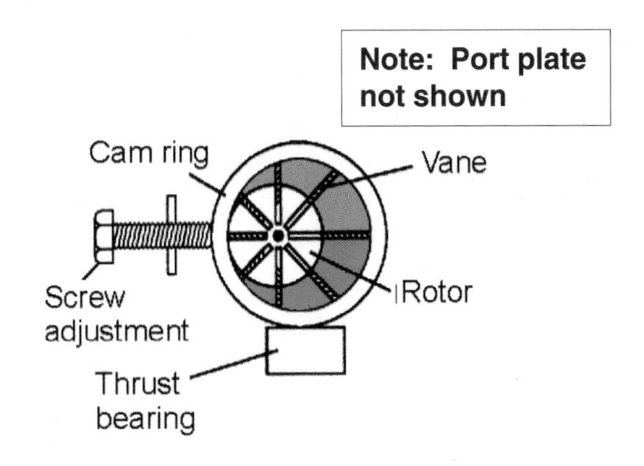

Fig. 3-25 Variable volume vane pump components

Variable volume vane pumps are unbalanced pumps. Their rings are circular and not cam-shaped. However, they are still referred to as cam rings.

Since the cam ring in this type pump must be free to move, the pumping mechanism does not come as a cartridge assembly.

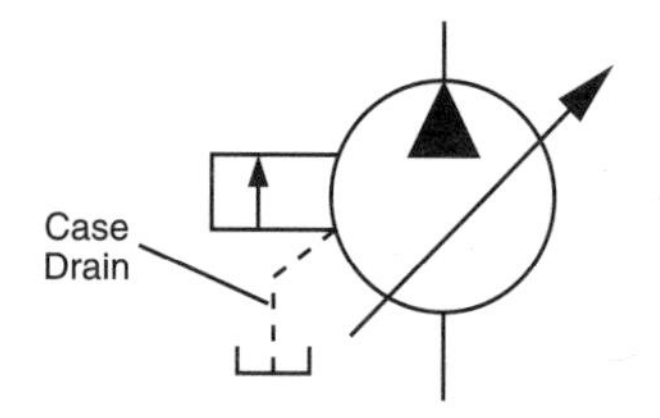

Fig. 3-26 Case drain shown on left

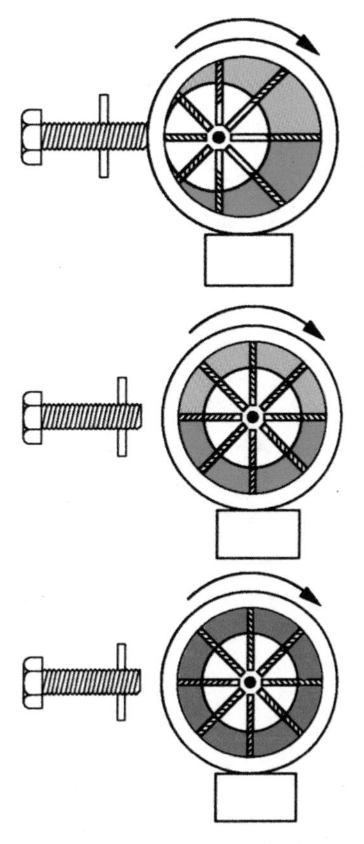

Fig. 3-27 Vane pumps can change output flow by means of the screw adjustment.

Case drain

All variable volume, pressure compensated pumps must have their housings externally drained (fig. 3-26). The pumping mechanisms in these pumps move extremely fast when pressure compensation is required. Any buildup of fluid within the housing would hinder their movement.

Also, any leakage which accumulates in a pump housing is generally directed back to the pump's inlet side. The leakage from a variable volume pump, while it is compensating, is generally hot. If it were diverted to the inlet side, the fluid would get progressively hotter. Externally draining the housing alleviates the problem.

The external drain of a pump housing is commonly referred to as a case drain.

How a variable volume vane pump works

With the screw adjusted in, the rotor is held off center with regard to the cam ring (fig. 3-27). When the rotor is turned, an increasing and decreasing volume is generated. Pumping occurs.

With the screw adjustment turned out slightly, the cam ring is not as off center to the rotor as before. An increasing and decreasing volume is still being generated, but not as much flow is being delivered by the pump. The exposed length of the vanes at full extension has decreased.

With the screw adjustment backed completely out, the cam ring naturally centers with the rotor. No increasing and decreasing volume is generated. No pumping occurs. With this arrangement a vane pump can change its output flow anywhere from full flow to zero flow by means of the screw adjustment.

How the pressure compensator of a variable volume vane pump works

The pressure compensator of a variable volume vane pump is the large adjustable spring which holds the cam ring off center against the volume adjustment (fig. 3-28). This causes an increasing and decreasing volume to form and pumping to occur. One half of the ring is under less-than-atmospheric conditions and the remaining half is pressurized.

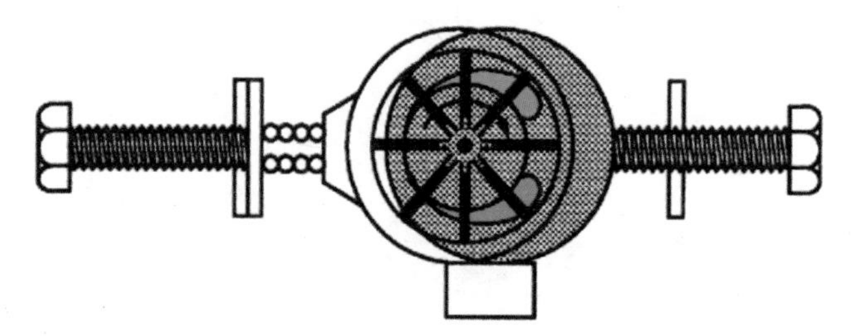
Fig. 3-28 Pump compensating

As pumping occurs, pressure generates a force on the inner surface of the cam ring. Because of the design of the kidney-shaped ports in the port plate, the generated force tends to move the ring in the main direction of, but not exactly at, the thrust bearing. A portion of the force is taken up by the compensator spring.

When pressure on the ring and force become large enough, the ring begins to move in the direction of the spring reducing discharge flow. The ring cannot move through the thrust bearing because it is a mechanical stop. Once the ring centers with the rotor, it stops and discharge flow ceases. The cam ring will not cross to the other side of the rotor because this would cause the pressure half of the pumping mechanism to form an increasing volume which is suction operated. Pressure would immediately drop resulting in the spring pushing the ring off center once again.

The more the pressure compensator adjustment is screwed in, the more the compensator spring is compressed and the greater the force holding the ring off center. To protect the pump from too high of a pressure adjustment, pressure compensated vane pumps are commonly equipped with over pressure stops.

Vane pump dual compensation

In a vane pump, a dual compensator consists of a solenoid valve, pilot valve, and orifice which senses system pressure (fig. 3-29).

With the solenoid valve closed, fluid pressure is allowed to accumulate in the spring chamber through the sensing orifice. This adds to the spring pressure biasing the cam ring against the volume adjustment. The maximum amount of fluid pressure is determined by the pilot valve setting. Compensator setting is a function of spring and fluid pressure.

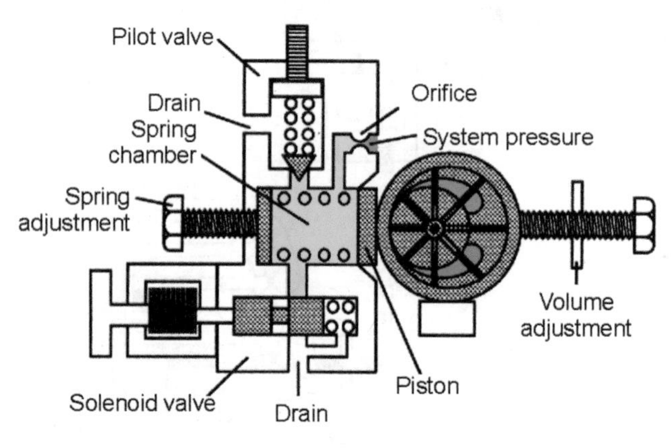

Fig. 3-29

If the spring were adjusted for 41 bar *(600 psi)* and the pilot valve were adjusted for 62 bar (*900 psi),* the compensator setting would be 103 bar *(1500 psi).*

A second compensator setting is achieved when the solenoid valve is energized. This action vents the spring chamber of fluid pressure. Spring pressure only holds the cam ring against the volume adjustment. Since this has a value of

41 bar *(600 psi)* in this example, the pump now compensates at 41 bar *(600 psi).*

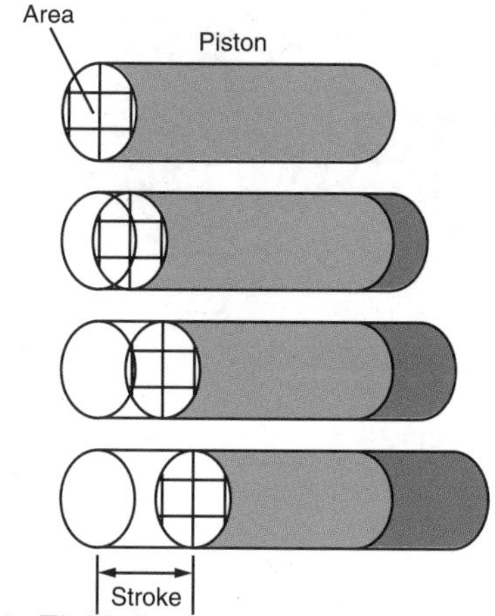

Fig. 3-30 Fluid displacement can be changed by varying stroking distance of pistons.

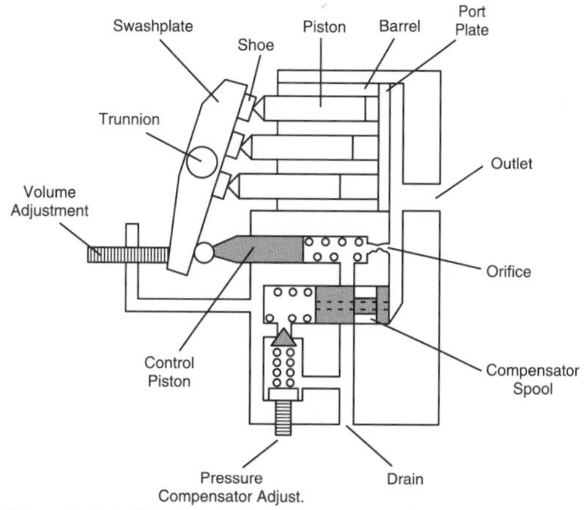

Fig. 3-31 Variable volume axial piston pump components

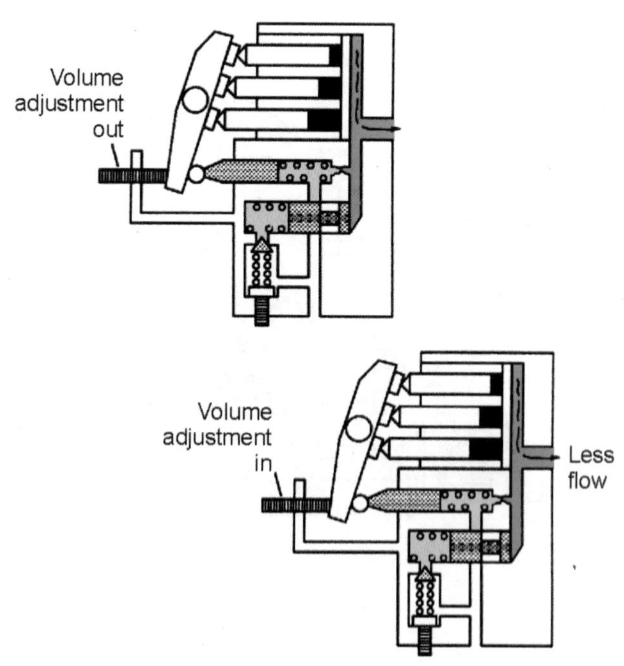

Fig. 3-32 Volume adjustment controls flow via the swashplate angle

Variable volume axial piston pump

The amount of fluid which an axial piston pump displaces per revolution is determined by the cross sectional area of the piston and the distance pistons are stroked within their cylinder bores (fig. 3-30). While pumping takes place, nothing can be done to change the cross sectional area of the pistons. But, a variable volume axial piston pump is designed so that the distance pistons are stroked can be varied.

What a variable volume axial piston pump consists of

The pumping mechanism of a variable volume axial piston pump basically consists of a barrel, port plate, pistons with shoes, an adjustable swashplate, shoeplate, maximum volume adjustment, control piston, orifice, compensator spool, pilot valve, and various bias springs (fig. 3-31).

Pistons are housed in the bores of the barrel. Attached to the barrel is a shaft which is coupled to an electric motor. The shaft can be connected at either the port plate or swashplate side.

How the volume adjustment of a variable volume axial piston pump works

The volume adjustment of a variable volume axial piston pump is a threaded rod which mechanically limits the swashplate angle (fig. 3-32).

With the volume adjustment backed completely out, the spring biasing the control piston positions the swashplate at an extreme angle. As the barrel is rotated by a prime mover, piston shoes are forced by mechanical bias to contact the swashplate surface resulting in the pistons reciprocating within the bore. In one half of the circle of rotation, pistons move out of the cylinder barrel generating an increasing volume; fluid fills the increasing volume through the inlet port. In the other half of the circle of rotation, pistons are pushed into the cylinder barrel generating a decreasing volume; fluid is pushed out to the system through the outlet port.

With the swashplate at its extreme angle, maximum volume discharges from the pump. If the threaded rod were screwed in slightly so that the swashplate had less of an angle, pistons would not reciprocate to such a great degree. This results in less fluid discharging into the system.

The more the volume adjustment of a variable volume axial piston pump is screwed in, the less flow is discharged.

How the pressure compensator of a variable volume axial piston pump works

The pressure compensator of a variable volume axial piston pump uses a combination of spring and fluid pressure.

In the pressure compensated vane pump seen previously, a single spring acted as the compensator. This was possible because the force moving the ring did not act directly opposite the spring, but at an angle. Most of the force was taken up by the thrust bearing. In a pressure compensated axial piston pump, this is not the case.

As a pressure compensated axial piston pump is operating, pressure acts on the ends of the pistons doing the pumping. With the centerline of the swashplate trunnion offset from the centerline of the barrel, the pistons under pressure attempt to push the swashplate to an upright position. This force can be considerable and is counteracted directly by spring and fluid pressure acting on the control piston.

A pressure compensator of a typical axial piston pump consists of a control piston, compensator spool, adjustable pilot valve, and bias springs (fig. 3-33). When pressure at pump outlet is high enough to overcome the setting of the pilot valve and the bias springs, the swashplate is pushed back to an upright position by the pressurized pistons.

Once the swashplate achieves a zero angle, it stops. The swashplate cannot be pushed to the other side of center since this would cause the pressure half of the pumping mechanism to form an increasing volume which is suction operation. Pressure would immediately drop resulting in the spring offsetting the swashplate once again.

The adjustable pilot valve, compensator spool and its bias spring are very similar in operation to a pilot operated relief valve. Pump discharge pressure is sensed through an orifice in the com-

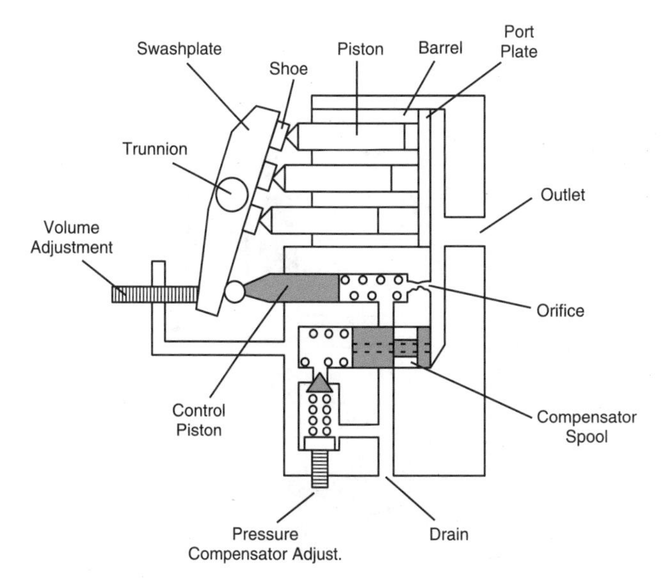

Fig. 3-33 Variable volume axial piston pump components

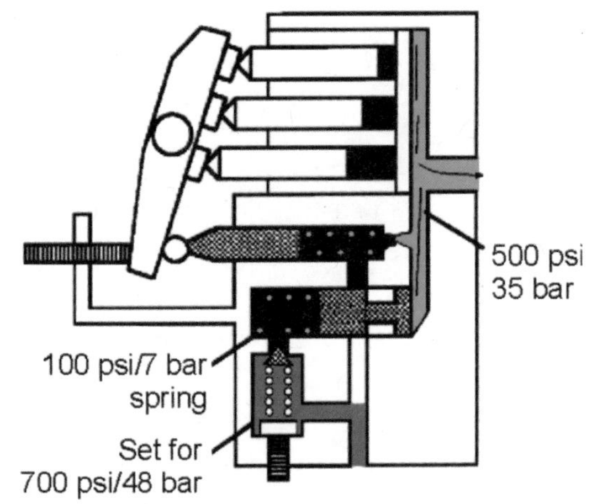

Fig. 3-34 Pressure in spring chamber limited to 48 bar (700 psi)

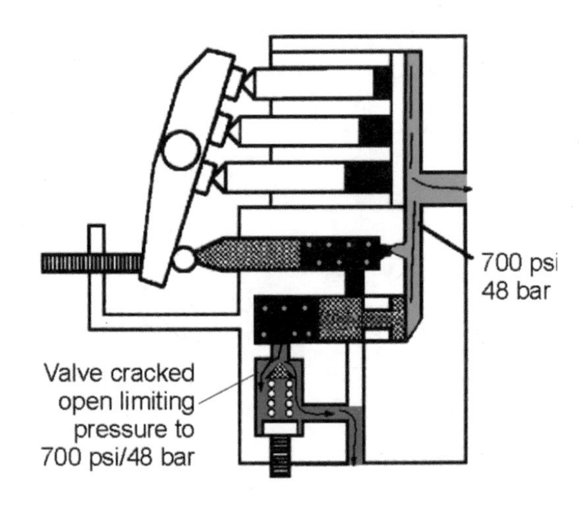

Fig. 3-35 Hydraulic pressure of 48 bar (700 psi) hold the spool

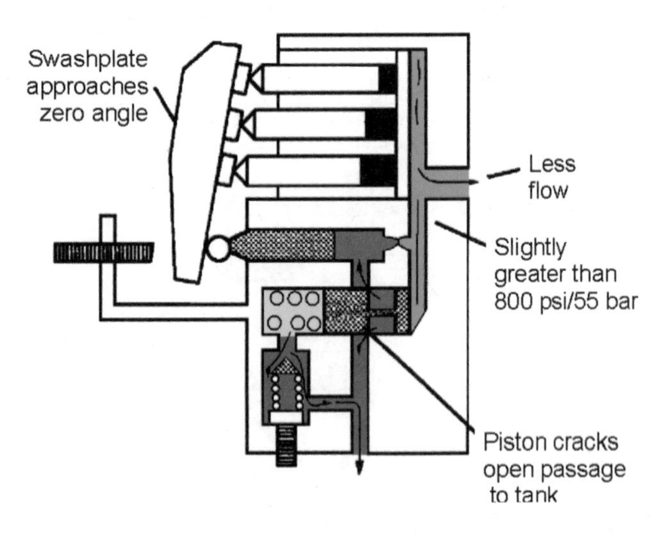

Fig. 3-36 Maximum pressure biasing the compensator spool is 55 bar (800 psi)

pensator spool. It is allowed to accumulate in the bias spring chamber of the spool and the amount of accumulated pressure is limited by the pilot valve. When pressure at the end of the compensator spool is sufficient to overcome the pilot valve setting and the bias spring, the compensator spool will move venting the control piston spring chamber.

Assume that the spring biasing the compensator spool has a value of 7 bar *(100 psi)* and that the pilot valve will limit pilot pressure in the spring chamber to 48 bar *(700 psi)* (fig. 3-34).

With a system pressure of 35 bar (*500 psi*), 35 bar (*500 psi*) acts to push the spool to the left. 35 bar (*500 psi*) is transmitted through the orifice to the spring chamber acting to hold the spool in place. Areas exposed to pressure on either side of the spool are equal; therefore, the spool is balanced except for the 7 bar (*100 psi*) spring. At this point, 35 bar (*500 psi*) attempts to push the spool to the left and a total hydraulic and mechanical pressure of 41 bar *(600 psi)* holds the spool in place.

With a system pressure of 41 bar *(600 psi)*, 41 bar *(600 psi)* at the right end of the spool acts to push the spool to the left. A total mechanical and hydraulic pressure of 48 bar *(700 psi)* acts to hold the spool (fig. 3-35).

When system pressure climbs to 48 bar *(700 psi)*, the pilot valve opens limiting the pilot/spring chamber to 48 bar *(700 psi)*. This means the maximum pressure biasing the compensator spool is 55 bar *(800 psi)* (fig. 3-36).

As pump discharge pressure climbs above 55 bar *(800 psi)*, the spool moves cracking open the control piston cavity. Pressure in this cavity begins to drop resulting in a smaller generated force at the control piston and an inability to keep the swashplate fully displaced. Consequently, discharge flow decreases. As pressure at pump outlet increases, the compensator spool moves more, resulting in less discharge flow (fig. 3-37). In this example, let's assume that as pressure climbs an additional 7 bar *(100 psi)* above 55 bar *(800 psi)*, the spring biasing the compensator spool will have been compressed, and the spool will have moved sufficiently to vent the control piston cavity, destroking or compensating the pump. The pump is fully compensated at 62 bar *(900 psi)*.

(The spring biasing the control piston is relatively light and can be neglected in calculating compensating pressure.)

When pressure drops at pump outlet, the compensator spool closes off the control cavity deventing the chamber. Pumping action resumes.

The more the pressure compensator pilot valve adjustment is screwed in, the more pilot pressure it allows to bias the compensator spool before compensation occurs.

With the operation of pressure compensated, variable volume vane and axial piston pumps in mind, we compare in the next section their operation with a fixed pump.

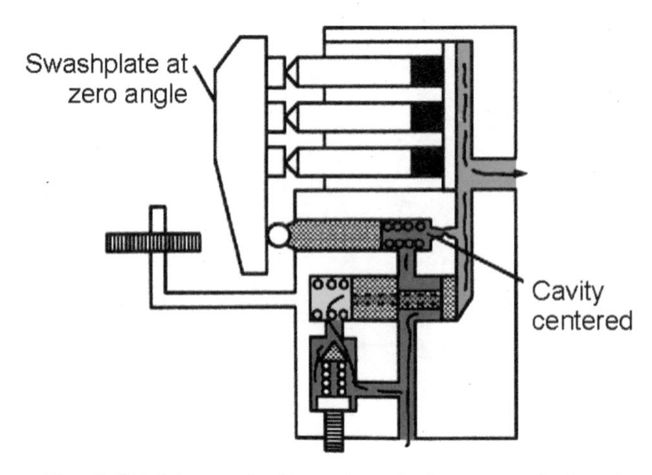

Fig. 3-37 The cavity is centered when swashplate is at zero angle.

Piston pump dual compensation

A dual compensator of a piston pump incorporates a solenoid valve, and another pilot valve (fig. 3-38).

With the solenoid valve closed, the pump compensates in its normal manner. Fluid pressure is allowed to accumulate in the compensator spool spring chamber. The amount of pressure is determined by the main pilot valve. This pressure adds to the compensator spring pressure biasing the compensator spool.

If the pilot valve were adjusted for 90 bar *(1300 psi)* and the compensator spring allowed pressure to climb an additional 14 bar *(200 psi)* before full compensation, the pump would be fully compensated at 103 bar *(1500 psi).*

Assume now that the auxiliary pilot valve is adjusted for 28 bar *(400 psi).* When the solenoid valve is energized, the auxiliary pilot valve connects with the compensator spring chamber. Maximum pressure is allowed to accumulate in the chamber is now 28 bar *(400 psi).* With the spring having a full compensation value of 14 bar *(200 psi)*, the pump now compensates at 41 bar *(600 psi).*

With a dual compensator accessory, generated power can more evenly match output power at actuators with different requirements.

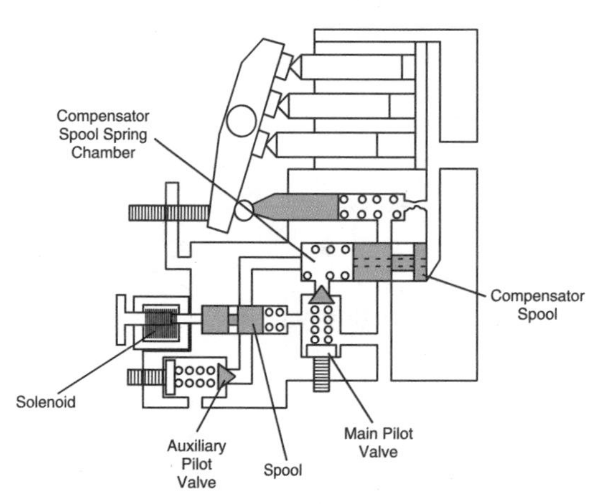

Fig. 3-38 Piston pump dual compensator components

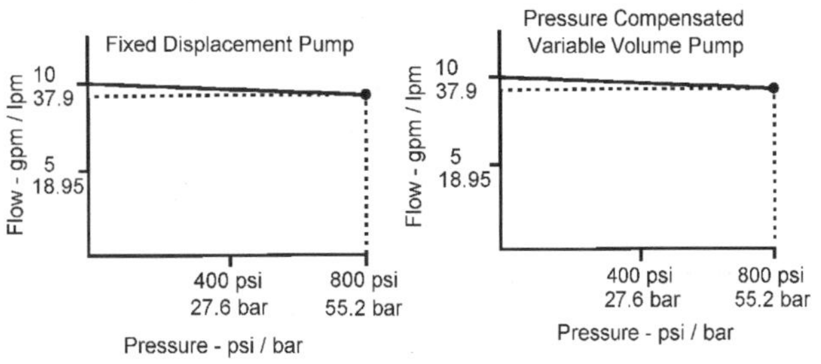

Comparing fixed and pressure compensated variable volume operation

Illustrated are two curves (fig. 3-39). One curve shows the discharge flow from a 37.9 l/min *(10 gpm)* fixed displacement pump as it operates between zero and 55.2 bar *(800 psi)*. The other curve shows the discharge flow from a 37.9 l/min *(10 gpm)* variable volume pump as it operates between the same pressures. Both curves are the same and they point out that the discharge flow rate from each unit at 55.2 bar *(800 psi)* is approximately 35.05 l/min *(9.25 gpm)*; this is 3.7 kW *(4.3 hp)*.

Fig. 3-39

Assume that the fixed displacement pump/ electric motor pressure is limited by a relief valve setting of 68.97 bar *(1000 psi)* (fig. 3-40). Referring to the fixed displacement curve between zero and 68.97 bar *(1000 psi)*, we see that at 68.97 bar *(1000 psi)* pump/electric motor is developing 34.11 l/min *(9 gpm)* which is 3.8 kW *(5.2 hp)*. This is the maximum power that the unit is allowed to develop; and as it passes through a relief valve, it is transformed into heat.

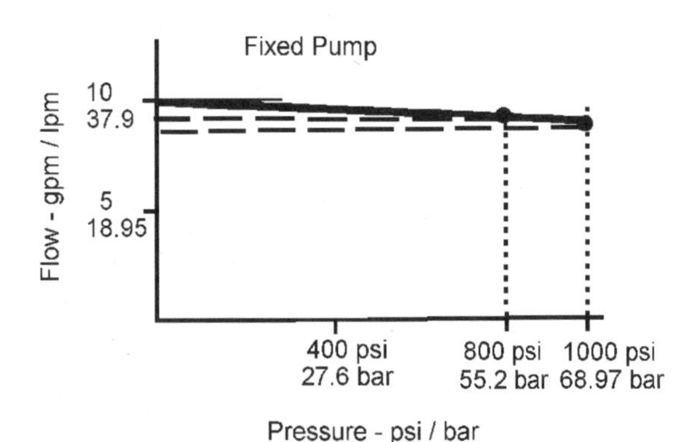

Fig. 3-40

Pressure compensated vane pumps are characteristically fully compensated within 13.8 bar *(200 psi)*. Piston pumps compensate fully within 6.9 bar *(100 psi)* or less.

Assume now that the pressure compensated variable volume pump has a compensator setting of 68.97 bar *(1000 psi)* (fig. 3-41). Referring to the pressure compensated pump curve between zero and 68.97 bar *(1000 psi)*, we find that at 55.2 bar *(800 psi)* and 34.86 l/min *(9.25 gpm)* (3.7 kW/*4.3 hp*), discharge flow begins to decrease. Consequently, generated power decreases as system pressure approaches 68.97 bar *(1000 psi)*. At a pressure of 62.1 bar *(900 psi)* discharge flow is 30.3 l/min *(8 gpm)* (3.1 kW/ *4.2 hp)*. At 67.85 bar *(950 psi)*, discharge flow is 15.16 l/min *(4 gpm)* (1.6 kW/*2.2 hp)*. At 68.97 bar *(1000 psi)*, discharge flow rate is zero l/min *(0 gpm)*.

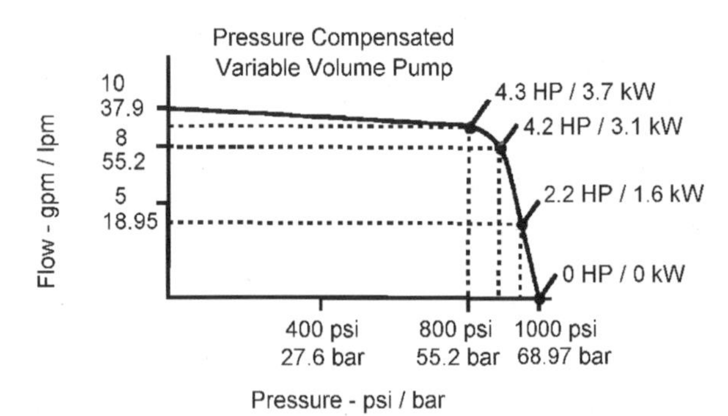

Fig. 3-41

With a pressure compensated pump coupled to a prime mover, generated hydraulic power is not only controlled by limiting pressure, but it is reduced by regulating discharge flow. This means less heat is generated in a system since generated power is not wasted over a relief valve.

We have seen that when a pressure compensated pump is fully compensated, the generated hydraulic power to the system is zero. However, the prime mover coupled to the pump must still develop power because of internal leakage and mechanical inefficiencies of the pump.

Overcenter axial piston pumps

As was illustrated, the displacement of an axial piston pump, and therefore its output volume, can be varied by changing the angle of the swashplate. It was also shown that the pump will develop no flow when the swashplate is centered with the cylinder barrel.

Some swashplates of axial piston pumps have the capability of crossing over center (fig. 3-41a). This results in the increasing and decreasing volumes being generated at opposite ports. Flow through the pump reverses.

From the overcenter axial piston pump illustrated, it can be seen that ports A and B can be either an inlet or outlet port depending upon the angle of the swashplate. This takes place with the cylinder barrel rotating in the same direction.

Overcenter axial piston pumps are often used in hydrostatic transmissions which will be covered in a later section.

Axial piston pumps can be variable displacement, pressure compensated and variable displacement, or variable displacement and overcenter. These combinations are also available with the bent-axis and radial design piston pumps.

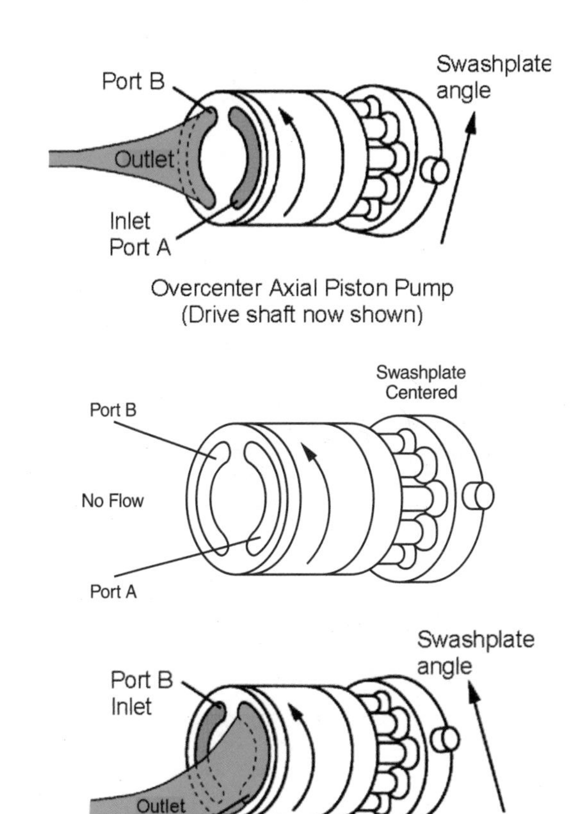

Fig. 3-41a Flow can be reversed by crossing over center

Load sensing theory of operation

The object of a load sensing system is to produce flow and pressure only upon demand, and only in amounts necessary to perform the needed work functions. The result is a system that wastes very little energy. A load sensing system also provides very precise control of hydraulic functions which would be difficult to achieve by any other means without converting significant amounts of energy into unwanted **heat**. This style of control requires a "load sensing" or flow control orifice at the pump outlet port (fig. 3-42). It is the pressure drop (ΔP) established across this orifice that determines the required pump displacement.

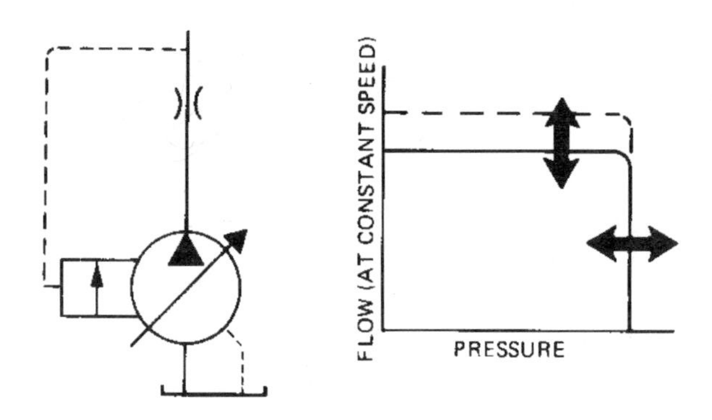

Fig. 3-42 Flow control orifice at pump outlet port senses load.

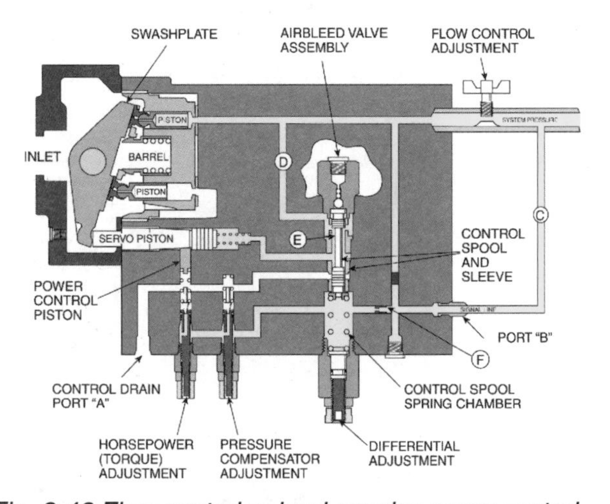

Fig. 3-43 Flow control or load sensing pump control

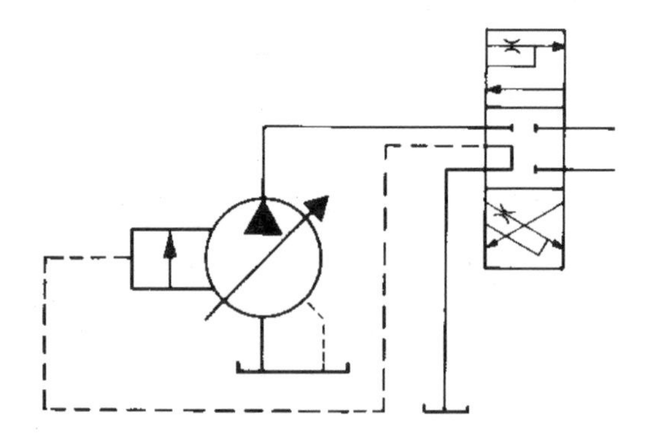

Fig. 3-44 A directional control valve may be used to meter flow.

LOAD PRESSURE	+	**CONTROL ORIFICE ΔP**	=	**PUMP PRESSURE**

Many variable displacement pumps used in a load sensing system have a control that will "destroke" or reduce the displacement of the pump whenever the difference in two pressures, supplied to the pump control mechanism, exceeds a preset value. Oppositely, the pump will "upstroke" or increase in displacement whenever the difference in the two pressures supplied to the pump control mechanism is less than the preset value. The pump displacement will reach its maximum value whenever the two pressures supplied to the control mechanism are equal (fig. 3-43).

In this manner, the pump will always try to maintain a constant pressure drop across the flow control orifice that is equal to the pump differential pressure control setting.

Since the pump control responds only to the difference between the two control pressures, the flow will be the same regardless of the magnitude of the pressure in the system. Also, since the pump control responds only to the pressure drop across the orifice, the flow will remain relatively constant regardless of the drive speed of the pump within the constraints of the pump's maximum displacement.

With a variable volume pump it is possible to control flow and still maintain pump pressure very near that required to move the load. This may require a signal line to feed pilot oil to the pump control section from a point downstream of an orifice such as a needle valve or directional control valve (fig. 3-44) being used to meter flow. The signal or pilot line is used to transmit the pressure sensed between the load and the orifice to the pump control mechanism which translates the signal into a flow required to maintain a certain pressure drop (ΔP) between the pump outlet and the sensing point which is downstream of the flow control orifice.

The principle benefit of load sensing is the ability to save energy by not requiring the pump to compensate at full pressure; instead, the pump will begin to compensate at some pressure close to the pressure required to move the load. Typically at the "load pressure" plus the flow control orifice pressure drop (ΔP).

Some examples will illustrate the benefit of a "load sensing" pump control. The first illustration is a constant pressure system and controlling flow at something less than the pump's total capability. The schematic (fig. 3-45) shows that the system is set up with a variable volume pump with its compensator set at 138 bar *(2000 psi)*. It is a 57 l/min *(15 gpm)* pump. The flow is restricted with a variable orifice (needle valve) such that the actuator receives only 30 l/min *(8 gpm)*.

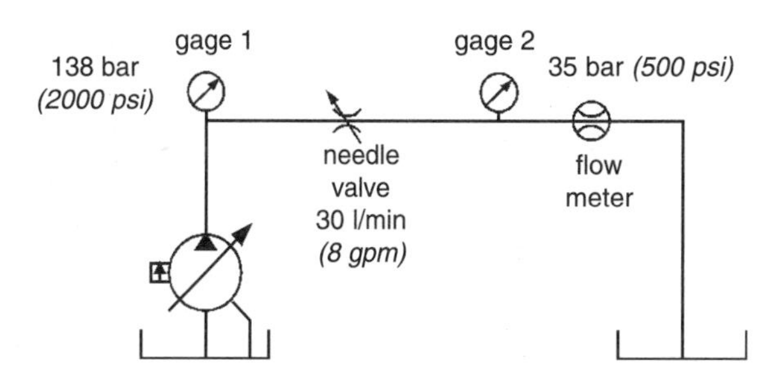

Fig. 3-45 Load sensing pump control circuit

The pressure necessary to move the load is 35 bar *(500 psi)*. The pressure at gage (1) would read approximately 138 bar *(2000 psi)* (the compensator setting of the pump). This can be seen by reviewing the operating (flow) curve for this style of pump (fig. 3-46). By following the 30 l/min *(8 gpm)* flow line to the pump operating curve and then down to the pressure axis, a pressure very near 138 bar *(2000 psi)* is revealed. The 138 bar *(2000 psi)* pressure would be the worst case and is therefore taken here.

This pump is operating in the partial compensation region, the very steep slope of the flow curve.

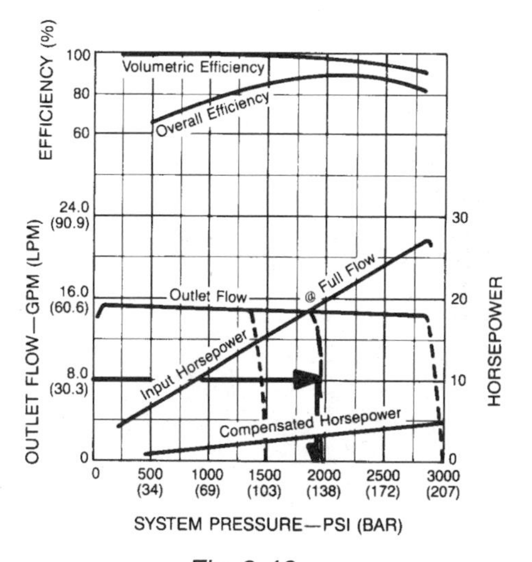

Fig. 3-46

The pressure at gage 2 downstream of the flow control orifice (needle valve) would read 35 bar *(500 psi)* (the resistance caused by the load).

Therefore, we have a pressure drop of 103 bar *(1500 psi)* across the variable flow control orifice 138 bar - 35 bar = 103 bar *(2000 psi - 500 psi = 1500 psi)*. Therefore, utilizing the hydraulic horsepower formula (hp = Q x psi x .000583) at 103 bar *(1500 psi)* and 30 l/min *(8 gpm)* flow, approximately a 5.22 kW *(7 hp)* loss is seen across the flow control orifice (needle valve).

The next system will have the same set of requirements. However, the pump will be controlled with a load sensing control system. The schematic (fig. 3-47) shows a set up for load sensing. By sensing the load pressure required and communicating that load pressure to the pump control section, the pump will maintain a preset pressure drop between the outlet of the pump and the load sense line. This pressure drop is normally between 7 bar *(100 psi)* and 14 bar *(200 psi)*, although it can be adjusted on some pumps to higher or lower values. Certain pumps are factory preset at a 10 bar *(150 psi)* differential; this value will be used for calculations. Now the pressure at gage 2 will still read 35 bar *(500 psi)* which is the load pressure. The pressure at gage 1 would read 35 bar *(500 psi)* plus the flow control orifice differential (ΔP) of 10 bar *(150 psi)* for a total of 45 bar *(650 psi)* (a detailed description of this type of control system is found later in this chapter). The pressure drop

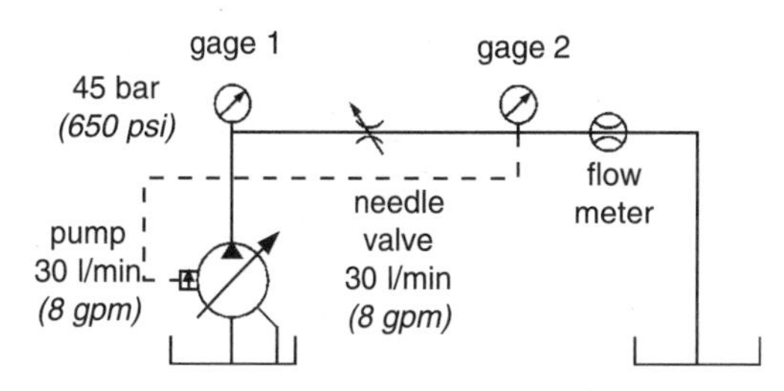

Fig. 3-47 Load sensing circuit

LOAD PRESSURE	+	**ORIFICE ΔP**	=	**PUMP PRESSURE**
35 BAR		**10 BAR**		**45 BAR**
(500 PSI)		***(150 PSI)***		***(650 PSI)***

across the variable flow control orifice (needle valve) at 30 l/min *(8 rpm)* would now be only 10 bar *(150 psi).* By using the horsepower formula, there is only .522 kw *(.7 hp)* wasted as heat across the orifice. This represents a tenfold heat savings factor.

Another important feature of a load sensing system is the ability to maintain flow regardless of changes in the required load pressure. Once the flow control orifice has been adjusted to provide 30 l/min *(8 gpm)* at the 35 bar *(500 psi)* load pressure should the load pressure increase during the work cycle to 69 bar *(1000 psi)*, the load sensing control will adjust the pump pressure to the preset flow control orifice differential. In this case, 10.3 bar *(150 psi)*, the pump outlet pressure will then be at 79.3 bar *(1150 psi).* Maintaining the same pressure drop across the orifice will give the same flow, and the actuator will not speed up or slow down as a result of a change in the actuator (load) pressure requirement. Also, it should be noted that the heat generation rate remained constant.

Load sensing works from both ends of the system. Not only will this control maintain a set flow rate regardless of load pressure required, it will also maintain a constant flow rate regardless of input speed to the pump (fig. 3-48). If a 31.25 cc *(2^3)* per revolution pump is driven with a variable speed drive, operating between 1000 rpm and 3000 rpm, the following situation would be observed: as the prime mover speed increases, the pump attempts to put out more flow at a given degree of displacement. As this flow tries to go through the preset flow control orifice, the pressure on gage 1 increases. However, with the load remaining the same, the load sense line now sees a greater flow control orifice (ΔP) differential than the original preset condition. This will cause the pump to destroke and go to a lower degree of displacement. The net effect is that flow remains the same regardless of input shaft speed into the pump. That is, within the displacement limitations of the pump.

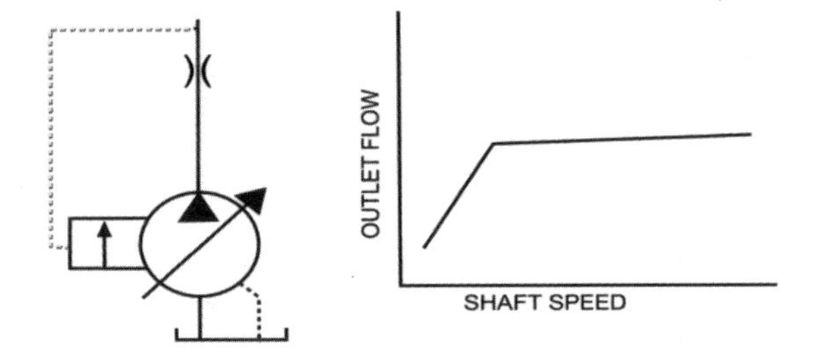

Fig. 3-48

Another useful function of the load sensing control is the ability to go to low pressure standby. The schematic (fig. 3-49) shows how by plumbing a valve into the load sense line and bleeding this line to tank, low pressure standby is achieved. Note that a load sensing directional control valve may automatically vent the load sensing line when in the neutral position.

The four key elements in the choice of a "load sensing" pump control are:

1) constant flow through the orifice regardless of the load pressure requirement
2) constant flow through the orifice regardless of the prime mover speed
3) ease of adapting to low pressure standby
4) less heat generation than standard methods

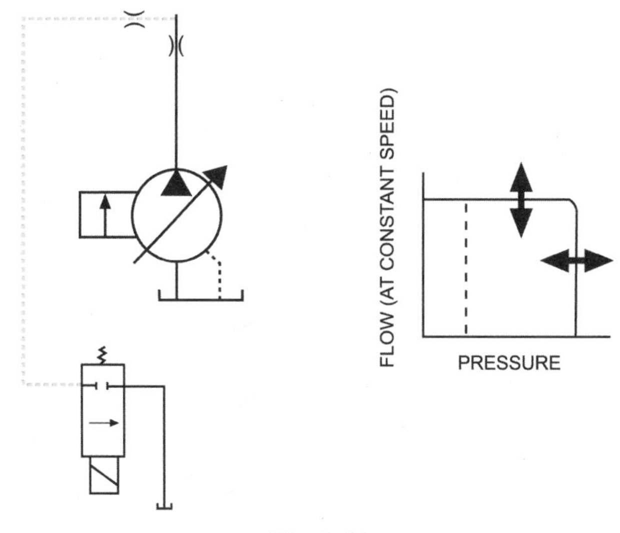

Fig. 3-49

The above description of the operation of a load sensing system included a needle valve (flow control orifice) creating the pressure drop to activate the pump control. Since the differential pressure control (ΔP) regulates a given pressure drop across an orifice, the flow through the orifice can also be varied in direct proportion to the area of the orifice[1]. Thus, by changing the size of the flow control orifice, the demand for flow from the pump can be varied. This permits the operation of a system where the demand for flow is determined by the size of the flow control orifice and the pressure losses across the orifice are limited to the setting of the pump differential pressure control mechanism, normally within the range of 7 bar *(100 psi)* to 14 bar *(200 psi).*

$$Q = 100A_O\sqrt{\Delta P}$$

[1]*Based on work by H. E. Merritt in the text,* Hydraulic Control Systems, *Herbert E. Merritt, John Wiley & Sons, copyright 1967.*

A special closed center, directional control valve could be used in a load sensing system performing the function of the variable flow control orifice in a simple system of fig. 3-50. It would be a special valve because it provides the ability to send the actual load pressure signals to the pump control. It is a closed center valve because the pump flow is blocked or reduced to zero in the center or neutral position.

In most practical load sensing systems, the pressure upstream of the control orifice is measured at the pump outlet. The pressure downstream of the control orifice must be measured in the valve and a "load signal" port is provided which must be connected to the pump. This is schematically shown in fig. 3-50.

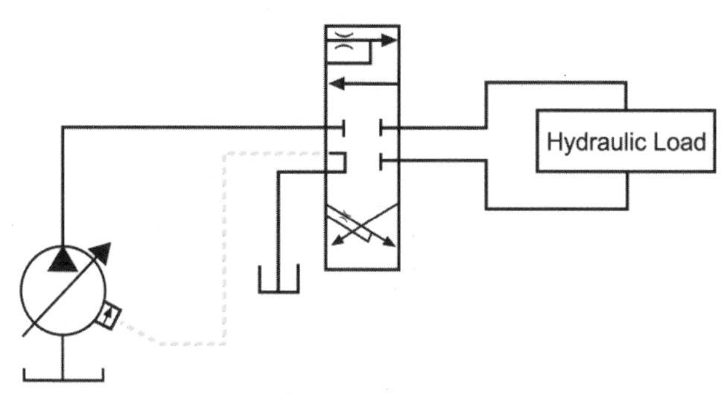

Fig. 3-50 Load sensing system

If simultaneous operation of several work functions requiring several valves is desired, then the load sensing valve may require additional control components. First the pump has only one control port to sense pressure downstream of a given control orifice. If the highest downstream pressure is used, then it is possible to supply flow to all lower pressure parallel functions up to the limits of the maximum displacement of the pump. This can be accomplished by a series of "logic" check valves that select the highest load

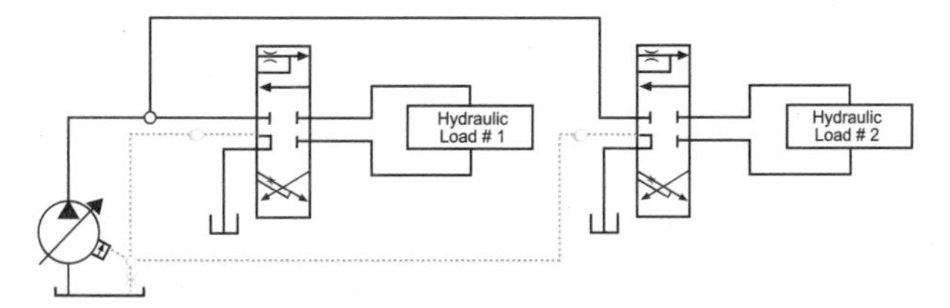

Fig. 3-51 Multiple function system schematic

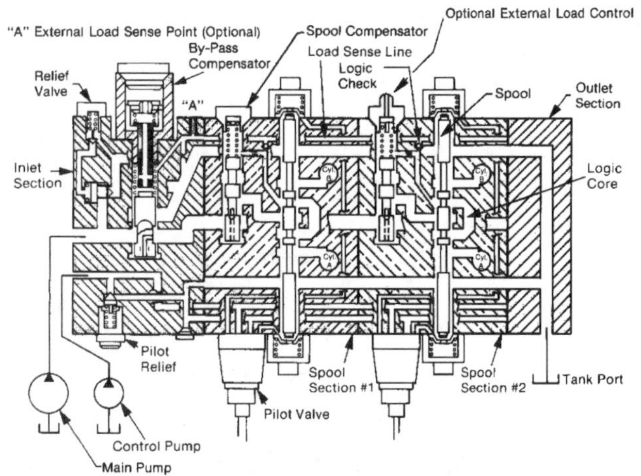

Fig. 3-52 Multiple function (spool) system components

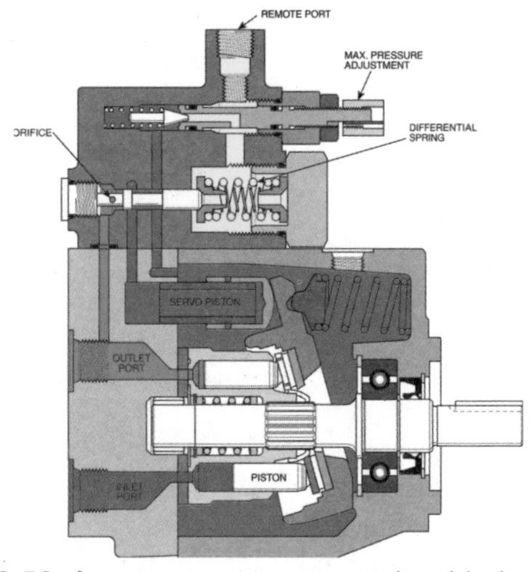

Fig. 3-53 A pressure compensator is added to limit the maximum operating pressure.

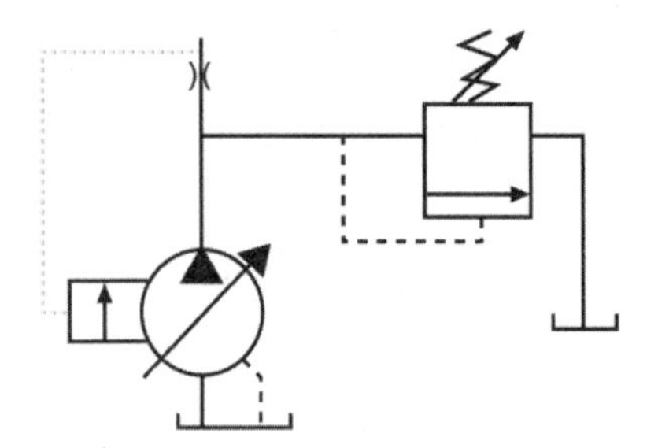

Fig. 3-54 A relief valve acts as a safety valve

pressure to operate the pump differential pressure (ΔP) control. Such a multiple function system is shown in fig. 3-51.

With the addition of the check valve logic circuit, it becomes necessary to provide a means for relieving the pressure that may become trapped between the check valves and the pump when the valve spools are returned to their center position. Therefore, a small orifice is required in parallel to the signal line and bleeding pilot flow to the reservoir. This "low signal bleed orifice" will thus vent any pressure that becomes trapped in the signal line when all spools are in the neutral position.

Another feature of the multiple function (spool) system is that the flow to the lower pressure functions will be limited only by the area of the orifice formed by the individual valve spools and the difference in pressure between the particular lower load pressure and the highest pump pressure. Each function will then experience a different pressure drop (ΔP), flow and possible excessive heat generation. In an effort to limit this effect, compensators (fig. 3-52) can be added to each valve section which will serve to limit the maximum flow to each function. There are innumerable variations in the form and function of these compensators and most are proprietary designs covered by patents. They are all designed to compensate for this effect due to simultaneous operation of multiple work functions at differing pressures.

In addition to the differential pressure control mechanism, load sensing pumps should have a means of limiting the maximum operating pressure of the pump. A "pressure compensator" (fig. 3-53) is added to the pump control package to perform this task. This device reduces the effective pump flow to zero when a preset maximum pump pressure is achieved. In addition to the pressure compensator, system (pump) relief valve (fig. 3-54) shall be a part of the system. The relief valve acts as a safety valve in case of a pump pressure compensator malfunctions. This relief valve also relieves excess pressures that may occur more quickly than the pump pressure compensator can react too, to limit.

SAFETY NOTE: All positive displacement hydraulic systems shall have a full flow system relief valve positioned as close as possible to the pump outlet. This "safety" should be set slightly higher than the pump compensator

setting. But not higher than the weakest (pressure rating) component in the system.

With a basic understanding of a load sensing system obtained in the previous material, following how a typical variable volume pressure compensated pump with load sensing controls actually operates will be much easier.

Let us now examine in detail a final example of the benefits of a "load sensing" circuit. This can best be accomplished through the comparison of three circuits performing the same operation. The sequence of operation, flow and pressure requirements of the circuit are as follows:

Sequence	Flow **(l/min - *gpm*)**	Pressure **(bar/*psi*)**
Advance	37/*10*	138/*2000*
Hold	0/*0*	207/*3000*
Return	45.5/*12*	41/*600*
Idle	0/0	Minimum

All three of the circuits will be driven at a constant speed of 1800 rpm. Each pump is capable of 56.9 l/min, 31.45 cc *(15 gpm, 1.92³/rev.)* or flow rate at 1800 rpm. Each circuit will have a 207 bar *(3000 psi)* maximum pressure capability. For simplicity, all leakage, inefficiencies and back pressures have been neglected in the calculations.

The first circuit to be examined is the "open center" circuit shown in fig. 3-55. The heat generated during operation of each system will be calculated using the following formula: BTU/hr = watts x 3.419 *(BTUs/hr = 1.5 x gpm X psi).* In this circuit the primary heat generators are the directional control valve (DV-1) and the relief valve (RV-1).

Simplified heat calculations for the "open center" circuit are found in Table 3.1. It can be seen from Table 3.1 that this circuit adds a total of 162,825 BTU's/hr/cycle in heat.

The second circuit (constant pressure) replaces the fixed displacement pump with a variable volume pressure compensated pump and replaces the open center directional control valve with a closed center directional valve (fig. 3-56).

Simplified heat calculations for this circuit are found in Table 3.2.

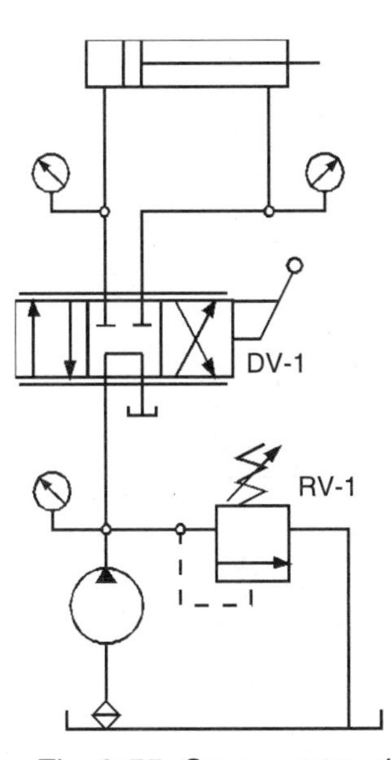

Fig. 3-55 Open center circuit

Table 3.1 Open Center Heat Calculations

DV1	**gpm X psi X 1.5 = BTU/HR**
1	10 x 1000 x 1.5 = 15,000
2	0 x 3000 x 1.5 = 0
3	12 x 2400 x 1.5 = 43,200
4	15 x 50 x 1.5 = 1,125
TOTAL	**BTU/hr/cycle: 59,325**

RV1	**gpm X psi X 1.5 = BTU/HR**
1	5 x 3000 x 1.5 = 22,500
2	15 x 3000 x 1.5 = 67,500
3	3 x 3000 x 1.5 = 13,500
4	0 x 0 x 1.5 = 00000
TOTAL	**BTU/hr/cycle: 103,500**

TOTAL HEAT (BTU/hr/cycle): 162,825

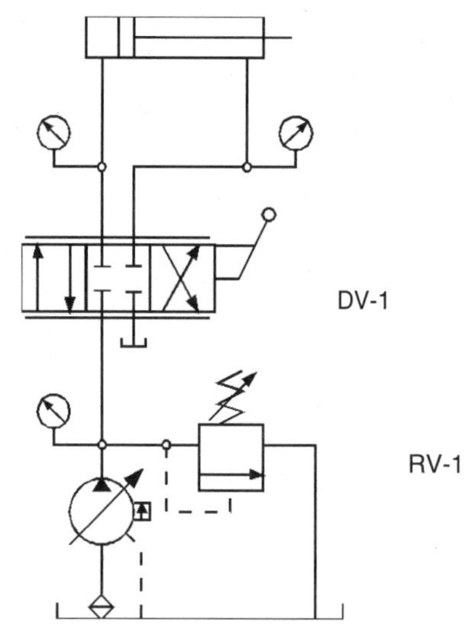

Fig. 3-56 Constant pressure circuit

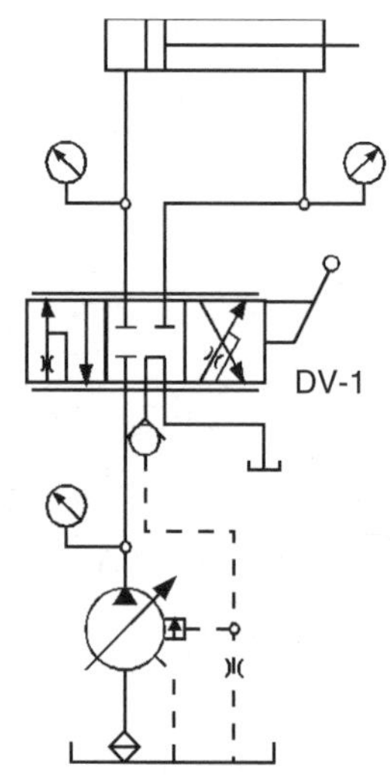

Fig. 3-57 Load sensing circuit incorporating pump and directional control valve

Table 3.2 Constant Pressure Heat Calculations

DV1	gpm	X	psi	X	1.5	=	BTU/hr
1	10	x	1000	x	1.5	=	15,000
2	0	x	0	x	1.5	=	00000
3	12	x	2400	x	1.5	=	43,200
4	0	x	0	x	1.5	=	00000
TOTAL BTU/hr/cycle:							**58,200**

Control Drain	gpm X psi X 1.5	=	BTU/hr
1	1 x 3000 x 1.5	=	4,500
2	1 x 3000 x 1.5	=	4,500
3	1 x 3000 x 1.5	=	4,500
4	1 x 3000 x 1.5	=	4,500
TOTAL BTU/hr/cycle			**18,000**

TOTAL HEAT (BTU/hr/cycle): 76,200

From this table it should be noted that only 76,200 BTUs/hr/cycle is added to the system. Comparing Tables 3.l & 3.2, a net savings of 86,625 BTUs/hr/cycle would be realized by utilizing a pressure compensated pump.

The third circuit incorporates a "load sensing" pump and directional control valve. The following are the circuit assumptions (fig. 3-57):

1) pump compensator control set at 138 bar *(2000 psi)*
2) minimum compensator setting is 14 bar *(200 psi)*
3) other assumptions are the same as constant pressure circuit

Simplified "load sense" heat calculations are found in Table 3.3:

By comparing Tables 3.1 & 3.3 a total savings of 146,925 BTUs/hr/cycle is seen.

This comparison points out the energy efficiency of the "load sensing" system.

Now that the energy savings of "load sensing" has been detailed, the next section will examine "load sensing" or "flow control" options on different style variable volume pumps.

Load sensing (flow control) with a variable volume pressure compensated axial piston pump

Flow control is achieved by placing a flow control orifice (fixed or adjustable) in the pump outlet port. The pressure drop (ΔP) across this flow control orifice is the governing signal that controls the pump's output flow as explained below (fig. 3-57).

Whenever the pressure drop at the flow control orifice increases (indicating an increase in output flow), the pump attempts to compensate by decreasing the output flow. It does this by sensing the lower pressure on the downstream side of the flow control orifice via line (C), which is balanced against the pump pressure via passage (D), on the control spool. The control spool then has reduced pilot pressure from line (C) and a spring on one end (lower end) and higher system (pump outlet) pressure on the upper end of the spool. The control spool is forced to the lower end against the control spool spring by differential pressure. This vents the servo piston cavity, destroking the pump to a point where the set pressure drop across the orifice is maintained and the set flow is obtained (fig. 3-58).

The converse of this is also true whenever the pressure drop decreases (indicating a decrease in output flow). In this case, the control spool is forced upward. This increases pump displacement in an attempt to maintain the predetermined pressure drop or constant flow. Several things should be noted. The pump is still pressure compensated and destrokes at the selected pressure setting. The pressure compensator control will override the flow control whenever the pressure compensator control setting is reached.

Orifice "F" is in position between the pump control section and the system pressure, line (C). In this position it will protect the control section (control spool chamber, pressure compensator dart, etc.) from saturation, an overflow of pilot oil into this area.

In order for this compensator to function, pilot oil must "flow-into" the control spool chamber. That is, a small amount of oil bleeds from the system into and through the control section and back to tank via control drain port (A). The amount of flow through this section may amount to only 1.14 l/min *(.3 gpm).*

To isolate the pump outlet pressure from the lower end of the control spool (control spool chamber), a plug has been positioned in the parallel pressure passage (to passage D) and before signal port (C).

If, instead of measuring the pressure drop across the flow control orifice in the pump outlet port, it is measured downstream of a directional control valve, a constant pressure drop will be maintained across the valve spool. This results in a constant flow for any given opening of the directional control valve regardless of the work load downstream or the operating speed of the pump.

The pump "senses" the amount of pressure necessary to move the load and adjusts output flow to match the valve opening selected and pressure to overcome the load plus the preset ΔP across the valve spool.

The benefits of this arrangement are that excellent, repeatable metering characteristics are achieved, and considerable energy savings are

Table 3.3 Load Sensing Heat Calculations

DV1 gpm X psi X 1.5 = BTU/hr	
10 x 200 x 1.5 =	3,000
0 x 0 x 1.5 =	0000
12 x 200 x 1.5 =	3,600
0 x 200 x 1.5 =	0000
TOTAL BTU/hr/cycle	**6,600**
Control Drain gpm X psi X 1.5 = BTU/hr	
1 x 2200 x 1.5 =	3,300
1 x 3000 x 1.5 =	4,500
1 x 800 x 1.5 =	1,200
1 x 200 x 1.5 =	300
TOTAL BTU/hr/cycle:	**9,300**
TOTAL HEAT (BTU/hr/cycle):	**15,900**

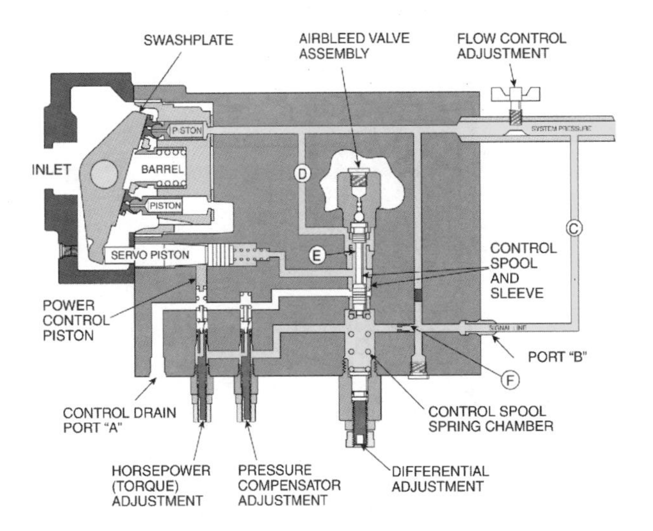

Fig. 3-58

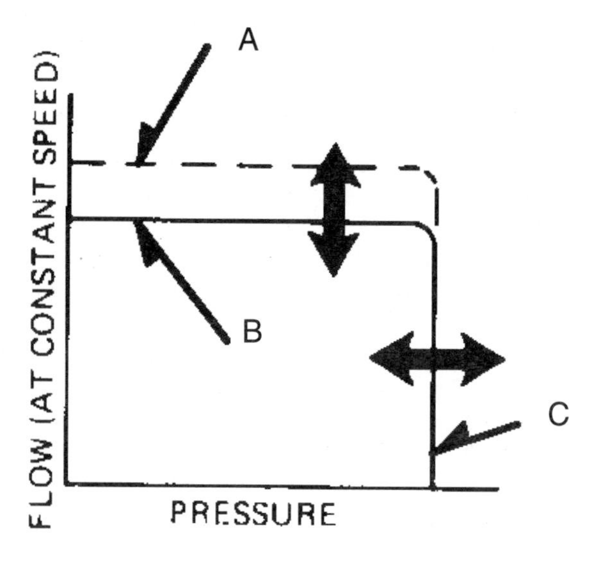

Fig. 3-59

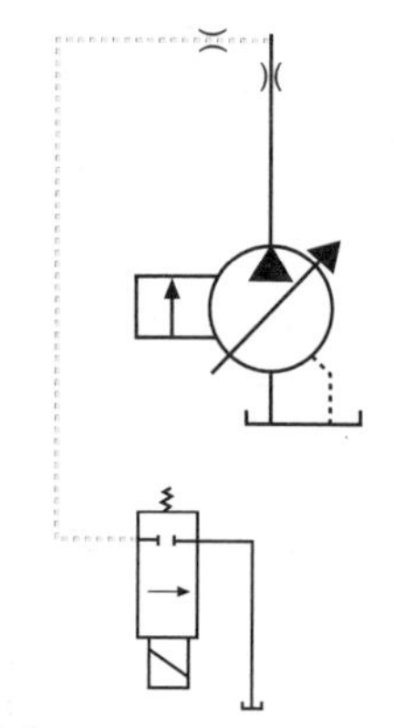
Fig. 3-60 Low pressure standby circuit

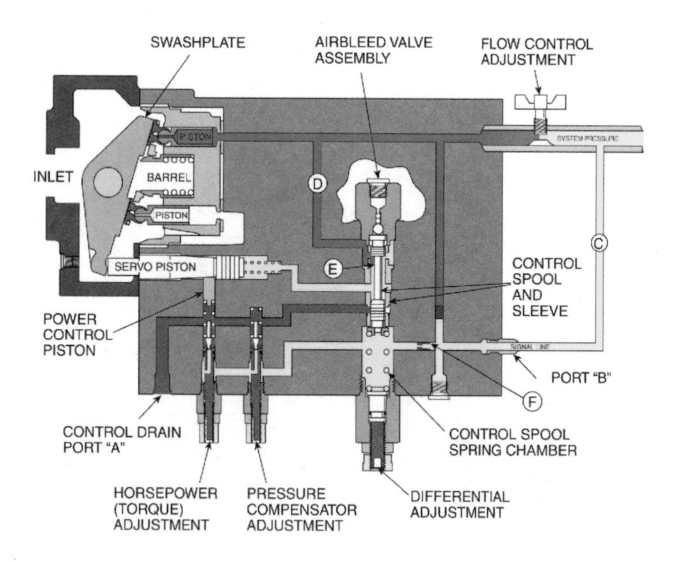

Fig. 3-61 Load sense pump

realized while metering, compared to using a straight pressure compensated system.

The graph (fig. 3-59) shows that this configuration provides for establishment of maximum flow (A), maintenance of operating flow at a selected level (B) and compensation of pressure deviations from a selected set point (C).

Low pressure standby

Low pressure standby can be provided by venting the "load sensing" port through a simple on/off valve suitable for flows of 4-8 l/min *(1-2 gpm)*. When flow or pressure is required, this valve is closed allowing system pressure to build behind the control spool and bringing the pump on stroke (fig. 3-60).

Operating modes

Once a particular load sense (flow control) pump has been chosen, various operating modes must be examined.

The first mode or condition is to see what effect closing the flow control orifice will have on the pump and system.

With this particular load sense pump (fig. 3-61), if the needle valve is closed, the pump outlet is blocked and pump outlet pressure begins to increase in passage (D). At the same time, system pressure in signal port line (C), depending on the circuit, has decreased to zero. Under these conditions, then the only thing holding the control spool in position is the 10-14 bar *(150-200 psi)* control spool spring. Pump pressure in passage (D) would increase to the above value, shifting the control spool venting the servo piston cavity allowing the pump to destroke to zero displacement. At this point the pump is at zero flow at 14 bar *(200 psi)* pressure, "low pressure standby." This condition generates minimum heat and requires minimum horsepower from the electric motor (prime mover).

Another operating mode of interest and possible concern would be when the load stalls when an excessive load resistance is encountered. If the load should stall, this would cause the pressure drop across the flow control (needle valve) to decrease toward zero. Under this condition pressure in the "load sense" line (C) and internal pilot passage (D) are approximately equal. With equal pilot pressure on both ends of the control spool, the control spool is held in position by the control spool spring. This forces a portion of the pilot oil in passage (D) through orifice (E) and into the servo piston cavity. This will cause the servo piston to extend and move the swashplate to an extreme angle, increasing the pump displacement.

This increased flow (trying to reestablish the set pressure drop across the flow control orifice) and resulting pressure increase will be seen in passages (D) and (C). Pressure in these passages continues to increase until the setting of the pressure compensator control is reached. At this point the compensator control is in command and controls the pump action by controlling the pilot pressure in the control spool chamber. Pressure in this chamber is limited by the pressure compensator dart venting pilot oil to tank via control drain port (A).

As pressure in passage (D) continues to increase it exerts a force on the control spool. This force shifts the control spool. This is possible due to the venting action of the pressure compensator dart. The control spool shifting will vent the oil in the servo piston cavity. This action will allow the swashplate to move toward the vertical or no flow position. With this type of control section connected in the load sense configuration, should the load stall, the pump will go to zero flow at the pressure compensator setting.

These two operating modes should be examined with each style of load sense pump control. Another example will more vividly point out why these checks are important.

At first glance the control section of this pump (axial piston pump) looks very similar to the previous example. However, let's see what examination of the two operating modes reveals (fig. 3-62).

It should be noted that the operation of the swashplate in this pump is somewhat different than the previous example. In this design a small bias piston is used to drive the swashplate to the vertical position. Through internal passages this piston is connected to the pump outlet port.

The larger servo piston is internally connected, via pilot passages, to the pump outlet pressure and compensator control section. This position having a larger area, is utilized to move the

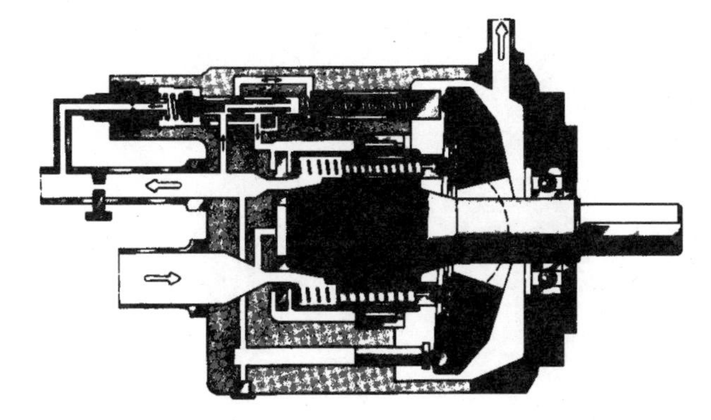

Fig. 3-62 Load sense pump control section

swashplate to the extreme on maximum displacement position. A spring in the servo piston holds the swashplate at some displacement at initial start up.

The force balance between the two pistons (bias and servo) determines the swashplate position and thus pump displacement (fig. 3-63).

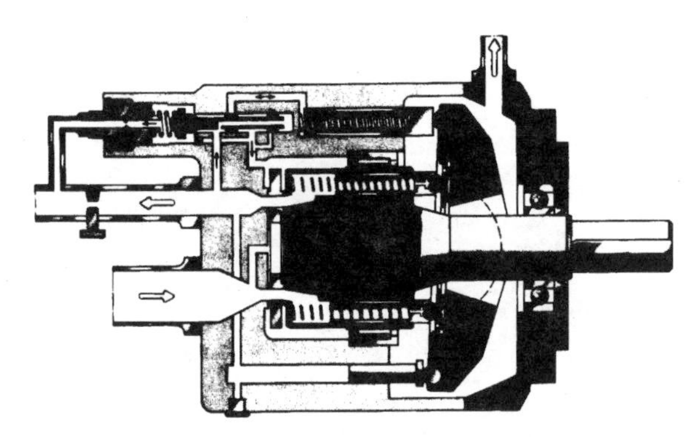

Fig. 3-63 Pump with flow control compensator

The first operating mode that needs examination is what happens when the flow control is closed off (shut). Under this condition, the pilot pressure downstream of the orifice would be zero. Therefore, the only force holding the compensator spool to the left is supplied by the 10 bar *(150 psi)* compensator spool spring. The right hand of the compensator spool is subjected to pump outlet pressure via a pilot passage drilled down the center of the spool. When the pump outlet pressure equals the compensator spool spring [which has a fixed value of 10 bar *(150 psi)*], the spool moves or shifts to the left thus venting the pilot oil in the servo piston cavity. This action then allows the swashplate to return to the vertical (zero or no flow) position. That is, no flow @ 10 bar *(150 psi)* or a "low pressure standby" condition, minimum horsepower required.

The next mode of operation that is of concern is when the load stalls. In this case, the pressure drop across the flow control is approximately zero. Therefore, the pilot oil pressure on each end of the compensator spool is approximately equal. This leaves the spool spring to hold the compensator spool fully shifted to the right. In this position the spool allows pump pressure (pilot oil) to be directly connected to the servo piston cavity. This causes the pump to move to maximum displacement. As flow exits the pump pressure will increase, the hydraulic balance will remain on the compensator spool and the spool spring will continue to hold the spool shifted to the right. This means that in this situation the pump stays at maximum flow at an infinite (dangerous) pressure which could result in equipment damage and/or personal injury. Therefore, a relief valve **must** be used to control system pressure.

SAFETY NOTE: All hydraulic systems with a positive displacement pump(s) must have a system relief valve.

Additional axial piston pump load sensing controls are described below (fig. 3-64).

The load sense port is connected downstream of a flow control orifice. Load pressure is sensed through this line, through the integral orifice and into the differential spring chamber. Load pressure and the differential spring bias the control spool to the left. This allows the swashplate spring to hold the swashplate at a nonzero displacement position. When pump outlet pressure, sensed internally to the left end of the control spool, begins to exceed the combined load pressure and differential spring value, the control spool moves to the right. This movement permits system pressure to enter the servo piston cavity and move the swashplate toward the zero displacement position.

Examining the two critical operating situations, the following conditions are found.

With the flow control orifice closed (non-passing), pressure downstream decreases toward zero. The only force remaining to hold the control spool to the left is that of the differential control spring. Therefore, when system pressure (pump outlet pressure) on the left end of the control spool begins to exceed the differential spring value, the control spool is moved to the right connecting the pump outlet port to the servo piston cavity. The result is the servo piston moves the swashplate to the zero displacement position.

Should the load stall, the second critical operation, the pump outlet port pressure and the load sense pressure in the differential spring cavity are essentially the same. The differential spring holds the control spool to the left until the load sense pressure approaches the pressure limiting control (pressure compensator) setting. The setting of this control limits the maximum pump outlet pressure and controls the movement of the control spool which, in turn, control the servo piston which drives the swashplate to zero displacement at maximum pressure.

In the next example, as with the other load sensing piston pump controls, a flow control orifice (needle valve) is required. A load sense line is then used to connect control spring cavity and the system downstream of the control orifice (fig. 3-65).

The results of examining the two critical operating conditions are as follows:

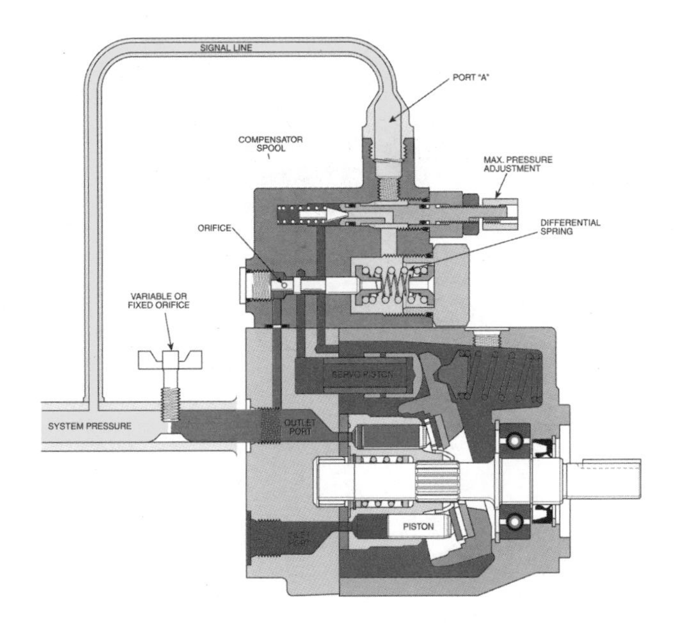

Fig. 3-64 Axial piston pump load sensing controls

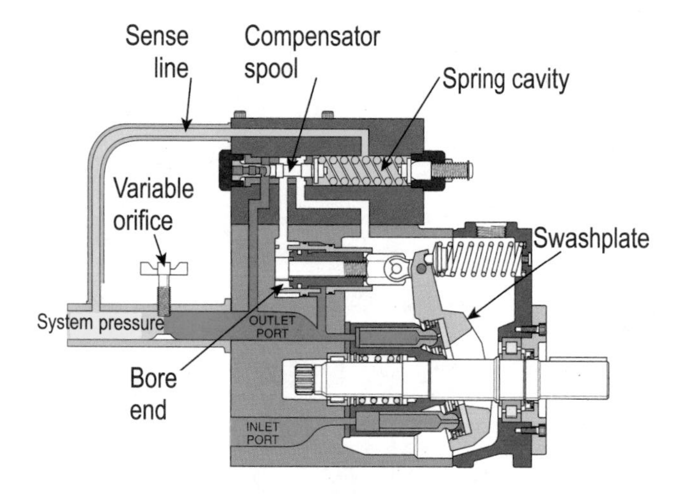

Fig. 3-65

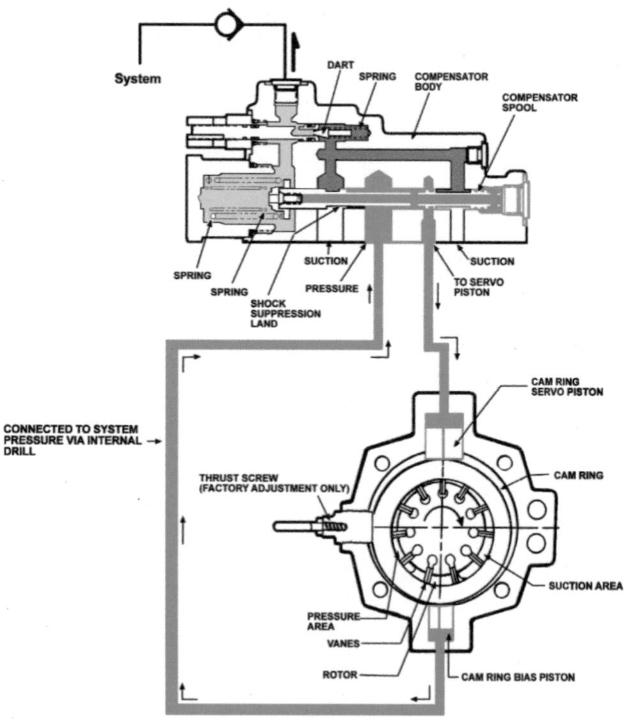

With the flow control valve closed, pump outlet port pressure moves the compensator (control) spool to the right against the differential spring. Load sense line pressure would be virtually zero. Under these conditions, the pump servo piston drives the pump to zero displacement at the differential spring setting.

However, when the load stalls, the compensator (control) spool is **hydraulically** balanced. Load sense pressure in the differential spring cavity and the pump outlet port pressure are equal. Under these conditions, the differential spring holds the compensator (control) spool to the left. This allows the pump to stay at full displacement, (full flow) at an infinite (dangerous) pressure that could result in equipment damage and/or personal injury. Therefore, a relief valve **must** be used to control system pressure.

SAFETY NOTE: All hydraulic systems with a positive displacement pump(s) must have a system relief valve.

Load sensing (flow control) is done on many styles of variable volume pumps, not just on the axial piston type described.

Another style of pump that can be adapted to the load sense (flow control) configuration is the variable volume vane pump. This configuration is possible if the pump has a remote control (pilot controlled) style compensator (fig. 3-66).

Fig. 3-66 Variable volume vane pump with remote control style compensator

The first control configuration to be examined will be the “flow-out” style remote compensator. In this configuration the pilot control oil “flows-out” of the compensator’s “remote control port.” A “load sense” pilot line and check valve must be connected from this port to the downstream side of the flow control orifice (flow control). Thus, under normal operating conditions a force balance is established on the compensator spool. On one end (left end) a fixed spring and “load” pressure is balanced by pump outlet pressure on the opposite end (right side) of the compensator spool.

The ΔP (pressure drop) is communicated back to the compensator via the “load sense” (flow control) line. It is this signal that adjusts the pump output (displacement) in an effort to maintain a constant flow through the orifice.

This can be seen by examining the orifice flow equation for a circular orifice:

$$Q = Kd_o^2\sqrt{\frac{\Delta P}{\rho}}$$

From this equation, it can be seen that flow through an orifice is affected by the size of the orifice (d_o^2). In the case of the above example the orifice size is fixed.

The ΔP, pressure drop, across the orifice is kept constant by the compensator control as described above.

Therefore, the only variable to affect flow through the orifice is the viscosity or density of the fluid. This is a temperature related variable that is not controlled by the "load sense" (flow control) configuration.

However, as with the axial piston pumps, two operating modes must be examined.

The first is to examine the pump/system operation with the "load sense" (flow control) orifice closed.

Under these conditions the pump outlet pressure is fed internally to the right end of the compensator spool.

With the system downstream of the "load sense" orifice now receiving no oil, the pressure in this portion of the system would be at a very low value, perhaps a few bar *(psi).* Under these conditions the compensator pilot oil passes through the internal orifice and out the remote control port, through the check valve and out into the low pressure system.

This then leaves just a fixed spring (13 bar, *200 psi)* and pilot oil pressure equal to the check valve spring, on the left end of the compensator spool.

Pump outlet pressure would then build to a value equal to the fixed spring (13 bar, *200 psi)* and the check valve pressure drop, at this point the compensator spool would move to the left allowing the pump to stroke to zero displacement at a few bar/*psi* over the fixed spring rate, "low pressure standby" condition. This would require minimum horsepower from the prime mover.

The second mode of operation would be when the load would slow and stall. Under these conditions, the pressure in the "load sense" (flow control) line would increase as the load pressure increased due to the slowing and stalling of the load. This pressure is communicated back into the compensator control. This pressure, plus the fixed spring, force the compensator spool to the right which in turn causes the pump to stroke toward full flow displacement.

The pressure in the system and ultimately in the compensator control section increases until the pressure reaches the dart setting of the compensator.

At this point the pressure is limited by the dart setting and the pump strokes toward zero displacement.

Under these conditions, the pressure in the system and the pressure in the compensator controller are virtually the same. This balances the hydraulic force on the check valve allowing the check valve spring to close the check valve.

The check valve is needed in this situation to prevent system pressure oil from entering and saturating (overflowing) the compensator dart section. If the check valve were not there, saturation would occur, resulting in loss of pump control.

As discussed earlier "flow-out" is only one type of pump compensator control; there is also a "flow-in." With a "flow-in" compensator control, the internal pilot flow is blocked. Control pilot oil must "flow-into" the compensator for control.

A consideration with this type control is the possibility of saturating the control section with too much pilot flow and losing pump control. To prevent this, a fixed orifice is required in the "load sense" (flow control) line; replacing the check valve (fig. 3-67).

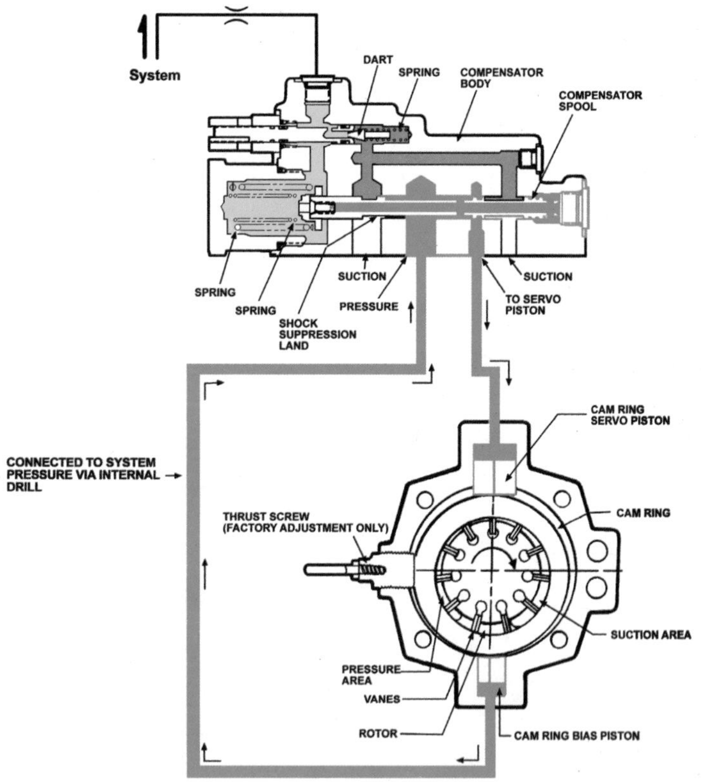

The two critical modes of operation must now be examined. As with the other examples, the first mode of operation will be with the system flow control closed.

Under this condition, downstream flow is nonexistent and pressure is extremely low. Therefore, there is no pilot flow or pressure in the "load sense" (flow control) line. There is no pilot oil signal into the compensator. The only force holding the compensator spool to the right is the fixed spring (13 bar, *200 psi).*

Fig. 3-67 A fixed orifice replaces the check valve to prevent losing pump control

The pump outlet pressure is internally connected to the right end of the compensator spool. When pump pressure reaches the value of the fixed spring, the spool will begin to move to the left allowing the pump to stroke toward zero displacement. Thus the pump will be in a "low pressure standby" condition, minimum horsepower demanded from the prime mover.

The second mode of operation is when the load stalls.

When the load stalls the pressure in the "load sense" (flow control) line and the internal passage to the right end of the compensator spool are virtually the same. The fixed spring holds the compensator spool to the right.

As pressure builds, it approaches the compensator dart setting. This dart limits the maximum pilot oil pressure in the compensator.

At this point, the pump pressure on the right of the compensator spool begins to work against the fixed spring and pilot oil pressure moving the compensator spool to the left. This allows the pump to destroke to zero displacement at the compensator pressure setting.

Generally speaking, when the control section of a variable volume pump (such as those just discussed) has a built-in compensator dart pressure control, the pump will destroke to zero displacement at the compensator dart setting if the load stalls.

Without the built-in compensator dart control, the pump will stroke to full displacement at **relief valve** setting.

On the other hand, if the flow control orifice (needle valve) in the system is closed, the pump will go to zero displacement at minimum pressure, i.e., "low pressure standby."

Horsepower (torque) limiting control

Horsepower (torque) limiting is a control(s) that sets a maximum horsepower or torque that the pump can demand of the prime mover, thus helping to prevent an overload condition. That is, an electric motor would draw too much amperage tripping circuit breakers, burning out the motor windings or stalling the motor. With an internal combustion engine as the prime mover, an overload condition would stall the engine.

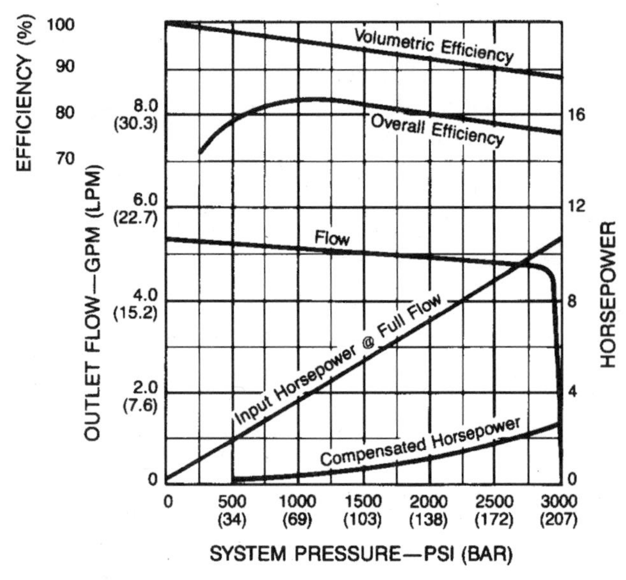

Fig. 3-68

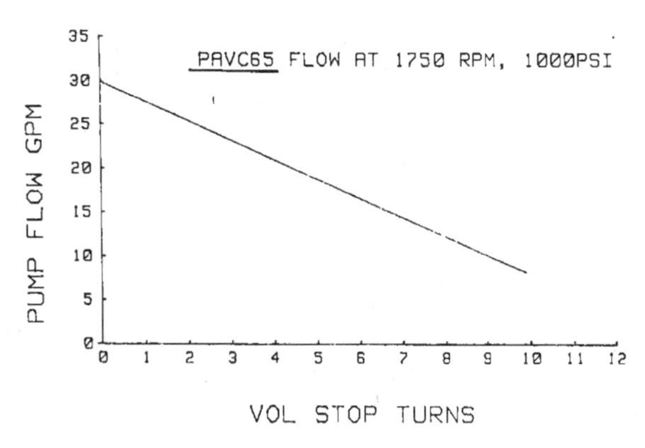

Fig. 3-69

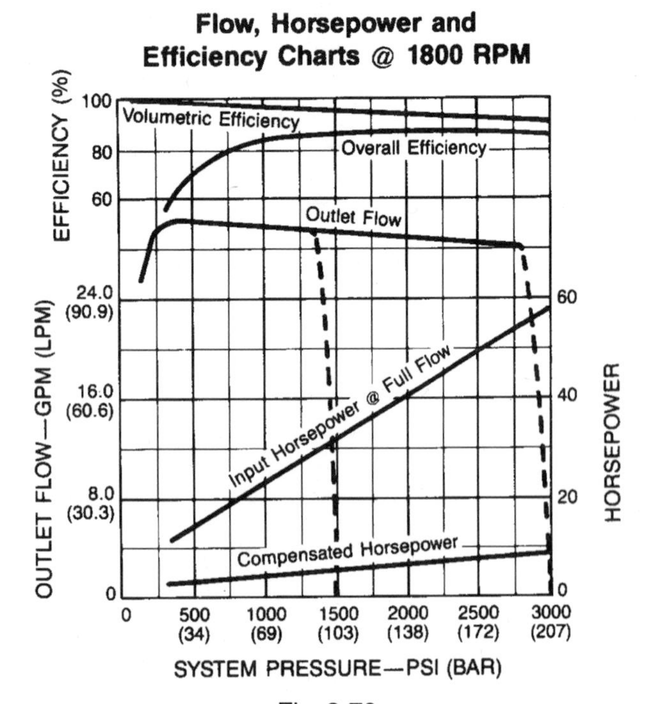

Fig. 3-70

The following formula shows the relationship between the horsepower into (horsepower from the prime mover) the pump and the horsepower out of the pump:

$$W_{in} = \frac{Q_m P_m \text{ x.}166 \text{x} 10^{-4}}{\text{pump overall efficiency @ } P_m}$$

$$hp_{in} = \frac{Q \text{ x } psi \text{ x} (.000583)}{\text{overall efficiency @ } psi}$$

It can be seen by reviewing the above formula that three variables affect the horsepower "in" from the prime mover; they are:

1. pump flow (Q_m, *Q*)
2. pump pressure - bar (*psi*)
3. pump overall efficiency at a given pressure

However, in reviewing the set of typical operating curves (fig. 3-68) for a variable volume pressure compensated pump, it can be seen that over a given pressure range the overall efficiency remains relatively constant, thus having less of an affect on the "horsepower in" than the combination of flow (Q) and pressure (bar/*psi*).

Knowing these facts then, a crude horsepower limiter could be fashioned utilizing the pressure compensator adjustment and maximum volume stop on a variable volume pressure compensated pump. Both of these adjustments have been described in detail in previous chapters. Empirically or through manufacturer's data, a flow versus volume stop turns relationship can be established. Such a relationship, in graph form, is shown at the left (fig. 3-69).

Utilizing the known input horsepower limit, the pump operating curve and the above curve (flow vs. turns) the maximum flow for a given horsepower can be determined.

Determining limiting flow

As an example at the left (fig. 3-70) is the set of operating curves for a variable volume pressure compensated piston pump. Maximum input horsepower can be as high as about 41.8 kW *(56 hp)*. Limiting the input horsepower to 20 (14.9 kW) would require the following steps:

1. Rearrange the "horsepower in" formula to solve for flow (Q):

$$Q_m = \frac{W \times \text{overall efficiency@} P_m}{P_m \times .166 \times 10^{-4}}$$

$$Q = \frac{hp_{in} \times \textit{overall efficiency@ psi}}{psi \times .000583}$$

2. Substitute into the formula the known values for the variables:

Maximum power set by maximum system pressure

$$Q_m = \frac{14{,}920 \times .85}{10.3 \text{ MPa} \times .166 \times 10^{-4}}$$

Maximum system pressure (compensator setting)

$$Q = \frac{20 \times .85}{1500 \times .000583}$$

3. Solve for flow (Q):

$$Q_m = \frac{12{,}682}{171} = 74 \text{ l/min}$$

$$Q = \frac{17}{.875} = 19.5 \text{ gpm}$$

4. Locate 74 l/min (*19.5 gpm*) on the flow vs. turns graph (fig. 3-71) and draw a horizontal line from that point on the vertical axis to the curve.

5. Draw a vertical line down to the horizontal axis. This point of intersection shows how many turns the maximum volume stop of this particular pump must be turned in (4.5 turns).

This procedure will limit the pump to a maximum of 74 l/min *(19.5 gpm)* at 103 bar *(1500 psi)* or approximately 14.9 kW *(20 hp).*

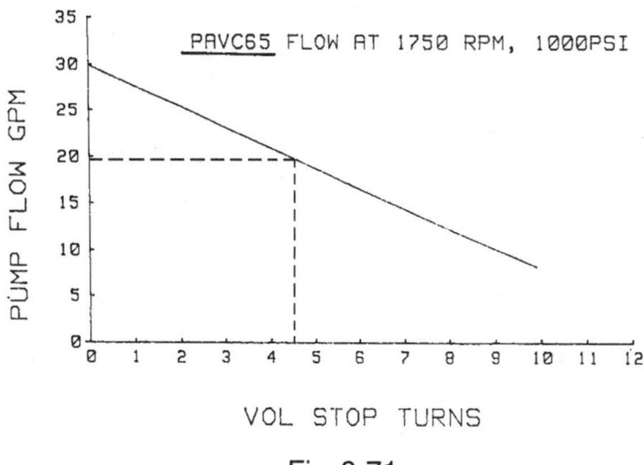

Fig. 3-71

What would happen to the input horsepower requirement of the pressure increases to 138 bar *(2000 psi)*?

Examining the "W_{in}" *(horsepower in)* formula it can be seen that:

fixed by max. volume stop
now 138 bar *(2000 psi)*

$$W_{in} = \frac{Q_m \times P_m \times .166 \times 10^{-4}}{\text{pump overall efficiency@ } P_m}$$

$$hp_{in} = \frac{Q \times psi \times .000583}{\text{overall efficiency@ } psi}$$

approximately the same as before

... the torque *(horsepower)* would increase; a possible overload condition. The flow (maximum volume stop) would need readjustment to a reduced flow. This would return the torque *(horsepower)* limit to 14.9 kW *(20 hp)*.

Each time the maximum pressure changes, the flow (maximum volume adjustment) must be reset.

There are two other ways of conveniently and inexpensively limiting the input torque *(horsepower)*. These two methods are hydromechanically and electronically. The electronic method will be discussed in the chapter on electronic control of pumps.

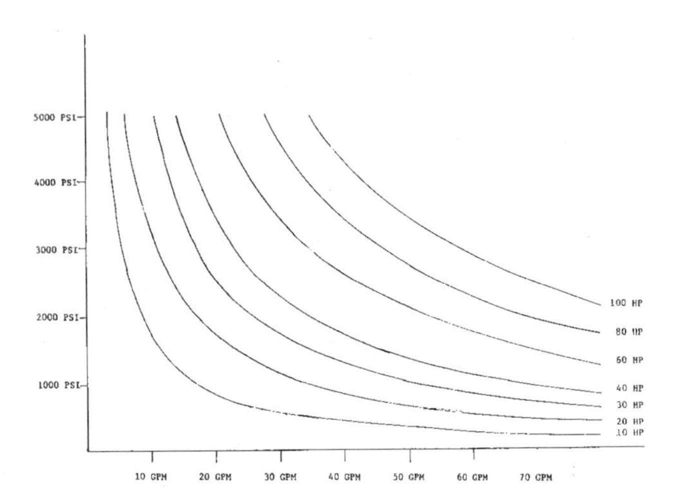

Fig. 3-72 Horsepower curves

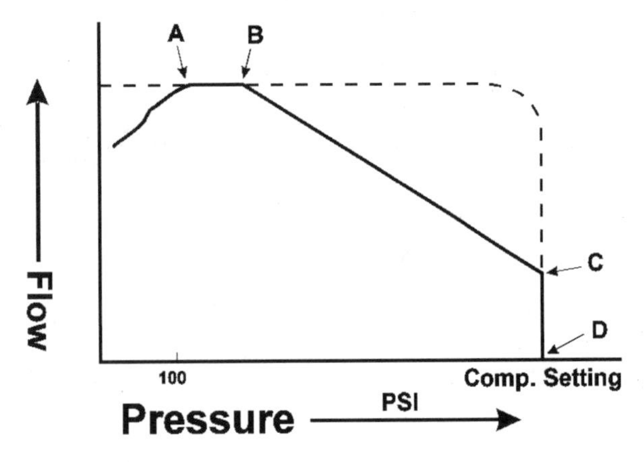

Fig. 3-73 Pump operating curve

Hydromechanical control

As previously shown, the "Watt in" *(horsepower in)* is an interplay between flow (Qm, *Q*) and pressure Pa *(psi)*. A graphical look (fig. 3-72) at this interplay, for a fixed horsepower, would reveal a curve similar to the one at the right; a hyperbola. Each curve represents a fixed horsepower.

These curves can be closely approached hydromechanically, but at a very high manufacturing cost. However, upon close examination, the vast majority of the constant horsepower applications require only a rough approximation or one point on the constant horsepower curve.

This being the case greatly simplifies the hydromechanical pump controls. Such a simplified control scheme would yield a pump operating curve (flow vs. pressure) shown in fig. 3-73.

Below point "A" the system resistance is below the minimum required by the pump and therefore, full pump flow may not be achieved.

From points "A" to "B" system (pump) pressure is sufficient to drive the pump to full displacement. At point "B" the simplified constant horsepower control begins to vary this displacement (decreases) inversely with the pressure (increasing). This continues until point "C" is reached.

Between points "C" and "D" the pump output (displacement) is controlled, reduced, by the pump compensator control as the pump pressure remains relatively constant.

Hydromechanical horsepower limiting control

A typical hydromechanical horsepower torque limiting control will now be examined (fig. 3-74).

The horsepower (torque) control is sensitive to the position of a control or a servo piston. When the servo piston is to the right, the swashplate causes low flow and the power control servo piston develops maximum spring pressure on its companion dart (mechanical feedback). When the servo piston is to the left and the flow is high, the power control servo piston reduces spring pressure on the dart. This allows it to open under less pressure in the control spool spring chamber, thereby venting some of the pressure in the control spool spring chamber. As with the operation of the pressure compensator control, this allows the control spool to move left, venting the servo piston cavity and causing the servo piston to move right. This reduces output flow and thereby power.

As indicated in the pictorial drawing, pressure in the control spool spring chamber is affected by both the pressure compensator control and the power (torque) control. The resultant pressure in this chamber is a function of the set points of these two controls. Both set points are adjustable.

As shown in the graph at the right (fig. 3-75), this arrangement provides for establishment of various flow vs. system pressure levels and controls the power (torque) level to reduce power consumption. (Avoids the standard pressure compensated corner horsepower.)

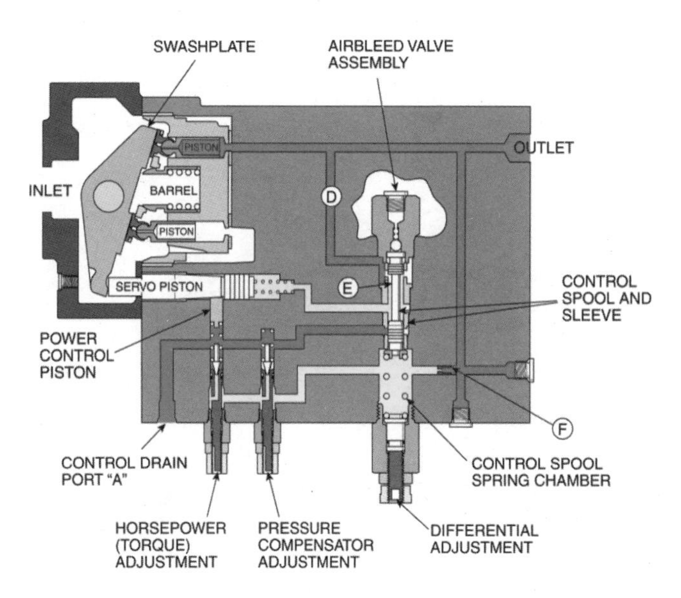

Fig. 3-74 Hydromechanical horsepower torque limiting control

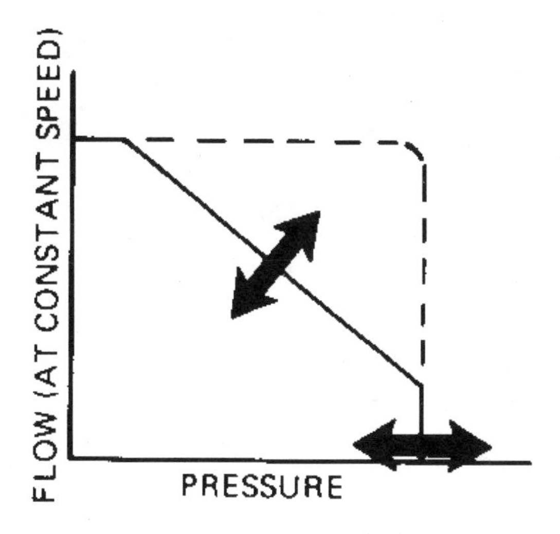

Fig. 3-75

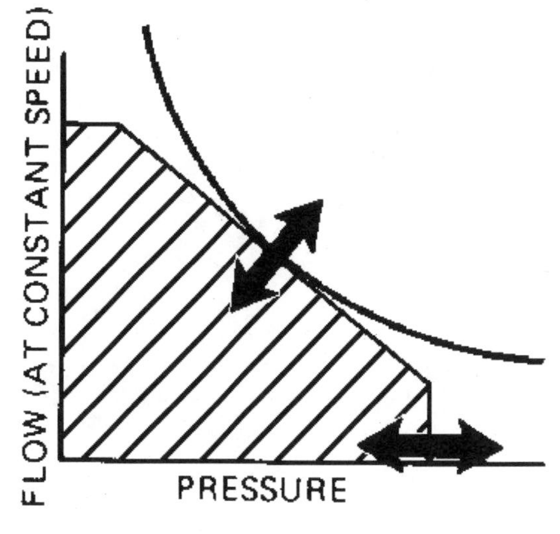

Fig. 3-76

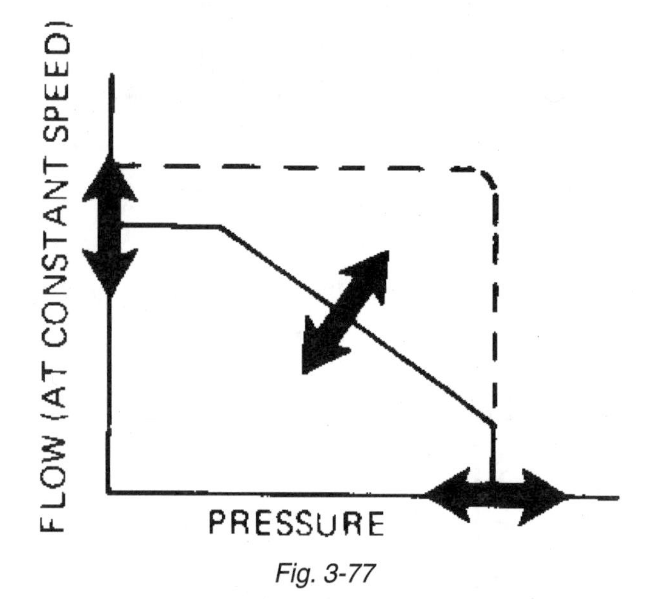

Fig. 3-77

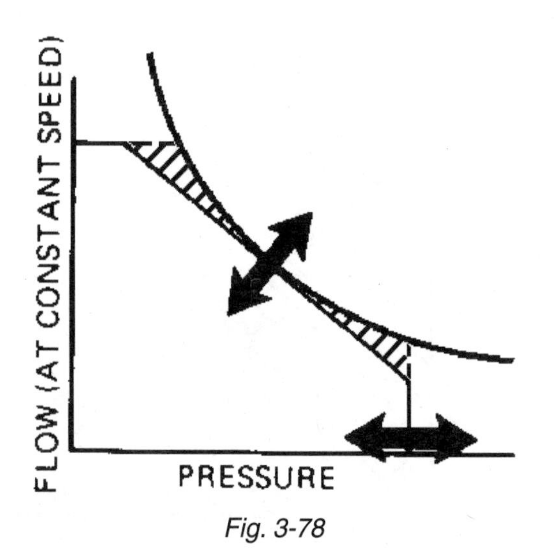

Fig. 3-78

The constant horsepower curve can be translated diagonally via the horsepower adjustment (fig. 3-76). This allows the constant (diagonal) horsepower curve, torque *(horsepower)* limit, to touch (tangent) the true constant horsepower curve at a single point. Under this condition any pump flow and pressure combination will be *equal to* or *less than* the horsepower limit setting (shaded area of graph).

It should be noted that the angle or slope of the pump horsepower curve is determined by the angle or slope machined into the servo piston body. It is this slope that determines the *rate of change* of the spring tension (via the power control piston) of the torque *(horsepower)* adjustment.

This semiautomatic spring adjustment determines the rate of change of the pressure in the *control spool* spring chamber. This in turn affects the movement or rate of movement change of the control spool (compensator spool). The movement of this spool then controls the position of the servo piston via controlling the rate at which pilot oil enters or leaves the servo piston cavity. In short, the rate of change of the horsepower (torque) adjustment spring affects the rate of change of the servo piston position which relates directly to displacement or flow and to horsepower (torque) demand.

A close examination of the pump operating curve with the pump horsepower (torque) curve tangent to the constant horsepower curve, reveals two areas (crosshatched areas) where the pump horsepower (torque) adjustment limits the combination of flow and pressure well below the constant horsepower curve (fig. 3-77). This is in part due to the fact that one curve is linear [pump horsepower (torque) adjustment] and the other is non linear (constant horsepower curve).

To regain a portion of this lost flow and pressure, it would be necessary to cut multiple slopes or steps on the servo piston instead of the single slope. This special machining would add greatly to the cost of the component; therefore it is done in special cases only.

Horsepower (torque) limiting with load sensing (flow control)

Depending upon the design of the pump control section it may be possible to combine the load sensing (flow control) control with the horse-power limiting control.

With this combination of controls the position of the compensator control spool is a function of the interplay of three controls; horsepower (torque) adjustment (2), pressure compensator (3), and load sensing (flow control).

This combination of controls will cause the pump to operate within the limits of the envelope shown by the graph at the right (fig. 3-78). This envelope limit is established by the three controls; horsepower (torque), pressure and flow. These three controls set the maximum limits of the pump operating envelope.

With this combination of controls, the pump will operate on flow (load sensing) control until the combination of flow (Q or l/min - *gpm*) and pressure (bar or *psi*) begins to exceed the set point of the horsepower (torque) limiter; it then becomes the controlling element. When the pressure side of the horsepower (torque) control begins to exceed the pressure compensator adjustment set point, it then takes control establishing the maximum attainable pump pressure.

In reviewing the previous example of horsepower (torque) control, it can be seen that control is achieved through control of the pilot oil pressure in the compensator control spool spring chamber (fig. 3-79).

Other styles of horsepower (torque) control utilize the above method by incorporating a control dart, in parallel with the pressure compensator dart, and a means of varying the spring tension on the dart. Several examples will be briefly discussed below.

In certain designs access to the pilot oil in the control spool spring chamber may require external plumbing. Two such examples are shown at the right.

The first piston pump control shown (fig. 3-80) functions similarly to that described in the **hydro-mechanical horsepower limiting control** section. However, a cam affixed to the swash-plate trunnion is required to vary the tension on the horsepower (torque) control dart.

The other control example (fig. 3-81) incorporates an additional **control spool**. This spool can be thought of as an extension of the servo

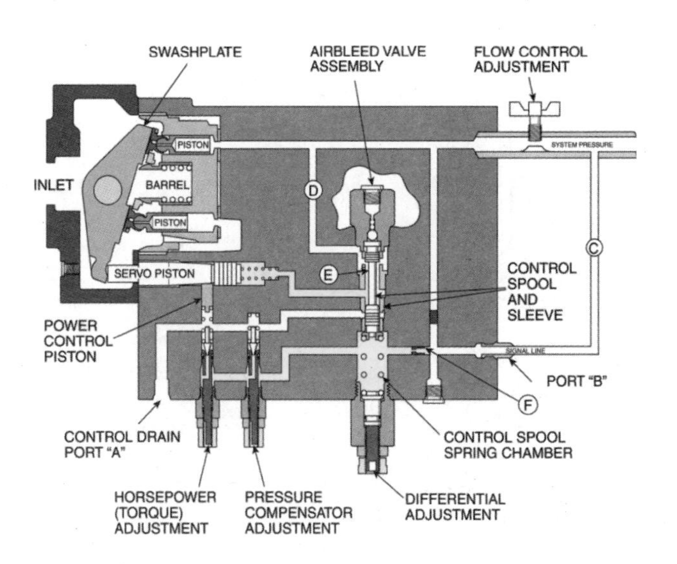

Fig. 3-79

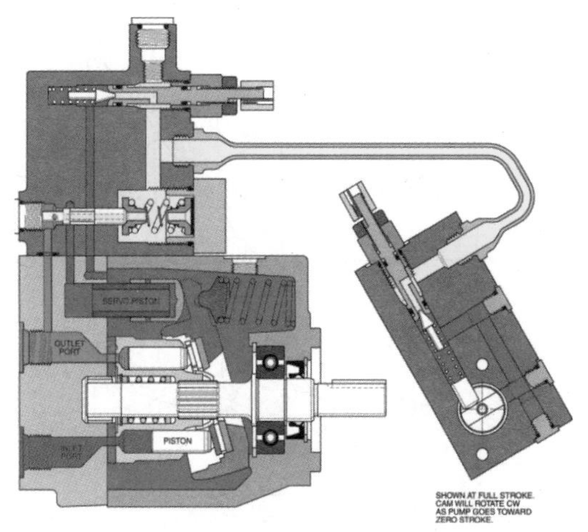

Fig. 3-80 Control dart

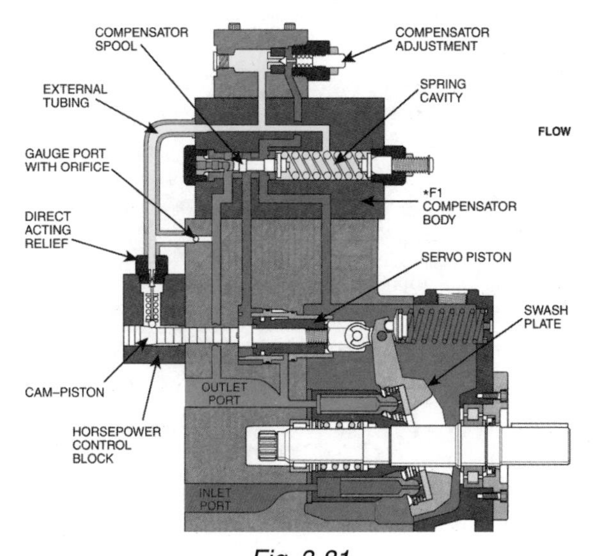

Fig. 3-81

piston. It is the spool that has a slope machined on it. This slope determines the *rate of change* of the spring tension on the torque *(horsepower)* control adjustment.

A similar style of control can be seen incorporated into the pressure compensated variable volume vane pump at the right (fig. 3-82). Its operation would be identical to that of the above piston pump control.

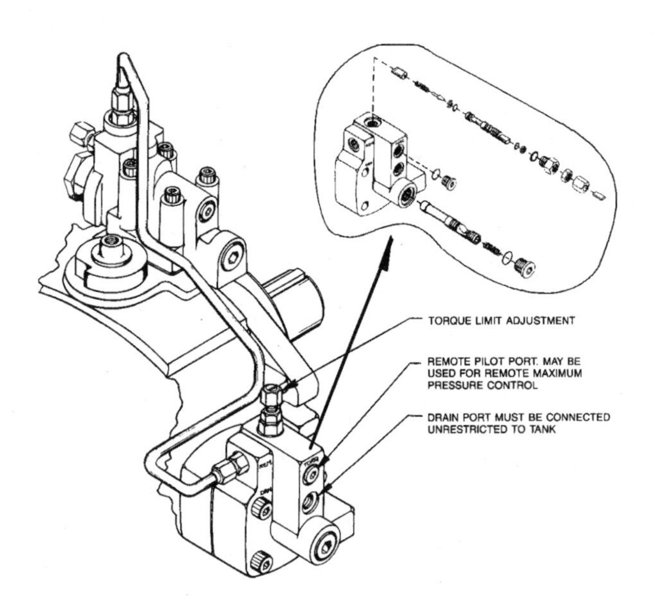

Fig. 3-82

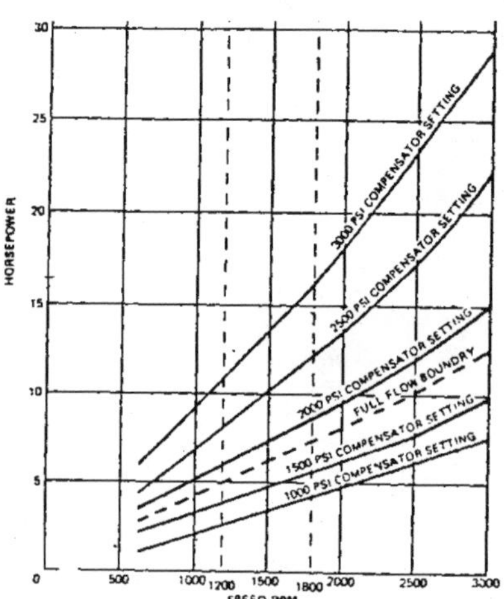

Fig. 3-83

Minimum horsepower full flow boundary

Under certain conditions, a variable volume pressure compensated pump, equipped with a horsepower limiter control, may not achieve full displacement at low pump pressures.

The performance data presented at the right (fig. 3-83) graphically illustrates the above condition. The graph relates the pump shaft speed (rpm), horsepower limiter setting to the pump pressure compensator setting. All points (compensator settings) above the *full flow boundary* curve are achievable; that is, full pump flow is attainable. With compensator settings below the full flow curve full pump flow *is not* attainable at low pressures.

The above pump performance data will be utilized in the next examples. The system conditions require the pressure compensator be set at 138 bar *(2000 psi).* The pump shaft speed is 1800 rpm and the horsepower control is set to limit the maximum horsepower demand to *15* (11.2 kW). The 11.2 kW (*15 hp)* line intersects the pump shaft speed of 1800 rpm. Therefore, from this it is known that all points above the curve are achievable; full flow is attainable.

This means that the interplay caused by the adjustment or tension on the springs which control the horsepower (torque) and pressure controls are such that they do not allow pilot oil to exit the control spool chamber. Therefore, the compensator (control) spool remains to the right holding the pump at full displacement (full flow).

Now, if the horsepower control is set to a lower value, say 3.7 kW *(5 hp),* the horsepower line and pump shaft speed (1800 rpm) intersect below the *full flow boundary* curve. This means that the pump will not achieve full flow at low pressures.

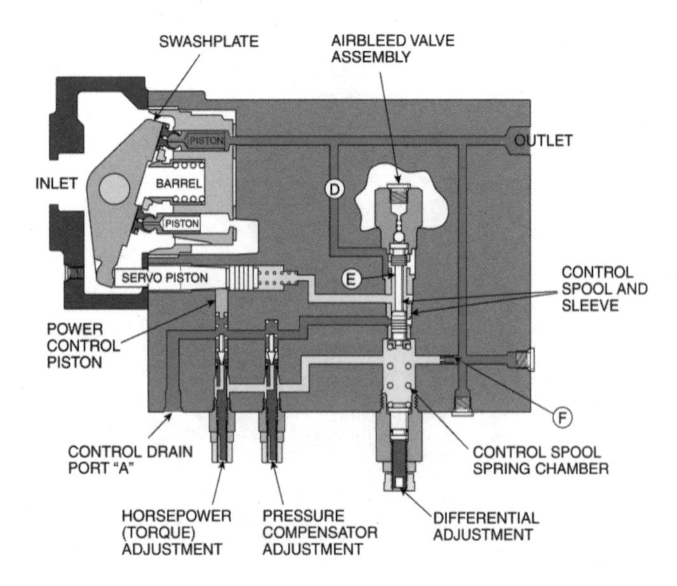

Fig. 3-84

The reason this occurs can be seen by reviewing the pump control configuration at the right (fig. 3-84). As the horsepower control is reduced, the tension on the horsepower control spring is reduced. This reduced tension will allow the horsepower control dart to pass a small portion of pilot oil from the control spool chamber. The compensator (control) spool must now move to the left until a new force balance is established across the spool. In turn this establishes a new balance or position of the servo piston at a point *less than* full flow.

Examining circuit conditions, in detail, prior to incorporating a horsepower controlled pump will minimize the chance of a problem due to the *less than* full flow situation.

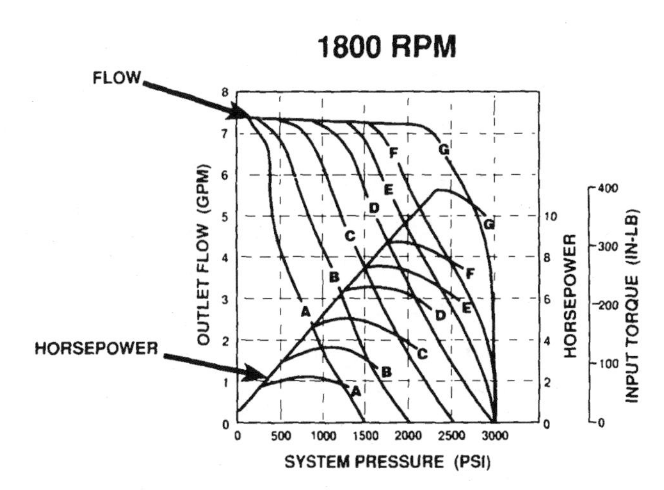

Fig. 3-85

Horsepower, flow and pressure curves

Each flow and corresponding horsepower curve represent a particular torque (horsepower) control setting (fig. 3-85). The maximum attainable "horsepower in" (torque) for a particular setting is shown at the apex of the horsepower curve. Flow at a particular torque *(horsepower)* setting will follow the flow vs. pressure curves.

As an example, a torque *(horsepower)* setting equivalent to operating curve "C" will be used. Pump flow will follow the flow curve labeled "C" and with zero flow (deadhead flow) will occur at 172 bar *(2500 psi)*. This will occur even if the compensator setting is above 172 bar *(2500 psi)*. It can be seen that at pressures above 35 bar *(500 psi)*, flow will decrease to less than full flow. Pump flow at 69 bar *(1000 psi)* will be a little more than 23 l/min *(6 gpm)*.

By examining the horsepower curve "C" it can be seen that maximum "horsepower in", for this particular setting, will be 3.7 kW *(5 hp)* or approximately 20 N-m *(180 in-lb)* of torque, at 86 bar *(1250 psi)*.

Setting a horsepower limiter control

There are two ways to set the typical horsepower limiter control. Each is described below.

As an example, suppose a maximum of 3.7 kW *(5 hp)* is to be drawn from a 5.6 kW *(7.5 hp)* electric motor.

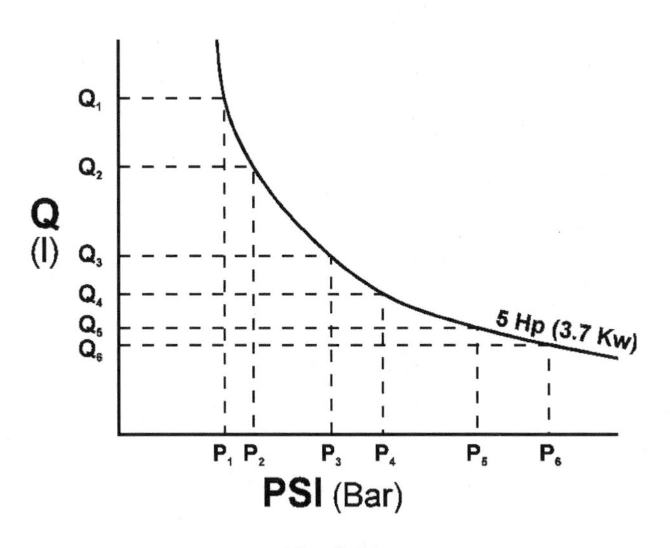

Fig. 3-86

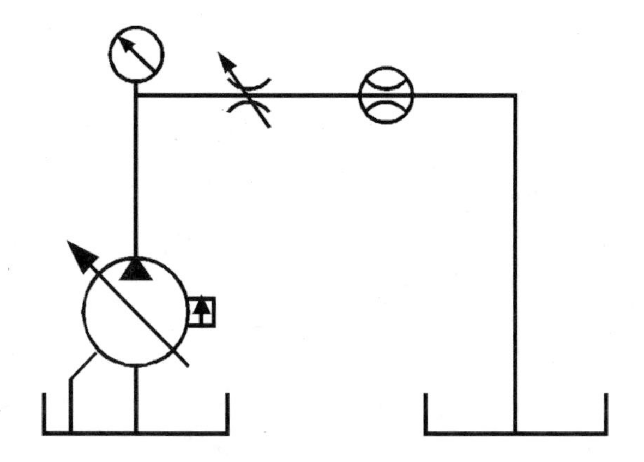

Fig. 3-87 Third step - connecting needle valve, pressure gage and flow meter

The first step in setting the horsepower control is to plot a 3.7 kW *(5 hp)* curve on standard graph paper.

Utilizing the horsepower formula

$$W = \frac{Q_m P_m \; x .166 \, x 10^{-4}}{\text{pump overall efficiency @ } P_m}$$

$$hp = \frac{Q \; x \; psi \; x \, (.000583)}{\text{overall efficiency @ psi}}$$

At six randomly selected pressures; solving for flow (Q) the 3.7 kW *(5 hp)* curve can be approximated with an acceptable degree of certainty (fig. 3-86).

The next step is to set the horsepower control at a value below the required limit but slightly above its minimum setting.

The third step is to connect a needle valve, pressure gage and flow meter between the pump outlet and the tank (fig. 3-87).

Next with the pump running, the needle valve is adjusted and flow restricted until each of the randomly selected, pressures from above, is reached. These pressures and corresponding flows are plotted on the above graph. This will show how close the horsepower control is adjusted to the required setting.

If the resultant curve is below the 3.7 kW *(5 hp)* curve, then the horsepower control must be adjusted to increase the horsepower demand. On the other hand, if the curve is above the 3.7 kW *(5 hp)* curve, the adjustment must be reduced.

The pressures and flows at the pre-selected points must be graphed to again verify how close the horsepower adjustment is to the required setting. This may require three or four attempts before the horsepower adjustment is set. As a note, if a flow meter is not available, the timing (speed) of an actuator could be used to calculate the pump flow at a selected pressure.

Another, and faster, way of setting the horsepower control requires the use of an ammeter. The ammeter could be permanently wired into the electric motor circuit or a clamp-on ammeter could be used. Either style of meter would register the amperage draw of the motor. The current draw corresponding to the 3.7 kW *(5 hp)* setting could be calculated or obtained from the motor manufacturer. In the case of the above example, the current draw would be somewhat less than that shown on the nameplate for the amperage draw at the maximum rating of the 5.6 kW *(7.5 hp)* motor being used.

Chapter 3 exercises
Exercise 1 - torque (horsepower) limiting

INSTRUCTIONS: Answer the following questions or solve the following problems as required.

1. A maintenance person has asked for your help. This individual has been instructed to limit the maximum pump horsepower of a particular pump. The pump **is not** equipped with a built-in horsepower limiting control. The maintenance person has supplied you with the following information: PAVC 65 PUMP, OPERATING SPEED IS 1750 RPM, OPERATING PRESSURE IS 69 BAR (*1000 PSI),* OVERALL EFFICIENCY IS 85%. The maximum torque is to be limited to 5.6 kW *(7.5 HP).* What flow setting (number of volume stop turns) is required? (HINT: use the graph in fig. 3-71)

2. A true fixed horsepower curve is a hyperbola, as shown in fig. 3-72. However, the typical hydromechanical horsepower limiting control **cannot** duplicate this curve. Why is this?

3. Production has informed you that they are having trouble with the horsepower limiting control on one power unit. At low pump pressures the pump **does not** seem to achieve full displacement. They supply you with the following information: OPERATING SPEED 1800 RPM, TORQUE SETTING IS 4.4 KW *(6 HP)* THE PUMP COMPENSATOR SETTING IS 86 BAR *(1250 PSI).* What is the problem?

4. A new machine has just arrived at your plant and must be placed into production as quickly as possible. However, the ammeter (on the power unit) that you had planned on using to set the pump's horsepower limiting control was damaged in transit. The plant superintendent has asked you to find another way to set the control. What procedure will you follow?

Chapter 3 exercises (cont'd.)
Exercise 2 - load sensing exercise

Instructions: In this assignment the answers are already given. You must supply the questions.

ANSWER 1. Pump drain line is restricted

ANSWER 2. A flow control orifice is required at the outlet port of the pump

ANSWER 3. Flow - in

ANSWER 4. Low signal bleed orifice

ANSWER 5. To prevent compensator control section saturation

ANSWER 6. Low pressure standby

ANSWER 7. Two operating modes

ANSWER 8. System relief valve

Chapter 4

Hydraulic motors and hydrostatic drives

Hydraulic motors convert the working energy of a hydraulic system into rotary mechanical energy. Hydraulic motors operate by causing an imbalance which results in the rotation of a shaft. This imbalance is generated in different ways depending upon whether the motor is a vane, gear, piston or geroller type.

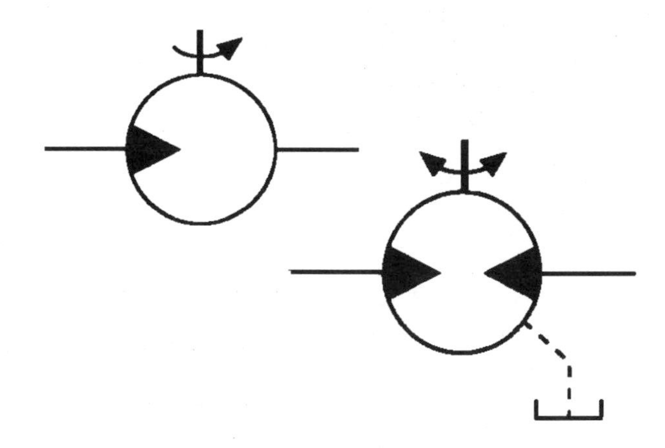

Fig. 4-1 Hydraulic motor symbols that depict unidirectional and bidirectional rotation.

Hydraulic motors are positive displacement unidirectional rotation or bidirectional rotation devices as illustrated in fig. 4-1. That is, as a motor receives a constant flow of fluid, the motor speed will remain relatively constant regardless of the pressure.

Hydraulic motors as discussed in mobile hydraulic systems in this text will be divided into three categories: Implement or auxiliary drive motors, wheel and or track drive motors, and swing drive motors.

Implement or auxilary drive

Motors that fit into this category will be used to drive such mobile implements as paddle wheels on combines and excavators, trenchers, cable drums, etc. The main discussion in this area will focus on fixed displacement vane, gear and piston type motors.

Vane motors

A vane motor is a positive displacement motor which develops an output torque at its shaft by allowing hydraulic pressure to act on vanes which are extended.

What a vane motor consists of

The rotating group of a vane motor basically consists of vanes, rotor, ring, shaft, and a port plate with kidney-shaped inlet and outlet ports (ports not shown) as shown in fig. 4-2.

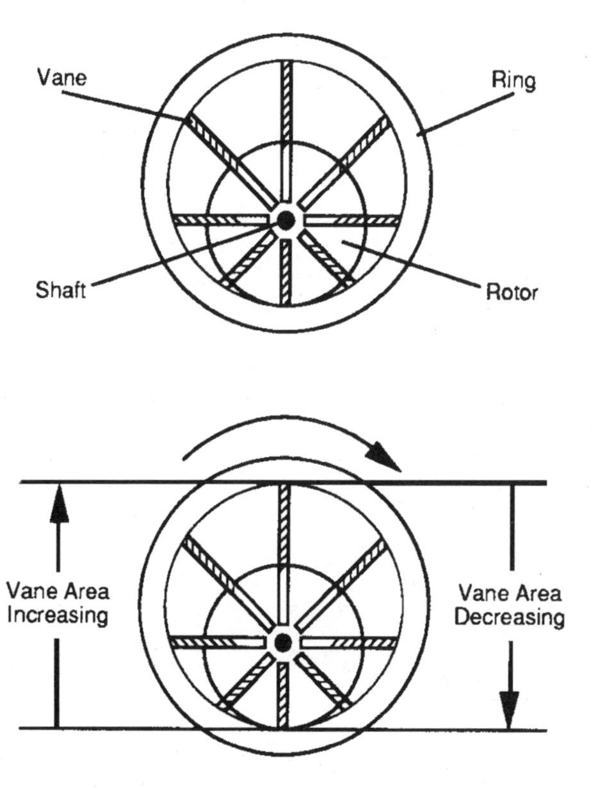

Fig. 4-2 A typical representation of a fixed displacement vane motor

How a vane motor works

All hydraulic motors operate by causing an imbalance which results in the rotation of a shaft. In a vane motor, this imbalance is caused by the difference in vane area exposed to hydraulic

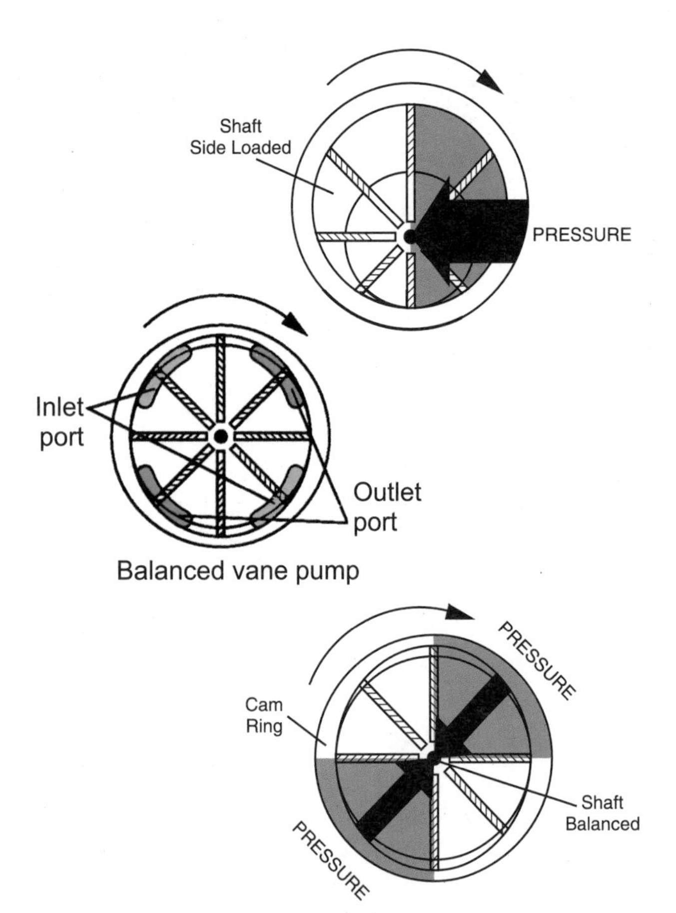

Fig. 4-3 Balanced vane motor

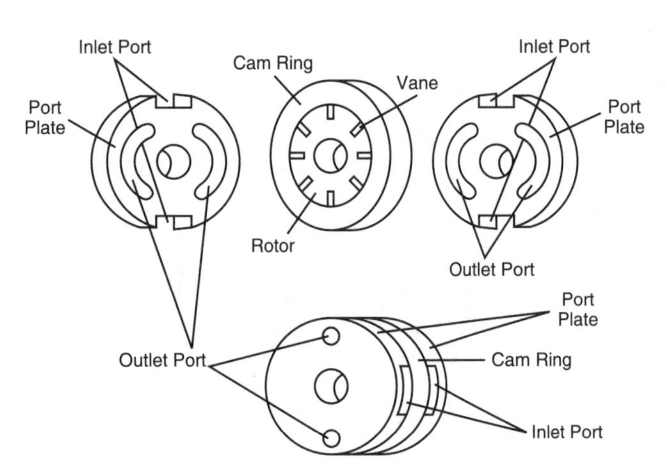

Fig. 4-4 Fixed displacement vane motor cartridge assembly

pressure. In fig. 4-2, with the rotor positioned off-center with respect to the ring, the area of the vanes exposed to pressure increases toward the top and decreases at the bottom. When pressurized fluid enters the inlet port, the unequal areas of the vanes result in a torque being developed at the motor shaft.

The larger the exposed area of the vanes, or the higher the pressure, the more torque will be developed at the shaft. If the torque developed is large enough, the rotor and shaft will turn.

Balanced vane motor design

In a hydraulic motor, two different pressures are involved — system working pressure at the inlet and tank line pressure at the outlet. This results in side loading the shaft which could be severe at high system pressures (fig. 4-3). To avoid shaft side loading, the inner contour of the ring is changed from circular to cam-shaped. With this arrangement, the two pressure quadrants oppose each other and the forces acting on the shaft are balanced. Shaft side loading is eliminated.

A balanced vane motor consists of a cam ring, rotor, vanes, and a port plate with inlet and outlet ports opposing each other. (Both inlet ports are connected together, as are the outlet ports, so that each can be served by one inlet or one outlet port in the pump housing).

Vane motors used in mobile hydraulic systems are generally of the balanced design.

Cartridge assembly

The rotating group of mobile vane motors is usually an integral cartridge assembly (fig. 4-4). The cartridge assembly consists of vanes, rotor, and a cam ring sandwiched between two port plates.

An advantage of using a cartridge assembly is easy motor servicing. After a period of time when motor parts naturally wear, the rotating group can be easily removed and replaced with a new cartridge assembly. Also, if the same motor is required to develop more torque at the same system pressure, a cartridge assembly with the same outside dimensions, but with a larger exposed vane area, can be quickly substituted for the original.

Extending a motor's vanes

Before a vane motor will operate, its vanes must be extended. Unlike a vane pump, centrifugal force cannot be depended on to throw-out the vanes and create a positive seal between cam ring and vane tip. Some other way must be found.

There are two common means of extending the vanes in a vane motor. One method is spring loading the vanes so that they are extended continuously. The other method is directing hydraulic pressure to the underside of the vanes.

Spring loading (fig. 4-5) is accomplished in some vane motors by positioning a coil spring in the vane chamber.

Another way of loading a vane is with the use of a small wire spring. The spring is attached to a post and moves with the vane as it travels in and out of the slot.

In both types of spring loading, fluid pressure is directed to the underside of the vane as soon as torque is developed.

Another means of extending a motor's vanes is with the use of fluid pressure, see fig. 4-6. In this method, fluid is not allowed to enter the vane chamber area until the vane is fully extended and a positive seal exists at the vane tip. At this time, pressure is present under the vane. When fluid pressure is high enough to overcome the spring force biasing the internal check valve, fluid will enter the vane chamber and develop a torque at the motor shaft. The internal check valve in this instance performs a sequencing function.

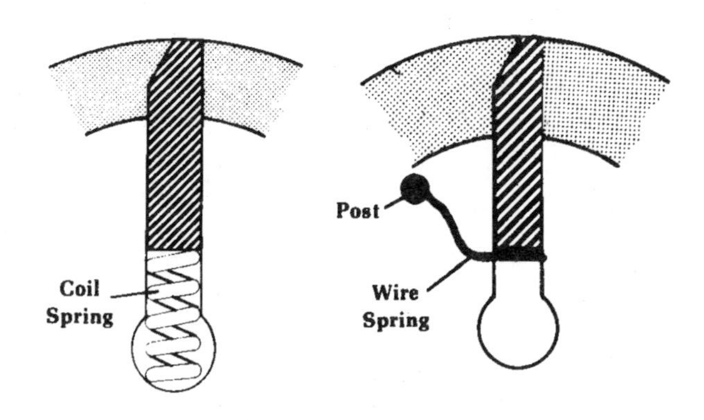

Fig. 4-5 Spring loaded vane assemblies

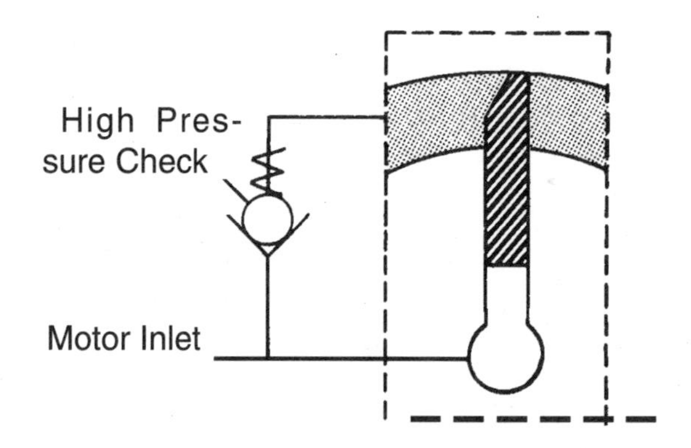

Fig. 4-6 Hydraulic pressure loading of vanes in a vane motor

Freewheeling of pressure loaded vane motors

When the load attached to a motor's shaft is allowed to freewheel, the load is allowed to coast to a stop. A motor, which uses hydraulic pressure to extend its vanes (see fig. 4-7), requires a 4 bar *(65 psi)* to 8 bar *(120 psi)* check valve in the tank line if the load is allowed to freewheel. The back pressure, which is generated because of the tank line check valve, keeps the vanes from retracting. This slows down the load more quickly.

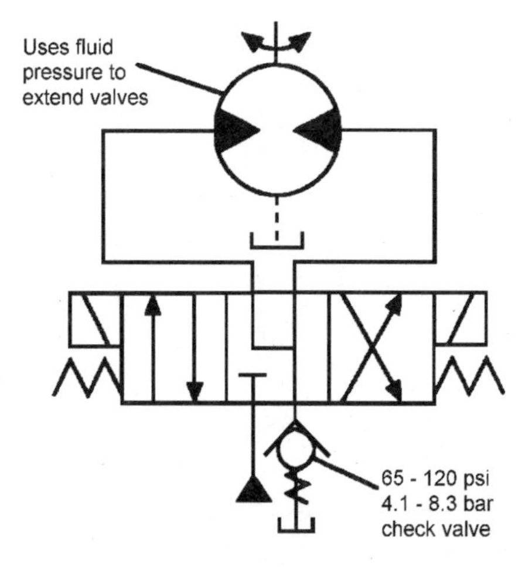

Fig. 4-7 Freewwheeling vane motors with pressure loaded vanes require a 4 - 8 bar (65 - 120 psi) backpressure.

Gear motors

A gear motor is a positive fixed displacement motor which develops an output torque at its shaft by allowing hydraulic pressure to act on gear teeth. One type of gear motor design is an external gear motor; that is, both meshing gears have teeth on their outer circumferences. The type of gear used in this motor is most commonly a spur gear.

The other type of gear motor that will be discussed is an internal gear motor. Consisting of one external gear which meshes with the teeth on the inside circumference of a larger gear. A popular type of internal gear motor in industrial systems is the gerotor motor. A second version of this internal gear motor is a geroller.

What an external gear motor consists of

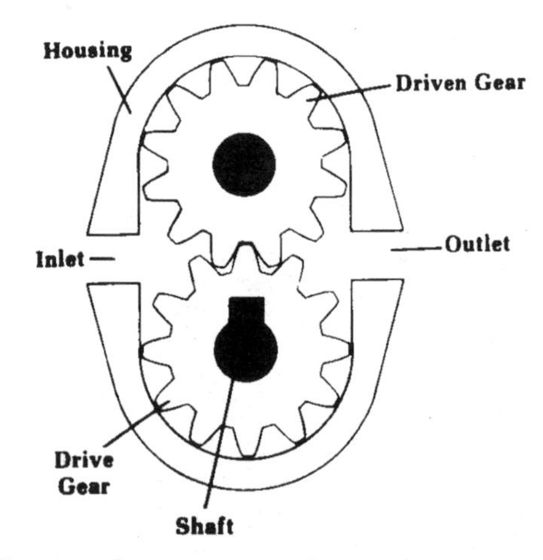

Fig. 4-8 Components of a typical gear motor

A gear motor, fig. 4-8, basically consists of a housing with inlet and outlet ports, and a rotating group made up of two gears. One gear, the drive gear, is attached to a shaft which is connected to a load. The other gear is the driven gear.

How an external gear motor works

Hydraulic motors operate by causing an imbalance which results in the rotation of a shaft. In a gear motor, this imbalance is caused by gear teeth unmeshing.

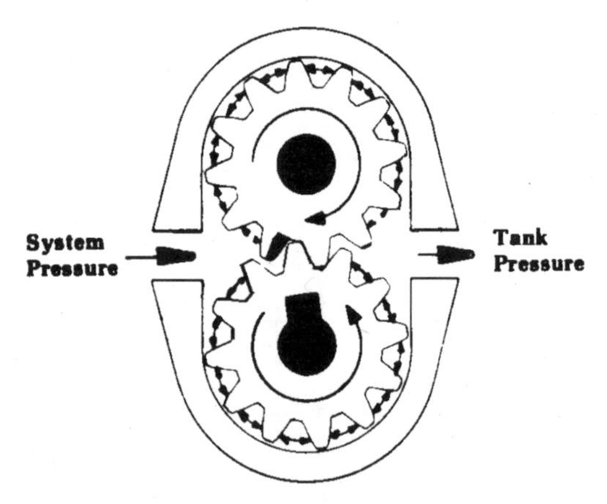

Fig. 4-9 The drive gear is rotating counterclockwise and the driven gear rotates clockwise creating an increasing volume at the inlet and a decreasing volume at the outlet.

In fig. 4-9 of the gear motor, the inlet is subjected to system pressure. The outlet is under tank pressure. As gear teeth unmesh, it can be seen that all teeth subjected to system pressure are hydraulically balanced except for one side of one tooth on one gear (shaded area). This is the point where torque is developed. Consequently, the torque developed by a gear motor of this type is a function of one side of one gear tooth. The larger the gear tooth or the higher the pressure, the more torque is produced.

You may wonder why the gear teeth do not turn in the opposite direction. To rotate in the opposite direction, gear teeth would have to mesh instead of unmesh. Gears which mesh generate a decreasing volume which pushes fluid out of the housing. The gears have no other choice but to unmesh.

Gerotor motor (an internal gear)

A gerotor motor is an internal gear motor with an inner drive gear and an outer drive gear which has one more tooth than the inner gear. The inner gear is attached to a shaft which is connected to a load.

The imbalance in a gerotor motor is caused by the difference in gear area exposed to hydraulic pressure at the motor inlet. In the gerotor motor, fig. 4-10, the exposed area of the inner gear increases at the inlet.

Fluid pressure acting on these unequally exposed teeth, results in a torque at the motor shaft. The larger the gear, or the higher the pressure, the more torque will be developed at the shaft.

Fluid entering the rotating group of a gerotor motor is separated from the fluid exiting the motor by means of a port plate with kidney-shaped inlet and outlet ports.

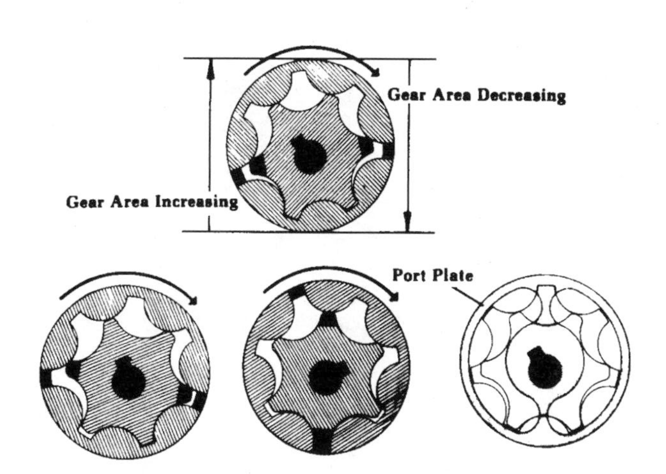

Fig. 4-10 The gerotor motor is an example of an internal gear motor.

Geroller motor

The geroller motor though similar in design to the gerotor motor, there are some significant differences in its construction and operation. The geroller, seen in fig. 4-11, consists of a locating ring with check ball holes, an inner rotor with sealing rolls, an orbiting outer element, anti-rotation rollers, and an output shaft.

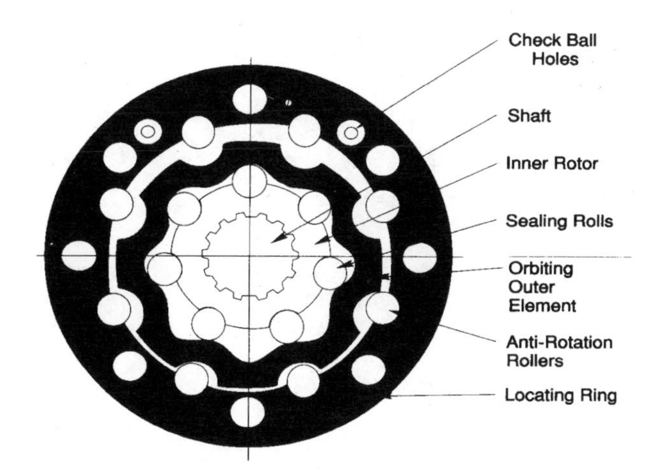

Fig. 4-11 Power element of a geroller type motor

How a geroller motor works

As in the gerotor motor both the outer and inner gears rotate as fluid pressure and flow enter the motor's inlet. And the inner gear makes an orbital rotation. But, in the geroller motor, the outer element does the orbiting while the inner roller has a circular rotation.

During motor operation, when high pressure enters the inlet, the sealing rolls, which are free to rotate within their location between the inner rotor and outer element, will seal between high pressure and low pressure chambers in essence, they act like check valves. The higher the pressure, the tighter the seal.

Two speed geroller motor

To achieve multiple speeds from a fixed displacement gear motor, a geroller motor (see fig.

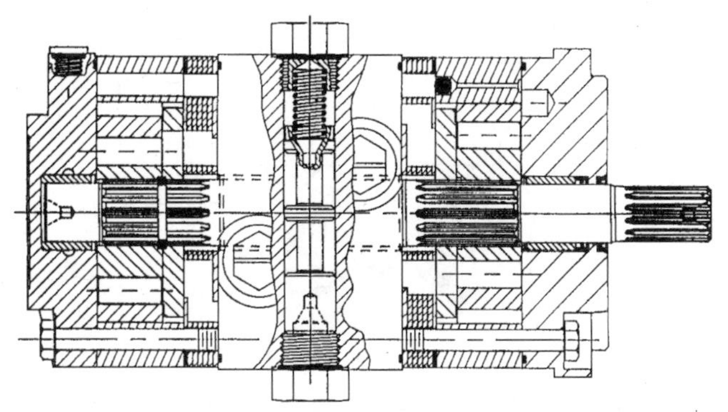

Fig. 4-12 *Two-speed geroller motors have two power elements mounted on a common shaft.*

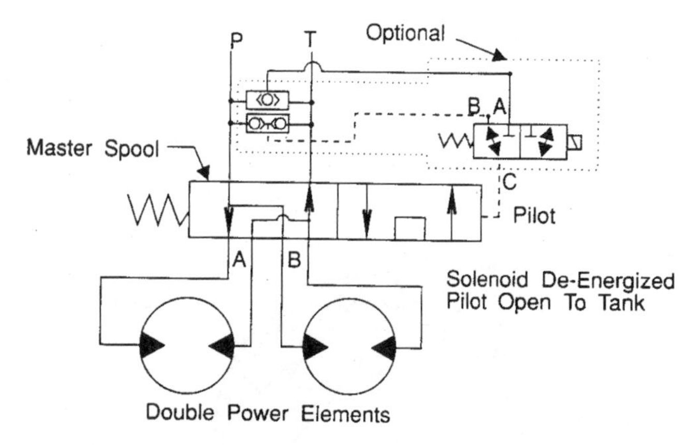

Fig. 4-13 High torque, low speed mode in a two-speed geroller motor has power elements supplied in a parallel flow.

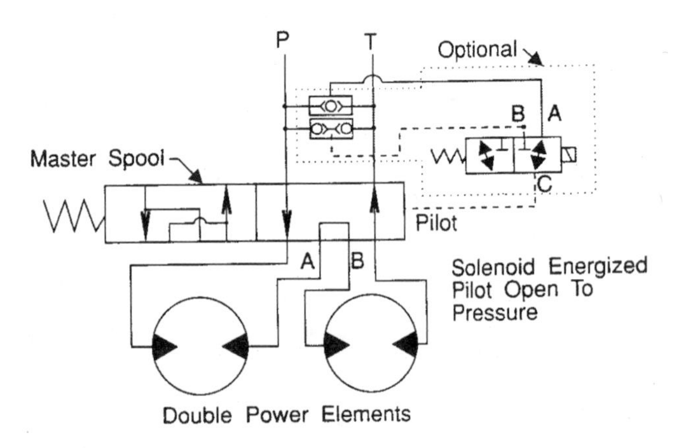

Fig. 4-14 High speed, low torque mode in a two-speed geroller motor has power elements supplied in a series flow.

4-12) can be designed to house two geroller elements into one body separated by a valving mechanism. The two elements are mounted to a common output shaft. These elements can be selected to operate either in a series or parallel mode.

The integral selector valve shifts between high torque, low speed (parallel) operation and high speed, low torque (series) mode. The selector valve can be either open center or closed center and may be actuated by external pilot pressure or by a solenoid valve.

Parallel mode in a two speed geroller motor

Fig. 4-13 shows graphically the operating principle of the two speed motor with the selector valve being controlled by a solenoid operated 2-position, spring offset, 3-way valve. With the solenoid valve in its de-energized condition the motor will operate in the high torque, low speed mode.

As high pressure and fluid flow enter the motor at port "P", It takes several paths within the motor housing. In the solenoid valve circuit, denoted by the dotted enclosure line, a shuttle valve will shift to allow system pressure to port "A" of the solenoid valve where it is deadhead. System pressure will also close the left side check of the double check valve. With the solenoid de-energized the pilot pressure at the right side of the master spool is open to tank through the solenoid valve and the right side check of the double check valve.

The master spool is held in its at-rest position by its bias spring. Through the flow path of the master spool, pump flow is directed out ports "A" and "B" to both motor elements. The parallel flow provides equal pressure to both elements and a reduced flow rate to each element resulting in the motors ability to generate high torque forces at slow rotational speed.

Series mode in a two speed geroller motor

When the solenoid valve is energized as shown in fig. 4-14, pilot pressure is directed from port "A" out port "C" of the solenoid valve to the right side of the master spool. With the master spool shifted against its spring, the flow path through the valve directs pressure and fluid flow out port

"A" only to one geroller section. The outlet of this first section is directed by the master spool out port "B" to the second geroller section. As fluid leaves the second geroller section it reenters the master spool and is directed back to tank. The two gerollers are now in series connection which allows the motor to automatically achieve a high speed, low torque mode.

Piston motors

The last motor in this section on implement motors is a positive fixed displacement piston motor which develops an output torque at its shaft by allowing hydraulic pressure to act on pistons. Two types of piston motors are commonly used in these types of applications, axial piston and bent axis piston.

What an axial piston motor consists of

The rotating group of an axial piston motor, fig. 4-15, basically consists of swashplate, cylinder barrel, pistons, shoeplate, shoeplate bias spring, port plate, and shaft.

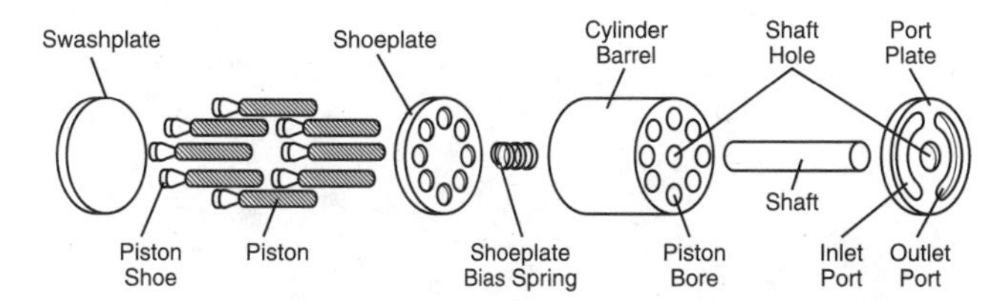

Fig. 4-15 Components of a typical fixed displacement axial piston motor

The pistons fit inside the cylinder barrel. The swashplate is positioned at an angle and acts as a surface on which the shoe side of the piston travels. The piston shoes are held in contact with the swashplate by the shoeplate and bias spring. A port plate separates incoming fluid from the discharge fluid. A shaft is connected to the cylinder barrel. In our example, it is attached at the port plate end.

How an axial piston motor works

To illustrate how a piston motor works, fig. 4-16 illustrates the operation of one piston in the cylinder barrel of an axial piston motor.

With the swashplate positioned at an angle, the piston shoe does not have a very stable surface on which to position itself. When fluid pressure acts on the piston, a force is developed which pushes the piston out and causes the piston shoe to slide across the swashplate surface. As the piston shoe slides, it develops a torque at the shaft attached to the barrel. The amount of torque depends on the angle of slide caused by the swashplate and the pressure in the system. If the torque is large enough, the shaft will turn.

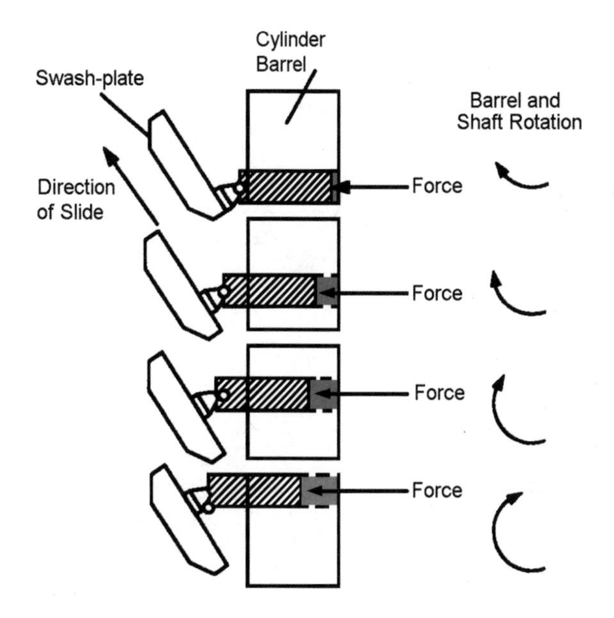

Fig. 4-16 Single piston of an axial piston motor operation

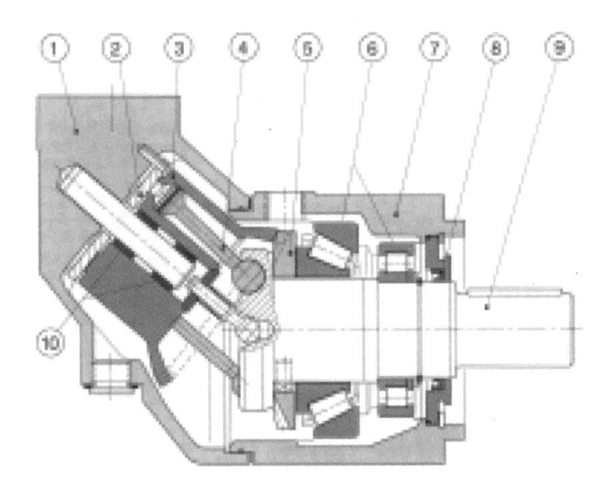

Fig. 4-17 Typical fixed displacement bent axis motor

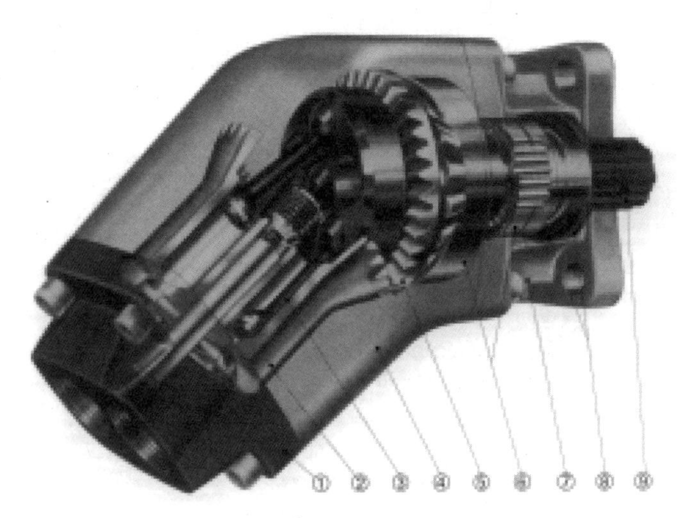

Fig. 4-18 Piston barrel and timing gear mesh to convert hydraulic energy into mechanical energy.

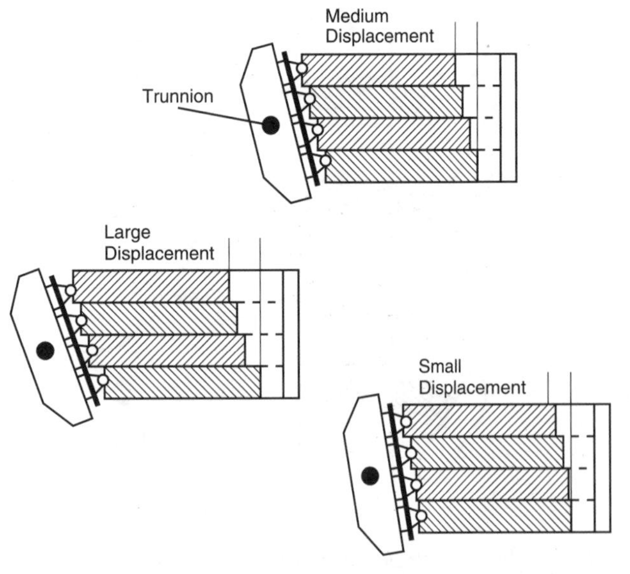

Fig. 4-19 Motor displacement is determined by the swashplate angle.

Torque continues to be developed by the piston as long as it is pushed out of the cylinder barrel by fluid pressure. Once the piston passes over the center of the circle, it is pushed back into the cylinder barrel by the swashplate. At this point, the piston bore will be open to the outlet port of the port plate.

A single piston in a piston motor develops torque for only half of the full circle of rotation of the cylinder barrel and shaft. In actual practice, a cylinder barrel of a piston motor is fitted with many pistons. This allows the motor shaft to continuously rotate as well as obtain maximum torque.

Bent axis piston motor

The fixed displacement bent axis piston motor shown in fig. 4-17, consists of the barrel housing (1), a valve plate (2), cylinder barrel (3), pistons with piston rings (4), a timing gear (5), roller bearings (6), bearing housing (7), shaft seal (8), the output shaft (9), and needle bearings (10). The angle of the piston barrel to the output shaft determines the displacement of the motor.

How a bent axis piston motor works

The view of a bent axis piston motor in fig. 4-18 shows the mechanical connection between the pistons and the output shaft. As pressurized fluid enters the inlet port in the barrel housing, the piston(s) at the inlet are force back in the barrel and with the pistons attached to the output shaft and thus the timing gear, the output shaft and barrel will rotate.

Of the vane, gear, and piston motors which have been described, only piston motors are available as variable displacement.

Variable displacement axial piston motors

The displacement of an axial piston motor, or any piston motor, is determined by the distance the pistons are reciprocated in the cylinder barrel.

Since the swashplate angle controls this distance in an axial piston motor (fig. 4-19), we need only to change the angle of the swashplate to alter the piston stroke and motor displacement.

With a large swashplate angle, the pistons have a long stroke within the cylinder barrel.

With a small swashplate angle, the pistons have a short stroke within the cylinder barrel.

By varying the angle of the swashplate then, the motor's displacement and consequently its shaft speed, and torque output can be changed.

Overcenter axial piston motors

Some swashplates of axial piston motors have the capability of crossing overcenter. A motor of this type is able to reverse its shaft rotation without changing the direction of flow through the motor.

In the overcenter axial piston motor, see fig. 4-20, the motor shaft is not shown. But, you can imagine it as being attached to the cylinder barrel at the port plate end through the swashplate side. We can see from the illustration that changing the angle of the swashplate by crossing overcenter results in a different direction of slide for the pistons. Consequently, cylinder barrel and motor shaft rotate in the reverse direction. This takes place with fluid flow passing through the motor in the same direction.

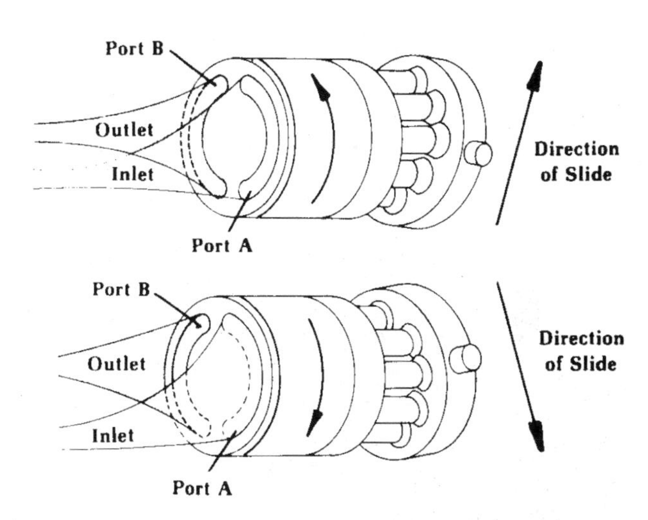

Fig. 4-20 Overcenter axial piston motors can reverse their rotation by reversing the angle of the swashplate.

Variable displacement bent axis piston motor

A typical variable displacement bent axis piston motor (see fig. 4-21) consists of: a servo piston (1), servo valve with adjustment screw (2), end cap with integrated displacement control (3), valve segment (4), pistons (5), cylinder barrel (6), synchronizing shaft (7), bearings (8), mounting flange (9), and output shaft (10). The amount of displacement of the bent axis piston motor is controlled by the servo piston and valve.

More detail on the controlling of the displacement angle for this type motor will be covered in the hydrostatic drive section.

Hydraulic motors in a circuit

We have seen that the torque developed by a hydraulic motor is the result of hydraulic pressure acting on the motor's rotation group. In these situations, we made the assumption that no hydraulic pressure was present after the motor. Even though this side of the motor is generally drained to tank, tank line pressure, or back

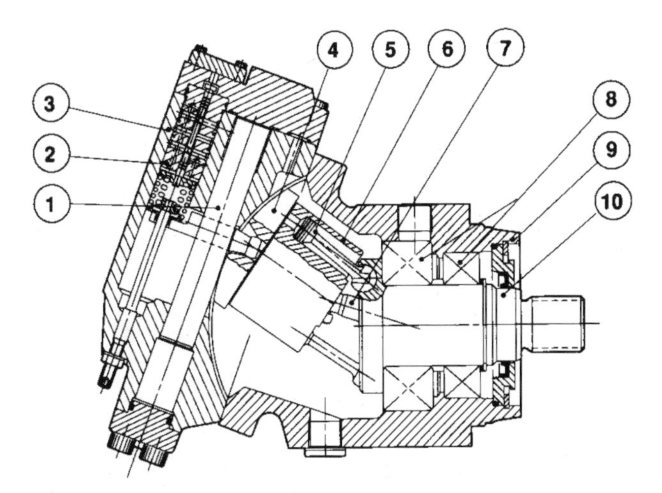

Fig. 4-21 Typical bent axis variable displacement motor. Angle of displacement is controlled by an integral servo valve.

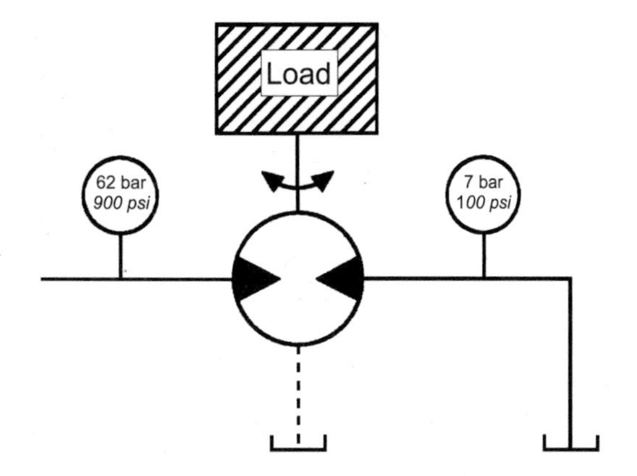

Fig. 4-22 Backpressure affects the torque capability of a hydraulic motor.

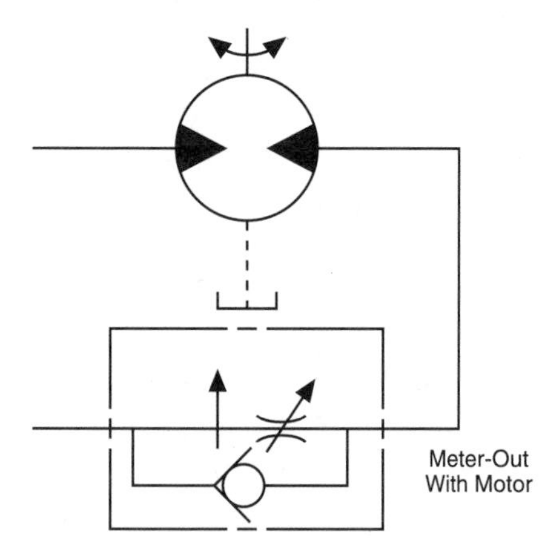

Fig. 4-23 To keep flow rate through the flow control constant, install a pressure compensated flow control valve.

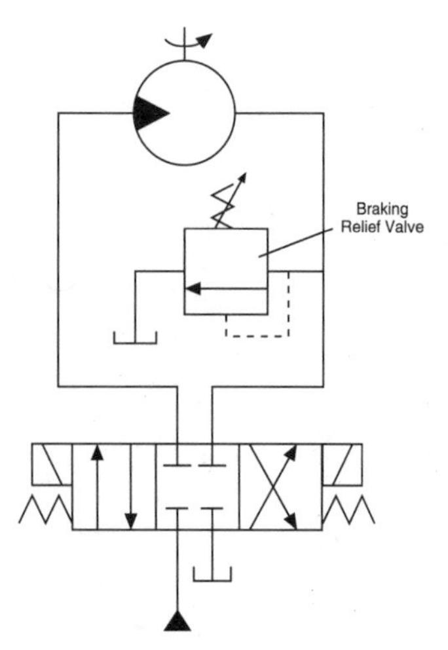

Fig. 4-24 A 3-position, closed center directional valve is used to brake the load.

pressure, can be as high as 6.9 bar (100 psi) in some systems. The force generated by the back pressure on the rotating group must be overcome before a load can be turned.

In fig. 4-22, the load attached to the motor's shaft can be turned, theoretically, with a pressure of 55 bar *(800 psi)* at the motor inlet. The back pressure in the tank line is 7 bar *(100 psi)*. With this condition, the load will not turn. An additional 7 bar *(100 psi)* must be present at the motor inlet to equal or offset the 7 bar *(100 psi)* backpressure. With 62.1 bar (900 psi) at the inlet, 62 bar (900 psi) is used to develop and turn the load. 7 bar *(100 psi)* offsets the back pressure. The 55 bar *(800 psi)* which was calculated to turn the load actually indicated the required pressure differential.

Hydraulic motors are generally externally drained. This means a portion of the flow entering the motor ends up as leakage. As the torque requirement and the pressure at the motor increases, more flow runs out the drain. As a result, motor shaft speed decreases.

To accurately control the speed of a hydraulic motor, a meter-out circuit is used. A meter-out circuit, fig. 4-23, controls the flow as it discharges from the motor and is not concerned with leakage. This is the only circuit which can control a motor's shaft speed accurately regardless of load.

One of the major concerns in motor circuits is the control of the load attached to the motor shaft. We have seen previously that a brake valve will keep a load from running away as well as allow the motor to develop full torque.

A brake valve senses the load. It automatically responds to the load's demand. Many times, the braking function is required to be a matter of choice rather than automatic. For example, in a conveyor system, which has no overrunning load and requires braking only periodically, a directional valve is used to select the braking function in a unidirectional motor circuit, see fig. 4-24.

Braking is performed by shifting the directional valve, usually to its center position, and blocking the flow out of the motor. When the pressure at the motor outlet increases to the braking relief valve setting, the valve opens and brakes the motor.

If the motor requires braking in both directions, see fig. 4-25, a braking relief valve can be connected through check valves to both motor lines. No matter which way the motor is rotated, braking is performed by the same valve.

In some applications, two braking pressures are necessary, fig. 4-26. For example, a conveyor, which is loaded in one direction and unloaded in the opposite direction, would require two different braking pressures to make most efficient use of its cycle time.

When two different braking pressures are required, two braking relief valves are connected in the motor lines. Each valve handles flow in different directions. Braking relief valves applied in this manner can also be used to achieve approximate starting and stopping positions with dissimilar loads in opposite directions.

A braking relief valve is a common, ordinary relief valve placed in a motor line. It is not a special valve. The setting of a braking relief valve is higher than the setting of the system relief valve.

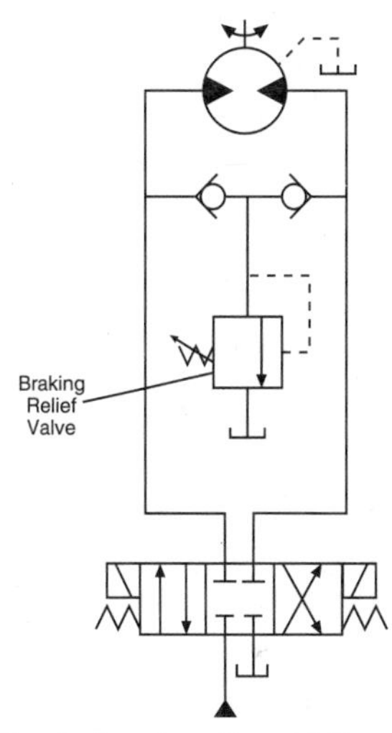

Fig. 4-25 Dual checks allow bidirectional braking.

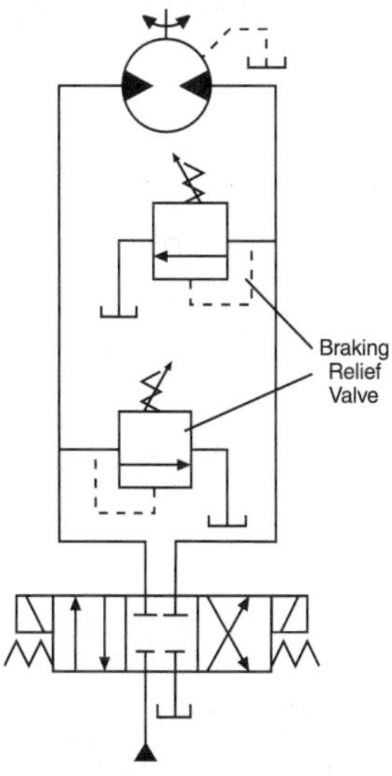

Fig. 4-26 Braking a motor in bidirectional rotation at different pressure requirments done with two braking valves.

Preventing motor cavitation

All of the motor circuits shown so far do not take into account that hydraulic motors can cavitate. They will cavitate just as a pump if a sufficient supply of fluid is not received at its inlet port while turning. This means that anytime a motor is braking, the motor inlet must not be blocked.

In a unidirectional motor circuit, this requirement can be met by allowing the motor inlet to be connected to tank through the center position of the directional valve as seen in fig. 4-27. When braking occurs, any less-than-atmospheric pressure at the motor inlet will result in fluid being drawn from the reservoir.

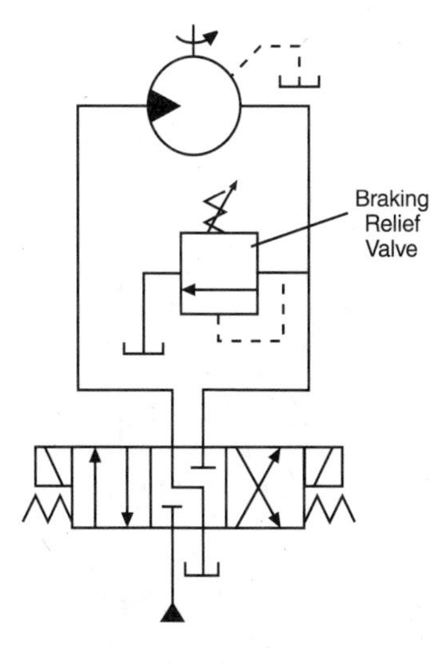

Fig. 4-27 Preventing motor cavitation in a unidirectional motor.

Makeup check valves

In a bidirectional motor circuit, supplying liquid to the motor inlet during braking is usually done with low pressure (.34 bar/*5 psi* or less) check valves positioned in each line. These are known as makeup check valves.

Crossover relief valves

A bidirectional motor circuit, using braking relief valves in both directions, can be designed so that the discharge from the relief valves is con-

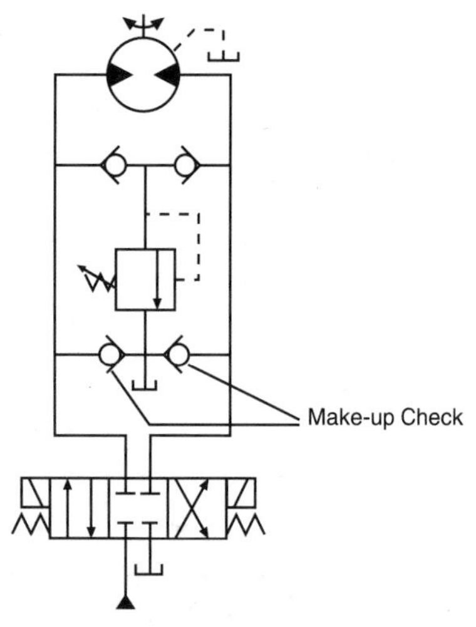

Fig. 4-28 Makeup checks help prevent cavitation in bidirectional motors.

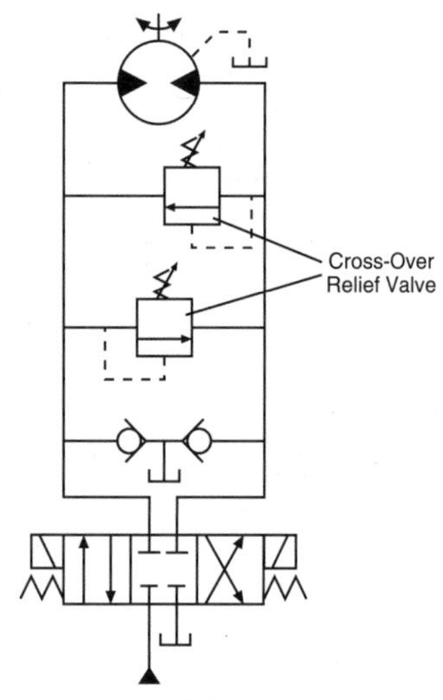

Fig. 4-29 Crossover relief valves prevent cavitation in bidirectional motors.

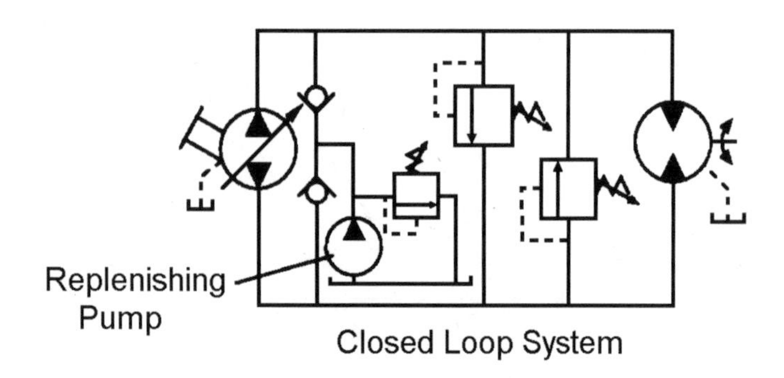

Fig. 4-30 Hydrostatic transmissions provide various speed ranges without having to have mechanical gearing.

nected to the opposite motor lines. From first glance, it may appear that these "crossover" relief valves would keep the motor's inlet well supplied since the motor's discharge fluid is redirected to motor inlet. However, makeup checks are still required since some fluid is lost through the motor drain and leakage across the directional valve. This arrangement of crossover relief valves and makeup checks is a very common bidirectional motor circuit.

Hydrostatic drives

Any hydraulic motor used in combination with various pumps is termed a hydrostatic drive. All of the previous circuits are considered hydrostatic drives. The hydrostatic drives circuits previously discussed all represent what are known as open loop systems. Hydrostatic drives can be classified based on torque range requirements; whether the drive is integral or split; is an open loop or a closed loop system; and whether constant or variable torque and or horsepower is needed.

Open loop

An open loop hydrostatic drive has the motor inlet connected to pump outlet and the motor outlet connected to tank. The motor rotation is stopped or reversed with a directional valve. The speed of the motor depends on pump flow rate and motor displacement.

Closed loop

A closed loop hydrostatic drive has motor inlet connected to pump outlet and motor outlet connected to pump inlet. The closed loop schematic shows motor rotation in either direction with variable pump input which will vary motor speed and direction. Any leakage in the system is made up by the replenishing pump. A small reservoir is used in this system since most of the system fluid is carried and stored in the system piping. Closed loop hydrostatic drives are compact.

Hydrostatic transmission

In common terminology, anytime a variable displacement pump or motor is used in a pump-motor circuit, the system is labelled a hydrostatic transmission.

In the closed loop hydrostatic transmission, fig. 4-30, the variable displacement, overcenter pump can vary the speed of the motor shaft as well as reverse shaft rotation. In closed loop systems of this nature, a small pump, called a replenishing pump, is used to make up for any leakage which occurs in the system.

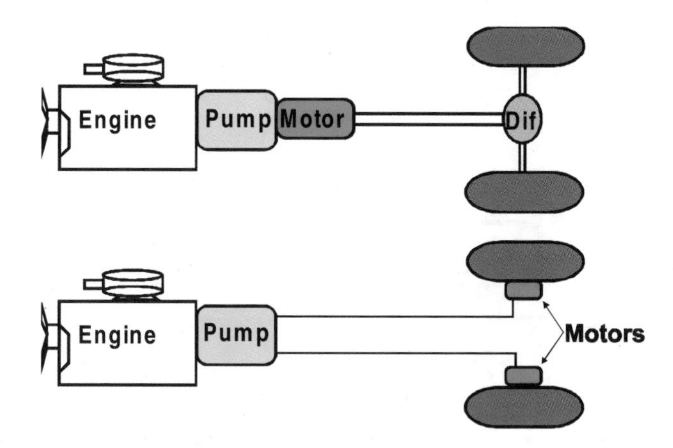

Fig. 4-31 Integral hydrostatic transmissions combine the pump-motor as a single unit. Split hydrostatic transmissions use individual pump(s) and motors.

Closed loop hydrostatic transmissions are compact systems. The reason being the reservoir is small and flow controls and directional valves are not needed to reverse or control the speed of shaft rotation.

Integral and split hydrostatic transmission

In a mechanical drive system for a wheel vehicle, the power train would consist of the engine, transmission, driveline(s), gear box, differential(s). fig. 4-31 illustrates a drive system where the transmission is replaced by a pump and motor combination which have their shafts in-line with each other. This represents an integral hydrostatic drive.

The mechanical drive train can be replaced completely with a split hydrostatic drive transmission which couples the pump to the engine and the motors to the drive wheels of the vehicle. By using one pump to drive two or more motors the system is considered a dual drive.

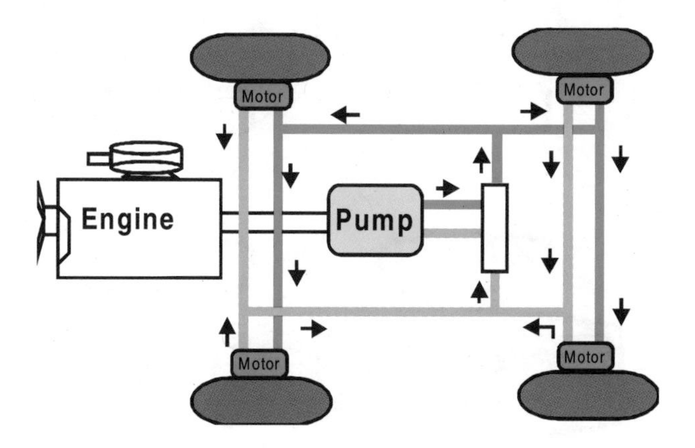

Fig. 4-32 Parallel circuit for a hydrostatic transmission.

Parallel hydrostatic transmission

As illustrated in fig. 4-32, the motors are connected in parallel. The outlet of the pump is connected to all four motors delivering the same pressure to all the motors at the same time. Since all motors receive the same pressure, the maximum pressure at any time is determined by the drive wheel that has the least amount of rolling resistance. A parallel system can also have problems while the vehicle is in a turning condition since the outer set of wheels will track faster than the inner set.

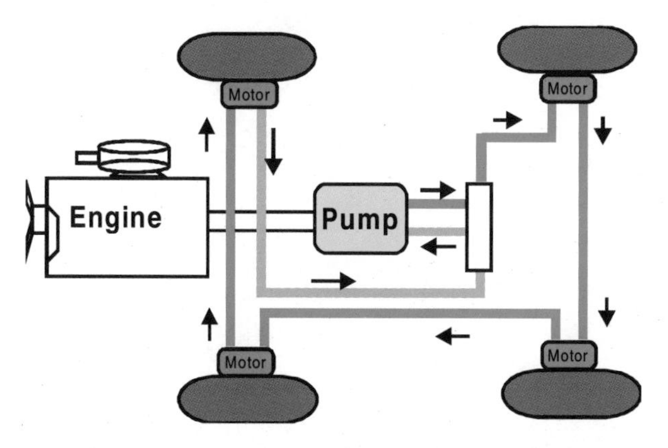

Fig. 4-33 Series circuit for a hydrostatic transmission

Series hydrostatic transmission

By connecting the motors as shown in fig. 4-33, a series hydrostatic transmission is formed. In a series circuit, the pump's output is directed into one motor and the outlet of that motor is connected to the inlet of the next motor, etc. Since each motor will receive the total flow from the pump, in turn, higher vehicle speeds can be achieved.

If a combination of parallel and series are used on one vehicle three speed ranges can be created for the vehicle in both forward and reverse motion.

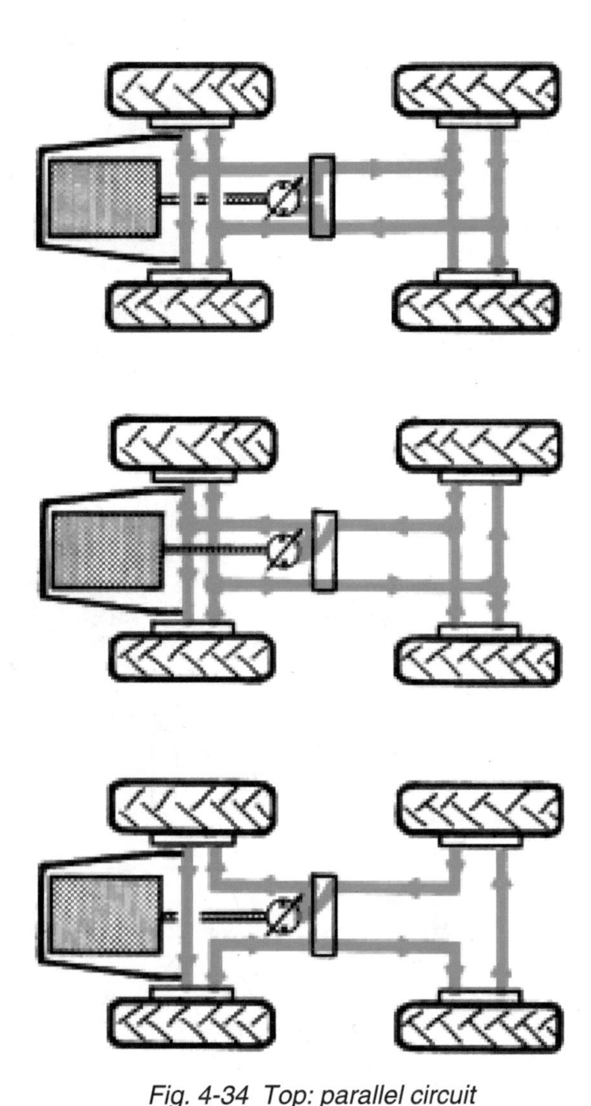

Fig. 4-34 Top: parallel circuit
Middle: paired series circuit
Bottom: series circuit

Reprinted from The Off-Road Vehicle Volume 1, courtesy of CPPA

Constructing a hydrostatic transmission

A rubber tired vehicle is to be changed from a mechanical drive to a hydrostatic transmission drive with three speeds forward and three speeds in reverse with a top speed of 30 km/h *(19 mph)*. The engine of the vehicle delivers about 75 kW *(100 hp)*. The pump chosen is a variable displacement piston pump delivering approximately 300 l/min *(80 gpm)* at a maximum of 393 bar *(5700 psi)*. All four motors selected have 156 mL/rev *(9.51 in^3/rev)* and can absorb the total pump output at the maximum pressure. To obtain the three speed ranges proper controls are to be selected so that for the lower speed range, the motors are coupled in a parallel mode as seen in fig. 4-34. To obtain the middle speed range, the controls will be shifted to place the motors in paired series mode. And to get the top speed range, the controls will be positioned to place the motors in series mode or to only have two wheel drive.

When reverse motion is needed, the pump flow is reversed within the system and the controls will give the same combination of operating modes. Therefore, the same speed ranges can be achieved in reverse.

Another way to achieve the goal of obtaining three different speed ranges would be to use motors that had variable displacement controls at all four wheels. By placing all four motors at their maximum displacement the low speed range would be obtained. The middle speed range would require the motors to be set at 50% displacement. And the top speed range would have two motors set at 50% displacement and two motors disengaged and free wheeling.

Preventing wheel spin

In the section on parallel mode hydrostatic transmission connections it was stated that the wheel with the least rolling resistance would determine the system pressure. It is also true that the wheel with the least rolling resistance could also consume all of the pump flow and cause the vehicle to slow down or completely stop. To prevent wheel spin in an all mechanical drive train, a differential(s) can be equipped with “lockup”. So too can the hydrostatic transmission drive be equipped to prevent or lessen wheel spin.

Fig. 4-35 illustrates a hydrostatic transmission equipped with fixed orifice, pressure compensated flow control valves. Through the proper placement of the flow controls, the flow to the individual motors or to pairs of motors can be accomplished. If it is desired to eliminate the “lockup” feature, at some point in the vehicle operation, directional control valves are used to bypass the flow controls.

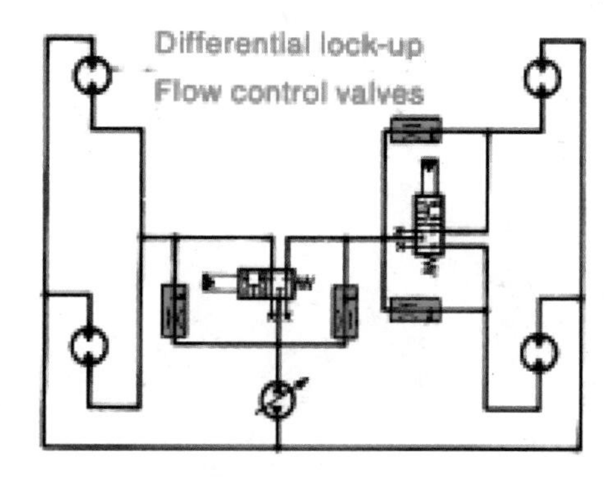

Fig. 4-35 A “lockup” circuit to prevent wheel spin Reprinted from The Off-Road Vehicle Volume 1, Courtesy of CPPA

Braking a hydrostatic transmission drive

Stopping the vehicle can be accomplished either hydraulically or mechanically. Hydraulic braking could be done by placing the pump into a zero flow condition and allowing the motor(s) to, in effect, drive the pump. This would in turn generate back pressure that would force fluid over the pressure limiting valve(s) (fig. 4-36). This pressure would create a force that is equal and opposite to the driving force of the vehicle therefore, stopping the vehicle.

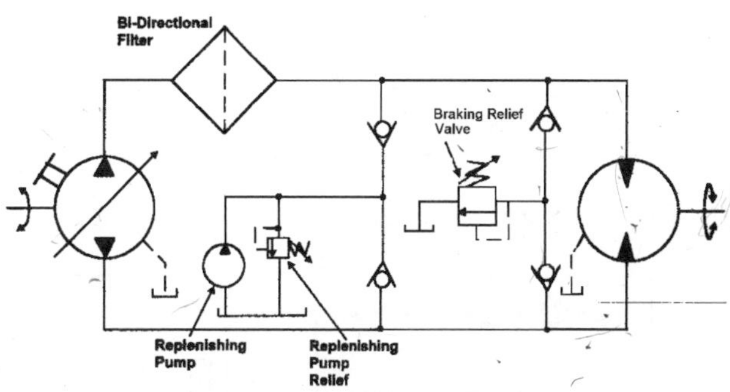

Fig. 4-36 Typical hydrostatic transmission braking circuit

However, if pressure were lost in the system due to a hose or fitting leak the hydraulic brake method would be inoperable. Therefore, a mechanical method of braking would be more reliable not mention safer. Some wheel motors have brakes built into the motor (fig. 4-37) or the brake is combined with the motor and the wheel hub.

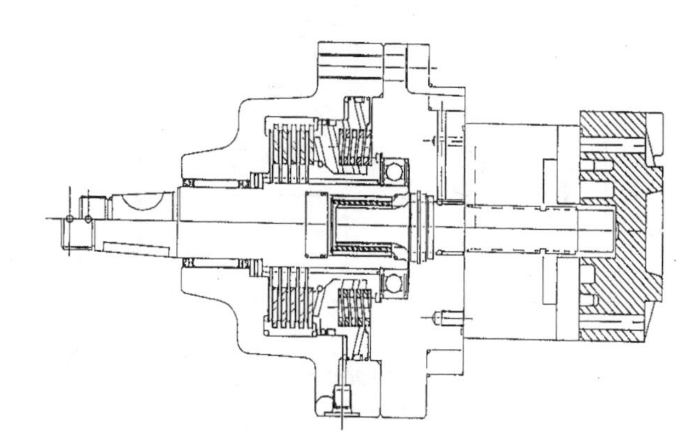

Fig. 4-37 Integral brake motor

Wheel motors

Wheel motors are designed to permit easy mounting, (fig. 4-38) of hubs and wheels onto self-propelled vehicles. They generally mount directly to the vehicle frame and become the axle and drive motor for the vehicle.

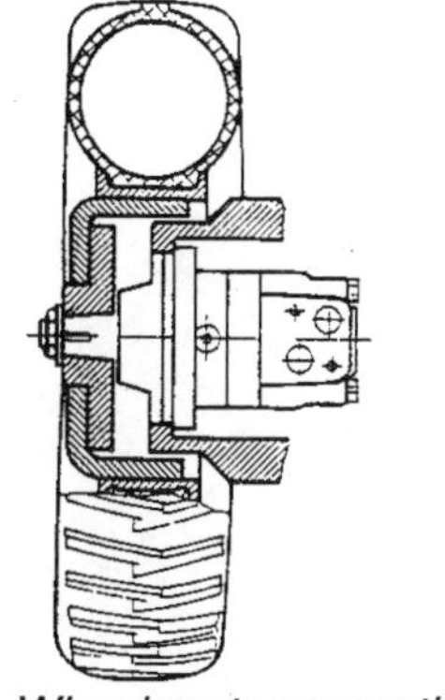

Fig. 4-38 Wheel motor mounting

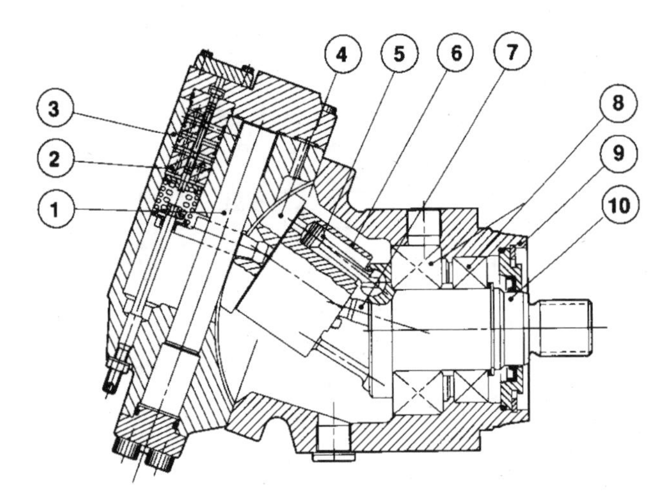

Fig. 4-39 Bent axis piston motor

Variable displacement bent axis wheel motor

Fig. 4-39 illustrates a bent axis variable displacement motor. The motor consists of: (1) a servo piston, (2) servo valve with adjustment screw, (3) an end cap with integrated displacement control, (4) valve segment, (5) pistons, (6) cylinder barrel, (7) synchronizing shaft, (8) bearings, (9) mounting flange, and (10) output shaft.

Displacement controls

Some variable displacement motors will have their displacement controls as an integral part of the motor. The following sections will be a discussion of some of the controls that might be used on a bent axis variable displacement wheel motor.

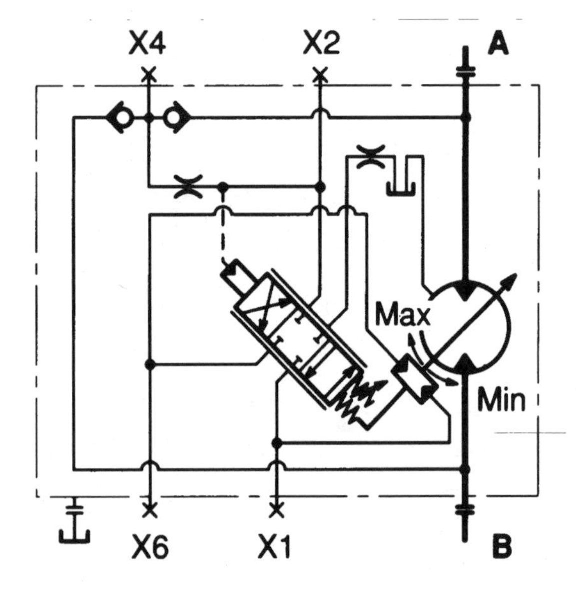

Fig. 4-40 Pressure compensated control schematic Internal pilot control of servo valve control

Pressure compensator control

Schematically represented in fig. 6.40 is a pressure compensator. This control automatically adjusts motor displacement to the output torque requirement.

The motor is normally positioned at its minimum displacement. When there is a demand for additional torque i.e. when the vehicle enters an upgrade, the displacement increases (providing more torque) while the motor shaft speed decreases proportionally.

The gauge ports denoted in the schematic are: “X1” Servo piston pressure (max displ.), “X2” Servo supply pressure (after orifice), “X4” Servo supply pressure (before the orifice), “X6” Servo piston pressure (min displ.). During motor operation, the servo valve bias spring positions the servo valve so that system pressure is directed through the parallel arrow position of the valve to the servo piston moving the pump angle to the minimum displacement position.

The threshold pressure, where displacement starts to increase is an adjustable pressure set by the adjustment screw on the servo valve spool spring at 150-400 bar *(2200-5800 psi)*. The actual pressure can be read at gauge port “X4”. In order to move the motor to maximum displacement, an additional modulator pressure from 15-50 bar *(220-725 psi)* must be felt at gauge port “X2” to force the servo valve completely into the crisscrossed arrow position, thus forcing the servo piston to the maximum displacement position.

Pressure compensator with brake defeat

The pressure compensator in fig. 4-40 is modified in fig. 4-41 to incorporate a solenoid controlled override function and a brake defeat function.

The override consists of a piston built into the end cover and an external electrohydraulic solenoid valve. When the solenoid is energized, system pressure is directed to the piston which in turn pushes the spool of the servo control valve. This causes the motor to lock in the max displacement position, irrespective of system pressure.

The brake defeat valve is also a part of the compensator control and consists of a two-position, three-way spool. The two ports "X9" and "X10" should be connected to the corresponding ports "A" and "B" respectively

The brake defeat function prevents the motor outlet port pressure to influence the pressure compensator. If, for example, port "A" is being pressurized when driving 'forward', pressure in port "B" during braking will not cause the motor to increase its displacement. Likewise, when driving in 'reverse' (port "B" pressurized), any braking pressure in port "A" will not influence the control.

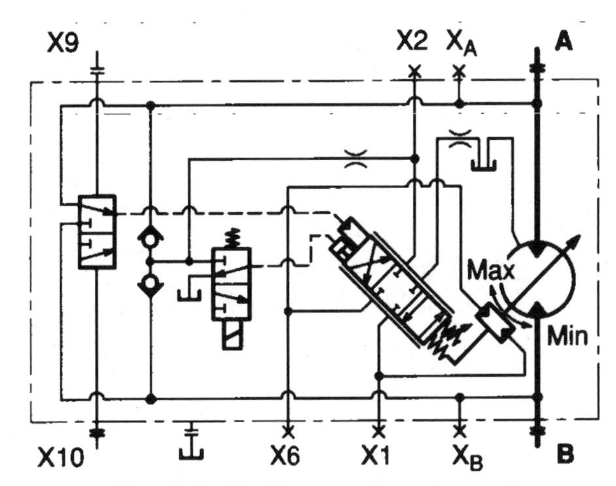

Fig. 4-41 Braking control schematic

Electrohydraulic proportional control

When continuously variable shaft speed is required in a vehicle the control of the motor can be performed by an electrohydraulic proportional DC solenoid valve as seen in fig. 4-42. The motor is normally in its maximum displacement position. When the solenoid current increases above its threshold value, the servo piston starts to move from the max to the min displacement position.

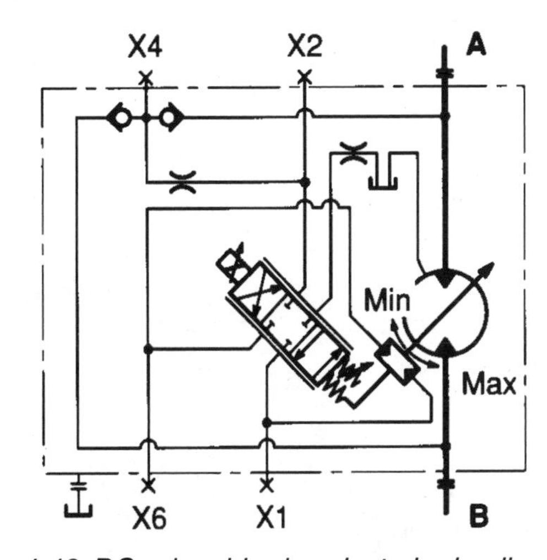

Fig. 4-42 DC solenoid valve electrohydraulic control schematic

Fixed displacement motor controls used in a hydrostatic transmission

If a fixed displacement motor is used in a vehicles hydrostatic transmission drive system various integral controls can be employed to control its operation. Fig. 4-43 shows a schematic representation of a flushing valve used on a fixed displacement motor.

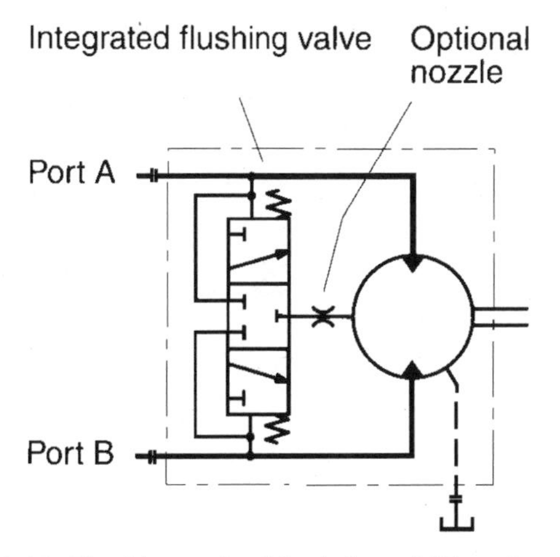

Fig. 4-43 Flushing valve block for additional cooling of fixed displacement motor

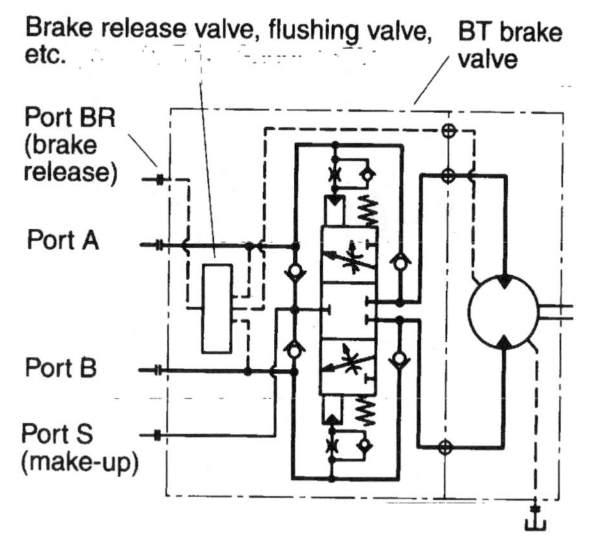

Fig. 4-44 Braking valve schematically represented

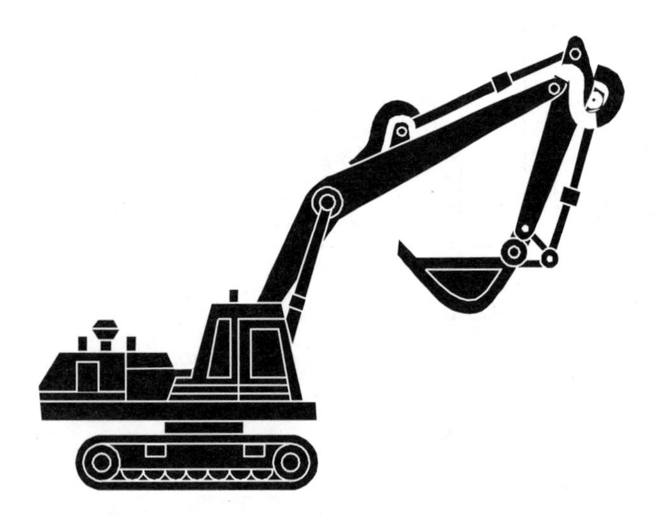

Fig. 4-45 Swing drive motors control the rotation of large shovels.

The flushing valve provides the rotating parts with an additional cooling flow, required when operating at high speeds and power levels. The valve also ensures that part of the main circuit fluid is being removed and replaced by cool filtered fluid from the charging pump.

Braking valve

When a fixed displacement motor is used in a vehicles hydrostatic transmission, the motor may operate faster than what corresponds to the available pump flow e.g. in a steep down hill. This causes motor cavitation and loss of braking power.

A brake valve, shown schematically in fig. 4-44, prevents cavitation by throttling the return flow from the motor as soon as the pressure level in the inlet port decreases to a low value e.g. 35 bar *(508 psi)*. At the same time, motor braking is provided when the pump flow decreases or is shut off.

Swing drive motors

Hydraulic motors that have been discussed up to this point can all be used in the third motor classification of swing motors. A swing motor application shown in fig. 4-45 might be a heavy equipment shovel. The swing motor will rotate the top half of the vehicle around the tracks in 360° of rotation.

Selection of a motor

The following equations are used to determine the required horsepower, torque, and flow rate for most motors used as implement motors. (The subscript "m" indicates metric.)

$$w_{in} = Q_m \times P_m \times .166 \times 10^{-4}; HP_{in} = \frac{QP}{1714}$$

$$w_{out} = (N)(T)(.1047); HP_{out} = \frac{NT}{63025}$$

$$T_m = \frac{D_m \Delta P_m e_m (10^{-6})}{2\pi}; T = \frac{D \Delta P e_m}{2\pi}$$

$$Q_m = \frac{D_m N}{10^3 e_v}; \ Q = \frac{DN}{231 e_v}$$

where:

w_m, *HP* = watts *(horsepower)*
Q_m, *Q* = Flow, l/min *(gpm)*
P_m, *P* = Pressure, Pa *(psi)*
ΔP = Press. differential across motor
T_m, *T* = Torque, N-M *(lb in)*
D_m, *D* = Motor displacement, cm^3/rev (*in^3/rev)*
N = Shaft speed, rpm
e_m = Mechanical efficiency
e_v = Volumetric efficiency

Note: 1 *psi* = 6895 Pa = 0.7 bar

Side Load

Side loads are imposed upon the shaft of a motor by:

- driving the load through a pulley or gear
- support the weight of a vehicle or other load on the shaft
- or both.

As an example in fig. 4-46, if the load requires a torque T Newton-meters *(pound-inches)* and is driven with a pulley on the motor shaft with a radius of R meters *(inches)*, the side load imposed on the motor shaft is T/R Newtons *(pounds)*. If the motor shaft is connected to a sprocket for a chain drive, R is one half the pitch diameter of the sprocket. If an external load with a weight of W Newtons *(pounds)* is also being supported by the motor shaft, the total side load on the shaft is determined by the equation:

$$\text{Side load (Newtons }/lb) = \sqrt{W^2 + (T/R)^2}$$

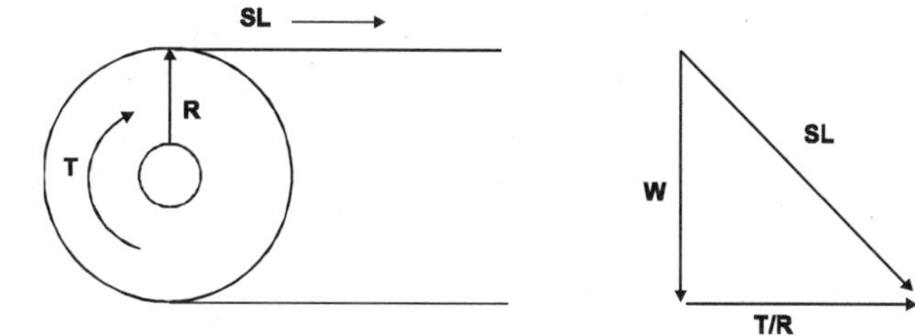

Fig. 4-46 How to calculate shaft side load

Sizing for vehicle propulsion systems

Hydraulic motors are often used to drive off-highway vehicles, either directly or through gear reducers. The power required to propel the vehicle, called "tractive effort," is supplied by the hydraulic motor(s). It is normally expressed in Newtons *(pounds)* and is the sum of the forces below:

TE = (RR+GR+F+ΔP) x 1.1

Where: RR = Rolling resistance (Newtons/*lb*)
GR = Grade resistance (Newtons/*lb*)
F = Acceleration force (Newtons/*lb*)
ΔP = Drawbar pull (Newtons/*lb*)

Surface Type	Surface Condition	R Value
Concrete	Excellent	0.010 lb. (0.005 kg)
Concrete	Good	0.015 lb. (0.007 kg)
Concrete	Poor	0.020 lb. (0.0091 kg)
Asphalt	Good	0.012 lb. (0.0054 kg)
Asphalt	Fair	0.017 lb. (0.008 kg)
Asphalt	Poor	0.022 lb. (0.010 kg)
Macadam	Good	0.015 lb. (0.007 kg)
Macadam	Fair	0.022 lb. (0.010 kg)
Macadam	Poor	0.037 lb. (0.017 kg)
Cobbles	Ordinary	0.055 lb. (0.025 kg)
Cobbles	Poor	0.085 lb. (0.039 kg)
Snow	2 in.(5.08 cm)	0.025 lb. (0.011 kg)
Snow	4 in. (10.16 cm)	0.037 lb. (0.017 kg)
Dirt	Smooth	0.025 lb. (0.011 kg)
Dirt	Sandy	0.037 lb. (0.017 kg)
Mud		0.037 lb. (0.017 kg) to 0.150 lb. (0.07 kg)
Sand	Level/Soft	0.060 lb. (0.027 kg) to 0.150 lb. (0.07 kg)
Sand	Dune	0.150 lb. (0.07 kg) to 0.300 lb. (0.14 kg)

Fig. 4-47 Typical "R" values for rolling resistance

Grade (Percent)	Slope (Degrees)
1%	0° 35'
2%	1° 9'
5%	2° 51'
6%	3° 26'
8%	4° 35'
10%	5° 43'
12%	6° 54'
15%	8° 31'
20%	11° 19'
25%	14° 3'
32%	18°
60%	31°

Fig. 4-48 Relationship table between grade in percent and slope in degrees

Tractive effort (TE) definition

Tractive effort is the total linear force that a vehicle can exert on the ground. Sometimes called "rim pull," it is the axle torque divided by the distance from the axle to the surface it is traversing.

Rolling resistance definition

Rolling resistance is the force in pounds required to propel a vehicle at constant speed over level terrain. It varies with the weight of the vehicle and the type of surface it is traversing. Soft sand, for example, offers more resistance to movement than concrete. Rolling resistance can be found using the following equation:

$$RR = GVW \times R$$

where:

RR = Rolling resistance (Newtons/*lb*)
GVW = Gross vehicle weight (Newtons/*lb*)
R = Rolling resistance factor dependent upon type and condition of surface. Typical "R" values are shown in fig. 4-47.

Grade resistance (GR) definition

Grade resistance is the additional force required to move a vehicle up an incline. The grade of a slope is normally expressed as a percentage, and represents the number of feet of rise in 100 feet of length. A slope that rises 10 feet in 100 feet has a grade of 10%. The gradeability of a vehicle is defined as the maximum grade the vehicle can climb. Grade resistance is calculated using the following equation:

$$GR = 0.01 \times GVW \times G$$

where:

GR = Grade resistance (Newtons/*lb*)
GVW = Gross vehicle weight (Newtons/*lb*)
G = Grade (%)

Fig. 4-48 gives the approximate relationship between grade in percent and slope in degrees.

Acceleration force (F) definition

The force required to accelerate a vehicle from an initial speed V1 (in feet/second) to speed V2 in T seconds is the accelerating force in pounds.

If the acceleration is from rest, V1 is zero. The acceleration force equation is:

$$F_m = \frac{V \times GVW}{t \times 9.81}; \quad F = \frac{V \times GVW}{t \times 32.16}$$

where:

V = Change in velocity (meters/sec - *ft sec*) (Final velocity - Initial velocity)
GVW = Gross vehicle weight (Newtons/*lb*)
t = Time for velocity change (seconds)
F_m, F = Acceleration force (Newtons/*lb*)

*Note: To obtain velocity in feet per second when mph is known, multiply mph by 1.467".

Drawbar pull (ΔP) definition

Drawbar pull is the force a vehicle can exert on a load in addition to the force required to propel itself. Actual force to tow or push a load can be calculated based upon Rolling Resistance, Accelerating Force and Grade Resistance of towed or pushed load.

Motor torque defined

Once the tractive effort has been calculated, hydraulic motor torque required can be estimated by:

$$T_m, T = \frac{TE \times r}{G \times N}$$

where:

T_m, T = Hydraulic motor torque (N-m/*lb-in*)
TE = Tractive effort (Newtons/*lb*)
r = Rolling radius of driven tires (meters/*inches*)
G = Gear reduction ratio between hydraulic motors and driven wheels (if none, use a value of 1)
N = Number of driving motors

Slip torque definition

Slip torque is the torque at the motor shaft that will cause the wheels or tracks to break traction and skid. It is affected by the weight of the vehicle and the coefficient of friction between the wheels or tracks and the surface. The equation to use to determine slip torque is:

$$ST_m, ST = \frac{VW \times u \times r}{G \times N}$$

where:

ST_m, *ST* = Hydraulic motor slip torque N-m *(lb in)*

VW = Maximum weight on driven wheel Newton *(lb)* including allowable vehicle overload dynamic weight shift

u = Coefficient of friction between tire and ground. (A value of 0.6 is used for "normal" tires and an average road surface.)

r = Rolling radius of driven tires (meters/ *inches*)

N = Number of driving motors

G = Gear reduction ratio between hydraulic motors and driven wheels (if none, use a value of 1)

Rolling radius

The rolling radius should be based on actual application. Factors such as ply rating, rated load and inflation pressure can result in different values.

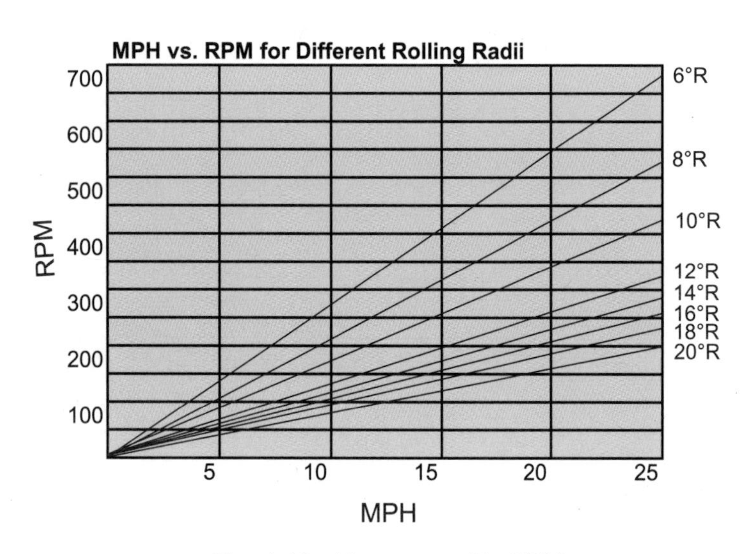

Fig. 4-49 - Motor speed in RPM

Hydraulic motor speed

The chart in fig. 4-49 will estimate the wheel rpm-vs-vehicle velocity for various rolling radii. For rpms not covered by the chart, the following equation can be used:

$$S_m = \frac{9.55\,(V)\,(G)}{r};\quad S = \frac{168 \times V \times G}{r}$$

where:

S_m, *S* = Required hydraulic motor speed (rpm)

V = Desired vehicle velocity (m/sec - *mph*)

G = Gear reduction ratio between hydraulic motors and driven wheels (if none, use a value of 1)

r = Rolling radius of driven tires (meters/ *inches*)

Chapter 4 exercises

1. **Situation:** A piece of mobile equipment is propelled by hydraulic motors. It requires high torque at low speed for one mode of operation.

   Problem: Design a system to meet the above requirements with the components listed below -- using the appropriate ANSI or ISO symbols.

   2 2-speed geroller motors
   1 selector valve
   1 solenoid operated 2-position, 3-way, spring offset valve
   2 shuttle valves

2. **Situation:** A hydraulic motor operates bi-directionally. The required braking pressure is different for each direction of rotation. Motor inlet makeup is of concern. A 4-way, 3-position, closed center DCV is being used to control the motor.

   Problem: Design a system to meet the above requirements with the components listed below - using the appropriate ANSI or ISO symbols.

   1 hydraulic motor, bi-directional
   2 braking relief valves
   2 check valves
   1 4-way, 3-position, closed center DCV

Chapter 4 exercise (cont'd.)

3. What is the difference between integral and split hydrostatic transmissions?

4. What is required in a hydrostatic transmission system to prevent wheel spin?

5. Utilizing ANSI or ISO symbols, draw the circuit for question no. 4 above.

6. What is the purpose of a flushing valve?

7. Using ANSI or ISO symbols, draw a flushing valve and hydraulic motor.

8. With a flow into a hydraulic motor of 129 l/min *(34 gpm)* at 155 bar *(2250 psi)*, what is the input horsepower to the motor?

9. What is the required hydraulic motor speed for the following case:

   Vehicle velocity - 16.7 kph *(10 mph)*
   Gear reduction ratio - 3:1
   Wheel radius - 19 cm *(7.5")*

Chapter 4 exercise (cont'd.)

10. What is the required motor torque for the following conditions?
    Tractive effort - 17,500 lbs.
    Number of motors - 2

Chapter 5

Hydraulic cylinders

Hydraulic working energy must be converted to mechanical energy before any useful work can be done. Hydraulic cylinders convert hydraulic working energy into straight-line mechanical motion (fig. 5-1).

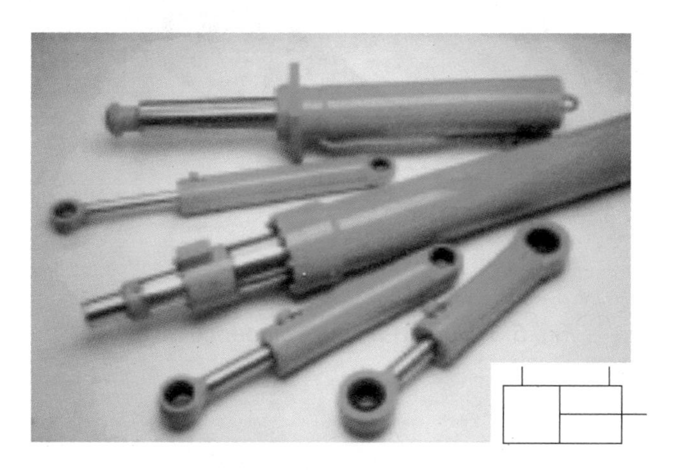

Fig. 5-1 Hydraulic cylinders and cylinder symbol.

Hydraulic cylinder types

Hydraulic cylinders are produced in various types (fig. 5-2). Common cylinders are described in the following:

- Single-acting - Plunger, ram or spring loaded cylinders; pressurized on one side only with a piston rod extending from one end.
- Double-acting - Double- or single-rod cylinder, in which fluid is applied to either side of the cylinder piston.
- Telescoping cylinder - Single or double acting, with nested multiple tubular rod segments, which provides a long working stroke in a short retracted envelope.
- Special types - Designed for specific customer applications.

In a cylinder, pressurized fluid flow is converted into a mechanical force and piston rod motion.

The following sections will show how rod speed is a function of flow, how mechanical force is affected by pressure, and how the piston and the piston rod area affect both speed and mechanical force.

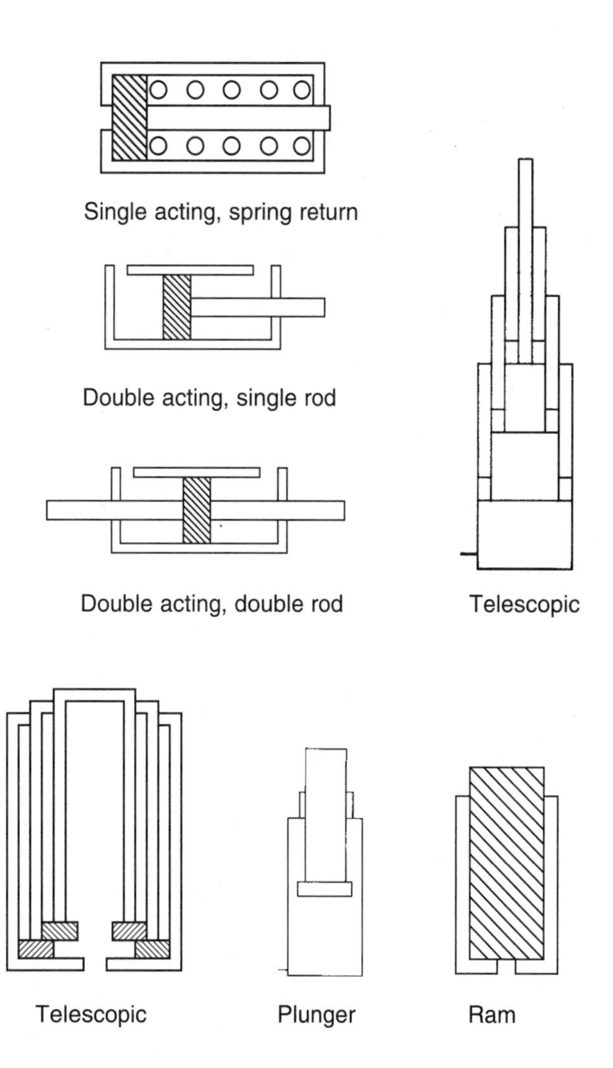

Fig. 5-2 Cylinder versions.

Single-acting cylinder

The single-acting cylinder works in one direction only, usually in the extend (commonly termed '+') direction. To reverse the direction, an outside force such as gravity or a spring force is required.

This cylinder could be a double-acting type with one side of the cylinder drained to tank or through the return line. If the non-pressurized side is not connected to tank, there is a risk of contamination entering the cylinder. This could result in damage to the cylinder barrel and seals.

The single acting cylinder can also be of a type called 'plunger cylinder' or ram. In this cylinder, the effective piston area consists of the piston rod, the plunger or the ram. The plunger cylinder or ram requires an extra long piston rod guide ring, external guidance or support to absorb transverse loads.

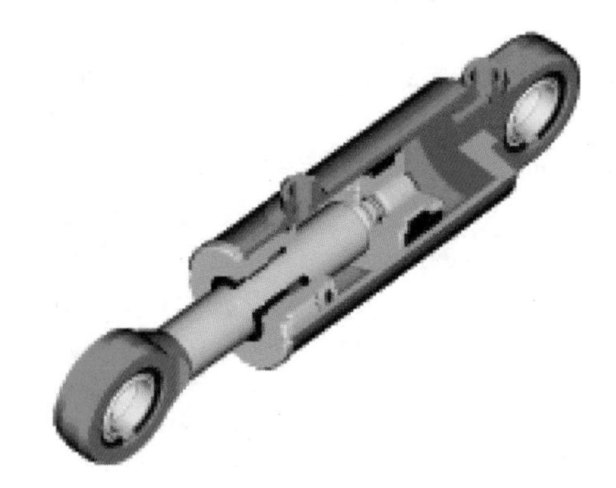

Fig. 5-3 Double-acting cylinder cutaway

Double-acting, single rod cylinder

The double-acting cylinder (fig. 5-3) is the most common of all cylinder types in industrial and mobile hydraulic applications. The double-acting cylinder works in both directions.

A double acting, single rod cylinder will have different speeds and different forces in the extend (+) and return/retract (-) directions. This is due to the presence of the piston rod, resulting in a smaller effective piston area (displacement) in the return/retract direction (fig. 5-4).

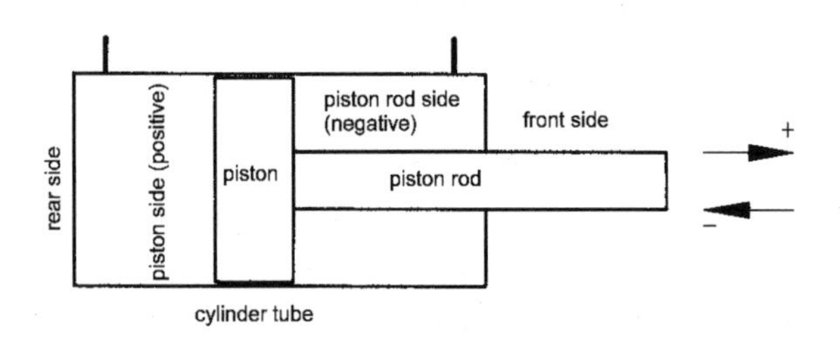

Fig. 5-4 Cylinder nomenclature

The same speed and force in both directions can be achieved by choosing a cylinder with a double piston rod or a 2:1 cylinder, controlled by a regenerative circuit (fig. 5-5). This circuit will be explained later in this chapter.

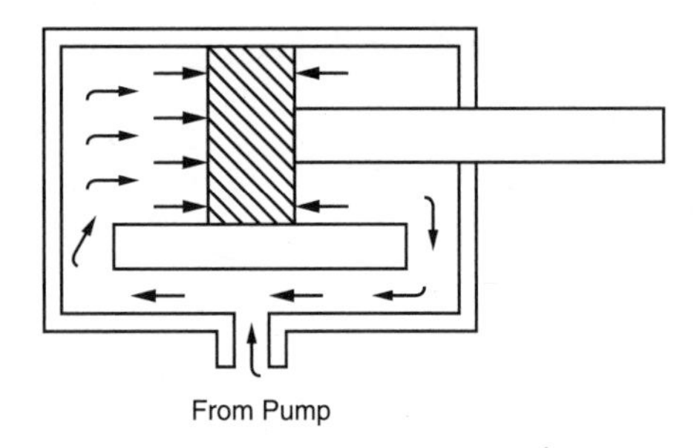

Fig. 5-5 A so called 2:1 cylinder

Telescoping cylinders

Telescoping cylinders (fig. 5-6) are used where a long stroke and limited space is required. Mostly these cylinders are of the single-acting type, but double-acting, telescoping cylinders are used in some applications.

A characteristic of the common telescoping cylinder is that force and speed are dependent upon which cylinder section, or 'step', is presently moving. This is due to the design of the cylinder where one section moves inside the next, larger section. The telescoping cylinder produces a comparatively large force at a low speed when extending, and low force at a high speed at the end of the stroke. This cylinder type is common on tipper or dump trucks.

If a constant speed is required during the entire cylinder stroke, a synchronizing, telescoping cylinder must be used. In this cylinder, all sections move simultaneously as the rod side of a section acts as a pump for the next, smaller section.

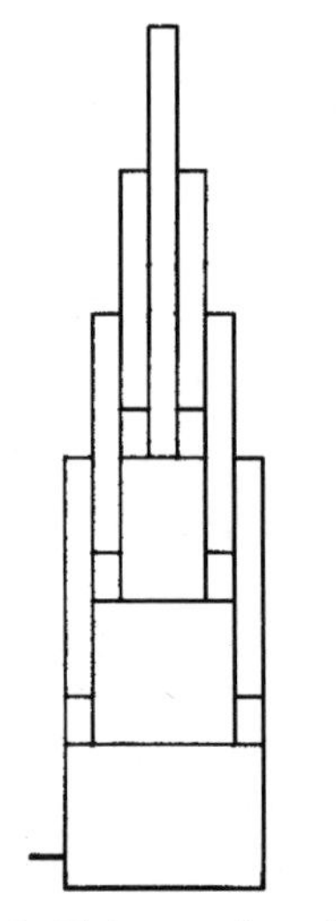

Fig. 5-6 Telescopic cylinder

Special types

Many cylinder manufacturers design and produce customized cylinders to meet specific requirements.

An example would be a cylinder that is an integral part of a machine. The cylinder may have special attachments, port connections and integrated valve functions (fig. 5-7).

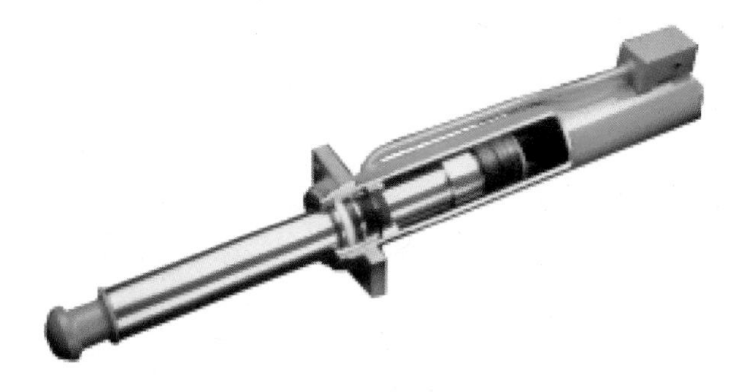

Fig. 5-7 Special type cylinder.

Cylinder designs

Mobile and industrial hydraulic systems have different cylinder requirements. Therefore, two design principles have developed. Mobile cylinders work with a much varying load requirement, while cylinders in industrial applications generally have a known work cycle, i.e. where the load is known and well defined.

Both design principles are optimized for their respective field of application. The principles overlap, however. Any cylinder design can be used within a wide range of applications, both mobile and industrial. The choice of cylinder is dependent upon the application and operating conditions, but can be influenced by convention and cost.

Cylinders for mobile applications

In mobile hydraulics, the requirement for compactness, low weight and flexible mounting are of vital importance besides cost (fig. 5-8).

Fig. 5-8 Typical cylinders for mobile applications

This often results in a welded cylinder design, where the attachment/mounting part (such as the 'ear') is welded to the cylinder barrel or piston rod. Normally, the rod end cover is attached to the barrel with threads or a locking ring, or bolted to the barrel.

The mobile cylinder can also be supplied with a built-in damper or cushion at one or both ends of the cylinder.

Design

The design of a hydraulic cylinder is based on the work that it will perform, i.e. a certain movement to a given load. This means that cylinders are the first components to be selected in a hydraulic system. If possible, all cylinders in a circuit should operate at approximately the same pressure level. The required pressure and flow to the various cylinders will then form the bases when sizing the pump, valves, fittings, tubing and hose.

Cylinder parts

A mobile hydraulic cylinder (fig. 5-9) consists of a cylinder barrel with end caps (including ports),

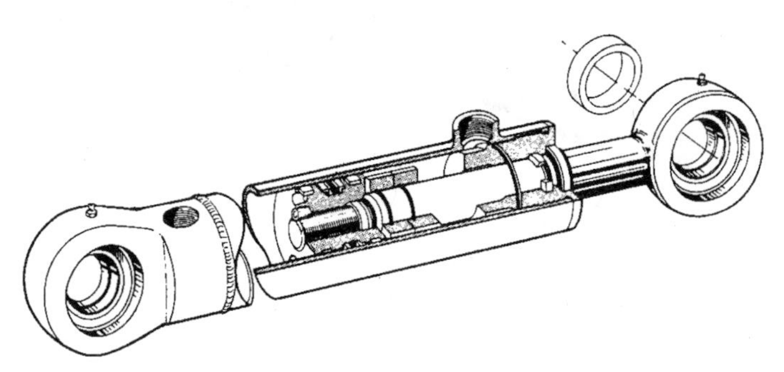

Fig. 5-9 Mobile hydraulic cylinder

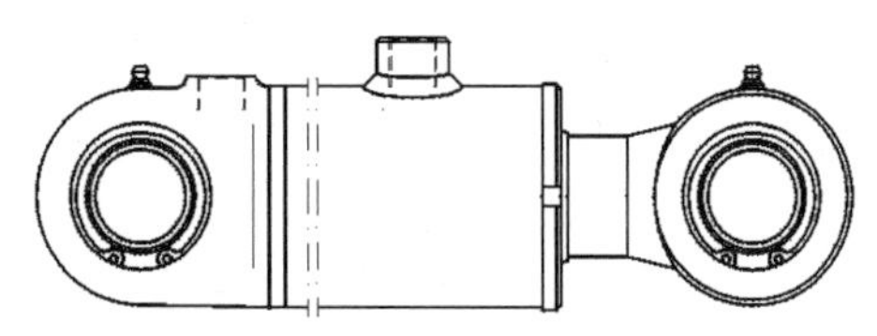

Fig. 5-10a

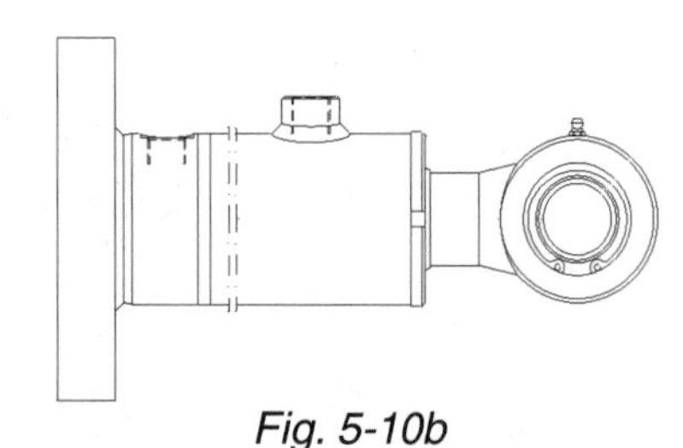

Fig. 5-10b

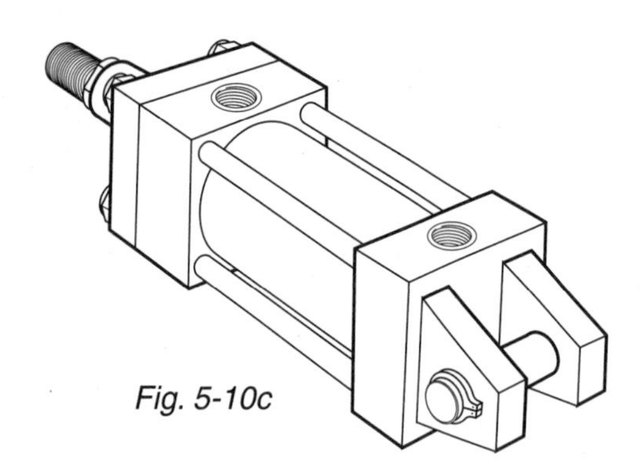

Fig. 5-10c

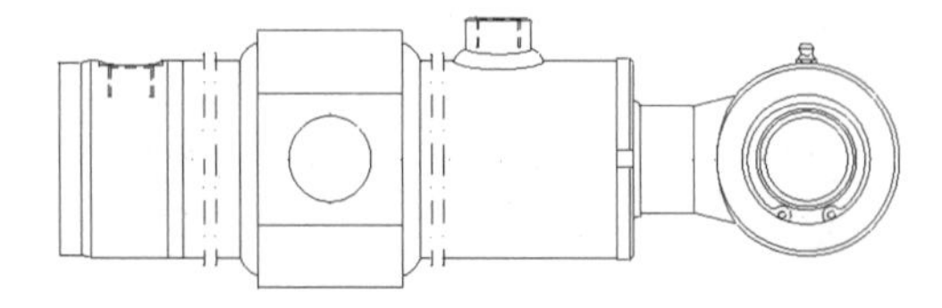

Fig. 5-10d

Fig. 5-10 Common welded mobile cylinders

and a piston with an attached piston rod. The rear end cap, facing the piston, is usually welded to the cylinder barrel. The front end cap, on the rod side, is attached by means of threads, a snap ring or tie rods. As the piston/piston rod moves back and forth, it is guided, and often supported, by a removable bushing called the rod gland or rod bearing.

Mounting styles

The force a hydraulic cylinder is able to develop is transferred to the mobile machine. How this is done directly affects the function, as well as the useful life, of the cylinder. An improper mounting style may cause unacceptable transverse forces or increase the risk of buckling.

The way the cylinder is mounted also determines what forces the cylinder barrel and its attachments will be subject to.

Common welded cylinder mounting types are (fig. 5-10):

- eye with ball-and-socket or sleeve (both ends) (a)
- flanged rear end cap (b)
- clevis mounting (c)
- threaded piston rod end (c)
- trunnion mounting (d)

In most cases, centerline mounting is recommended to minimize leakage caused by flexing of cylinder parts.

Mechanical motions

Cylinders convert hydraulic energy into straight line, or linear, mechanical motion. There are many ways of mounting the cylinder to the machine, such as:

- butt contact mounting
- movement in one plane
- movement in two planes

Butt contact mounting

When using a butt contact mounting, a very sturdy, non-sagging, machine is a prerequisite. Usually, this type of mounting is used for short linear movements. The piston rod must be properly, but not too sturdily, guided to meet reasonable setup/installation requirements.

There are two types of butt contact mountings, each resulting in a different force transmission:

- rear or front flange mountings transfer the cylinder force through the centerline (fig. 5-11).
- side lugs transfer the force outside the centerline (fig. 5-12).

Flange mounting is preferred. In order to minimize internal forces, the cylinder should be attached at the rear end cap if the cylinder is pushing, and at the front end cap if it is pulling. However, if a long cylinder is used for pushing, it may be favorable to attach it to the rod end, to reduce the possibility of buckling.

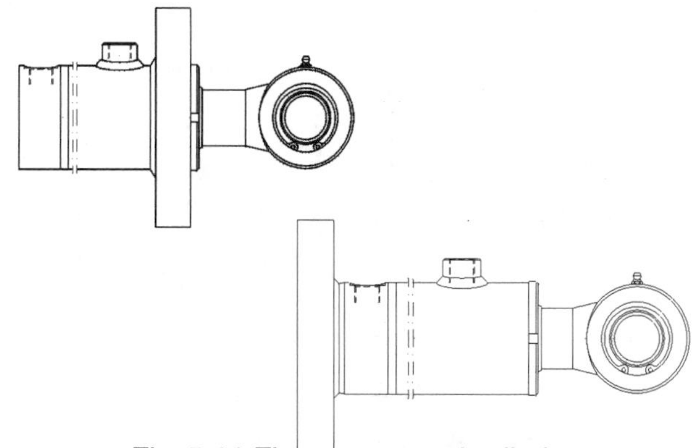

Fig. 5-11 Flange mounted cylinders

Long stroke cylinders should be supported at the free end. This support must be designed to allow the cylinder to move lengthwise to compensate for changes in length due to changes in pressure and temperature.

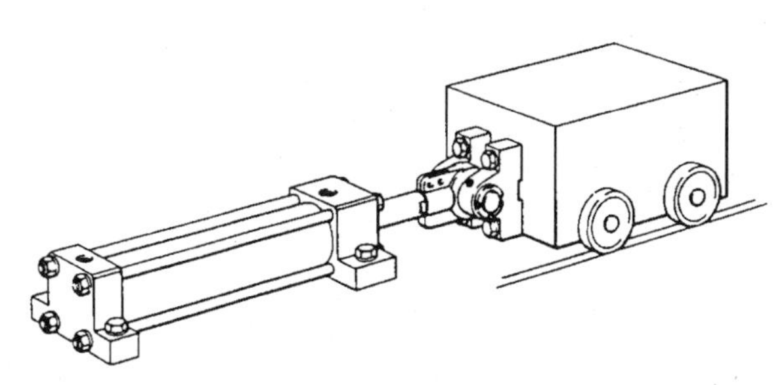

Fig. 5-12 Side lug cylinder

Side lugs, which transfer the cylinder force outside the centerline, often impose heavy loads on members/machine attachments. Otherwise the same, previous rules apply to this mounting method for either pulling or pushing cylinders, and for length changes. Thus, only attach the side lugs on one of the cylinder end caps.

Movement in one plane

If the load is moving in an arc, a mounting is required that permits the cylinder attachments to move accordingly. This requires a shaft or 'trunnion' mounting that can absorb the relative movement between the cylinder attachments and the machine. The shaft can either be mounted in an eye at the cap end, or consist of an intermediate, fixed trunnion which is journalled in bearings (fig. 5-13).

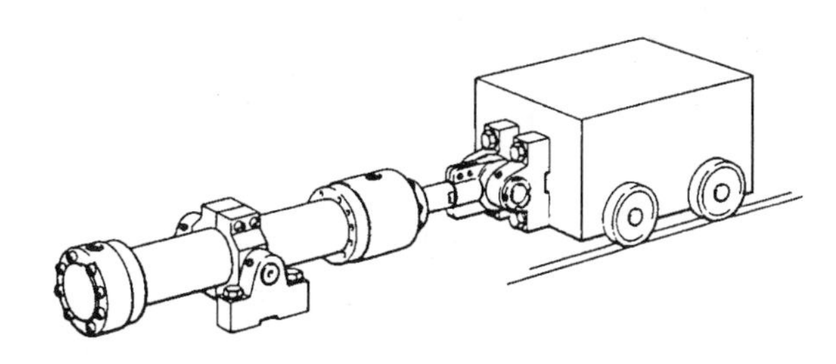

Fig. 5-13 Intermediate trunnion mounted cylinder

In order to reduce the bending stress in the shaft caps and bearings, this style mounting requires an accurate adjustment of the bearing bracket and the attachments in the butt contact plane. Certain types of bushings in the rear cap end and in the piston rod area can absorb an inclination of about 1° only (fig. 5-14).

Movement in two planes

In mobile equipment it may be necessary that the cylinder mounting style moves in two planes. The most common type is one with eyes on the cap end and with a piston rod with spherical bearing (fig. 5-14). This type of mounting requires a minimum of adjustment and relieves the external force on the cylinder within up to 15°.

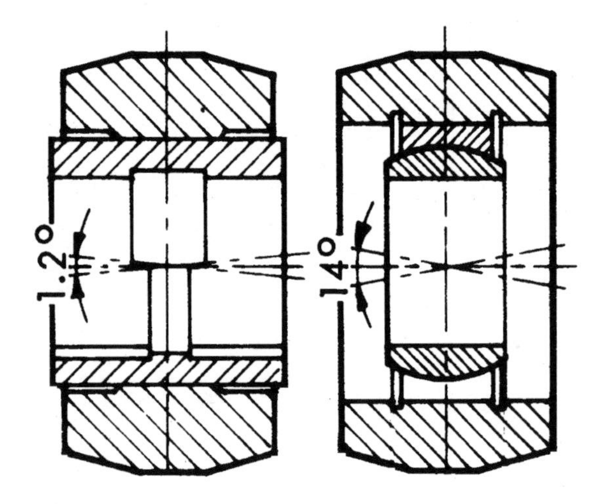

Fig. 5-14 Spherical mounting eye and bushing

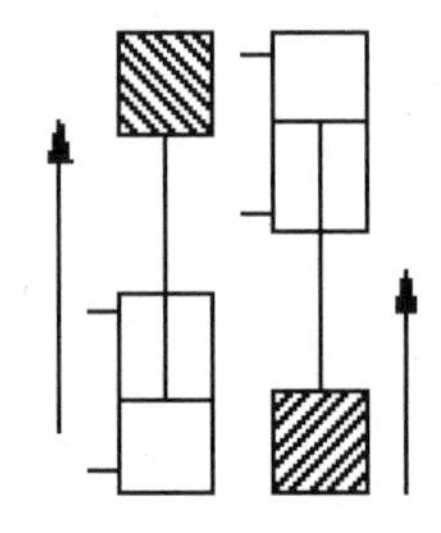

Fig. 5-15 Types of cylinder loading

It is important, for all types of movable mountings, that bearings and bushings be maintained and lubricated in a prescribed manner. Excessive friction in the ball-and-socket joints or bushings can otherwise cause a bending moment on the cylinder rod that adds to other stresses.

Types of cylinder loads

Cylinders can be used in an unlimited number of applications to move various types of loads. A load, which is pushed by a cylinder, is termed a '"pushing load"; a "pulling load" is pulled by the cylinder. These are also known as thrust (compression) and tension, respectively (fig. 5-15).

Forces and pressures affecting a cylinder

From an efficiency point of view, when several cylinders work simultaneously, all cylinders should operate at the same pressure. However, at the same time, it is often desirable that cylinders be standardized which, in turn, leads to compromises.

A cylinder is often required to lift or move a certain load. The required force depends upon several factors such as:

- mass
- gravity
- transmission ratio between cylinder and attachments
- frictional forces
- acceleration forces
- inertia

By choosing different working pressures, different cylinder dimensions/sizes can be used to do the same work. The higher the pressure, the smaller the cylinder dimension/size and also the pump, tubing, piping etc.

However, at the same product of pressure and cylinder area, the cylinder's mounting attachments and piston rod always transmit the same force. Thus, the chosen pressure level does not influence the mounting attachments and the rod. This means that for a constant load, if a high working pressure is chosen in order to get a smaller cylinder (with less flow requirement), the cylinder will require the same piston rod diameter as if a lower pressure and a larger cylinder diameter were chosen (fig. 5-16).

This, in turn, means that at higher pressures a greater difference will be seen between the force in the extend direction as compared to the retract direction, than at lower pressures (fig. 5-17).

Suppose the cylinder is working in both directions. When dimensioning/sizing the piston rod, static and dynamic forces must be examined. The required working pressure for a certain force is affected by the possible backpressure on the other side of the piston (fig. 5-18). This backpressure depends upon the resistance to flow within the system, e.g. in connections, piping, tubing, valves, filters, hoses, and cylinders connected in series or in a regenerative function.

The static forces are relatively easy to calculate, with the exception of frictional forces. In this case, previous experience must be relied upon.

It is considerably more difficult to estimate the dynamic forces the cylinder will be exposed to. An analysis of these forces begins by reviewing the duty or work cycle. A simple duty cycle can be divided into three phases or regions:

1. acceleration
2. constant speed corresponding to the added volume of fluid
3. retardation (deceleration)

Pump pressure in phases/regions 1-2-3

Phase/region 1: The main pressure relief valve limits the maximum pressure and the cylinder force (fig. 5-19)
Phase/region 2: The current load normally determines the force and pressure levels, but the system pressure is still limited by the main pressure relief valve.
Phase/region 3: The retardation/deceleration force may be limited by using pressure control valves in the cylinder ports.

The pressure control valves in phase 3 will, however, only help if the cylinder is not driven to the end of its stroke. If it is, it will be more difficult to estimate the retardation/deceleration forces.

As will be seen in the following section, there is a lot of information, about the cylinders and the rest of the system, that must be gathered in order to make the appropriate sizing decisions. A detailed discussion of this topic is beyond the scope of this text. However, certain items require mentioning.

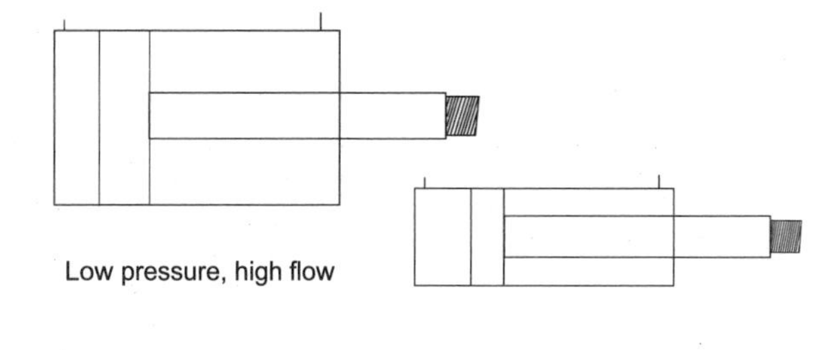

Fig. 5-16 Cylinder flow & pressure relationship

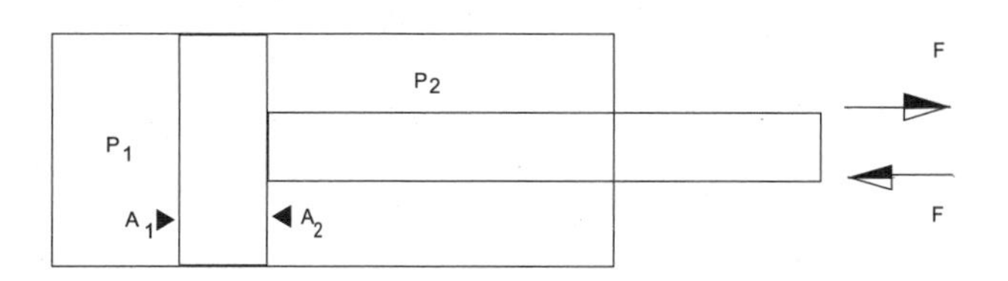

Fig. 5-17 Differential cylinder forces

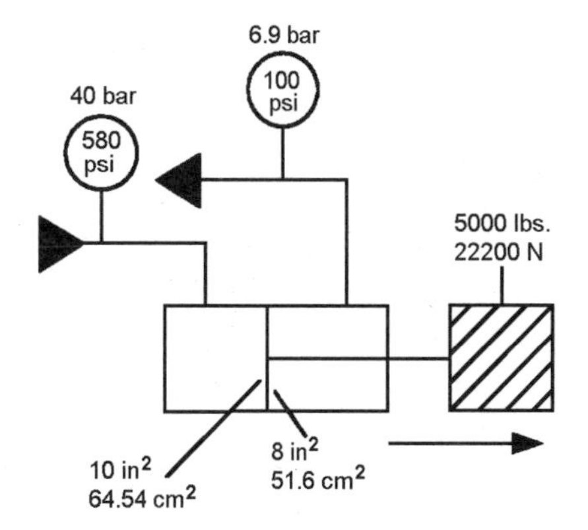

Fig. 5-18 Cylinder backpressure

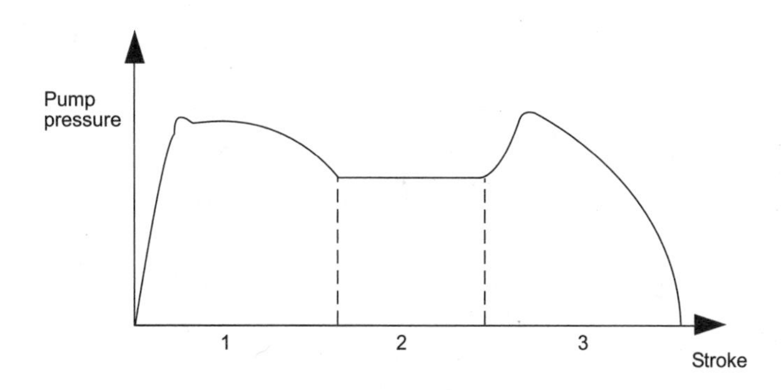

Fig. 5-19 Pressure regions

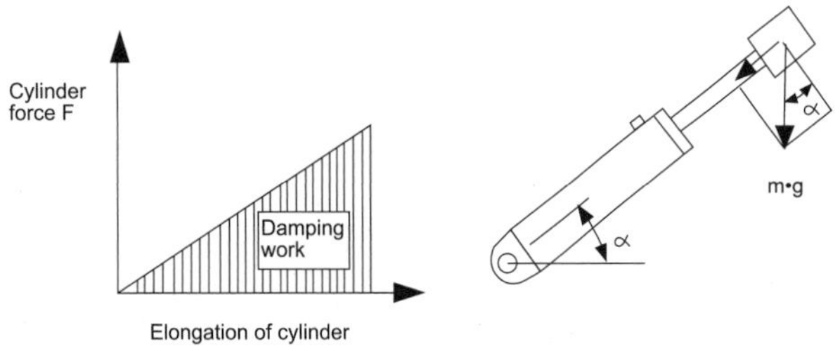

Cylinders without cushions

Assume that a working cylinder is moving a load/ mass (m) connected to the piston rod. If the piston rod is driven to the end of its stroke, the load/ mass will be retarded by a force (F) applied directly to the cylinder end cap.

This force will cause an elongation of the cylinder barrel and the piston rod, depending on the elastic stretching capability of the cylinder rod (fig. 5-20). A long cylinder will be more elongated by a certain load than a shorter cylinder, i.e. the elastic stretch will be greater and the retardation (stopping) force lower with the same kinetic energy for a long cylinder, compared to a short one.

Cylinder force F

Damping work

Elongation of cylinder

m•g

Fig. 5-20 Elongation of cylinder

Static forces are added to the retardation (stopping) force when a gravitational component is involved. In addition, the cylinder barrel is subjected to an additional load force. This is due to pressure in the cylinder which may reach a maximum pressure level at the end of the stroke.

Despite certain elasticity (stretching) of the machine structure, driving a cylinder full stroke should be avoided unless special measures have been taken to retard or stop its movement. Dampers, cushions or external shock absorbers are components designed to do just that.

Hydraulic shock

When the hydraulic working energy moving a cylinder's piston runs into a dead end (as at the end of a cylinder's stroke), the inertia of the moving liquid is changed into a pressure peak known as '"hydraulic shock". If a substantial amount of working energy is stopped, the shock may damage the cylinder.

Cushions

A cylinder can be equipped with a cushion to protect against excessive shock pressures (fig. 5-21). The cushion slows down a cylinder's piston movement just before reaching the end of the stroke. If required, cushions can be installed in both ends of a cylinder.

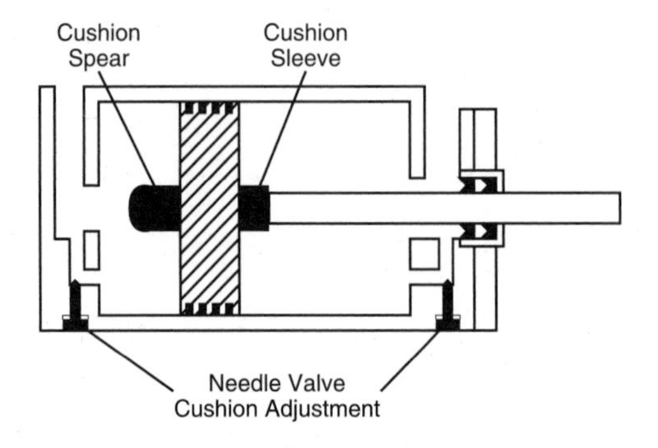

Fig. 5-21 Cylinder cushions

How a cushion works

As a cylinder piston approaches the end of its travel, a plug (spear or sleeve) gradually blocks the normal exit for the liquid and forces it to pass through a needle valve (fig. 5-22). At this point, some flow, on the opposite side of the cylinder, goes over the relief valve at the relief valve setting.

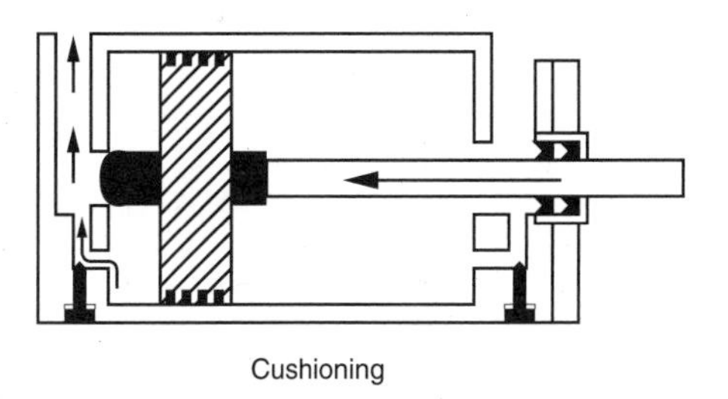

Fig. 5-22 How a cushion works

The remaining liquid between the end cap and cylinder piston is bled off through the needle valve and slows (cushion cavity) the piston. The adjustment of the needle valve determines the rate of deceleration.

In the reverse direction, flow bypasses the needle valve through a check valve within the cylinder.

Cylinders with cushions

Retardation (deceleration) of, and stopping, cylinder movement at the end of the stroke can also be achieved through the use of valves. These valves would be in the system and outside the cylinder. They could be used in combination with a position sensor or through an exterior mechanical stop, or an external shock absorber.

The cylinder is, however, often provided with a so-called damping/cushioning or an end of stroke brake, which slows the load during the very last portion of the cylinder stroke. Fundamentally, this slowing process is the same as that of a cushion and is provided by the creation of a small closed cavity. From this cavity oil is passed through a restriction in such a way that the piston speed is slowed to almost zero.

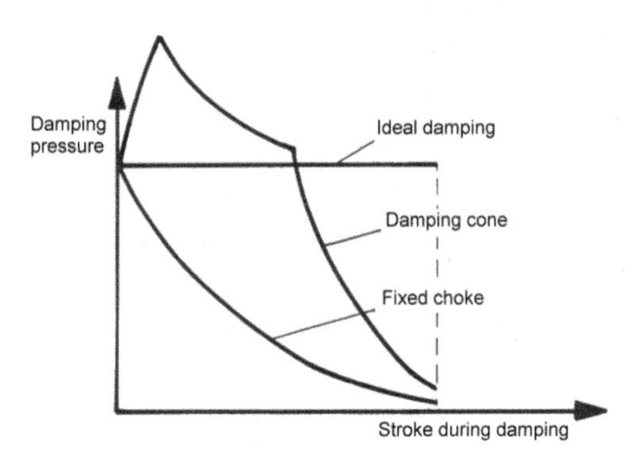

Fig. 5-23 Cylinder damping pressures

During the actual damping/cushioning process, the corresponding pressure in the cylinder is not allowed to exceed a certain maximum level. This is accomplished through proper design and sizing of the cylinder. Cylinder manufacturers should be consulted for guidance in this area.

The ideal damping/cushioning should be such that the pressure on the braking or cushioning side is kept at or below the maximum cushioning value (p_{max}) during the entire damping/cushioning process (fig. 5-23).

Some cushion designs present production problems such as eccentricity, narrow tolerances, etc. Another consideration in some designs is the narrow slot clearance between the cone and the hole which results in laminar flow (fig. 5-24). This laminar flow is affected by the size of the clearance and the fluid viscosity. Therefore, the damping/cushioning process becomes very sensitive to temperature.

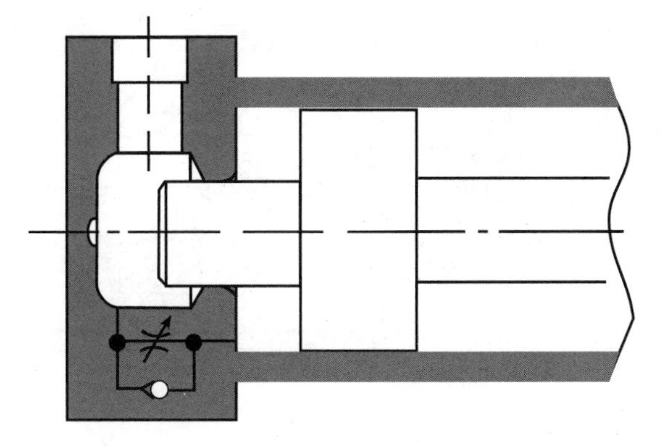
Fig. 5-24 Production considerations

It will be shown later in the text that the viscosity of common hydraulic fluids changes considerably with temperature. At low temperatures the cushion pressure may increase to a dangerously high level. At high fluid temperatures, the viscosity could be so low that the piston hits the end cap with almost full force.

To reduce the clearance and temperature problems, the cylinder is often equipped with a choking (needle valve) device which is placed between the damping/cushioning chamber and the cylinder port (fig. 5-24). The device is usually adjustable and may still allow the piston to hit the end cap. It always leads to a 'bottom stroke' i.e. the piston hits the end cap.

An example of a more efficient damping/cushioning device is shown in fig. 5-25. A moveable ring surrounds the cushioning pin/spear. A number of axial, tapered grooves are machined into the pin/spear, through which fluid can flow out of the damping/cushioning chamber.

By making the grooves in different lengths, specific cylinder damping performance can be obtained. If the performance envelope is chosen carefully for a certain load, a nearly constant damping pressure could be obtained.

The short length of the damping/cushioning ring flow path results in turbulent flow when combined with the grooves. Turbulent flow is independent of viscosity, and the damping/cushioning function is therefore independent of temperature variations.

The damping/cushioning pressure at the beginning of the cushioning process is affected only by the piston speed and the design of the damping/cushioning device. The process is, however, extremely dependent upon the mass of the load as well as of the operating pressures.

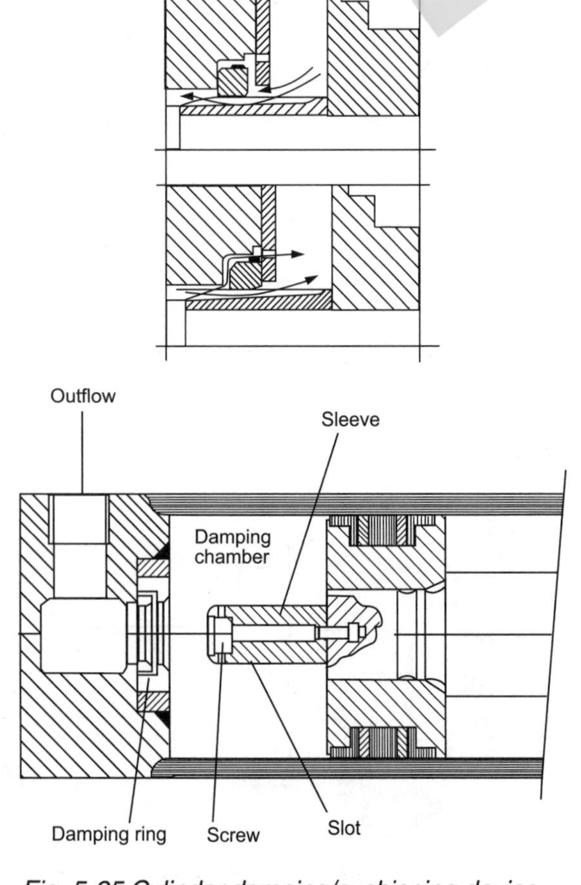

Fig. 5-25 Cylinder damping/cushioning device

Transverse forces

The capability of the cylinder to withstand forces acting at a right angle to the direction of movement, called "transverse or side forces", is limited. Examples of transverse forces are external loads acting on the piston rod and/or forces due to the weight of the cylinder. The weight of the cylinder may be an important consideration, especially if the cylinder is long and mounted horizontally.

The size of these transverse forces affects the useful life of the cylinder and can impose great pressures on the piston support rings and on the end cap gland.

In order to limit those forces on long-stroke cylinders, a spacer ring/stop tube may have to be installed. As a consequence, this will reduce the stroke of the cylinder. This reduces the load on the rod gland bushing (fig. 5-26).

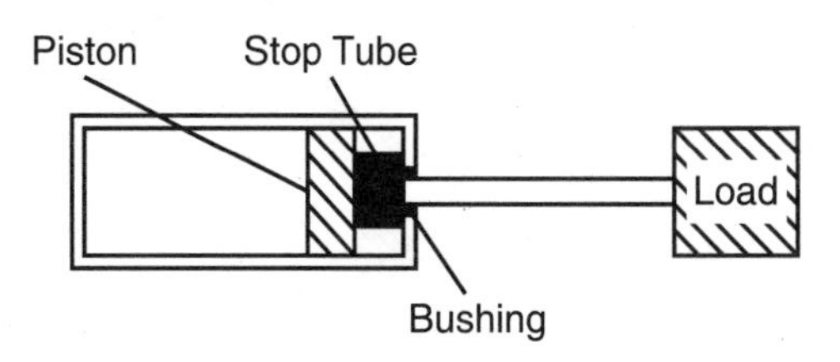

Fig. 5-26 Cylinder spacer ring/stop tube

However, most mobile cylinders are not equipped with spacers (stop tubes). To determine when a spacer (stop tube) is required, or of what length the spacer (stop tube) should be, consult the cylinder manufacturer.

Cylinder Rod Buckling

Despite the fact that a cylinder is designed for a certain pressure, i.e. to exert a certain force, it is not always possible to utilize it. Rod buckling may be a limiting factor.

Factors influencing rod buckling are:

- axial force (load)
- cylinder length (stroke)
- transverse forces (side forces)
- mounting style (freedom of movement)
- spacer/stop tube (affects initial deflection)
- piston rod material (has small effect)
- joint friction

For the cylinder illustrated in fig. 5-27 a certain initial deflection is observed due to the clearance between the piston and cylinder barrel, and between the piston rod and front end cap (rod gland bushing), and to eccentricity, etc.

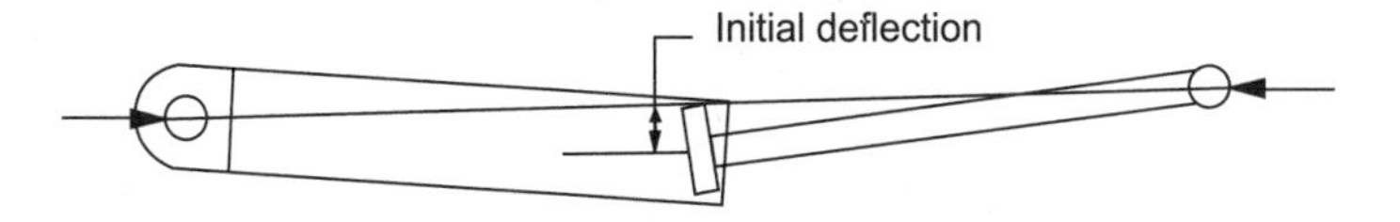

Fig. 5-27 Initial deflection

Initial deflection

The initial deflection forms a moment (lever) arm for the forces that load the cylinder. The initial deflection can be reduced if the cylinder is not driven full stroke, or if a spacer ring (stop tube) is placed between the piston and the front end cap.

When a cylinder is under pressure, the forces that act on the cylinder increase the initial deflection. Consequently, the moment (lever) arm of the axial forces increases with an increasing axial load. This deflection is counteracted by the force created by system pressure acting on the piston, cylinder barrel areas and mechanical re-

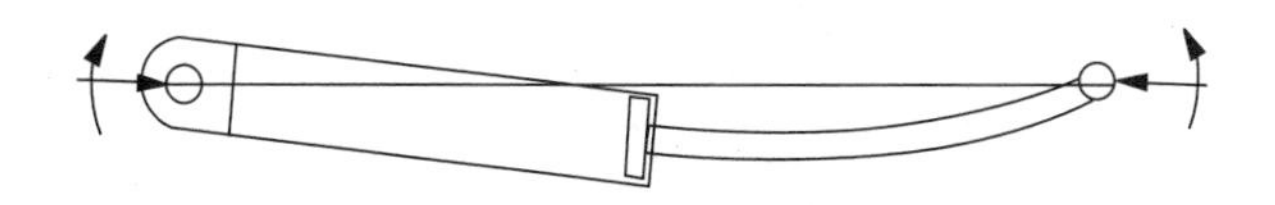

Fig. 5-28 Pressures and forces acting on the cylinder

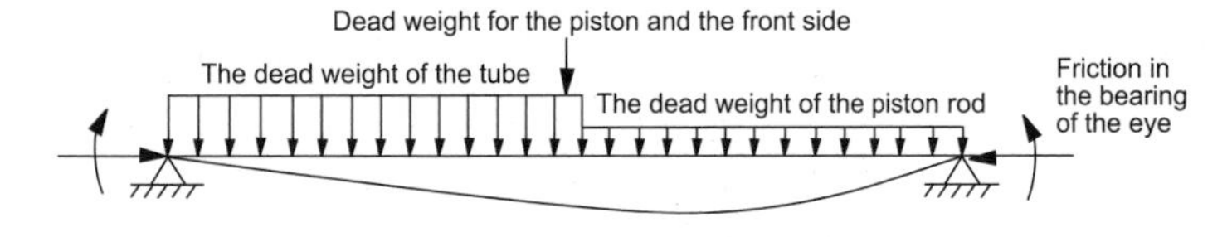

Fig. 5-29 Cylinder static loads

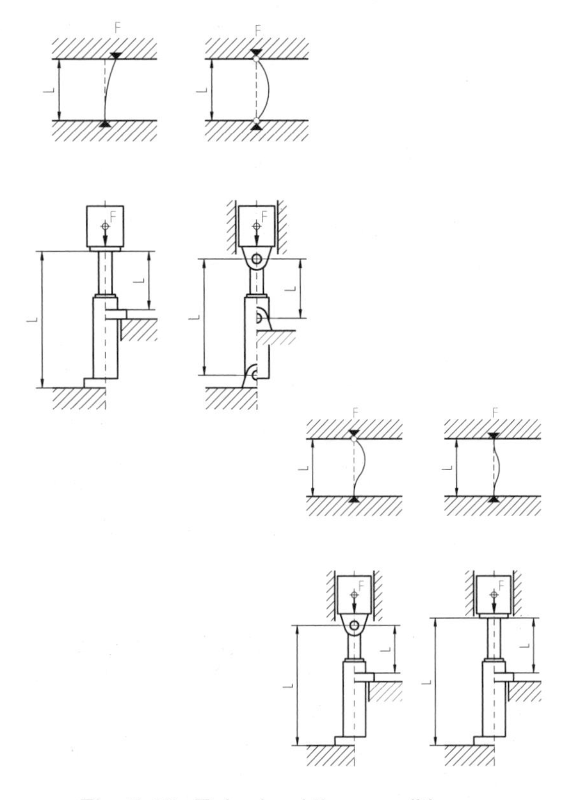

Fig. 5-30 Euler buckling conditions

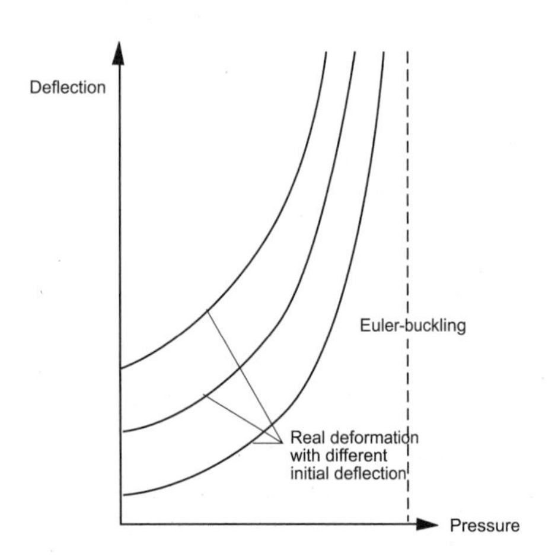

Fig. 5-31 Rod buckling graph

sistance of the piston rod and front end cap (fig. 5-28).

In certain cases the maximum allowable loading of the support ring (stop tube) and bushing (rod gland) will determine the maximum load the cylinder can withstand. For example, by using a spacer between the piston and front end cap, the dimensioning (sizing) of a cylinder could be reduced to that of a structural strength problem when calculating piston rod stress.

Static loads for a pushing cylinder (compression/thrust load)

The static forces affecting a cylinder are shown in fig. 5-29. In addition, large dynamic transverse forces may arise, especially in a mobile machine. The cylinder barrel is usually more rigid than the piston rod, and plastic deformation of the piston rod, because of buckling, may therefore occur.

There are four buckling conditions based on Euler equations (Leonhard Euler, 1744). Apart from the varying initial moments, typical mounting styles correspond to one of the Euler cases shown in the fig. 5-30.

According to the guidelines set by these cases, the highest allowable axial load before buckling occurs, is half as high for case two as compared to case four. The fourth case may, on the other hand, be subject to transverse forces acting on the cylinder possibly due to installation misalignment.

Initial deflection, among other things, will reduce the allowable rod buckling forces when compared with the corresponding Euler case. Buckling, as based on the Euler equations, is represented by the dotted line in fig. 5-31. True cylinder rod buckling is illustrated by the solid curved lines at selected, initial deflections.

The cylinder manufacturer normally defines, in the form of diagrams or tables, the highest allowable load as a function of cylinder diameter and mounting style. Often transverse forces, due to weight, are taken into account, while dynamic transverse forces and friction in the ball-and-socket joints are not.

Cylinder piston and piston areas

Piston and piston rod areas are generally defined as shown in fig. 5-32. The 'piston area' of a double acting, single rod cylinder indicates the piston area exposed to pressure on the cylinder cap end. The 'piston rod area' is the area of the piston, on the rod side of the piston, covered by the piston rod. The remaining area (piston-rod area) is the effective or annulus area. It is this area that is exposed to pressure. Since the rod covers a portion of the piston area, the annular area is always less than the piston area.

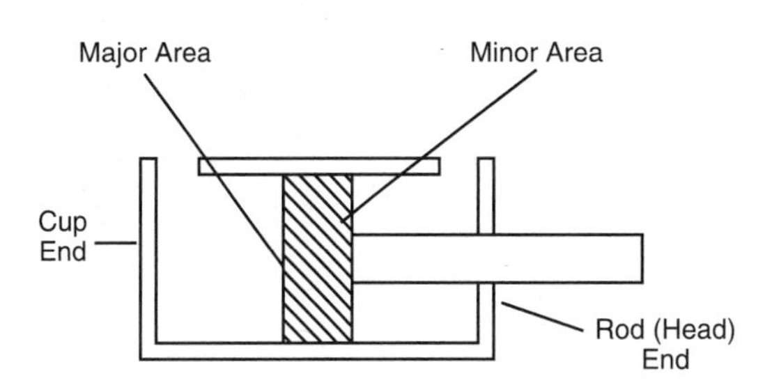

Fig. 5-32 Cylinder areas

Rod speed of an extending, double acting cylinder

The rod speed of a cylinder is determined by how quickly the volume behind the piston can be filled with liquid; the higher the flow a cylinder receives, the faster the rod extends. Rod speed can be calculated by using Formula 5-1.

Formula 5-1 Rod speed in an extending, double acting cylinder

$$v = \frac{Q_c}{A_p} \quad \text{(rod speed = flow/piston area)}$$

where: 'v' - rod speed [m/s]
'Q_c' - flow through the rear end cap [m^3/s]
'A_p' - piston area [m^2].

Discharge flow from an extending, double-acting cylinder

Flow entering through the cap end of a double-acting, single rod cylinder determines the rate at which a cylinder piston rod extends; at the same time, flow will be discharging from the rod end of the cylinder.

Discharge flow is an important concern in system design; it can be calculated by using Formula 5-2.

Formula 5-2 Discharge flow from an extending, double-acting cylinder

$$Q_r = v \times A_e \quad \text{(flow = rod speed x ring area)}$$

where:'Q_r'- flow out of the front end cap [m^3/s or *in^3/min*]
v - rod speed [m/s or in/min]
'A_e' - annulus aarea [$m/^2$ or *in^2*]; A_p minus the rod area A_r

While extending, the discharge flow from a single rod cylinder is always less than the flow entering through the cap end of the cylinder.

Rod speed in a retracting, double-acting cylinder

During retraction, when the pump flow is directed to the rod end of a single rod cylinder, a piston rod will retract faster than when extending (at the same flow). This can be calculated by using the previous expression for rod speed (Formula 5-1) if the piston area (A_p) is replaced with the annular area (A_e).

Discharge flow from a retracting, double-acting cylinder

Formula 5-3 Discharge flow from a retracting, double-acting cylinder

$$Q_c = Q_R \times \frac{A_p}{A_R}$$

where: 'Q_c' - flow out of the rear end cap [m^3/s]
'Q_R' - flow through the front end cap [m^3/s]
'A_p' - piston area [m^2]
'A_R' - ring area [m^2]

During retraction, when full pump flow is directed to the rod end of a single rod cylinder, discharge flow from the cap end will be greater than the flow going into the cylinder rod end. The discharge flow (out of the cap end) is calculated in Formula 5-3.

Typically when a double acting, single rod cylinder is retracting, more flow is discharging from the rear end cap than that entering the rod end. Therefore, pump flow is not necessarily the maximum flow in a system.

When a hydraulic system is designed, the amount of discharge (return) flow from retracting, single rod cylinders must be determined. This is one of the reasons piping, valving and filters on the return side of the system are sized larger than the corresponding components on the supply side of the system.

If component replacement is required on the return side of the system, ensure that smaller sized components are **not** being substituted for the original components.

Formula 5-4 Pushing force from an extending cylinder

$$F_p = p \times A_p$$

where: 'F_p' - thrust force [N]
'p' - pressure on piston [Pa]
'A_p' - piston area [m^2]

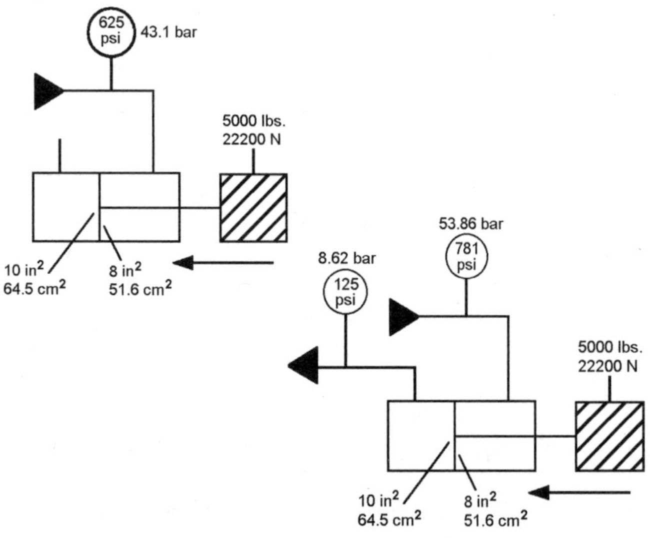

Pushing (compression/thrust) force from an extending cylinder

While extending, the mechanical force developed by a cylinder is the result of hydraulic pressure acting on the cap end piston area. This is expressed by Formula 5-4.

In the formula the assumption is made that pressure was zero on the rod side of the cylinder piston. Even though the annular area, A_e, is drained to tank while extending, tank line pressure, or 'back pressure', can be as high as 1 MPa *(150 psi)* or more in some systems.

This generates a force on the cylinder's annular area. This force, together with the resistance offered by the load, must be overcome before the cylinder can extend.

Anytime a load is moved by a cylinder, the force acting on the piston is the sum of the load resistance and the force acting on the annulus area because of the back pressure.

Fig. 5-33 Backpressure effects

Pulling (tension) force from a retracting cylinder

While retracting, the pulling/tension force developed by a cylinder is the result of hydraulic pressure acting on the annulus area as expressed in Formula 5-5.

If, for example, a load offers a resistance of 22 000 N *(5000 lbs)* and the annulus effective area of the cylinder piston is 50 cm^2 *(8 in^2)*, a hydraulic pressure of 4.4 MPa *(625 psi)* is required to equal the load (fig. 5-33) as calculated by the formula.

This also assumes that there is no backpressure acting on the piston. But, backpressure is even more pronounced while retracting than extending. This is a higher backpressure than seen during extension because discharge flow during retraction is usually more than during extension. While retracting a single rod cylinder, backpressure on the piston is usually higher than when extending.

It has been shown that the cylinder rod speed is affected by flow and piston area, and how the cylinder force is affected by pressure and piston area. In a later section, it will be shown what happens when one or more of these elements is restricted.

Double rod cylinder circuit

As shown previously, a double acting, single rod cylinder retracts faster than it extends. Some applications require, that the cylinder extends and retracts at the same speed.

One means of accomplishing this is with a double rod cylinder (fig. 5-34). Since a double rod cylinder typically has the same diameter rod on both sides of the piston, piston areas exposed to system flow are equal. With the rate of flow to each side remaining constant, rod speed is the same whether extending or retracting.

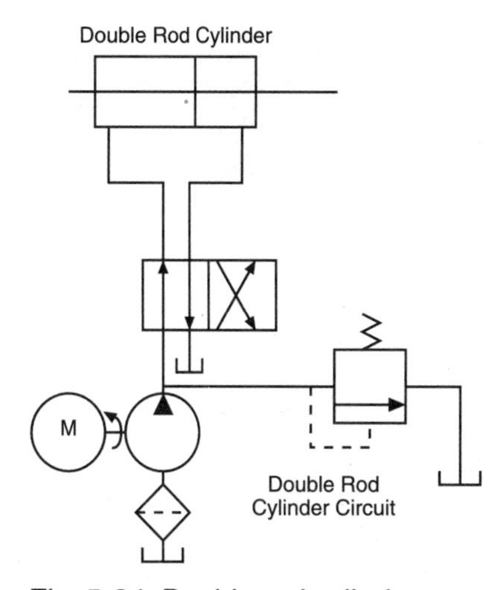

Fig. 5-34 Double rod cylinder

Regeneration with a 2:1 cylinder

Taking the discharge flow from the rod end of a cylinder and adding it to the flow into the cylinder's cap end increases a cylinder's rod speed (fig. 5-35). If a '2:1 cylinder' is used in the system, the cylinder's rod speed will be the same whether extending or retracting.

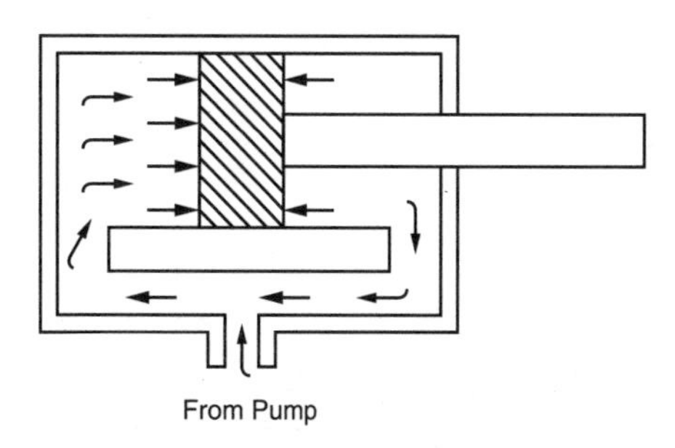

Fig. 5-35 Regeneration

A 2:1 cylinder has a rod with a cross sectional area equal to one half the piston area (fig. 5-36). In other words, the rod side of the piston has one half the area exposed to pressure as the cap side. (In practice, the rod area is not exactly one half the piston area due to the use of standard rod and piston material. However, it will be considered one half in order to facilitate calculations.)

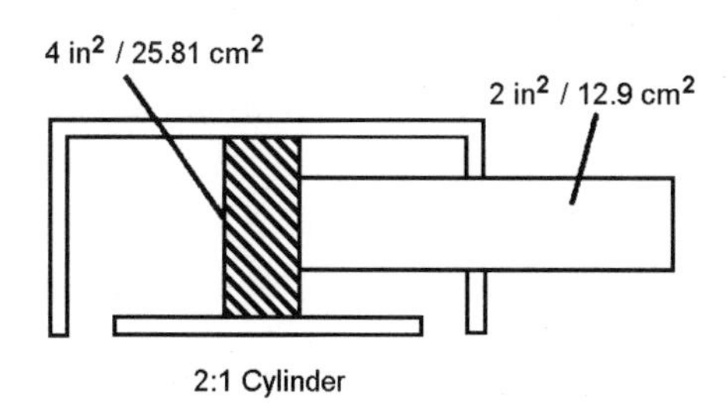

Fig. 5-36 2:1 area ratio cylinder

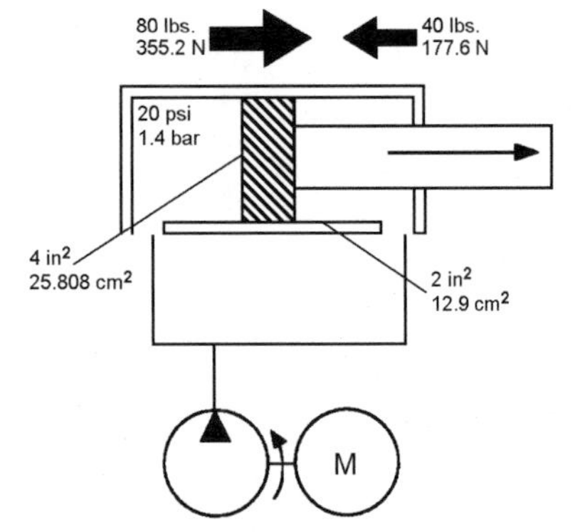

Fig. 5-37 Differential force during regeneration

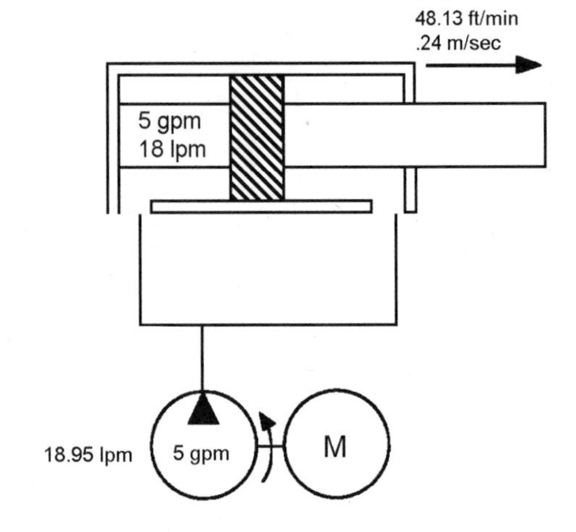

Fig. 5-38 Resulting regeneration movement

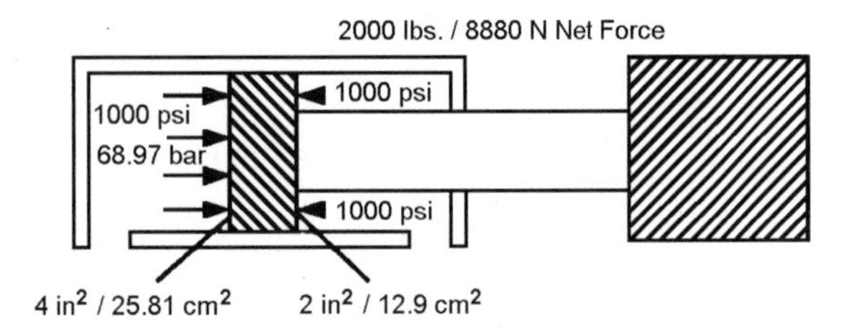

Fig. 5-39 Cylinder force during regeneration

With the cylinder in a circuit, flow and pressure are directed to both sides of the piston at the same time. At first it may appear that the cylinder is hydraulically locked. But, the difference in piston and annular areas exposed to pressure results in a larger force being generated on the piston area, extending the rod.

With 0.14 MPa *(20 psi)* on both sides of the piston, 350 N *(80 lbs)* of force would be generated to extend the cylinder rod. A 175 N *(40 lbs)* of force would be generated to retract the rod. This is a 175 N *(40 lbs)* differential in favor of extending the rod; the rod does extend (fig. 5-37).

As the piston rod moves out, fluid is being displaced from the rod end to the other side of the piston (cap end). This means pump flow is not required to fill the total volume behind the piston (cap end). Pump flow has to fill an area, only equal to the cross section area of the rod (fig. 5-38). With a 2:1 cylinder, this means the cylinder will extend twice as fast as normal with pump flow remaining the same.

Regeneration during rod extension

A characteristic of a 2:1 cylinder in a regenerative circuit is that the extending and retracting rod speeds are basically the same.

To retract the cylinder rod, the directional valve is shifted. The cap end of the cylinder is drained to tank. All pump flow and pressure is directed to the rod end of the cylinder.

Since the pump is filling the same volume as at the cap end (half cap end volume), the rod retracts at the same speed.

Cylinder force during regeneration

A design consideration of regeneration is that output force is reduced. Since fluid pressure is the same on both sides of the cylinder piston, the effective area on which the force is generated is the cross-sectional rod area.

In the example (fig. 5-39), as the cylinder contacts the workload, assume that the pressure climbs to a relief valve setting of 7 MPa *(1,000 psi)*. This pressure acts on a piston area of 25 cm^2 (4 in^2) which equals 17 500 N *(4,000 lbs)*. On the rod side, 7 Mpa (1000 psi) is acting on the annular area developing 8 750 N *(2000 lbs)*.

There is a net force of 8 750 N *(2,000 lbs)* to extend the rod.

Since areas on either side of the piston are balanced, except for the rod area, the net force is a result of pressure acting on a piston area equal to the rod cross-sectional area.

In calculations for speed and force during regeneration, the cross-sectional area of the rod is used, not the piston or annular areas.

Connecting a cylinder in a regenerative circuit results in a faster rod speed while extending, slower retract speeds and reduces the output force. Cylinder force is sacrificed for rod speed. And, with a 2:1 cylinder in a regenerative circuit, extending and retracting rod speeds are basically the same.

Sample regenerative circuits

Since a force reduction exists while a cylinder is in regeneration, regeneration is frequently employed only to extend the cylinder rod until it reaches the workload. When work is to be done, the rod side of the cylinder is drained (returned to tank) so that full force can be realized.

Illustrated in fig. 5-40 are two common examples of regenerative circuits in which the rod side of the cylinder can be drained (returned to tank) when necessary. In one circuit, this is accomplished through the centre position of a directional control valve. The other circuit employs an unloading valve.

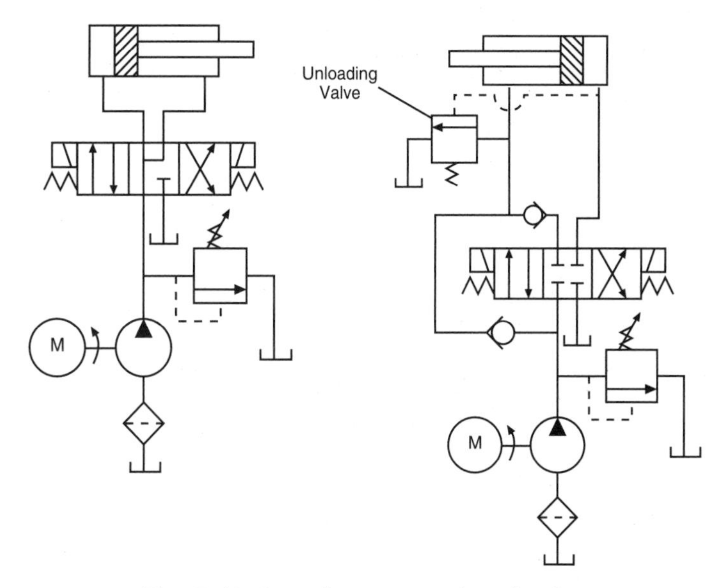

Fig. 5-40 Sample regeneration circuits

Previous sections have explained, how cylinders work in a circuit. The next section will show how cylinders are affected by wear.

Synchronizing two cylinders

One of the most difficult, if not impossible, things to accomplish in a hydraulic system is fully synchronizing the movement of two cylinders, even if the most sophisticated types of flow control valves are employed (fig. 5-41). Typical values for synchronization range from 3 to 1.6 mm ($^1/_8$" to $^1/_{16}$") depending upon the length of the cylinder stroke.

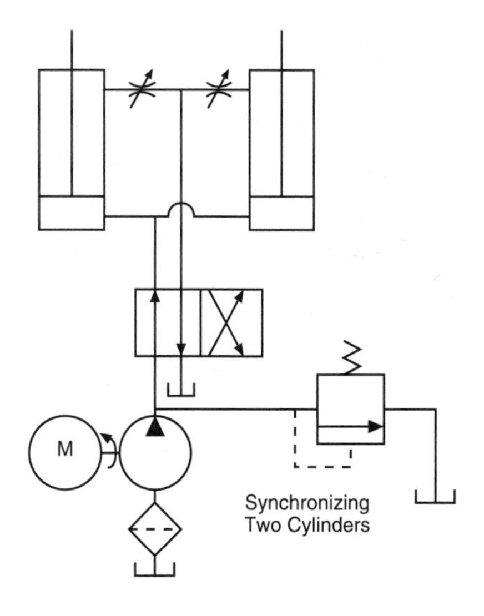

Fig. 5-41 Examples of synchronizing circuits

Even after the cylinders have been synchronized to within acceptable limits, they will, in a relatively short period of time, be out of synchronization because of the different wear characteristics of the cylinders and slightly different performance of the flow control valves to the same set of conditions.

In order to achieve a more positive control, some systems are designed so that the discharge flow from one cylinder is used as the input flow to another cylinder. Systems of this nature may still not be in perfect synchronization because of leakage.

A characteristic of these circuits is that they are equipped with makeup and replenishing lines between the cylinders.

NOTE: When two or more cylinders must work in parallel, it is recommended that the piston rods be mechanically connected. The connection must be rigid and could consist of a strong beam. Another alternative is to equip each cylinder with a position sensor and an electrohydraulic directional control valve. This method would permit tieing together of the cylinders electronically and the electronic monitoring of each cylinder's position relative to the other(s). This would be the most accurate but most costly solution to synchronizing multiple cylinders.

Seals

For proper operation, a positive seal must exist around a cylinder's piston as well as at the rod gland. Seals are therefore of crucial importance for the proper operation of a cylinder. The requirements may vary depending upon the application. There are no universal seals that can be used for all purposes. For different applications different materials and different seal designs must be chosen. The seal material, or compound, should be verified so as to be compatible with the system fluid and operating conditions. Every seal choice will be a compromise between various properties.

Factors influencing the choice of seals:

- pressure
- fluid
- temperature
- speed
- seal clearance
- friction
- sealing surface finish
- useful life

In the following section, a variety of seal types and materials are described as well as some of their typical properties.

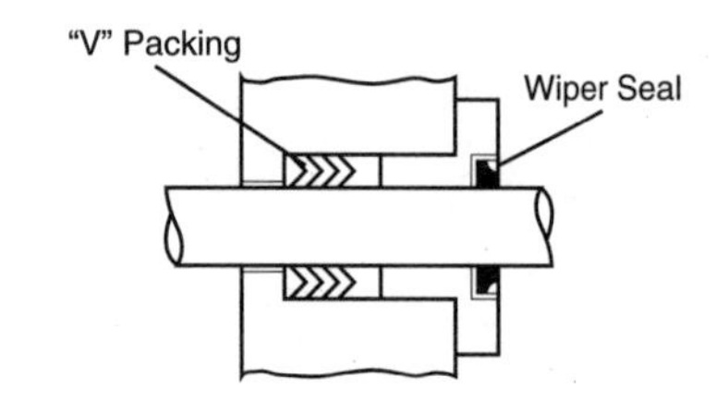

Fig. 5-42 Cylinder rod gland seals

Piston rod seal

Piston rod (gland) seals exist in several versions (fig. 5-42). Some cylinders are equipped with a 'V'-shaped or cup-shaped primary seal made of leather, polyurethane, nitrile (Buna-N®) or viton, and a wiper seal. Some designs thread into the rod end cap for easy maintenance.

In the past, the predominant piston rod seal has been made of nitrile rubber with a support or "back-up" ring. The support/back-up ring, which prevents seal extrusion, is usually made of acetal plastic (POM). Between those, a reinforced part of nitrile (Buna-N®) rubber is used to support the seal body and bridge the difference in hardness between the seal (nitrile rubber) and the support ring (acetal plastic).

A characteristic of this seal is its relatively low contact pressure (fig. 5-43) against the piston rod. However, a relatively thick oil film is transported/carried out on the piston rod if the piston rod is moving at low or no pressure or if it is exposed to vibrations. Even with higher pressure it will produce a relatively thick oil film, compared to other seal types.

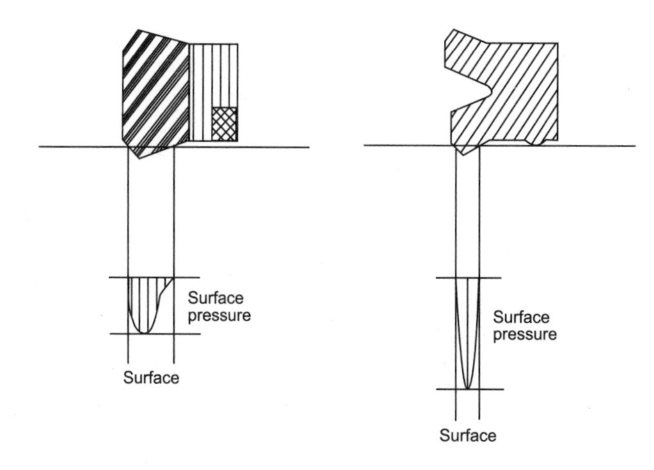

Fig. 5-43 Seal contact pressure

Nitrile (Buna-N) rubber has very good all-round properties. Elasticity, strength, and resistance to wear are good. Normal temperature range is -30°C to 90°C *(-22°F to 194°F).* There are also versions that are low temperature resistant, to be used for temperatures as low as -55°C *(-67°F).* This material can be used for many types of mineral oil, and for nonflammable (fire-resistant) hydraulic fluids of the HFA, HFB, and HFC types (refer to chapter 12, Hydraulic Fluids), and for water up to +60°C *(+140°F).* Working pressures up to 40 MPa *(5,800 psi)* and speeds up to 1 m/s *(200 ft/min)* are acceptable.

U-sleeve (lip seal)

A piston rod seal that is becoming more popular is the U-sleeve (fig. 5-44) made of polyurethane (PU). The U-sleeve (lip seal) has a sealing lip, which has a relatively small contact surface, with relatively high contact pressure. This design is better at scraping off the oil film (on the rod) under low pressure conditions than other designs.

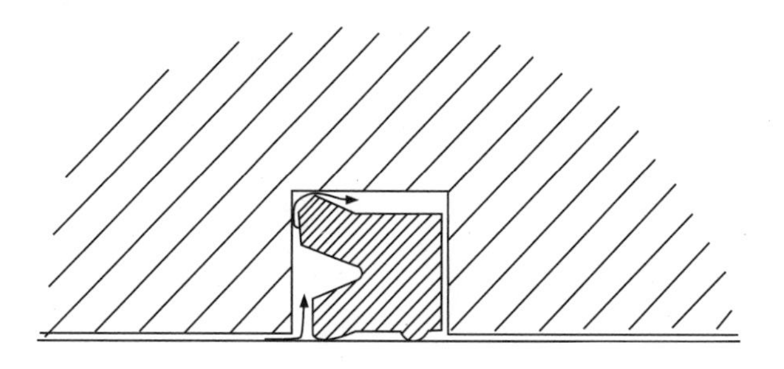

Fig. 5-44 Cylinder rod seal - U-sleeve/lip seal

Because of the high wear resistance of the polyurethane, long useful life can be obtained, despite the high surface pressure in the seal lip area.

Nevertheless, a small amount of fluid will pass the seal lip, in the form of an oil film on the rod. If this were not the case, the friction and the generation of heat and wear would shorten the seal life drastically.

Polyurethane has very good mechanical properties. Tensile strength, and tear and wear resistance are better than any other rubber type. However, the material is sensitive to hydrolysis (water absorption) and should not be used at temperatures higher than +50°C *(122°F)* in hydrous fluids (fluids containing water).

The normal temperature range is -30°C *(22°F)* to +90°C *(194°F)* for good material properties. There are also special versions, which support temperatures down to -40°C *(-40°F)* and higher on the upper range. Working pressure to 40 MPa *(6,000 psi)* and speeds up to 0.5 m/s *100 ft/min* can be tolerated.

One potential consideration with the polyurethane is that there are many different formulations of the material, and properties may vary greatly. Another consideration is that the temperature expansion coefficient is very high in relation to the coefficient for steel and grey iron. Therefore, at high temperatures, the seal is easily deformed, which at low temperatures (for example at a cold start) will cause the seal to leak between the seal and the groove in which it is installed.

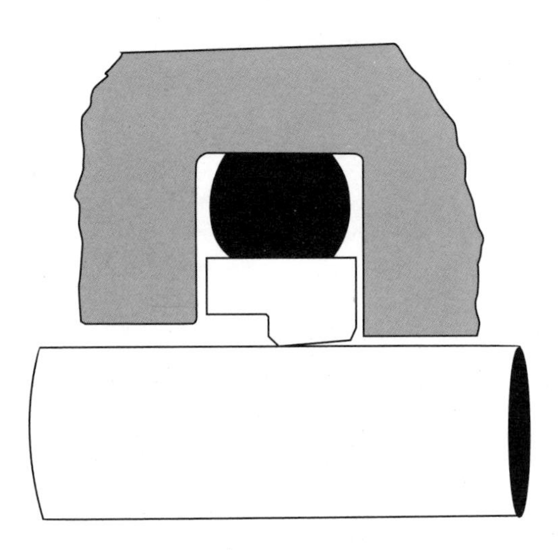

Fig. 5-45 Sliding ring seals

Slide ring seals

The slide ring seal consists of two parts: a slide ring made of PTFE (polytetraflourethylene) which seals against the piston rod. PTFE lacks flexibility, therefore, the required seal tension is provided by an ordinary o-ring (fig. 5-45).

PTFE or Teflon® has an extremely low coefficient of friction which is why the seal is used in applications where low friction and no stick-slip operation are required. The seal is also used in applications with speeds higher than 0.5 m/s *(100 ft/min)*.

A larger oil film, on the cylinder rod, is characteristic of this type of seal, as compared with those previously mentioned. Therefore, the PTFE seal will have specially designed lips. Double seals are often used to minimize external leakage.

PTFE has very good chemical resistance and is nonflammable. The normal temperature range is between -160°C *(-320°F)* and +250°C *(482°F)*. It is the choice of material for the o-ring that determines the temperature and pressure operating levels. Since PTFE has limited strength, the slide ring seal is filled or impregnated with graphite, glass, carbon or bronze to extend the seal life.

Besides the common rod seals just described, there are special seals. None can meet the sealing needs of today and tomorrow. Therefore, always discuss the seal choice with a cylinder manufactured, to obtain the best compromise solution for a given application and system demands.

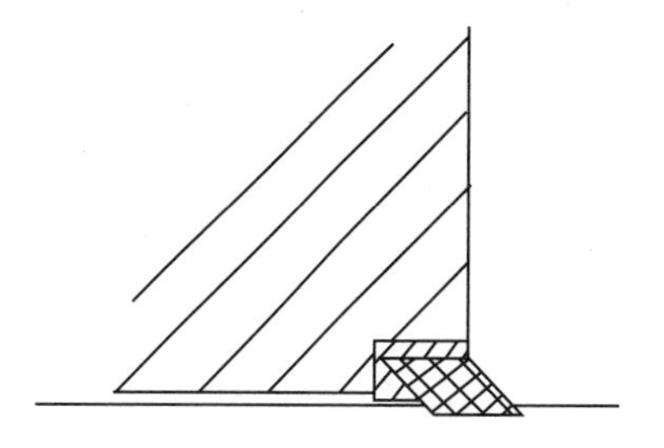

Fig. 5-46 Scraper rod seal (metal)

Scraper

The scraper (or "scraper ring") has a very important purpose, to prevent contaminants on the rod from entering the hydraulic system. Dirt must be prevented from entering the mobile cylinder. Thus the scraper must have a strong, but yet flexible, lip. Because of its favorable mechanical properties, polyurethane is a common material for scrapers.

The scraper can include a metal ring for installation with a tight fit in an open groove (fig. 5-46), or be all made of polyurethane, or, alternatively, nitrile (Buna-N) rubber for installation in a closed groove (fig. 5-47).

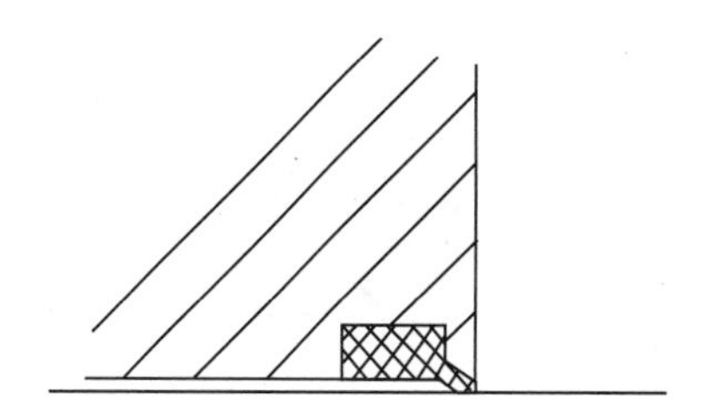

Fig. 5-47 Scraper rod seal (elastomeric)

Piston seal

Cast iron piston rings (fig. 5-48), lip seals (fig. 5-49), or a single bidirectional sealing elements were typically seals for pistons.

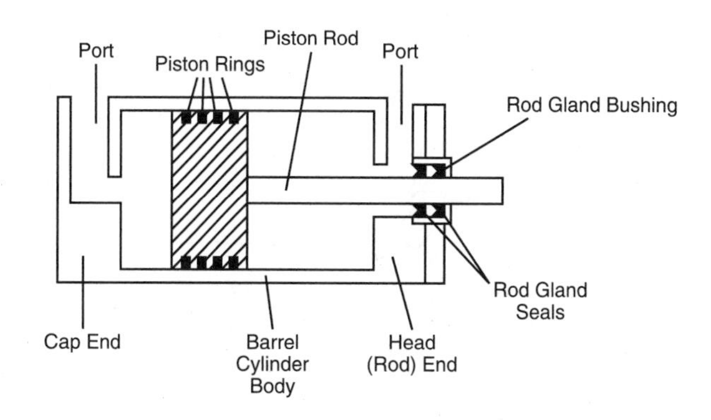

Fig. 5-48 Cylinder piston rings

Piston rings are durable, but exhibit some leakage under normal operating conditions. Lip seals and bidirectional seals offer a more positive seal, but may be less durable.

A common piston seal today is the double-acting compact seal (fig. 5-50), which is composed of several parts. The seal body is normally made of nitrile (Buna-N) rubber. On both sides are supporting rings, usually Hytrel (thermoplastic polyester), which help prevent leakage, and then guide/wear rings which absorb transverse forces on the piston. These guide/wear rings are normally made of polyamide (nylon) or acetal plastic.

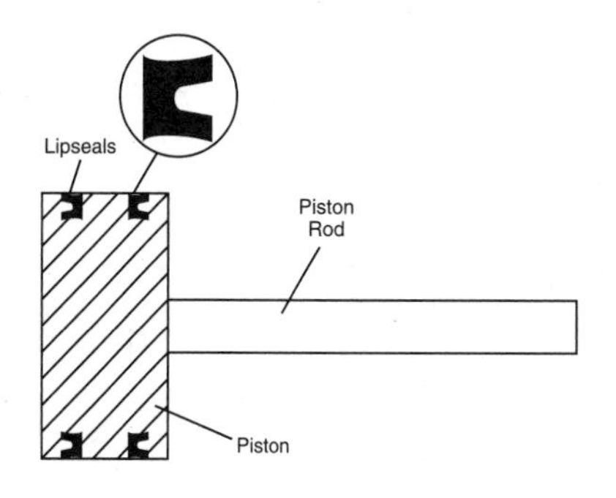

Fig. 5-49 Piston lip seals

The seal body is designed in such a way that, at start up, a relatively thick oil film between the seal and the cylinder barrel inner surface builds up. Therefore, the compact seal, in spite of its soft seal body, has good stick-slip properties and relatively low friction. When the cylinder is stopped, the seal presses the oil film out of the way, which makes static tightness very good.

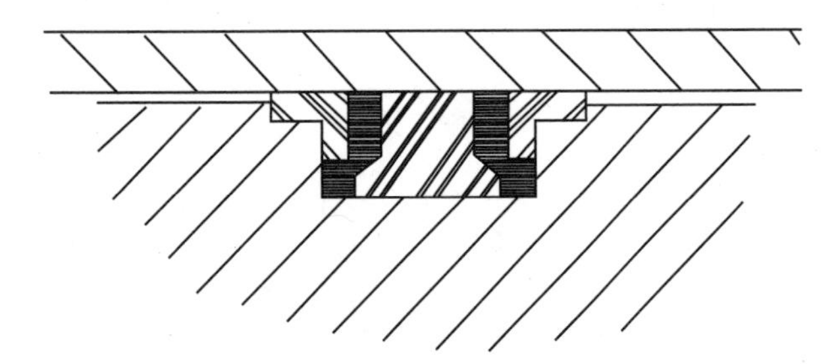

Fig. 5-50 Double-acting compact seal

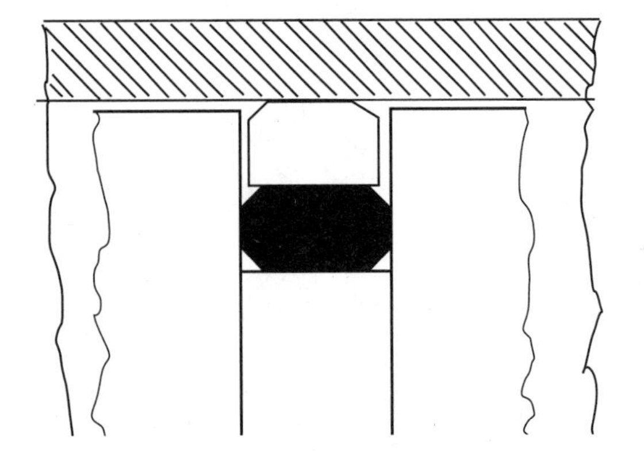

Fig. 5-51 Fluor rubber compact seal

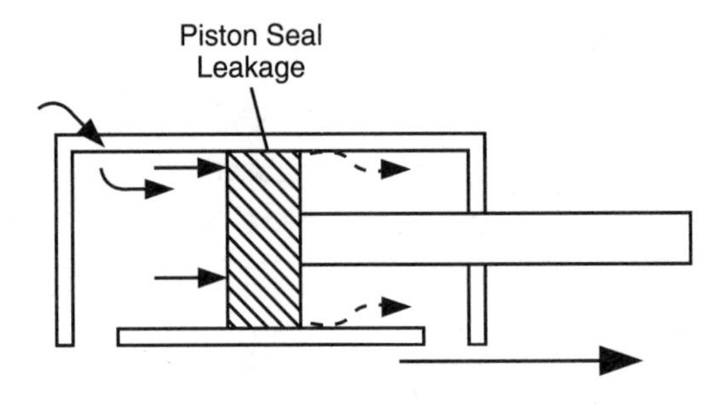

Fig. 5-52 Piston seal leakage

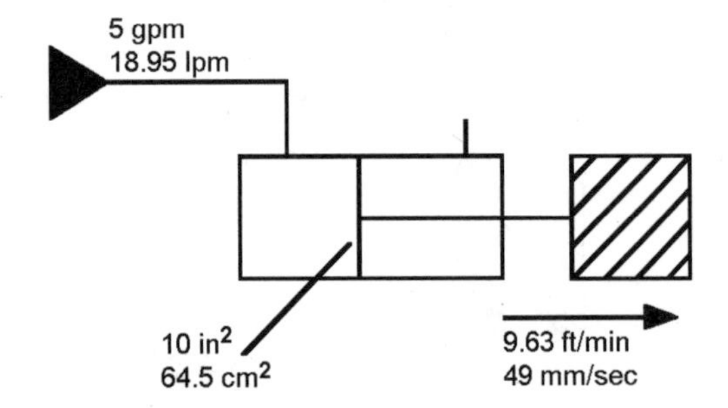

Fig. 5-53 Cylinder with good piston seals

The compact seal can be obtained with a seal body made of fluor rubber (fig. 5-51) instead of nitrile (Buna-N) rubber. Fluor rubber (FPM), best known under DuPont's trade name Viton®, is primarily intended for high temperature operation. The material is somewhat inferior to nitrile rubber at low temperatures. The recommended working range is between -20°C *(-4°F)* and 150°C *(302°F)*. Its water resistance is better than nitrile rubber. Fluor rubber can also be used for non-flammable fluids like chlorinated hydrocarbons and phosphates (HDF-fluids, but not Skydrol).

Piston seal leakage

As a cylinder operates, the cylinder seals wear, resulting in leakage at the piston rod and past the cylinder piston (fig. 5-52). Leakage at the rod seal results in a housekeeping problem and can readily be detected. Seal leakage past the cylinder piston is not easily detected. The next section shows that piston seal leakage causes the rod speed to decrease. It may even lead to pressure intensification in some cases.

Piston seal leakage affects rod speed

Piston seals are commonly made of cast iron or a resilient, synthetic compound. Leakage past the piston rings, or worn lip seals, can reduce the cylinder speed. However, a noticeable reduction in cylinder rod speed would require some 1.89 l/min *(0.5 gpm)* or more to compensate for the leakage. At this point, the cylinder probably has extensive internal damage.

In a clamping application, leakage past one piston may not be considered a problem. However, when there are many leaking cylinders on a machine, the clamping pressure may not be attainable, because the entire pump flow leaks past the cylinder pistons.

Rod speed of a cylinder is determined by how quickly pump flow can fill the volume behind a cylinder piston. This is the case, whether the rod is extending or retracting. A cylinder piston with an excessively worn seal allows fluid to bypass. This fluid does not fill the volume behind the cylinder piston and therefore does not contribute to rod speed.

In the illustration (fig. 5-53) a cylinder with a piston area of 60 cm^2 *(10 in^2)* receives 18.95 l/min *(5 gpm)* through the cap end. When extending, the rod speed (calculated according to Formula

5-1) is a function of fluid flow, filling the volume behind the piston.

An excessively worn piston seal would cause fluid to pass by the piston seal. This leakage, when subtracted from the pump flow, would indicate the decrease in rod speed.

Assume that the worn piston seal in the cylinder (fig. 5-54) allows 1.89 l/min *(0.5 gpm)* to bypass during extension. Consequently, 17.06 l/min *(4.5 gpm)* fills the volume behind the cylinder piston even though 18.95 l/min *(5 gpm)* enters the end cap. A speed calculation shows a 10% decrease in speed, the same as the ratio of the flows (leakage, pump) 2:20 (0.5:5).

Piston seal wear and leakage causes the cylinder rod speed to decrease even though full pump flow enters the cylinder. This means work will take longer to complete. At the same time, system operating temperature will increase because of the lost hydraulic power.

A cylinder usually leaks excessively in only a portions of the stroke (fig. 5-55) where it is primarily cycled. This is due to contaminants in the fluid, causing piston seal wear and cylinder bore scoring. Once the piston has passed the worn portion of the bore, bypass fluid is reduced and the rod speed increases.

Pressure intensification due to piston leakage

In some cases, piston seal leakage can cause pressure intensification. A cylinder, connected to a directional control valve and a pilot operated check valve (fig. 5-56) is required to raise and hold a load at mid-stroke. The load is 25 kN *(6,000 lbs)*; the cylinder piston has a piston area of 40 cm^2 *(6 in^2)* and a 35 cm^2 *(5 in^2)* annular area.

The cylinder in the circuit leaks excessively at the rod gland and past the piston. Since leakage at the rod gland is usually quite obvious, assume the rod gland seal is then replaced. This, however, still leaves the piston seal leakage uncorrected.

When the directional valve spool is shifted, flow at a pressure of 7 MPa *(1,000 psi)* enters the cap end of the cylinder, raising the load. As soon as the spool is returned to the neutral (center) position, the load immediately starts to fall because of the piston seal leakage and the rod

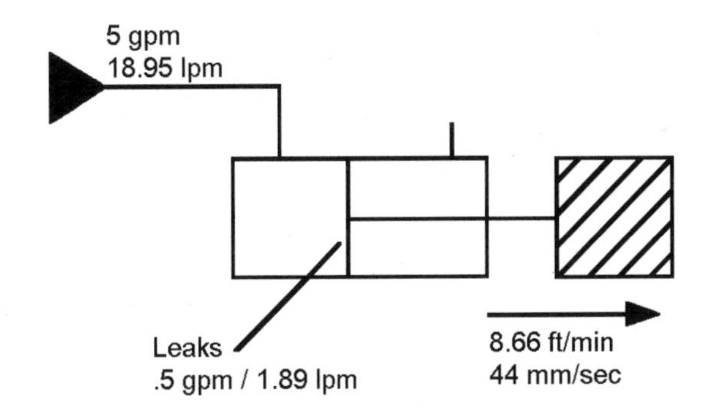

Fig. 5-54 Cylinder with worn piston seals

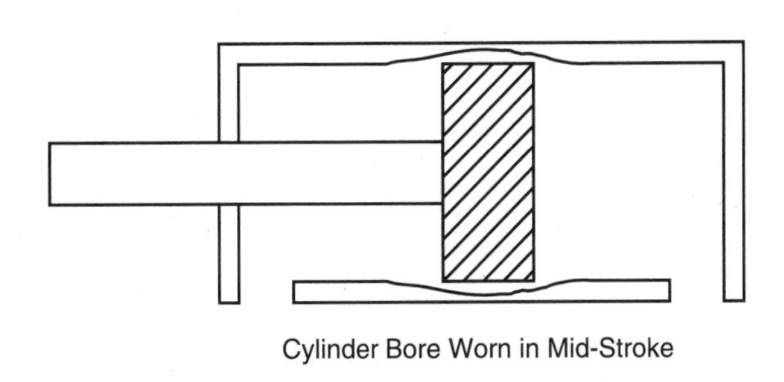

Fig. 5-55 Cylinder bore worn

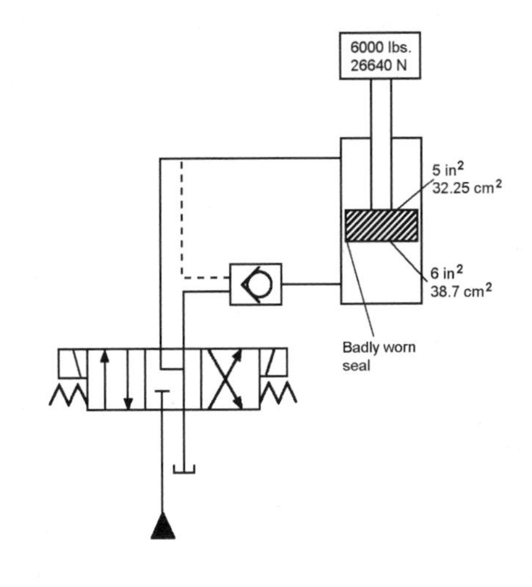

Fig. 5-56 Circuit resulting in pressure intensification

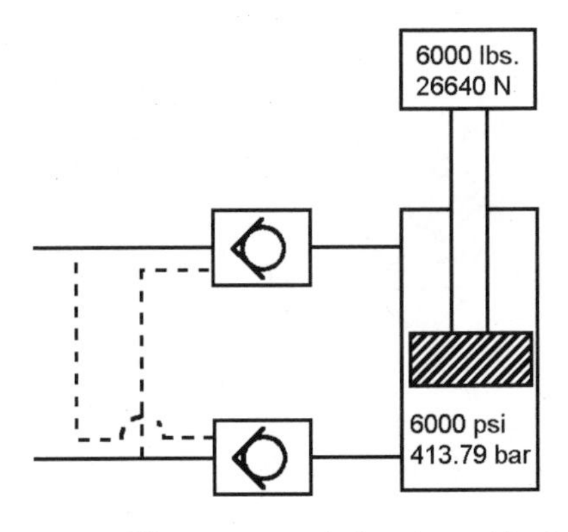

Fig. 5-57 Pilot operated checks added to the circuit

end connected to tank (return via the center position of the directional control valve).

To remedy this problem, pilot operated check valves are added to both ports of the cylinder (fig. 5-57). Now, when the directional valve is centered, fluid is cannot escape from the cylinder.

Because of the worn piston seal, fluid can still transfer from the high pressure side of the piston (cap end) to the opposite, low pressure side (rod end).

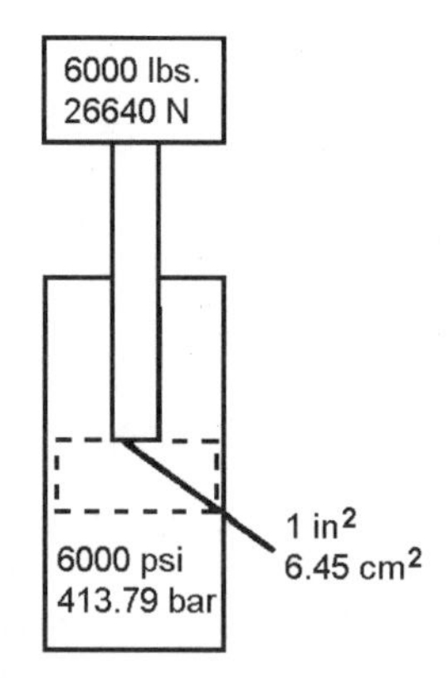

Fig. 5-58 Resulting pressure intensification

It may seem at first that, since the piston has in effect no seal, the load will drift down even though fluid is trapped in the cylinder. This is not the case and can be explained by an example.

A piston/piston rod is positioned halfway in a cylinder filled with liquid; it is obvious that drift is impossible as long as liquid does not leak out of the cylinder.

Returning to the cylinder with the worn piston seal, we find that the pressure is the same on both sides of the piston. Pressure acting on the piston area is offset by pressure acting on the annular area except for the rod cross-sectional area which supports the load.

Since the rod area is 5 cm^2 *(1 in^2)* and the load is 25 kN *(6,000 lbs)*, pressure generated in the cylinder would be 50 MPa *(6,000 psi)* (fig. 5-58).

The generation of high pressures in this manner can lead to ruptured seals, "blown" cylinder barrels, and external leakage. This process is known as pressure intensification.

Checking for piston seal leakage

Checking for piston seal leakage can be accomplished by seeing the effect of bypass flow on rod speed. To check for piston leakage, a needle or shut-off valve must be plumbed into the rod end cylinder line.

With the valve closed and the piston fully retracted, the cap end is subject to full system pressure. The needle valve is then cracked open, allowing the piston to move a short distance. The valve is then closed.

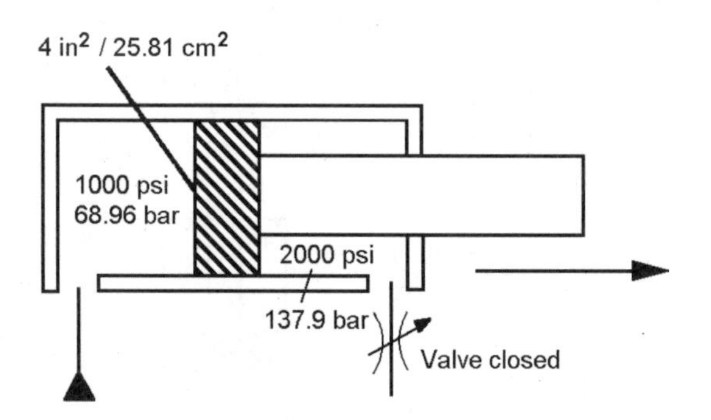

Fig. 5-59 Test for piston seal leakage

At this point, full system pressure is acting on the piston area, resulting in an increase in pressure on the annular area. In fig. 5-59 a 2:1 cylinder has a piston area of 26 cm^2 *(4 in^2)* and an annular area of 13 cm^2 *(2 in^2).*

With the relief valve set at 7 MPa *(1,000 psi)* and with the rod end port blocked, a force of 18 kN *(4,000 lbs)* is generated on the piston side to extend the rod. 18 kN *(4,000 lbs)* acts on the 13 cm^2 *(2 in^2)* annular area of the piston, which results in a 13.8 MPa *(2,000 psi)* backpressure on the rod side.

With 6.9 MPa *(1,000 psi)* at the cap end and 13.8 MPa *(2,000 psi)* at the rod end, any fluid leakage will transfer from the rod end to the piston cap, causing the piston rod to extend (fig. 5-60). This check is performed at intervals along the cylinder stroke.

As a piston seal check is performed, the rate at which the rod drifts out determines the reduction in rod speed as the cylinder operates in a system.

In the illustrated 2:1 cylinder (fig. 5-60), assume that, under test conditions, the rod drifts out at a rate of 30 cm/min *(1 ft/min)* as a 6.9 MPa *(1000 psi)* pressure differential existed across the piston. With the cylinder operating in a system at that differential, the same reduction in speed can be expected.

With the cylinder (fig. 5-61) receiving 19 l/min *(5 gpm)*, the piston rod would extend at a rate of 12 cm/s *(23 ft/min)*. If, instead, the piston seals were in perfect condition, the rod would extend at 12.5 cm/s *(24 ft/min)*. If the reduced rod speed, caused by piston seal leakage, cannot be tolerated, the cylinder should be repaired or replaced.

Keep in mind that cast iron piston rings can leak 16-49 cm^3 *(1-3 in^3)* of oil per minute at a pressure of 69 bar *(1,000 psi).* They are designed to leak somewhat for the purpose of lubrication. This flow should not be confused with leakage flow due to wear.

Leakage flow (bypass) should be considered when studying or troubleshooting a hydraulic system.

In the next section intensified pressure is present at the rod end of an unloaded, extending, single rod cylinder while flow is being metered (restricted) out of the rod end of the cylinder. This pressure can cause harm to a cylinder.

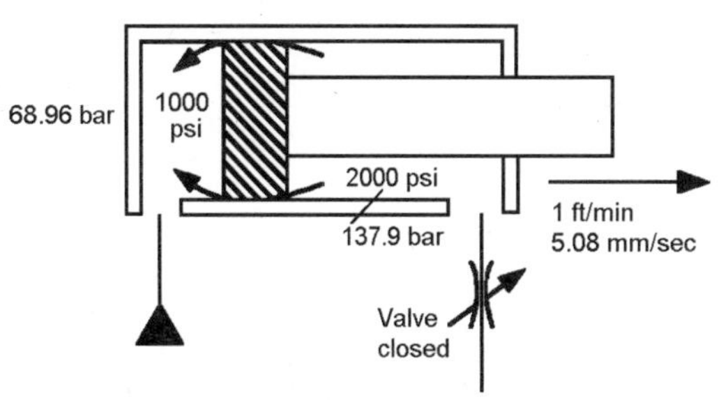

Fig. 5-60 Pressure intensification causing the cylinder to extend

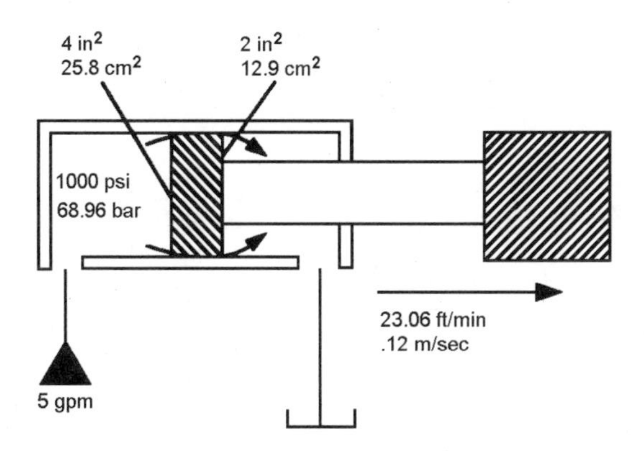

Fig. 5-61 Cylinder with leaking piston seals

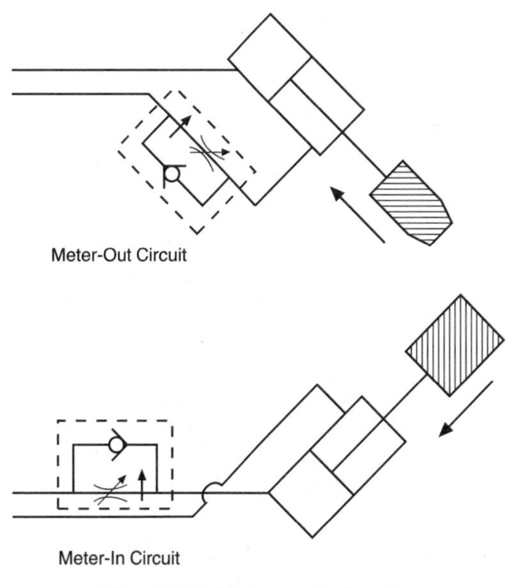

Fig. 5-62 Meter-out circuit

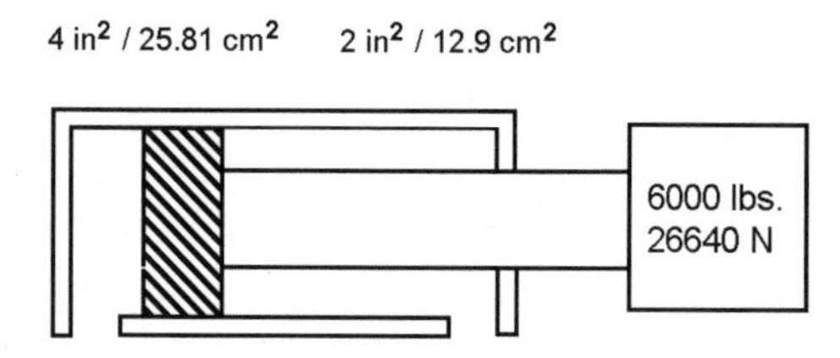

Fig. 5-63 Cylinder load

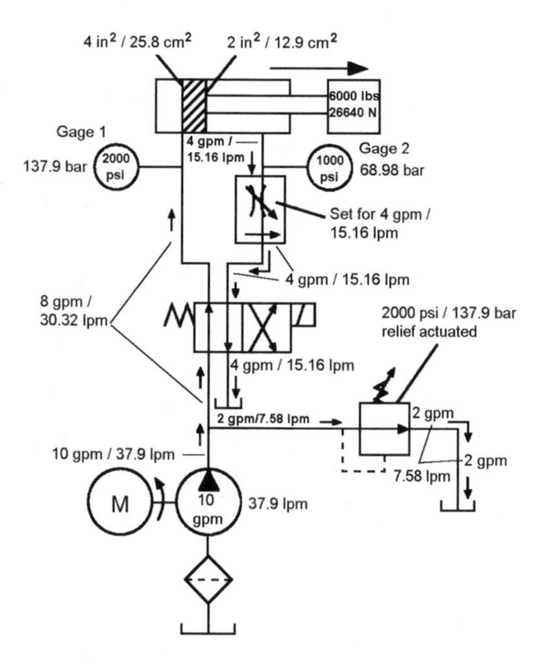

Fig. 5-64 Meter-out circuit

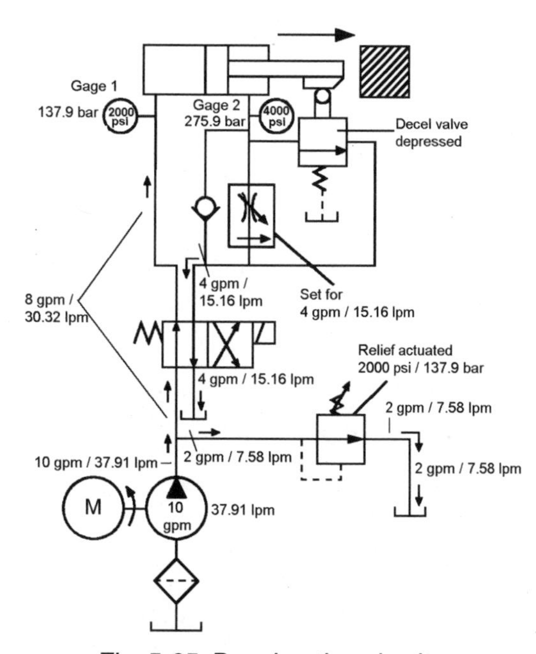

Fig. 5-65 Deceleration circuit

Intensification at the cylinder rod end

A flow control valve could be positioned at the rod side of a cylinder. This valve would then restrict flow from the cylinder and as a result the cylinder would not be allowed to runaway from pump flow. This is known as a meter-out circuit (fig. 5-62).

The flow control valve keeps the cylinder from running away by causing a back pressure on the annular or effective area of the piston. The resultant force keeps the piston and piston rod under control.

In any meter-out circuit in which a single rod cylinder is pushing out a load, whenever the pressure acting on the piston area and its resultant force is more than is required to equal the load, the excess force develops a back pressure on the piston annular or effective area.

In the illustration (fig. 5-63), a 2:1 cylinder is required to move a 26 kN *(6000 lbs)* load. The area of the piston is 26 cm² *(4 in²)* and the effective area is 13 cm² *(2 in²).* From the formula: Pressure = force/area, it can be calculated that 10 MPa *(1,500 psi)* must act on the piston area in order to equal the load.

In the circuit (fig 5-64), assume a pump flow of 40 l/min *(10 gpm),* relief valve setting is 14 MPa *(2000 psi)*, and the flow control valve is set at 16 l/min *(4 gpm)* out of the cylinder as the rod extends. With a 2:1 cylinder, if a pump flow enters the cap end one half of that discharges from the rod end. With the flow control set at 16 l/min *(4 gpm)*, only 32 l/min *(8 gpm)* is allowed into the cylinder.

When pump/engine is turned on, pressure begins to build in the system. (This happens almost instantly.) When pressure reaches 10 MPa *(1,500 psi),* the load has been equalled. But, the pump cannot push its flow of 40 l/min *(10 gpm)* into the system at 10 MPa *(1,500 psi)*. More pressure is developed. When pressure reaches 13.8 MPa *(2,000 psi)*, the relief valve cracks open enough to pass 8 l/min *(2 gpm)* to tank. At this point, gage 1 at the cap end indicates 13.8 MPa *(2,000 psi).*

This pressure acts on the piston area resulting in a force of 35 kN *(8000 lbs)* extending the piston rod. 75% is used to equal the load and the back pressure acting on the annular or effective area offsets 25% in excess. The backpressure is generated as a result of the flow control valve restriction. Gage 2 indicates 6.9 Mpa *(1000 psi)*.

If the same situation occurred with no load, the restriction of the flow control valve would cause an extremely high pressure to be generated in the rod end of the cylinder. In the circuit illustrated (fig. 5-65), the 2:1 cylinder extends rapidly to the workload. At a point near the work, a deceleration valve is closed off forcing fluid to pass through a flow control valve as it exits the cylinder rod side. The rod continues to extend for a short distance and then contacts the workload.

Between the time the deceleration valve closes and the load is contacted, a maximum pressure is generated on the piston area, but a load is not present to make use of it. This is therefor absorbed by increased back pressure acting on the annular piston area. For a 2:1 cylinder resulting in double the pressure. Gage 2 indicates 27.6 MPa *(4,000 psi)* and continues to indicate that until the load is contacted.

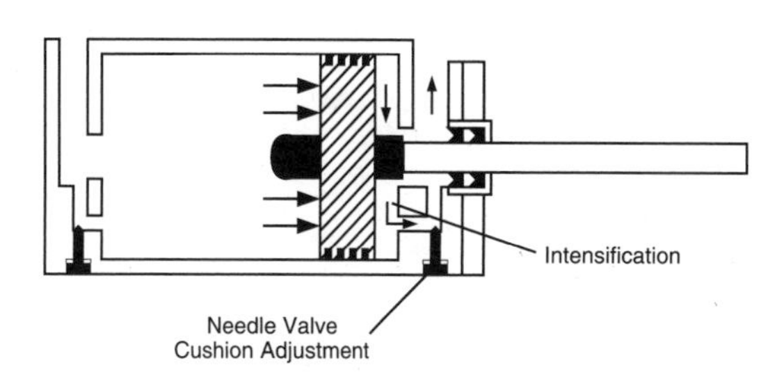

Fig. 5-66 Pressure intensification with a cylinder cushion

Pressure intensification at the rod side of a cylinder can cause rod seals to leak or rupture. Intensification can be expected to occur in the above manner any time an extending, single rod cylinder is being metered out without a load. Since cylinder cushions (fig. 5-66) are also meter-out restrictions, pressure intensification will occur any time an extending, single rod cylinder goes into the cushion. This same condition will occur if a mobile directional control valve is used containing a restricted or meter-out type spool. This does not affect the rod seal, but can cause leakage fluid to discharge from the needle valve cushion adjustment.

From previous illustrations, it has been shown that discharge flow from pump is not necessarily the maximum flow rate in a system. The above example points out that a relief valve setting is not necessarily the maximum pressure in a system.

Guide (bearing) rings/wear rings

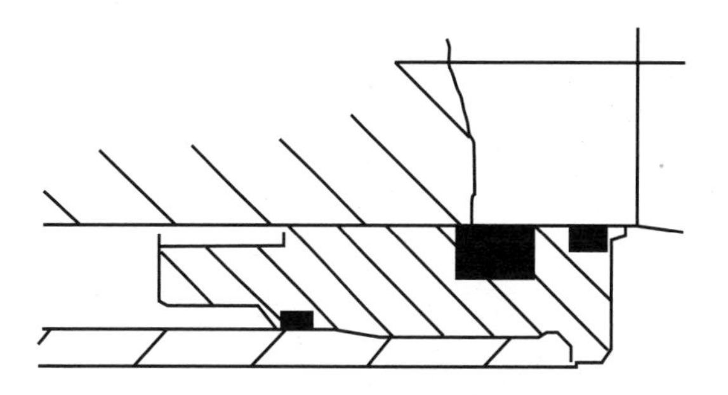

Fig. 5-67 cast iron guide/wear ring

The clearances between the movable and the fixed parts of the cylinder have a crucial effect on the function and the useful life of the cylinder. Wrong bearing material and too narrow clearances, may cause a break down due to parts welding together. On the other hand, too wide a

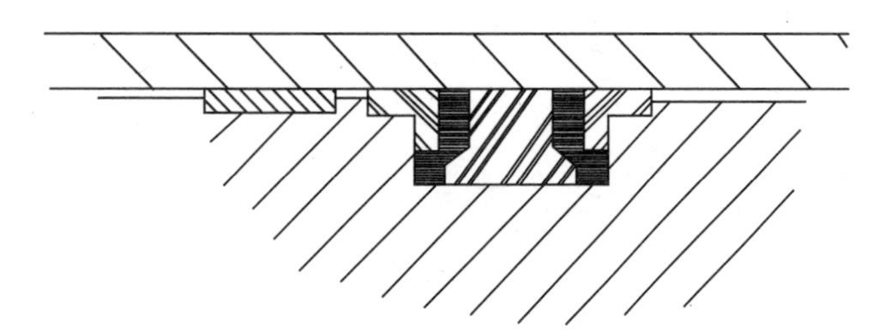

Fig. 5-68 Additional guide/wear rings

clearance may cause leakage due to the seals wearing out.

A solution is a piston rod guide/wear ring made of cast iron (fig. 5-67). Grey iron in combination with a hard-chromium plated piston rod has proven to withstand heavy loads. In combination with oil, grey iron has very good lubrication properties. Certain types of plastics have also proven to have good low friction and wear properties.

Under hard operating conditions, such as with transverse/side forces on the cylinder, standard bearings/wear rings are not adequate to handle such forces. In such cases, additional bearings/ wear rings can be added (fig. 5-68) or the standard bearings/wear rings can be replaced with others made of a material more resistant to surface pressure.

Bearings/wear rings made of polyamide and acetal plastic, reinforced or not, with glass-fibre, are available. Under very high load conditions, bearings/wear rings made of fabric reinforced phenol compound plastic, coated fabric, are often used.

Conditions for useful life

In addition to the requirements already mentioned concerning not allowing contamination to pass the scraper and enter the cylinder, there are other factors to be considered when designing a hydraulic system.

Some particles of contamination will enter the cylinder from the outside, and also certain small wear particles may be formed inside the cylinder. To avoid these staying in the cylinder and damaging the machined surfaces and seals, it is important to move the oil out of the cylinder.

To do this, the tubing between cylinder and directional control valve must not be very long, otherwise the same oil will constantly pass in and out of the cylinder, without ever being filtered. All oil in the system should pass through a filter, allowing the contaminates to be removed. In addition, the cylinder temperature will rise dangerously high if the same oil passes in and out of the cylinder continuously.

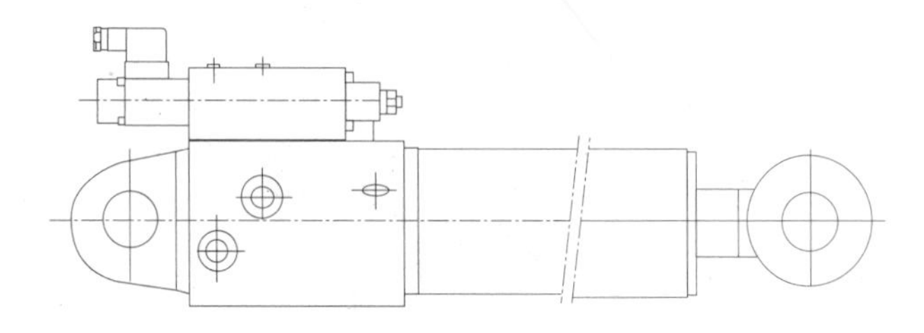

Fig. 5-69 Integrated cylinder

Integrated cylinders

Cylinders are often available with built-in or optional valving features. The choice of these fea tures is determined, in part, by the following factors:

- security requirements
- leakproof / tightness requirements
- environmental requirements
- lack of space, etc.

It is quite common in today's mobile industry to have a complete cylinder positioning unit, i.e. a cylinder with a servo or directional control valve and control system all built into or attached to the cylinder (fig. 5-69).

To prevent an unintentional load movement resulting from a hose burst or an internal leak, a load-holding valve is often employed . For improved security/safety, the valve must be connected to the cylinder by a pipe/tube, a flange or be built into the cylinder end cap (fig. 5-70). The built-in feature provides a neater and cleaner design and presents less risk for external leakage problems.

When choosing the load holding valve, it is extremely important to consider the entire system as well as the valve type. If the wrong combination is chosen, the cylinder is likely to be overloaded.

Many mobile applications are subject to legal requirements for hose burst securing/safety. Examples of widely used valve types to meet these requirements are the following:

- flow sensitive check valve
- choking with or without pressure compensation
- pilot controlled check valve
- pilot controlled pressure reduction (overcenter type)
- sleeve valve, seat valve or a combination of these two types

Extensive testing may be required to find a workable solution, as control problems may be encountered. For example, an overcenter valve in a hydraulic system can cause instability and create system oscillations.

A hydraulic pressure control valve, used to limit shock, may be built-into the cylinder cap or very near it. This valve helps minimize the shock pressure in the cylinder.

A cylinder supporting a static load (fig. 5-71) must always have a pressure limiting (shock) valve connected to it to limit the pressure on the

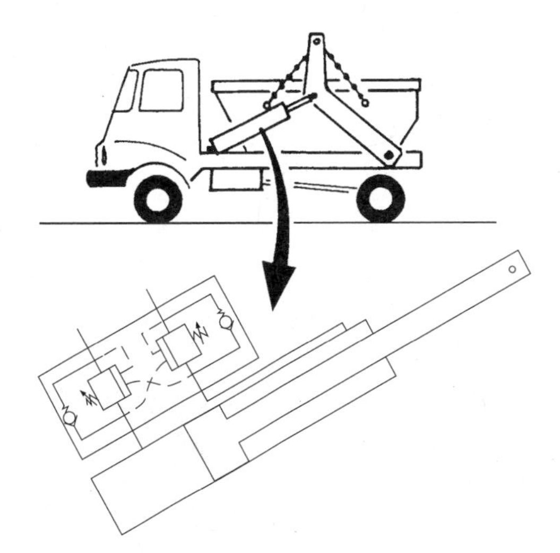

Fig. 5-70 Close coupled pressure control

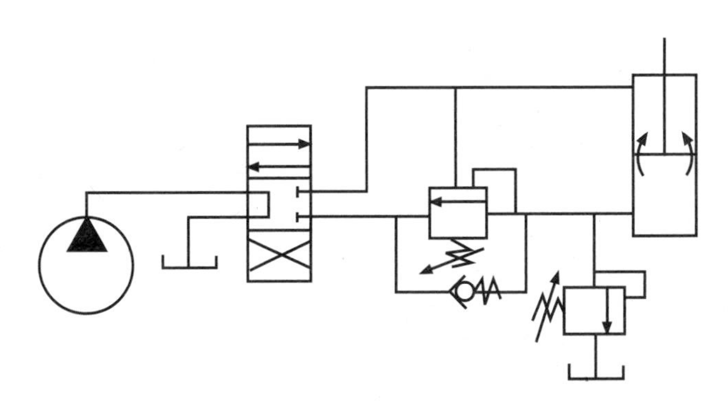

Fig. 5-71 Cylinder supporting a static load

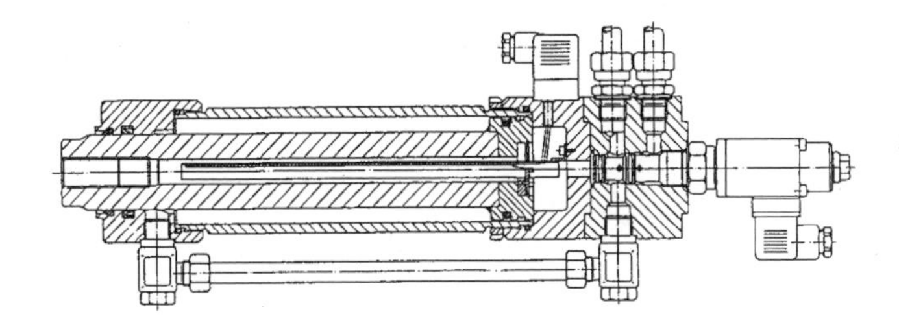

Fif. 5-72 Cylinder with integral position sensor

piston side of the cylinder. This valve will also limit this pressure in case of internal leakage past the piston seal.

Quite frequently in today's mobile system designs, cylinders are equipped with valves that are built in or bolted on. The cylinder is often combined with a position sensor to form a complete "closed loop" positioning unit (fig. 5-72). There are several advantages to this:

- faster and more accurate automatic control
- hose burst protection
- simplified hydraulic system
- pump and tank tubing in series circuit
- easier troubleshooting
- simplified service.

The same technology is often used on automatic drilling rigs in highly automated mines. In this case the concept offers advantages such as simplified tubing, better automatic control accuracy, lower system costs and reduced leakage.

Fig. 5-73 Cylinder for "closed loop" system

Cylinders for "closed loop" systems

Cylinders with built-in position sensors are becoming more and more common (fig. 5-73). Until just a few years ago these devices were very expensive, now the cylinder sensor is found in mobile as well as industrial applications.

What has spurred the development of these components are:

- increased security/safety
- continuous monitoring
- automation
- rationalization
- simplified mechanics and better ergonomics

A wide variety of sensors are available today; they can be divided into two groups: absolute and incremental sensors.

Absolute sensors keep indicating position relative to the end position and recover a cylinder's unique position signal even after a power failure. A voltage or current from the sensor is always proportional to the position of the cylinder. Some examples of absolute sensors are:

- potentiometer
- inductive sensor
- capacitive sensor
- acoustic sonar sensor
- absolute encoder

The potentiometer is often a hybrid type, a plastic-coated, thread-wound, straight/linear potentiometer.

The other types require a great deal of space in the rear end cover of the cylinder. With the use of this style the use of ball-and-socket style mounting is not possible. The position indicators have very good accuracy and reliability. The price level of the different types varies, from 0.5 to 5 times the price of the cylinder.

The hybrid potentiometer has been successful because it is easy to build into the cylinder and its low price.

Incremental (encoder) sensors only count pulses. A certain cylinder position is not unique to the sensor signal, thus it can be set to zero in any position. At e.g. a power failure, the cylinder does not know where its position is and must therefore be referred to a mechanical zero point. Incremental sensors can be optical or magnetic.

Selecting a mobile cylinder

The hydraulic cylinder converts hydraulic energy, in the form of pressure and flow, into mechanical energy - a linear movement, accurately controlled and with great power.

The requirements of a hydraulic cylinder are wide and varying. When selecting a cylinder, it is important to know the application as well as the properties of the different cylinder types.

Answering the questions below will establish a good starting point for the selection of the right cylinder:

- How much force is the cylinder to provide?
- What is the piston stroke?
- At what pressure will it work?
- Will it work simultaneously with other cylinders?
- Will it work together with other cylinders?
- At what frequency will it work?
- What is the piston speed?
- How large a mass will be connected to the piston rod?
- Is a damper/cushion required at one or both ends of the cylinder?
- What transverse forces, static or dynamic, will it be exposed to?
- Is any influence in the form of shock, heating, cooling, corrosion etc., to be expected?
- Are there any requirements of the hydraulic fluid or seals?
- What type of cylinder and attachments are appropriate for this application?

Chapter 5 exercises

Instructions: Solve the problems

1. **Problem:** Connect the hand pump, cylinder and reservoir so that with each stroke of the hand lever (forward and backward), the load is raised the same amount. The load must not be allowed to drop when the hand pump is not pumping. To solve the problem, three check valves are needed. These are the only components which may be added to the circuit.

   Do not be concerned with lowering the load or refilling the reservoir.

   In the hand pump, the rod side of the piston has half the area as the cap end side.

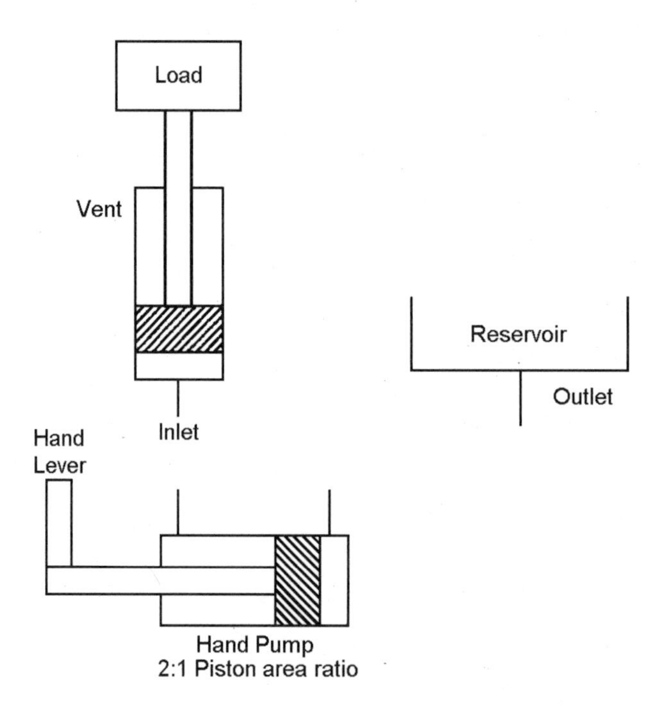

Chapter 5 exercises (cont'd.)

2. In situation A, the cylinder rod is pushing out the load. In situation B, there is no load on the cylinder rod.

   The system relief valve is set at 138 bar *(2000 psi).* The flow control valve is metering flow out of the cylinder.

   What do the gages read in each case?

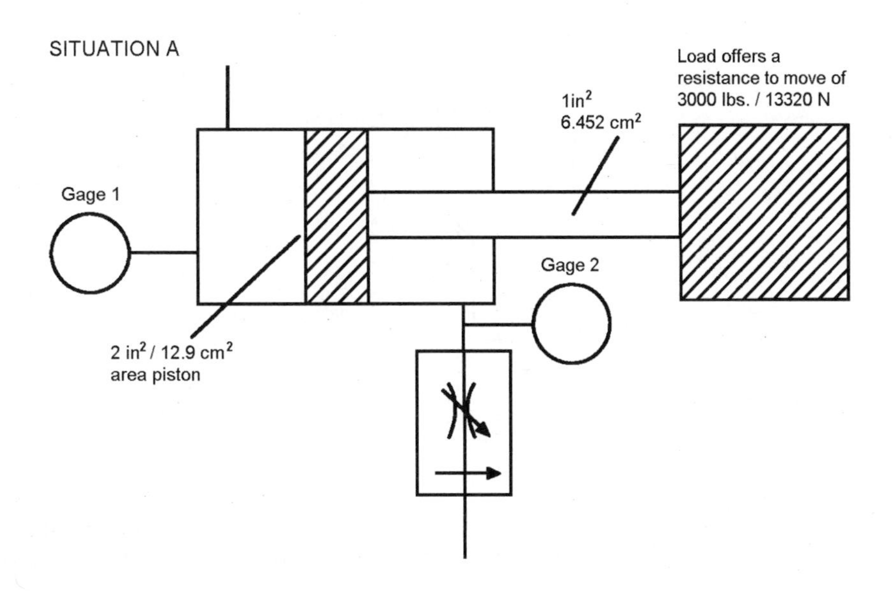

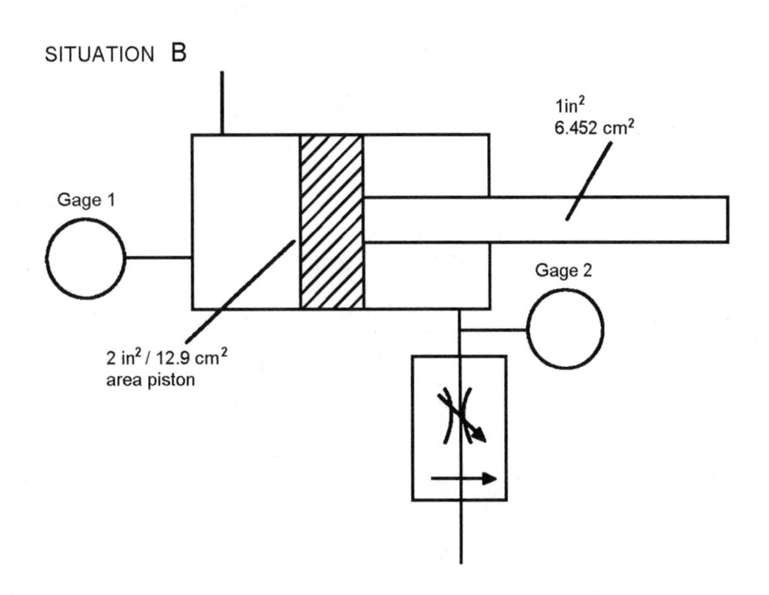

Chapter 5 exercises (cont'd.)

3. **Situation:** The 26 640 N *(6000 lb)* load must be suspended, but excessive leakage across the piston causes the load to drift down. A pilot operated check valve was placed in each cylinder line to control the drift.

   Problem: The cylinder's rod seal was blown out. Why?

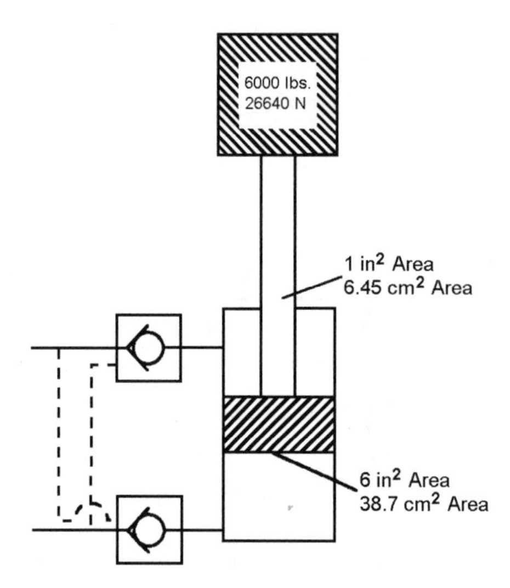

4. Problem: In the regenerative circuit below, the components are connected so that the cylinder piston rod travels at the same speed extending and retracting. System pressure = 69 bar *(1000 psi);* flow = 38 l/min *(10 gpm).*

   If the cylinder rod has a 5 cm *(2")* diameter, what is the rod speed? ______________

   What is the maximum force which can be developed by this cylinder? ______________

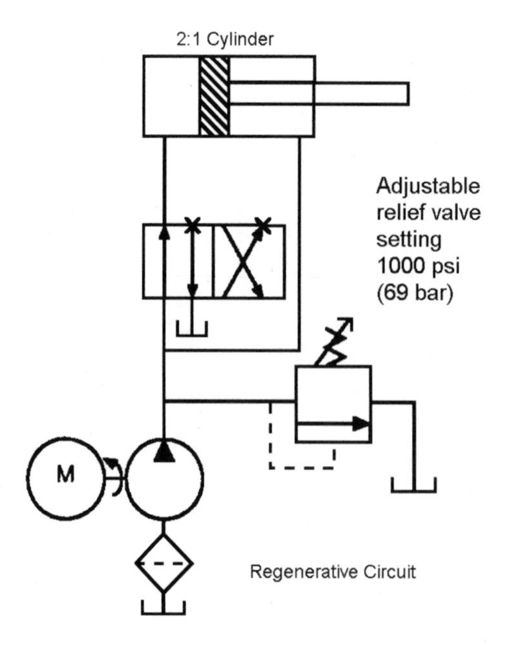

Regenerative Circuit

Chapter 5 exercises (cont'd.)

5. A cylinder with a 7 cm *(3")* bore and a stroke of 41 cm *(16")* receives 68 l/min *(18 gpm)*. What is the piston rod velocity?

6. A cylinder with a 20 cm *(8")* bore and a 1 m *(36")* stroke must extend in one minute. How much flow is required?

7. A cylinder with a 25 cm *(10")* bore and a .6 m *(24")* stroke must move a 38 718 N *(78,540 lb)* load through its stroke in three seconds. How much hydraulic horsepower must be delivered to the cylinder?

Chapter 6

Pressure control valves

In mobile hydraulic systems, the pressure is controlled with pressure control valves. There are basically five types which can be either direct or pilot operated:

- Pressure relief valves (limit the maximum pressure in a system or in part of it).
- Pressure reducing valves (reduce the system pressure on the outlet side to a constant value regardless of fluctuations in the main system above the selected pressure).
- Pressure compensators (maintain a constant pressure difference across e.g. a variable restrictor).
- Sequence and unloading valves (open or close a flow path at a preset pressure level).
- Overcenter valves (allow cavitation free load lowering preventing the actuator from running ahead when pulled by the load).

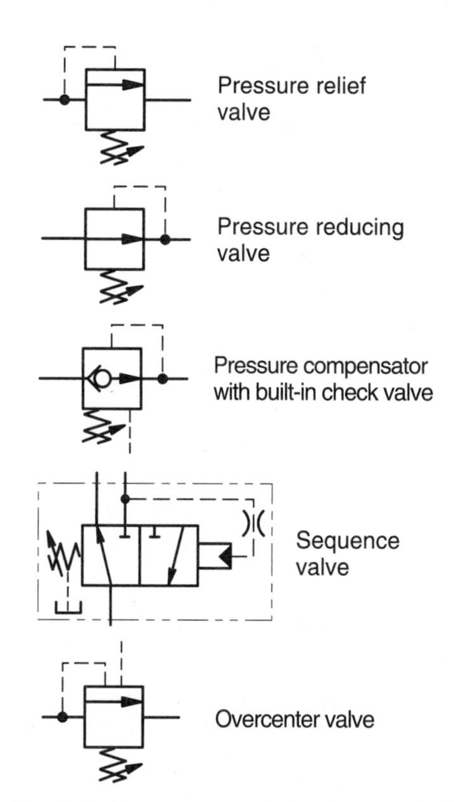

Fig. 6-1 Pressure control valve symbols

Direct acting system relief valves

A direct acting pressure relief valve limits the maximum system pressure. It can be an integral part of the main hydraulic directional control valve (DCV).

The relief valve consists basically of a poppet, a seat, a bias spring, and an adjustment screw.

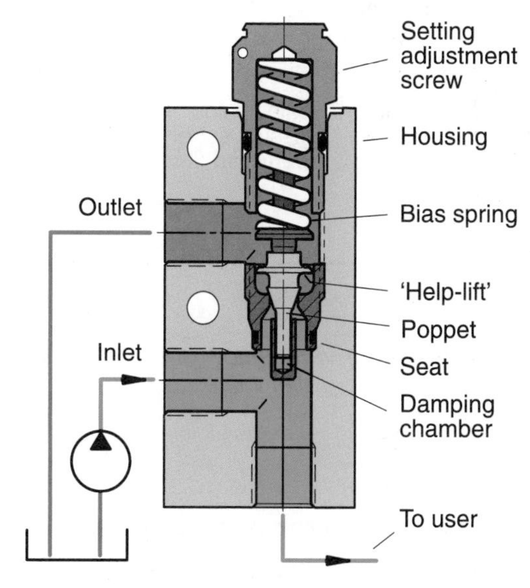

Fig. 6-2 Direct acting relief valve (cut-away)

Valve function

In fig. 6-2, the inlet (primary) port is connected to system pressure and the outlet (secondary) port to tank; the poppet is held against its seat by the bias spring. When the pressure overcomes the spring force the poppet is forced off its seat and opens the connection to tank. Flow from the pump is now diverted to tank and system pressure is limited to that of the bias spring setting.

The poppet opens more or less depending on the flow, keeping the pressure balance between the primary pressure and spring force. Hydraulic energy is thereby converted into heat energy (Formula 6-1).

Formula 6-1 Energy is a function of flow, pressure and time

$$W = Q \times \Delta p \times t$$

where: 'W' is energy
'Q' is flow
'Δp' is pressure differential
't' is time.

Valve characteristic

The opening of a direct operated valve starts when inlet pressure exceeds the spring setting. This 'cracking' pressure is normally a little bit

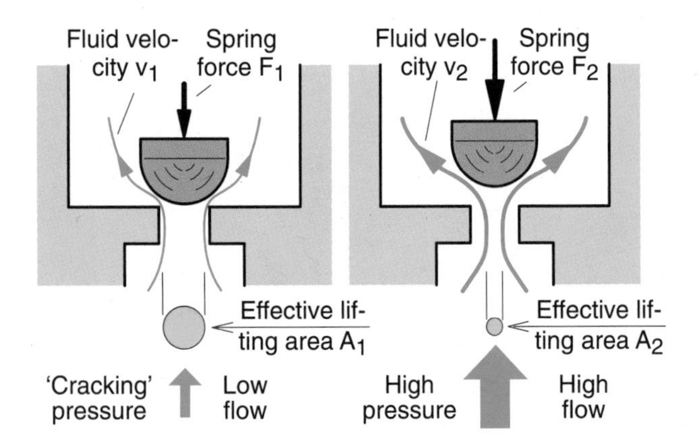

Fig. 6-3 Effective lifting (pressurized) area.

> *Formula 6-2 Spring force is a function of pressure and area.*
>
> $$F = p \times A$$
>
> where: 'F' is spring force
> 'p' pressure acting on the the poppet
> 'A' area on which pressure is acting.

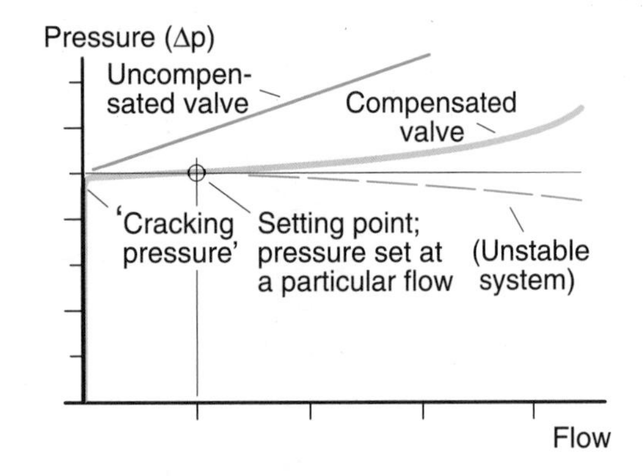

Fig. 6-4 Pressure-versus-flow of a direct operated relief valve.

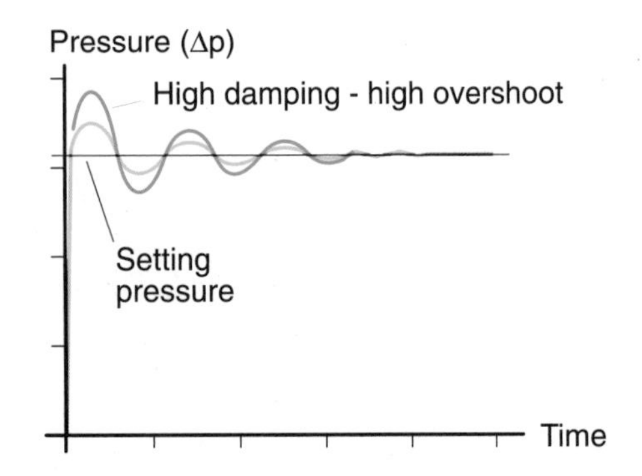

Fig. 6-5 Diagram showing pressure 'overshoot' (when the relief valve opens suddenly)

lower than the 'set pressure'. The valve acts in this manner because of the compression of the (quite stiff) spring due to the increasing flow passing the poppet and the decreasing effective inlet area.

When the valve starts to open at the cracking point, the hydraulic static pressure is acting on area 'A1' of the poppet (fig. 6-3). The poppet is lifted a certain distance, compressing the spring. In order to pass a higher flow through that restricting orifice, a higher fluid pressure is therefore required.

The higher fluid velocity, passing through the orifice between the seat and the poppet, decreases the effective lifting area ('A_2' is smaller than 'A_1') because the static pressure drops.

When force 'F' increases (due to the compression of the spring) and the effective lifting area 'A' decreases, a higher hydraulic pressure is consequently required; refer to Formula 6-2. The result is that pressure increases with increasing flow as shown in fig. 6-4. This increase can be reduced to a certain degree by allowing the impulse force of the fluid to act on the 'help-lift' edge area of the valve poppet (shown in figs. 6-2 and 6-6).

An increasing characteristic improves the margin of stability of the valve. To cope with the above contradiction, the spring characteristic must be carefully chosen. If not, the poppet starts to oscillate, causing high pressure spikes and intolerable noise; refer to the 'unstable system' curve in fig. 6-4.

The main reason why direct operated relief valves are used as shock valves in actuator ports (fig. 6-8) is that they open much faster, have a lower overshoot (fig. 6-5) and have less leakage than pilot operated valves.

Relief valve damping

The dynamic stability of a valve in a system is dependent of the weight and shape of the poppet and the seat as well as of the valve spring characteristic. Mechanical or hydraulic vibrations may propagate to the relief valve and generate poppet oscillations that must be damped properly. Fig. 6-2 shows a piston type damper on the inlet (primary) side. The fluid volume from the damping chamber must pass a small clearance between the valve poppet piston and the corresponding boring.

The degree of damping depends on the size of the clearance between the poppet 'piston' and the boring in the seat, as well as of the viscosity of the fluid; the higher the degree of damping - the higher the overshoot (fig. 6-5).

Pressure relief valve applications

The pressure relief valve has many uses in a hydraulic system. It can be utilized as a system relief/safety valve, or as a service line 'shock valve', protecting e.g. an actuator from excessive, external forces.

Actuator relief valves

A pressure control valve, protecting a hydraulic cylinder or motor by absorbing system pressure spikes/ 'shocks', is often referred to as an actuator line relief valve or shock valve. It may be set above or below system pressure.

The valve shown in fig. 6-6 is a cartridge assembly. It is usually an integral part of the main directional valve (DCV), located between the actuator port and the tank gallery. It can also be installed in a separate housing.

Normally these valves are combined with an anti-cavitation function which allows fluid flow in the opposite, makeup direction, reducing the risk of cylinder or motor cavitation.

As shown, the valve consists of a seat, poppet, bias spring, adjusting nut or screw (when adjustable), a check valve spring, and cap.

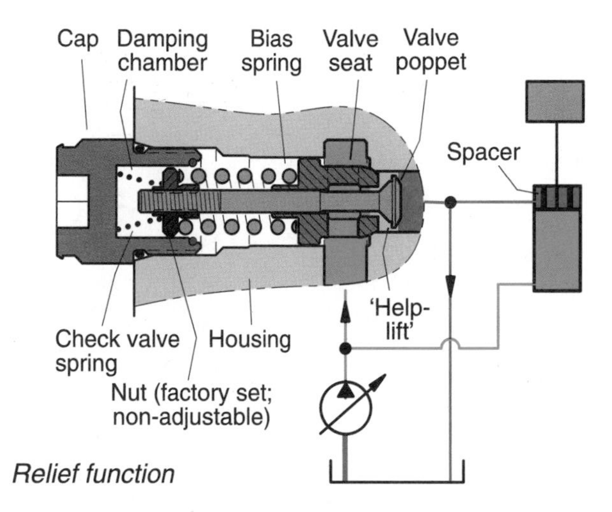

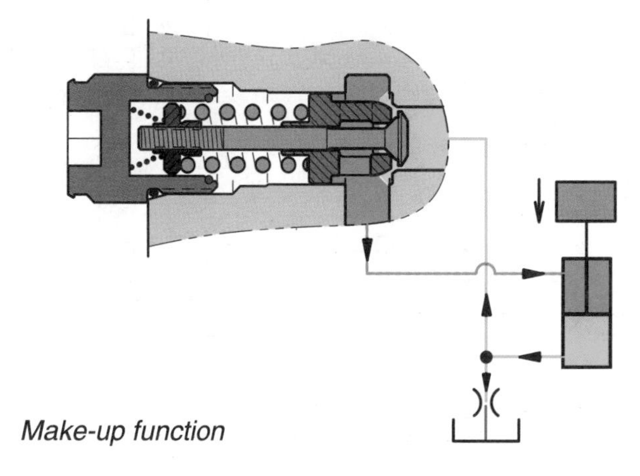

Fig. 6-6 Cartridge type relief valve (with makeup check valve function)

Relief valves in a circuit

Fig. 6-7 illustrates a circuit with a main pressure relief valve but without actuator port relief (shock) valves. Lacking reliefs, the ports connected to the cylinder are blocked when the DCV spool is in the center position. If, during the operation of the system, an external force should cause the pressure to rise in either port 'A' or 'B', the pressure increase could cause damage to the cylinder or the lines leading to it.

As can also be seen, the system relief valve is isolated from the actuator ports by the spool's centre position. It, therefore, does not protect the cylinder from possible pressure spikes.

A DCV equipped with port reliefs, fig. 6-8, protects the cylinder against high pressure spikes

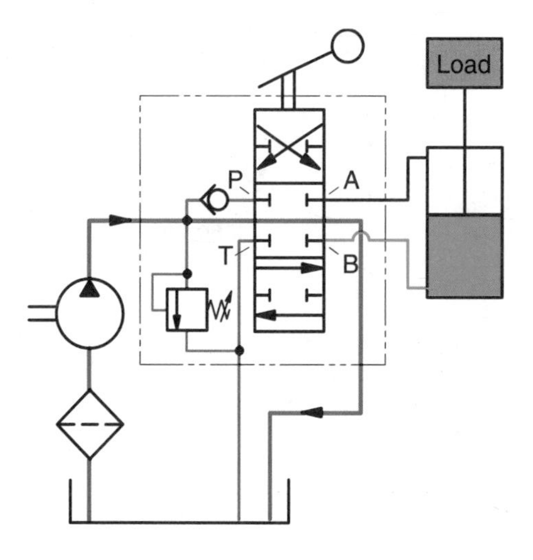

Fig. 6-7 Hydraulic circuit with main pressure relief valve

caused by e.g. external forces. Each actuator line relief valve can be adjustable or have a fixed pressure setting.

If an external force causes the pressure to rise in one of the cylinder ports, the pressure forces the corresponding valve to open to tank, thus limiting the pressure. When the relief valve opens the cylinder piston moves down as fluid is being diverted to tank.

The movement of the piston may, if not counteracted, cause a condition known as *cavitation* on the opposite, rod side of the cylinder. To prevent this damaging phenomenon, check valves are installed in the actuator ports to allow a makeup flow.

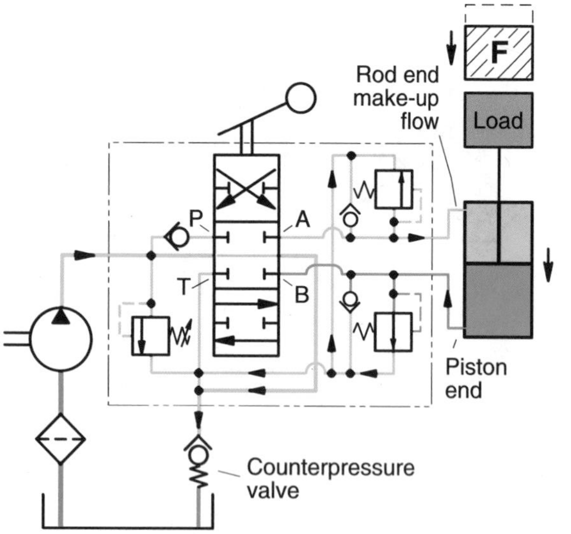

To explain the operating principle in greater detail, suppose port ‘B’ is connected to the piston end of the cylinder and port ‘A’ to the rod end (fig. 6-8). An additional load, ‘F’, bearing down on the initial load, causes compression of the fluid in the piston end of the cylinder. If ‘F’ is large enough, a pressure spike is being created, which is higher than the setting of the relief valve, and the ‘B’ port relief opens.

Fluid from the piston side is partly diverted to tank, and partly through the ‘A’ port check valve to the rod side of the cylinder, thus reducing the risk of cavitation.

If, instead, a large pulling load is acting on the cylinder (fig. 6-9), the ‘B’ port relief valve would open. Because of the smaller rod side area (as compared to the piston side area), the flow pushed out of the rod side is insufficient to fill the increasing volume of the piston side (through the check valve in port ‘A’). For this reason, additional fluid is required from the tank gallery (supplied by the pump or from the tank).

NOTE: In order to obtain sufficient counterpressure for the makeup function, a counterpressure valve may have to be installed in the return line to tank (as shown in figs. 6-8 and 6-9).

Rod end make-up flow
F
Load
P
A
T
B
Piston end
Counterpressure valve

Fig. 6-8 Actuator line relief valve function (pressure shock at piston end)

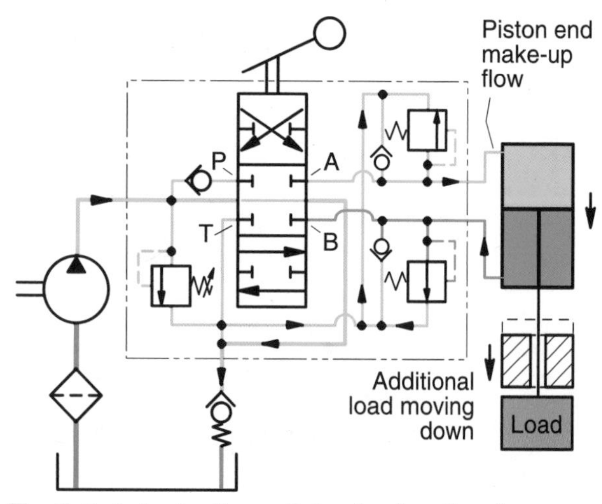

Fig. 6-9 Actuator line relief valve function (pressure shock at piston rod end)

Crossover relief valves

Crossover reliefs are often used in applications with hydraulic motors. When the motor is operating (in either direction) it needs to be protected from high pressure spikes caused by external torque loads.

An example is a motor that is being stopped abruptly (fig. 6-10). This may happen e.g. when a vehicle, propelled by a hydraulic motor, suddenly hits an obstacle. The entire pump flow would momentarily be directed over the crossover relief valve until the pump would have time to destroke or the main pressure relief (not shown) would act.

If the operator of the vehicle should suddenly shift the directional control valve to neutral (fig. 6-11), the pressure spike being created by the motor load inertia would be limited by the corresponding crossover relief, thus permitting the rotating mass to stop smoothly.

In most cases, crossover reliefs consist of two direct operated cartridge type relief valves with built-in check valve function. They are connected on their low pressure sides by a crossover passage as shown. When one of the relief valve opens, fluid passes over the check valve of the opposite cartridge to the low pressure side of the motor.

The relief valve cartridge is held against its seat by a light spring (conical type shown). When acting as a check valve, the entire cartridge moves against this spring, letting flow pass to the other motor port. The risk of cavitation on the low pressure side of the motor is then reduced.

NOTE: In order to concentrate on 'the message', certain components have been omitted in the schematics shown in this chapter, components that are essential for the proper functioning of a hydraulic system.

Pilot operated pressure relief valves

Principally, there are two versions of pilot operation. In fig. 6-12, pressure is sensed from the inlet (primary) side; this type of pressure sensing is known as *'direct pilot operation'*.

Pressure control valves can also sense pressure from another part of the hydraulic system by means of an external pilot line, fig. 6-13. This type of pressure sensing is known as *'remote pilot operation'*. Unlike a direct operated pressure control valve, where a spool or poppet is held biased by spring force only, a pilot operated valve has its spool biased by both fluid pressure and spring pressure.

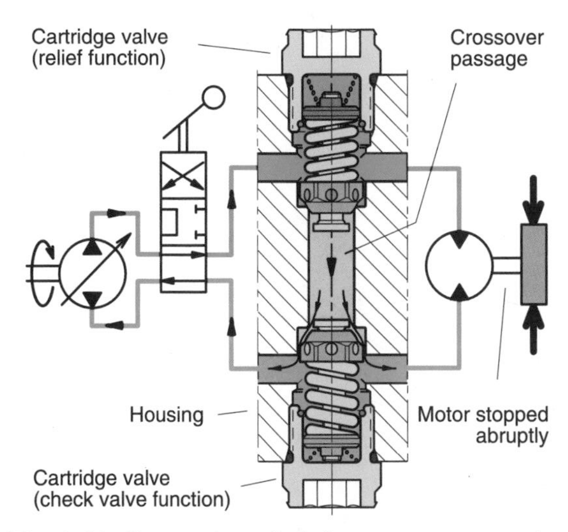

Fig. 6-10 Crossover reliefs in a pump-motor circuit (the motor is suddenly stopped)

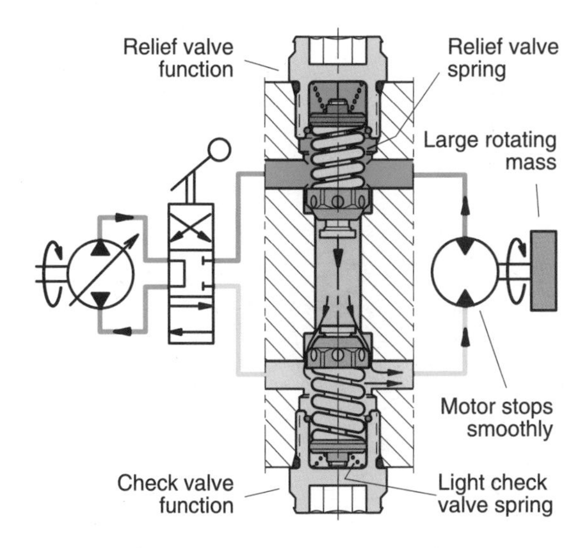

Fig. 6-11 Crossover reliefs in a pump-motor circuit (the directional valve shifted to neutral)

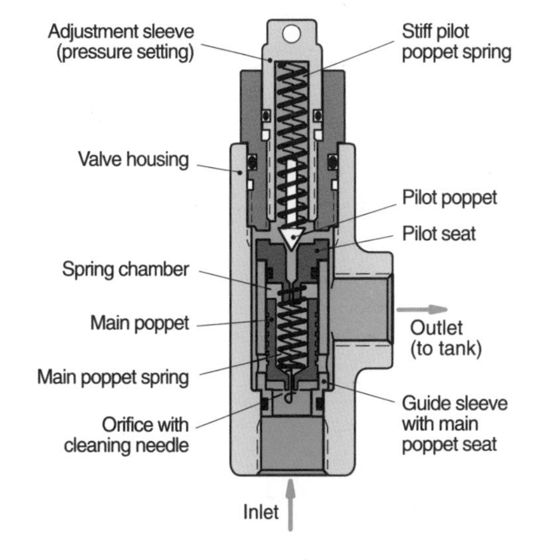

Fig. 6-12 Pilot operated, adjustable relief valve (cutaway)

Design

The pilot section of the relief valve (fig. 6-12) consists of a simple, spring-biased poppet, which handles small flows at a high pressure differential. The main section also contains a simple, spring-biased poppet/sleeve that handles large flows at a low pressure differential. By using the two together, large flows can be handled at high pressures.

The valve in the illustration consists of a main poppet with a fixed orifice and cleaning needle, main poppet bias spring, combined guide sleeve with seat, pilot seat, pilot poppet with bias spring and adjustment sleeve.

Valve function

In the pilot operated relief valve shown (fig. 6-12), severe increase in pressure at increased flow is eliminated by the light poppet (or spool) bias spring. Fluid pressure and the poppet spring bias the main poppet sleeve of the valve.

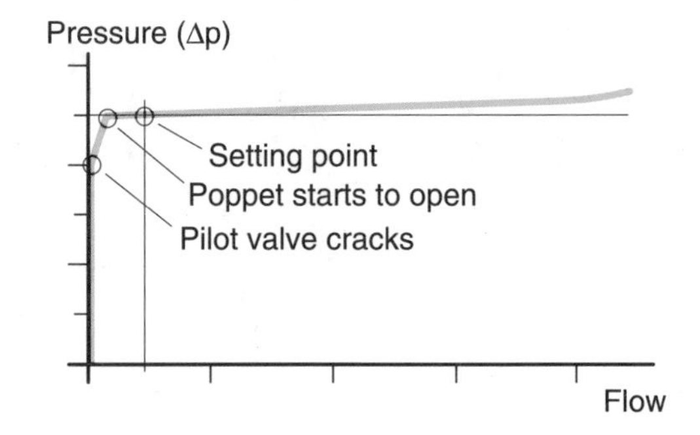

Fig. 6-13 Pressure -versus-flow diagram of a pilot operated relief valve.

When a certain system pressure is reached, the poppet is lifted off its seat. The slight increase in pressure at an increased flow is primarily due to the compression of the light spring and the flow forces acting on the poppet. Fig. 6-13 shows a typical pilot operated relief valve characteristic.

The maximum fluid pressure that is allowed to bias the spool is determined by the setting and characteristic of the pilot valve. To allow pressure to enter the spring chamber, an orifice is drilled through the main poppet. The cleaning needle prevents the orifice from being clogged.

As an example, assume the spring biasing the main valve poppet has a value of 7 bar *(100 psi)* and that the pilot valve limits the pilot pressure in the spring chamber to 140 bar *(2000 psi)*. With a system pressure of 140 bar *(2000 psi)*, acting at the bottom of the poppet to push the poppet up, a total mechanical and hydraulic pressure of 147 bar *(2100 psi)* acts to keep the main poppet down.

Since the pilot valve is set to limit the fluid pressure in the bias spring chamber of the main poppet to 140 bar *(2000 psi),* the pilot valve opens and fluid flows to tank. The total pressure acting down is still higher than the pressure acting up. As can be seen, the pressure at the cracking point is somewhat lower than the pilot stage opening pressure, which in turn is lower than the main poppet opening pressure.

The strong spring coefficient results in considerable pressure increase in the pilot stage (refer to fig. 6-13). The result is that a small flow must go through the pilot valve before the 'stationary stage' is reached, a flow that can be seen as leakage.

The fluid flow through the orifice in the main poppet increases, causing a certain pressure drop over the orifice, which results in a lower pressure above the main poppet. When the pressure drop is equal to the pressure setting of the light bias spring, the main poppet starts to open. Depending on the design of the orifice and the poppets, a different pressure drop and opening characteristic can be achieved.

The maximum pressure that can bias the main poppet in the 'down' position is 147 bar *(2100 psi)*. If the pilot valve were to have an ideal pressure-versus-flow characteristic, the pressure below the main poppet could only rise somewhat above 140 bar *(2000 psi)*; this is not the case with the illustrated design.

Note that the pilot operated pressure relief valve has at least twice as many potential leakage paths as the direct operated valve.

Remotely controlled, pilot operated relief valves

Remote adjustment of a pilot operated relief valve can be accomplished by utilizing a separate relief valve as shown in fig. 6-14.

When the pilot valve is mounted in a separate valve body, it can be installed in an easily accessible location for ease of adjustment by the machine operator or maintenance person. Note, however, that stability problems may develop in this setup; proper dampening is then required.

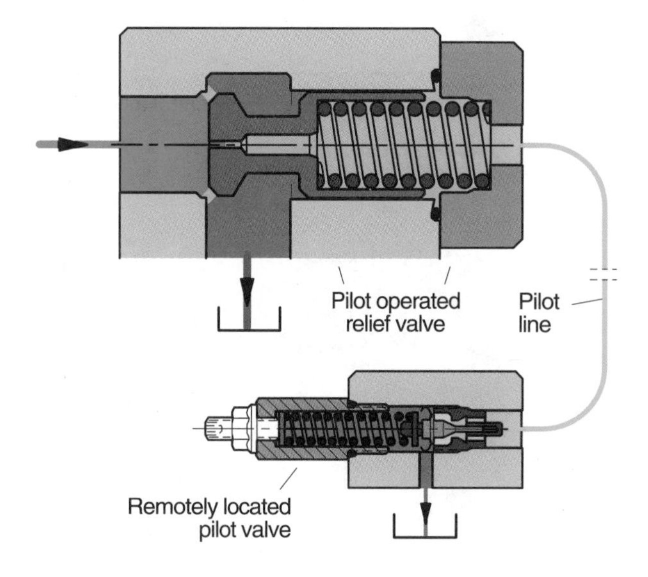

Figure 6-14 Pilot operated relief valve with adjustable pilot valve (cutaway)

Pressure reducing valves

Another type of direct acting pressure control valve is one which allows flow to pass through the valve while it is in its at-rest (normally open) position.

This valve type, commonly called *pressure reducing valve*, limits the pressure level in the outlet (secondary) port, and keeps it constant, inde-

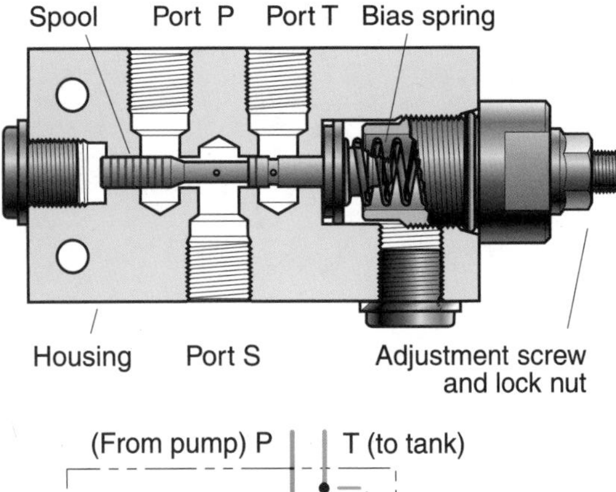

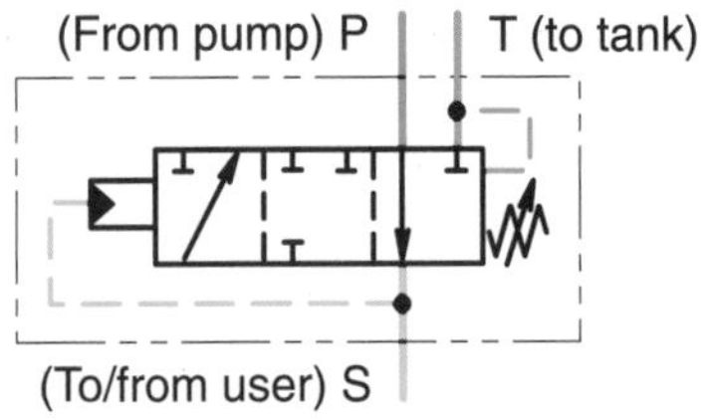

Fig. 6-15 Three-way pressure reducing valve (cutaway) with schematic

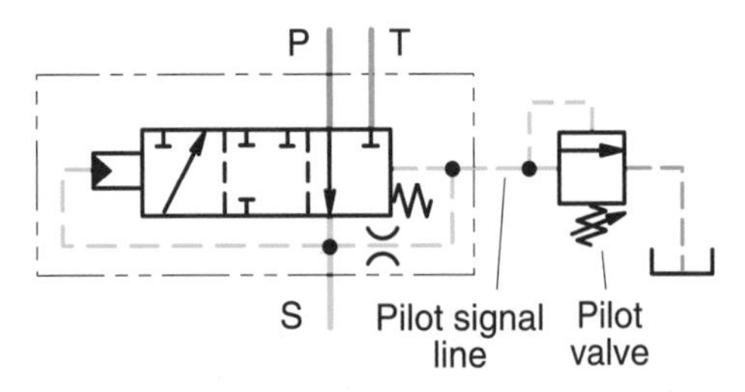

Fig. 6-16 Three-way pressure reducing valve, externally pilot operated

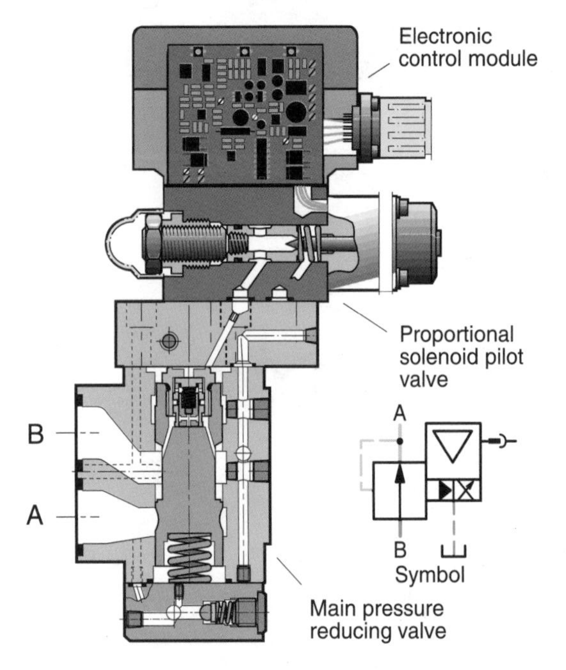

Fig. 6-17 Electronically controlled, pressure reducing valve (with symbol)

pendent of pressure variations in the inlet (primary) port. In order to maintain the pressure on the outlet side, the inlet pressure must be at least at the same level as the preset pressure.

Fig. 6-15 shows a direct acting pressure reducing valve. It consists of a valve housing, spool, bias spring, adjustment screw and lock nut. These valves are either two-way or three-way; the illustration shows a three-way valve.

The valve is open (P-to-S) as long as the pressure level on the secondary side (S) is lower than the bias spring setting. The spring chamber is internally drained to tank through a small hole in the spool. The secondary pressure is acting (through another small hole in the spool) against left end area of the spool . When the pressure in S increases to the setting value, the spool moves to the right and throttles (reduces) P-to-S. If the pressure on the secondary side tends to increase above the setting, the valve spool moves further (to the right) and connects S with T. Thereby the valve acts as a pressure relief valve on the secondary side.

The pressure reducing valve can easily be pilot operated by connecting an adjustable pilot relief valve to the spring chamber (fig. 6-16). This is quite similar to the pilot operated relief valve shown earlier.

If the pressure reducing valve has to handle large flows, it is usually pilot operated. Fig. 6-17 shows a pilot operated valve equipped with an integrated electronic control module that monitors a proportional solenoid pilot valve. This allows pressure setting (in port A) via a remote, electric signal.

The diagram in fig. 6-18 shows that, at a setting of e.g. 150 bar *(2200 psi),* the valve closely maintains the set pressure in port A, irrespective of variations in the inlet pressure in port B.

Sequence valves

Fig. 6-19 shows a 3/2 sequence valve (three ports, two positions) which is controlled with an internal pilot signal. The spool shifts position when pressure (in port P) increases to the bias spring setting pressure. The valve can be either normally closed (as in fig. 6-19) or normally open.

Sequence valves are available in many versions, e.g. 3/2 (shown), 2/2, and pilot (remote) operated valves. They are used e.g. in low flow pilot circuits where a need exists for closing or opening a connection at a preset level of a pilot signal from another circuit.

Another use of the sequence valve is in crane and forestry loader applications. The valve can be part of an overload protection system. It prevents any attempt to lift a load that will exceed max allowed lift moment (weight x 'arm') of the basic structure of the crane or loader.

Inlet compensators

An inlet pressure compensator/bypass valve can be integrated in a proportional DCV (directional control valve) to provide a flow that is independent of load pressure. The valve function is used in constant flow, closed circuit (CFC) systems with fixed displacement pumps. The speed of the function (actuator) is thereby directly proportional to the position of the lever (which determines the spool opening, or orifice, area).

The inlet compensator is a normally a closed pressure valve, located at the inlet of the DCV (fig. 6-20). It consists of a compensator spool, bias spring, relief valve and dampening orifice.

The unloading valve in a circuit

Fig. 6-20 shows the DCV spool in the neutral position. Pump flow entering port 'P' passes around the compensator spool and is communicated to the bottom of the spool. The bias spring cavity and relief valve (known as the load sensing circuit) are drained to tank across the load sensing groove on the DCV spool.

Therefore, any pump pressure above the bias spring rate, typically 6 bar *(85 psi),* will cause the compensator spool to shift 'up', dumping pump flow to tank at a 6 bar *(85 psi)* pressure differential.

When the DCV spool is moved either 'down' or 'up' (fig. 6-21), load pressure in actuator port 'A' or 'B' (whichever is connected to port 'P') is sensed in the four-way body, and communicated to the inlet compensator bias spring chamber and relief valve. The compensator maintains a constant pressure drop across the four-way spool while dumping any excess pump flow to tank.

The position of the DCV spool establishes a specific flow rate, and load pressure is sensed at the bias spring chamber. Load pressure and the bias

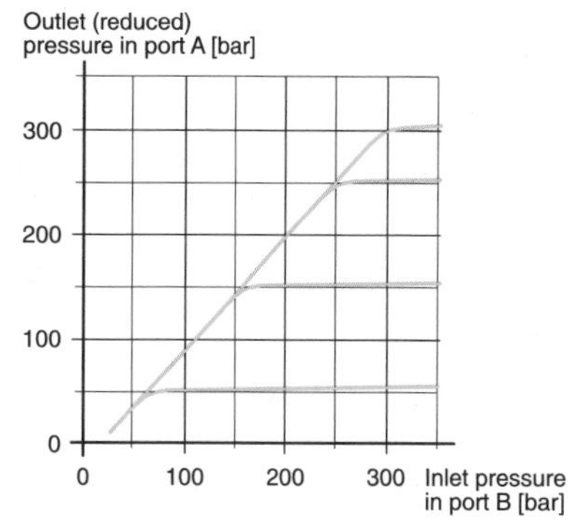

Fig. 6-18 Outlet versus inlet pressure diagram of the pressure reducing valve in fig.17

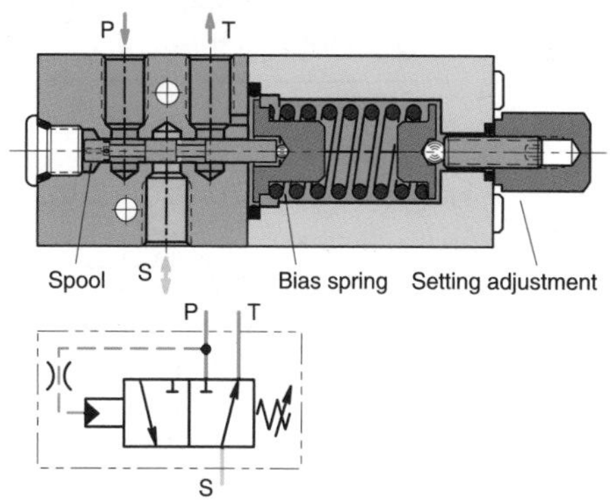

Fig. 6-19 Sequence valve cutaway (with symbol)

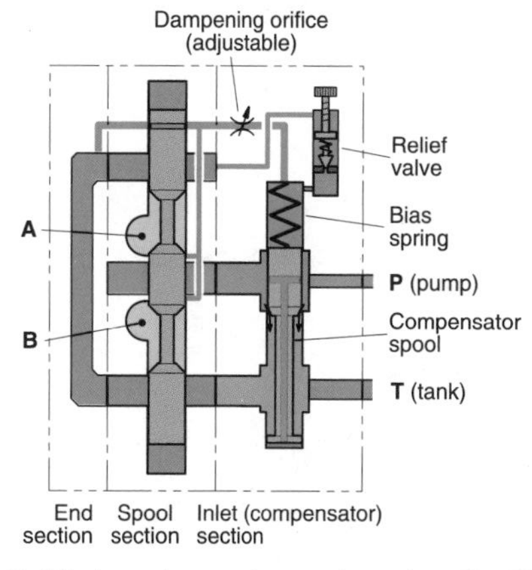

Fig. 6-20 Load sensing valve circuit with inlet pressure compensator (spool in neutral)

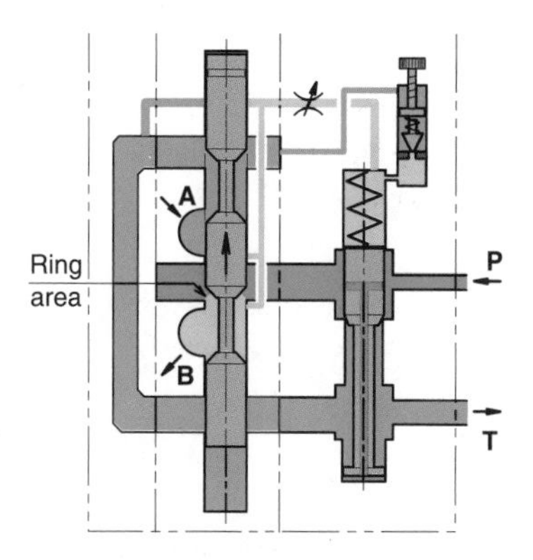

Fig. 6-21 Inlet compensator valve actuated (valve spool shifted 'up')

Formula 6-3. Pressure drop over an orifice.

$$\Delta p = C \times \frac{Q^2}{A^2}$$

where: 'Δp' is pressure drop over the orifice (= the ring area of the spool; fig. 6-21).

'C' a constant (which depends on the shape of the orifice and the specific gravity of the fluid)

'Q' flow through the orifice

'A' orifice area.

When the pressure drop over the orifice is kept constant, flow is directly proportional to the orifice area, or Q (flow) is a function of A (area).

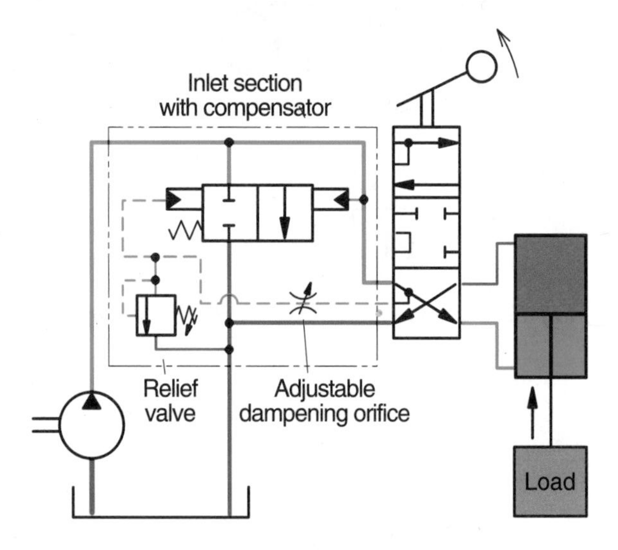

Fig. 6-22 Inlet compensator schematic

spring force push the compensator spool down, while, at the same time, pump pressure is sensed at the bottom of the spool. Flow will not pass from port 'P' out to the actuator port until pump pressure exceeds load plus bias spring pressure.

During valve operation, the compensator spool is continuously modulating, allowing only enough flow to the function, equivalent to the degree of spool shift at a pressure drop equal to the bias spring pressure. Because the compensator constantly compares load pressure with pump pressure (thereby maintaining a constant pressure drop across the four-way spool), the output flow from the four-way spool is constant, for a certain position of it, regardless of changes in load pressure; refer to Formula 6-3.

NOTE: If the output flow requirement tends to equal pump flow, the compensator blocks 'P' to 'T', and output flow cannot be compensated.

Control of maximum load pressure

To limit the load pressure, the inlet compensator valve incorporates a relief valve, figs. 6-21 and 6-22, which senses load pressure via the bias spring chamber. When load pressure exceeds the relief valve setting, the relief valve opens, limiting maximum pressure in the bias spring chamber to the relief valve setting. Therefore, any load pressure above relief valve setting will cause the compensator spool to dump pump flow to tank at a pressure drop equivalent to the relief valve setting plus bias spring rate.

The adjustable dampening orifice of the compensator is used to slow down the response of the compensator. By slowing down the response, the compensator will not "hunt" in trying to follow rapidly changing load pressures.

Pressure reducing compensator

A pressure reducing compensator, fig. 6-23, is used in the directional control valve (DCV) in conjunction with pressure compensated pumps, load sensing pumps, accumulator circuits, or when several DCV's are connected in parallel in a circuit.

In some valve designs, the compensator is used as an inlet compensator and in others as an individual valve section compensator. The compensator consists of a spool, bias spring, adjustable dampening orifice and, sometimes, a relief valve.

The major difference between the pressure reducing (normally open) compensator and the previously described unloading (normally closed) valve is that the compensator is held closed in its at-rest position.

The pressure reducing compensator performs three major functions:

- It blocks unused pump flow at the inlet, allowing pressure compensated or load sensing pumps to destroke.
- It provides a constant ΔP over the 4-way spool.
- It limits max load pressure.

When the DCV's spool is in the neutral position, the compensator blocks all pump flow at the inlet. This permits the pump to destroke or allow pump flow to be used elsewhere in the system.

The bias spring cavity is drained to tank as seen in fig. 6-23. Any pump pressure above the bias spring rate, typically 6 bar *(85 psi)*, causes the compensator spool to shift 'up', blocking any pump flow.

When the DCV's spool is shifted away from neutral (fig. 6-24), the pressure reducing compensator maintains a constant dp over the 4-way spool; any unused pump flow is blocked at the inlet. The load pressure, sensed at the at the bias spring cavity, and the bias spring rate act against the left side of the compensator spool while pump pressure acts on the opposite side.

During valve operation, the compensator modulates, allowing only enough flow to the function that corresponds to the degree of spool shift at a pressure drop equal to the bias spring.

Because the compensator constantly monitors the ΔP over the 4-way spool and modulates to maintain this pressure drop, the output flow to the function will be constant regardless of changes in load pressure.

The adjustable orifice is utilized to dampen compensator oscillations in case the load pressure fluctuates severely. The relief valve senses load pressure in the bias spring cavity. Any pump pressure above the relief setting causes the compensator spool to close, maximizing load pressure.

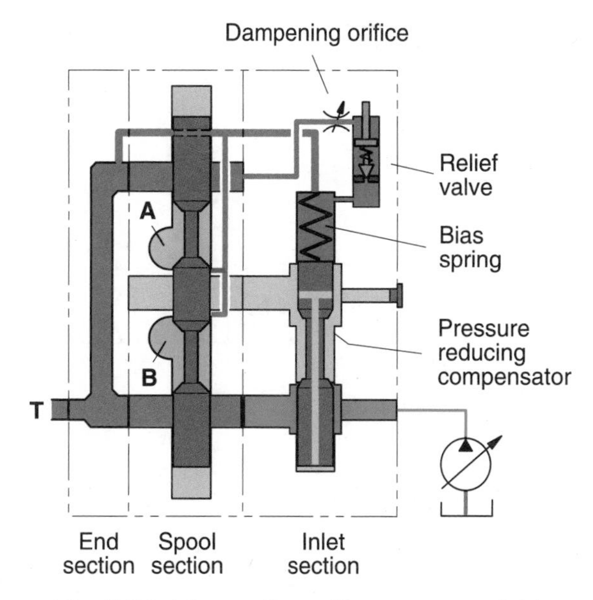

Fig. 6-23 Inlet section with pressure reducing compensator

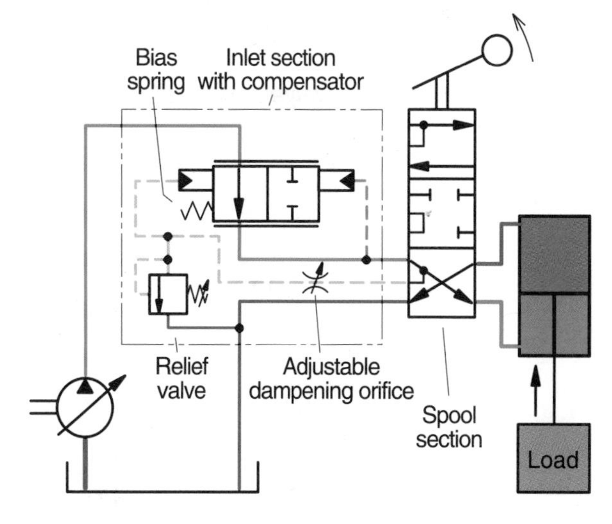

Fig. 6-24 Pressure compensator circuit (control valve spool shifted 'up')

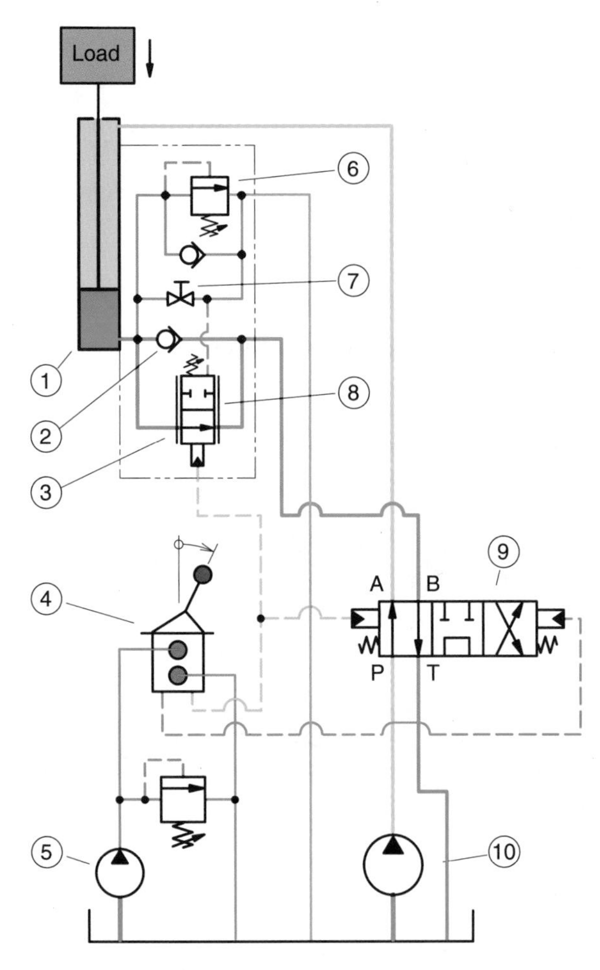

1. Hydraulic cylinder
2. Lift function check valve
3. Lowering function valve
4. Remote control valve
5. Pilot pump
6. Shock valve
7. Manual override
8. Lowering valve
9. Main DCV
0. Main pump

Fig. 6-25 Hose break valve circuit (lowering mode)

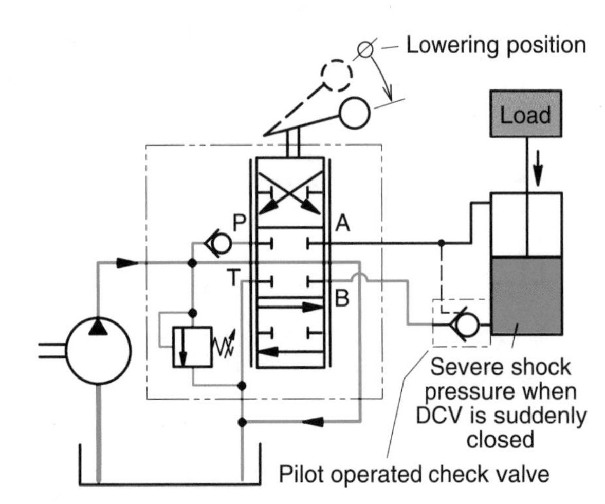

Fig. 6-26 Pilot operated check valve installed on a cylinder as a load holding valve

Load control valves

The following valves can be utilized to prevent e.g. a suspended load from dropping uncontrollably:

1) pilot operated check valve
2) counterbalance valve

A load control valve may be required to:

- lock up a load in case of a hose or pipe breakage
- prevent load lowering caused by leakage in the DCV
- provide smooth control in the lowering mode
- prevent actuator shock pressure in the event the DCV is suddenly closed
- require a minimum of power when lifting or lowering a light load

The counterbalance usually satisfies all the above requirements while the pilot operated check valve meets only the first two.

NOTE: As an alternative to the above valves, a so called *hose break valve* (fig. 6-25), flange mounted onto the cylinder, can be utilized. In case of a breakage in the line connected to port B, the load lowering movement can be stopped smoothly simply by returning the remote control handle to neutral; no shock pressure will be experienced.

In addition to the lowering spool valve, the hose break valve contains a manual override and a valve preventing shock pressures in case of a pilot signal line failure.

As shown in fig. 6-25, the hose break valve is used in connection with a pilot controlled directional control valve.

Pilot operated check valve

The simple, pilot operated check valve installs e.g. on a cylinder as shown in fig. 6-26. In the lowering mode, the valve is kept open by the pressurized pilot line. As soon as there is no pressure in port A, the check valve effectively locks the load in position (but cannot prevent movement caused by cylinder seal leakage).

Because the valve is either closed or open, a severe shock can develop in the cylinder when the DCV is abruptly returned to neutral from the lowering position; this can only be prevented by slowly moving the DCV handle.

High shock pressures can also develop when there is a rupture on the line between the check valve and port B. The pilot operated check valve may cause pronounced system pressure fluctuations (detected by high noise) and is therefore not normally installed in this type of application.

Counterbalance valve

In mobile hydraulics, the counterbalance (or overcenter) valve is mainly used as a safety device such as a hose burst or load holding valve on a boom cylinder.

The main purpose of the counterbalance valve is to allow cavitation free load lowering, preventing the actuator from running ahead when pulled by the load. It also acts as a pressure relief valve in one flow direction and a check valve for free flow in the opposite direction. In addition, it works as a hose rupture valve when installed directly on the hydraulic cylinder.

In mobile applications, the DCV spool controls the speed of the function. In some countries, however, an added counterbalance valve is mandatory equipment.

The counterbalance valve in a circuit

In the hydraulic circuit, fig. 6-27, the DCV directs flow to the rod end of the cylinder to lower a pushing load. The weight of the load forces fluid out of the piston end of the cylinder causing the load to drop uncontrollably (if the speed is not limited by the main DCV spool). Pump flow may not be able to keep up with the piston movement, causing cavitation in the rod end of the cylinder.

To avoid this situation, a counterbalance valve is installed in the line leading from the piston end of the cylinder to port B. A pilot line from the rod end now monitors the counterbalance valve.

A pilot pressure, typically 10 to 30 % of the pressure in the piston end, is required to open the valve. If the piston tends to increase its speed, pressure on the rod side will drop and the valve starts to restrict the flow.

A hydraulic winch application is shown in fig. 6-28. The counterbalance valve prevents the hanging load from running away as a certain pilot pressure is required to keep the valve open.

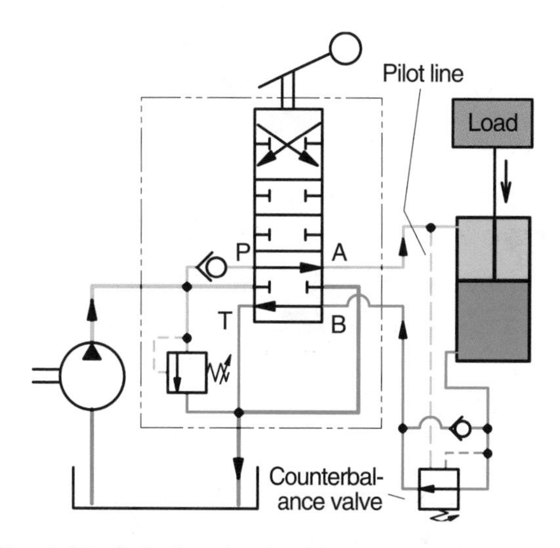

Fig. 6-27 Cylinder circuit with counterbalance valve

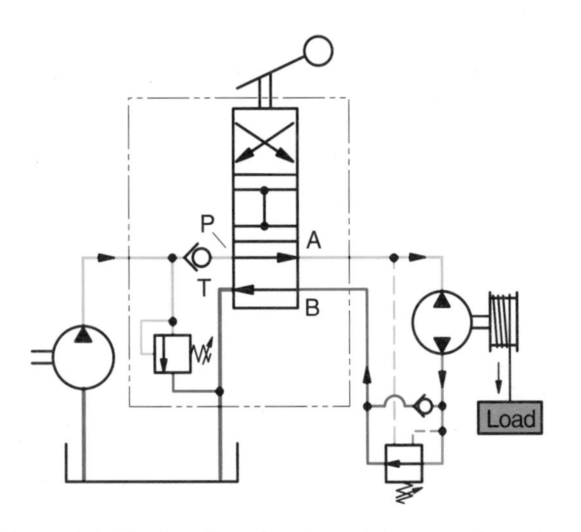

Fig. 6-28 Bi-directional motor with a counterbalance valve acting when lowering the load

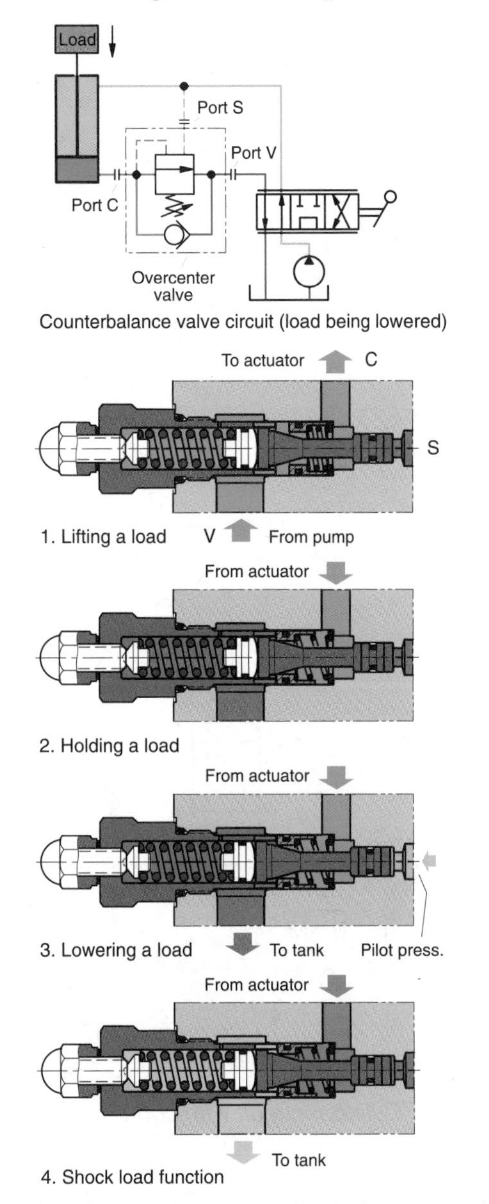

Fig. 6-29 Counterbalance valve in four operating modes

The pilot controlled, counterbalance valve has the following features:

- It locks the load (without leakage) when the DCV is inactive.
- The load movement is stopped also in case of a hose or pipe failure.
- It allows cavitation free load lowering, controlled by the pilot action over the lowering flow and by spool fed pressure, preventing the actuator from running ahead, when pulled by the load.
- It normally includes a check valve for free flow in the reverse direction.
- It includes a pressure relief function to cope with pressure shocks caused by external forces or overrunning loads (in case the DCV's spool has an open center).

External pressure relief valves are required in connection with a closed-center DCV. If not compensated for, a back pressure will develop between the main spool and the overcenter valve.

Valve function

Fig. 6-29 shows a cutaway of a common valve design (the corresponding symbol appears on top). The valve has the following operating modes:

1. load lifting
2. load holding
3. load lowering
4. shock load function (in the load lowering mode)

The pressure relief function of the counterbalance valve must be able stop any lowering (or return) flow at maximum allowed actuator load (corresponding to a maximum pressure, p_{max}). The relief valve is commonly set at least 30% above p_{max} (Formula 6-4).

When load pressure is applied to the actuator port, the valve stays closed until the pilot port is sufficiently pressurized (holding a load). Pilot and load pressures act together to open the valve. When this combined pressure reaches the valve setting, the load starts to move.

If the load attempts to pull away from the supply (pump) flow, pressure in the pilot line decreases. The relief valve will then throttle (reduce) the flow and catch up with the load. In the opposite direction the flow passes freely through the check valve as shown (lifting a load).

The pilot pressure, required to start the movement of the load, can be calculated according to Formula 6-5. Note that the relief valve characteristic is always increasing with flow, which, in turn, increases the margin of stability. This means that the pilot pressure has to be set higher for larger flows.

Valves with various pilot ratios are available on the market. In the following are shown some general rules on how the pilot ratio affects the counterbalance valve function in a system.

- A *high* pilot ratio (typically 6 to 8) allows a lower pilot pressure and faster operation of the actuator. It is also less energy consuming and is best suited in applications where the load pressure is relatively constant.
- A *low* pilot ratio requires a higher pilot pressure and consumes more energy. It provides, however, a more precise and smooth control of the lowering function. A low pilot pressure is used in applications where the load pressure varies which can cause instability.

NOTE: Attention must be paid to stability aspects when overcenter valves are installed in a load-sensing system. This can require analysis of the system and adjustment of system components.

Formula 6-4 Pressure relief setting versus max load pressure

$p_s \bullet 1.3 \times p_{max}$ where:

'p_s' is the relief valve setting of the counterbalance valve

'p_{max}' max allowed load press

Formula 6-5 Required pilot pressure versus load pressure

$$p = \frac{p_s - p_C}{R}$$

where:

'p' is required pilot pressure

'p_s' relief valve setting

'p_C' pressure in port C (fig. 6-29)

R pilot ratio of the counterbalance valve (usually 3 to 8)

Other pressure control valves

The *counterpressure/backpressure* valve is used as a spring biased and/or pilot operated check valve when installed in the tank line of a circuit; refer to figs. 6-8 and 6-9). It can also be integrated in a directional control valve. This setup increases the counterpressure/backpressure in the return line, which improves makeup capabilities, thus reducing the risk of activator cavitation.

Accumulator relief valves

There are some specific pressure control valves used in mobile hydraulic systems. One is the accumulator relief valve (fig. 6-30) which is utilized in the steering and braking circuits of e.g. a front end loader. Its purpose is to regulate hydraulic fluid flow into and out of the accumulator, preventing reverse flow from the accumulator to the pump and relieve pressure surges in the steering system.

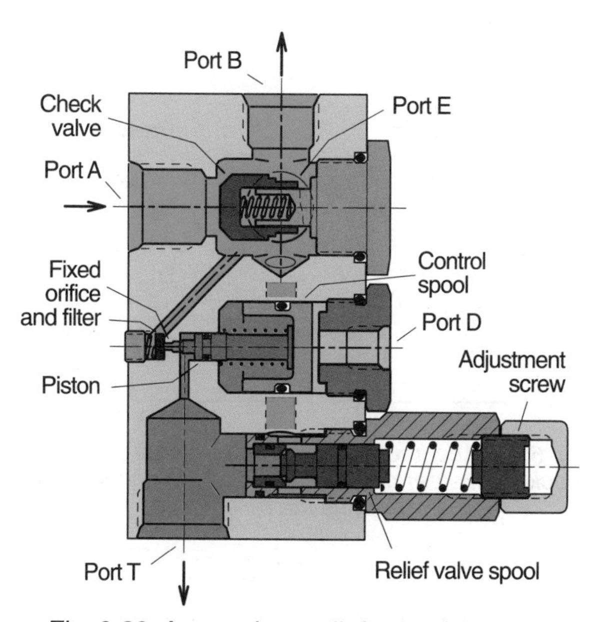

Fig. 6-30 Accumulator relief valve (shown with open, main check valve)

The valve contains a check valve, relief valve spool, adjustment screw, control spool, fixed orifice with filter, and a piston.

Valve function

According to OSHA (Occupational Safety and Health Administration), heavy equipment must have a means of steering in the event of an engine or steering pump failure. Because of this requirement, machine manufacturers often use an accumulator in the system to provide the necessary fluid flow and pressure.

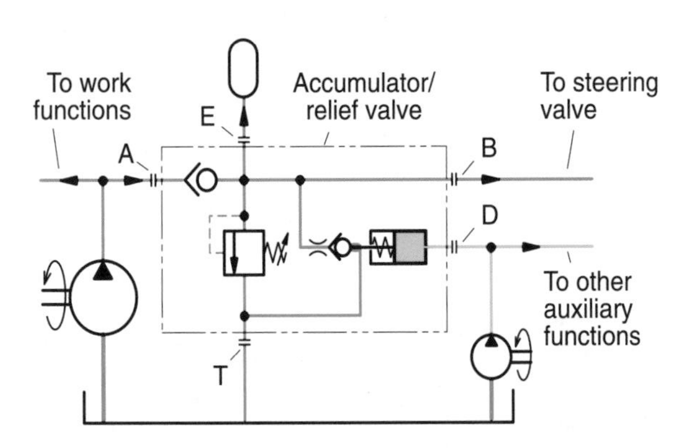

Fig. 6-31 Accumulator relief valve circuit (pump flow to steering valve and accumulator)

The schematic in fig. 6-31 illustrates how the accumulator relief valve is used to provide a method of isolating the accumulator in the system. At engine start-up, the hydraulic fluid from the pump enters the valve at port 'A', unseating the check valve.

The flow continues through port 'B' to the steering valve and to the accumulator through port 'E'. Fluid flow also travels through internal passages of the valve to the relief valve spool area where it deadheads.

The fixed orifice in the valve is blocked by pilot pressure entering the control spool area through port 'D', moving the control spool and piston to the left.

NOTE: The pilot pressure must be supplied from a separate source of fluid pressure, such as the transmission, steering pump or main hydraulic pump.

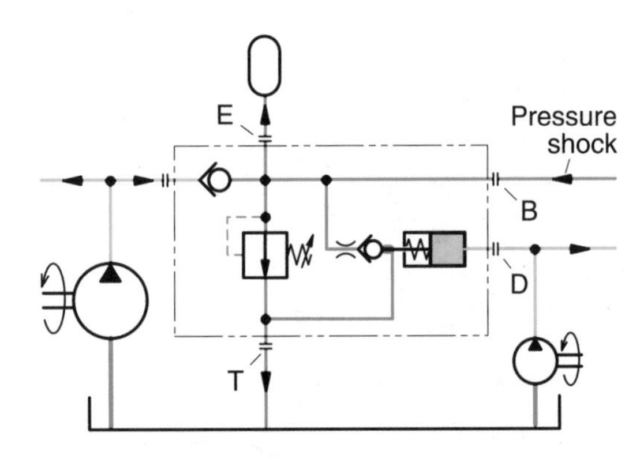

Fig. 6-32 Accumulator relief valve circuit (pressure shock from steering valve charges the accumulator and/or opens the relief valve)

If, during operation of the steering system, a shock pressure should develop in port 'B' (fig. 6-32), the check valve, shown in fig. 6-30, will be forced to the left, blocking off port 'A'. This pressure can often be absorbed by the accumulator through port 'E'.

When the accumulator is charged to its maximum capacity the surge pressure is felt at the relief valve spool which is set to open, in most cases, at a pressure 25 to 35 bar *(400 to 500 psi)* above the maximum system pressure. If the surge pressure exceeds the setting of the relief valve, it will open and allow excess pressure to exhaust back to tank.

Whenever the machine's engine is shut down, the pump flow at port 'A' stops, and the check valve closes, trapping accumulator pressure in the steering system. As a safety feature, to prevent accidental steering of the machine when the engine is shut down, the stored pressure in the accumulator is automatically relieved back to tank (fig. 6-33).

Pilot pressure in port 'D', that held the control spool and piston to the left, is allowed to dissipate whenever the engine is shut down. The bias spring plus pressure from the accumulator forces the control spool to the right. This opens the fixed orifice area and allows the accumulator to slowly bleed through the orifice and back to tank. The filter in front of the fixed orifice protects the orifice from becoming plugged by contamination.

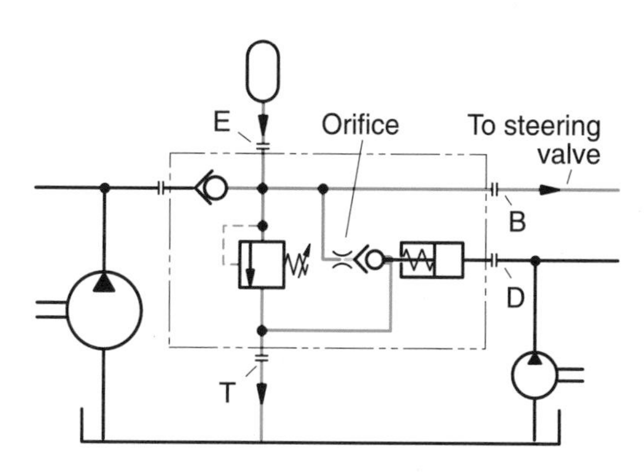

Fig. 6-33 Accumulator relief valve circuit (the accumulator flow feeds the steering valve and gradually bleeds off to tank)

Reducing energy loss during machine idling

When the machine is idling and no useful work is being done, it is an unnecessary waste of power to circulate a high pump flow through the hydraulic system. Lowering the pump flow can be done e.g. by:

- reducing the engine rpm
- venting system flow to tank by unloading an installed, pilot operated, main relief valve

Unloading a pilot operated relief valve

Unloading or 'venting' a relief valve refers to releasing the fluid pressure which is biasing the main poppet of a pilot operated relief valve as seen in fig. 6-34.

By releasing the pilot pressure, the only pressure holding the poppet closed is the light pressure of the bias spring. This results in the pump having to develop a relatively low pressure to open the poppet.

NOTE: The solenoid can be activated e.g. by pulling an 'emergency switch' on the instrument panel of the machine. When the operator hits (pushes) the switch, pump flow is diverted to tank and any engaged function will immediately stop moving.

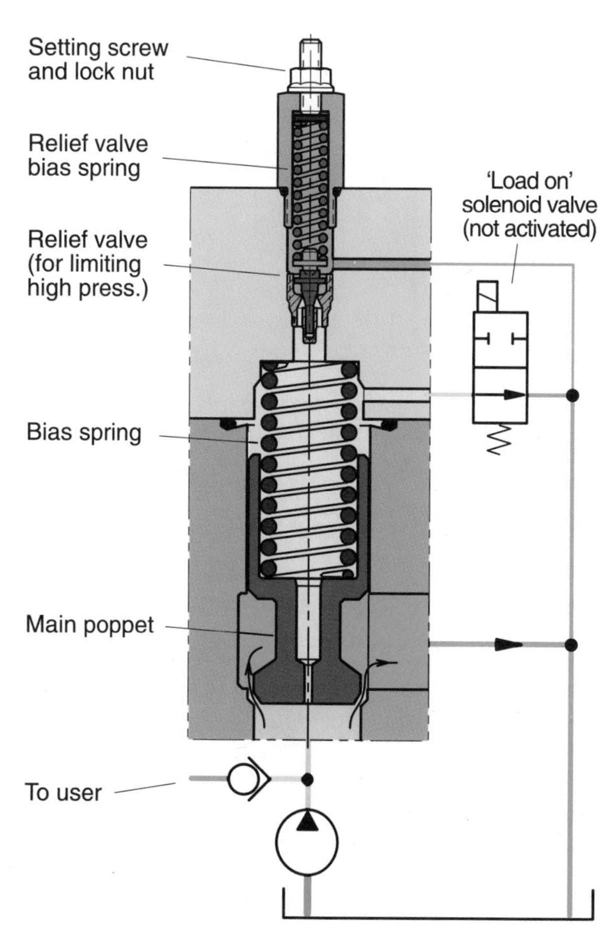

Fig. 6-34 By venting the pilot operated relief valve, the pump is unloaded

Unloading a fixed displacement pump in an accumulator circuit

In an accumulator circuit, when the accumulator is fully charged, pressure from the pump will rise until it overcomes the setting of the pilot operated relief valve. The circuit maintains system pressure at a constant value but generates a lot of heat, which may damage the pump and other components in the circuit. When an accumulator is used to develop system flow, it discharges its flow between a maximum and minimum pressure.

With an ordinary pilot operated relief in the circuit, the valve would bring in the pump as soon as the accumulator pressure drops below the relief valve setting; this may be an undesirable condition.

To unload the pump flow back to tank at a very low pressure and keep it unloaded until it is required to recharge the accumulator, an electrical pressure switch and a solenoid operated, two-way poppet valve can be used (fig. 6-35).

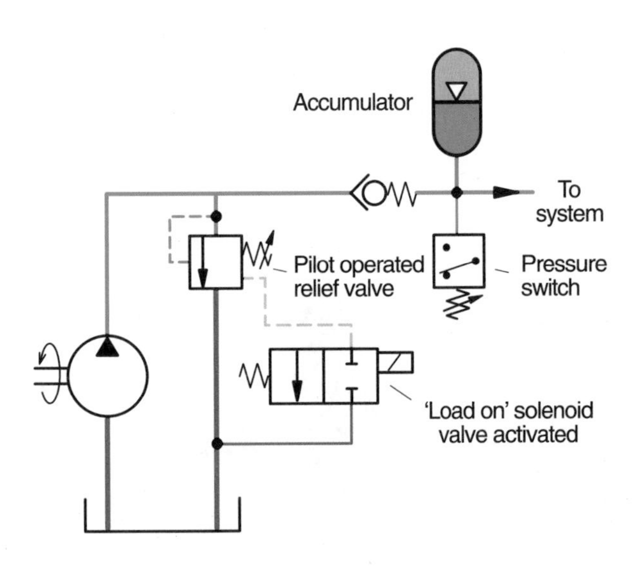

Fig. 6-35 Unloading a fixed displacement pump (typical solenoid method)

The pressure switch sends an electric signal to the solenoid operated two-way poppet valve, which is connected to the vent port of the pilot operated relief valve. When the accumulator is charged to the required maximum pressure, the pressure switch closes, sending a signal to the solenoid valve which vents the pilot operated relief valve (fig. 6-35).

When the accumulator requires recharging, at a lower pressure, the pressure switch opens, breaking the signal to the solenoid valve. This allows the return spring to return the poppet valve to the closed position, deventing the pilot operated relief valve.

The pilot operated relief valve closes off the return-to-tank flow path. The pump is allowed to recharge the accumulator and/or supply fluid to the system.

Summary

The pressure control valve is a very important component in the hydraulic system. It is utilized, in one form or the other, in every mobile hydraulic system. As the name implies, the main purpose of the valve is to limit pressure in a pump or actuator, thereby preventing an excessive pressure buildup that may interfere with the functioning of a component, or even damage or destroy it.

As a consequence, the pressure control valve may be the most important component in helping prevent personnel injury.

The most common pressure control valve is the *pressure relief* valve which comes in two versions: direct and pilot operated. The biggest advantage of the direct operated valve is that it opens very fast; this is the reason why it is utilized to limit chock pressures in an actuator.

The pilot operated valve maintains close to the set pressure irrespective of flow (up to max specified flow). Another advantage is that the pressure can be set from a distance by utilizing a small pilot valve.

The *pressure reducing* valve maintains a set pressure downstream of the valve. It is used in part of a hydraulic system that cannot use (or stand) the full system pressure (which may vary considerably).

The *sequence* valve is a small control valve used for various tasks, mainly in pilot control circuits.

Variants of the pilot operated pressure valves are the *inlet pressure compensator/bypass* valve and the *pressure reducing compensator*, built into a directional control valve. The first mentioned valve is utilized in a 'constant flow, closed circuit' system with a fixed displacement pump, and the last mentioned in a load sensing system with a pressure compensated pump.

The *counterbalance* (or *overcenter*) valve is used in connection with a cylinder or motor. It prevents e.g. a 'hanging' load from running away uncontrollably when being lowered.

An example of a specialized pressure control valve is the *accumulator relief* valve which is used e.g. in the steering circuit of an off-road vehicle such as a front end loader. In case of an engine or steering pump failure when the vehicle is travelling, the steering valve is connected to an accumulator that provides the pressurized fluid flow necessary for the steering until the vehicle is brought to a stop.

Chapter 6 exercise
Pressure control valves

1. **Situation:** The tilt cylinder must extend first and hold at a pressure of 35 bar (500 psi).

   Problem: Add the appropriate valves.

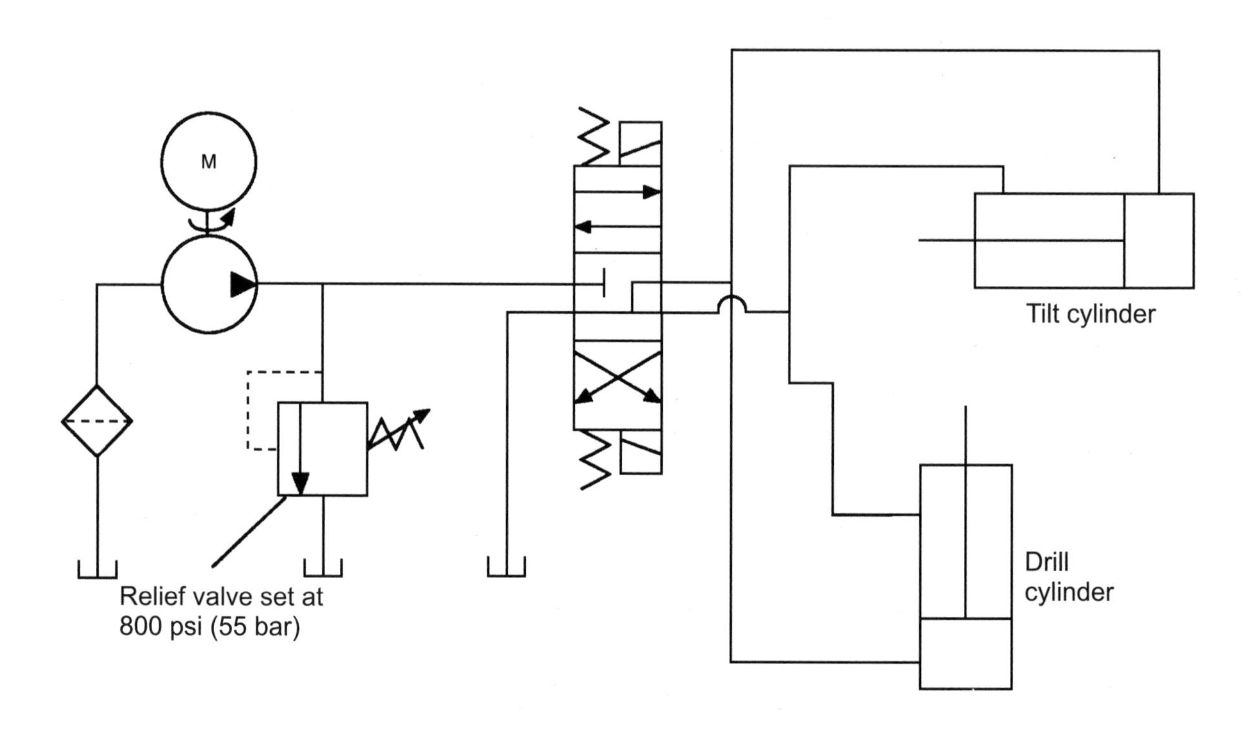

2. Cartridge type relief valves perform the functions of _____ and _____.
   a. relief, bleed-down
   b. relief, makeup
   c. relief, sequence
   d. relief, overcenter

3. Crossover reliefs are most commonly used with what hydraulic components? _____
   a. valves
   b. overcenter pumps
   c. double rod cylinders
   d. motors

4. _____ is one method of unloading a pump.
   a. Utilizing inlet compensators
   b. Adding a sequence valve to the circuit
   c. Incorporating pilot operated check valves
   d. Venting a relief valve

5. The main purpose of a _____ valve is to allow cavitation-free lowering, preventing the actuator from running ahead when pulled by the load?
   a. counterbalance
   b. sequence
   c. pressure reducing
   d. relief

6. Which of the following is not a function performed by the pressure reducing compensator? _____
   a. limits maximum load pressure
   b. unloads accumulator
   c. blocks unused pump flow at the inlet
   d. provides a constant ΔP over the 4-way spool

Chapter 7

Flow control valves

Speed control in a mobile hydraulic system is achieved through flow regulation. This can be done, with minimal energy loss, by changing the rotational speed of the fixed displacement pump or by adjusting the displacement of the variable displacement pump.

Flow control can also be obtained by using a flow control valve (throttle valve or constant flow valve) or by throttling the spool of a directional control valve; this cannot be done, however, without energy loss. Flow control valves (examples shown in fig. 7-1) will be examined in this chapter.

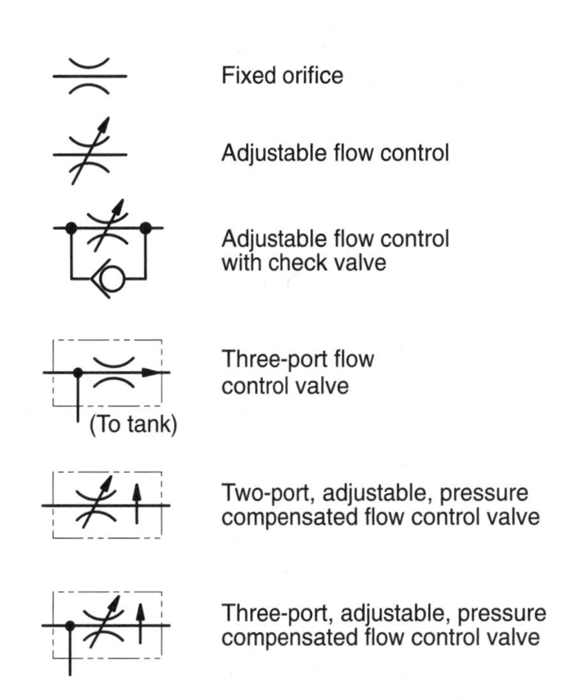

Fig. 7-1 Flow control symbols

The function of a flow control valve is to reduce the flow rate in part of a circuit. It performs its function by creating a restriction to the flow. If the flow after the flow control is insufficient, the restriction can be opened up (valid for adjustable flow controls) or system pressure upstream of the valve has to be increased.
When the flow control is supplied from a fixed displacement pump, some of the flow will have to take another path, usually through a relief valve, or be diverted to tank via a three-port flow control valve.

If a flow control valve, which restricts pump flow, is placed in an actuator line (fig. 7-2), the fixed displacement pump will attempt to push its total flow through the valve. If the restricting orifice is small enough, the pump pressure will reach the relief valve setting, and part of the flow will be diverted through this valve to tank (which, in the example will be 15 l/min *(4 gpm).*

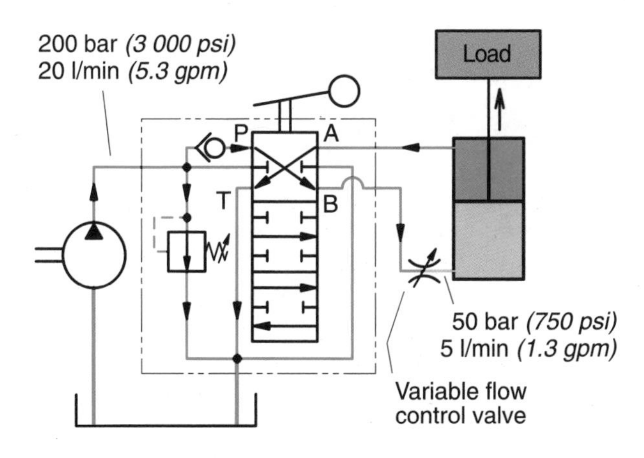

Fig. 7-2 Load speed restricted by flow control valve

The pressure loss across the relief valve and flow control valve will create a loss of energy, which, if lasting long enough, can lead to system overheating and malfunctioning.

In the above example, the energy loss in the flow control valve will be 1.3 kW *(1.7 hp),* and in the relief valve 5 kW *(6.7 hp).*

NOTE: In this chapter, pressure losses in directional valves and circuit lines are assumed to be negligible.

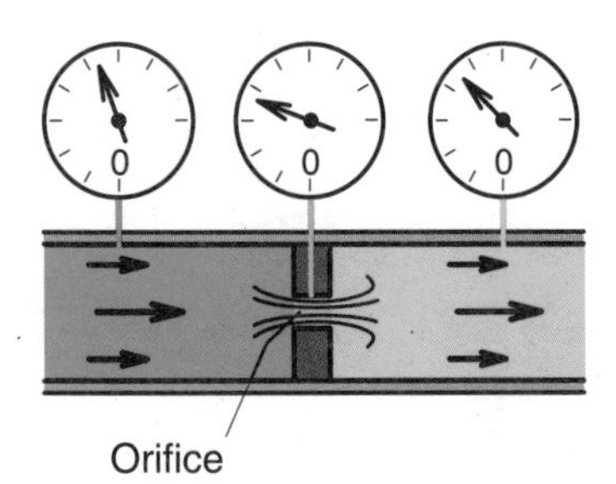

Fig. 7-3 Flow restricted by an orifice creates a pressure loss (Δp)

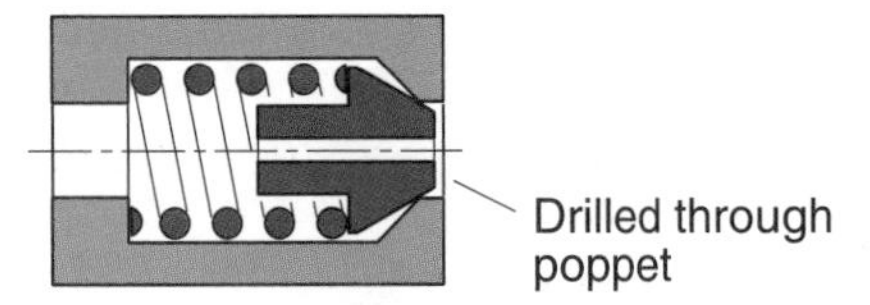

Fig. 7-4 Flow restricted by an orifice in a check valve poppet

Formula 7-1. Pressure drop across an orifice.

$\Delta p = C \times \frac{Q^2}{A^2}$ where: 'ΔP' is pressure drop
'Q' flow through the orifice
'A' orifice area

NOTE: The constant, 'C', varies with fluid viscosity and the shape of the orifice.

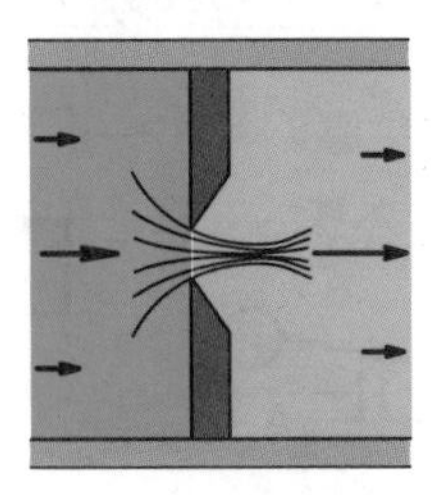

Fig. 7-5 Flow through a sharp-edged orifice

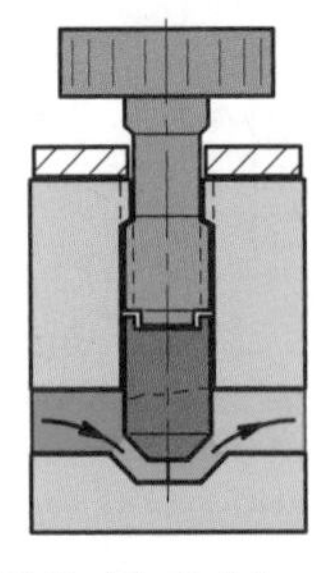

Fig. 7-6 Variable-orifice gate valve

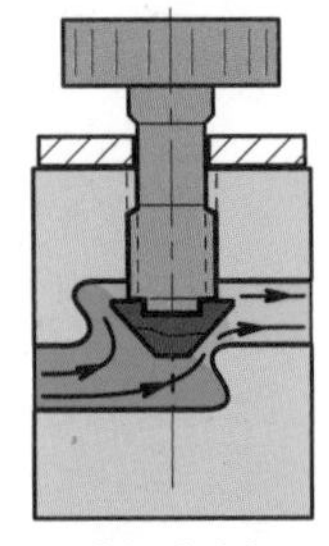

Fig. 7-7 Variable-orifice globe valve

Orifice

An orifice is a relatively small opening in a fluid's flow path (fig. 7-3). Flow through an orifice is affected by three factors (Formula 7-1):

1) orifice size (and shape)
2) pressure differential (ΔP) across the orifice
3) fluid viscosity

Orifice size affects flow

These three factors affect the flow rate through the orifice. An example is a garden hose which has sprung a leak. If the hole is small, the leak will be in the form of a drip, but if the hole is large, the leak will be substantial. In either case, the hole in the hose is an orifice, which meters a flow of water to the surrounding area.

Fixed orifice

A fixed orifice is a reduced opening of a nonadjustable size. Examples of fixed orifices used in hydraulics are drilled-through pipe plugs, throttle screws or check valves with a hole through the centre (fig. 7-4).

If the orifice is short and sharp, i.e. the length is smaller or is the same as the diameter (fig. 7-5), the flow through it is turbulent. This means, in practice, that it is not affected by changes in viscosity (due to temperature changes). In other cases the flow is laminar and strongly affected by the viscosity.

Variable orifice

Many times, a variable orifice is more desirable than a fixed orifice because of its degree of flexibility. Gate valves, globe valves and needle valves are examples of variable orifices.

Gate valve

A gate valve (fig. 7-6) has a flow path straight through the centre. Turning the handle, which positions a gate or wedge across the fluid path, changes the size of the orifice. Although gate valves are not designed to restrict flow, they are found in some systems where coarse metering is required.

Globe valve

A globe valve (fig. 7-7), does not have a straight-through flow path. Instead, the fluid must bend 90° and pass through an opening, which is the seat of a plug or globe. Changing the position of the globe changes the size of the opening.

Needle valve

The fluid going through a needle valve (fig. 7-8) must turn 90° and pass through an opening, which is the seat for a poppet with a cone-shaped tip. The position of the cone in relation to its seat determines the size of the opening. The orifice size can be changed very gradually because of the fine threads on the valve stem and the shape of the cone.

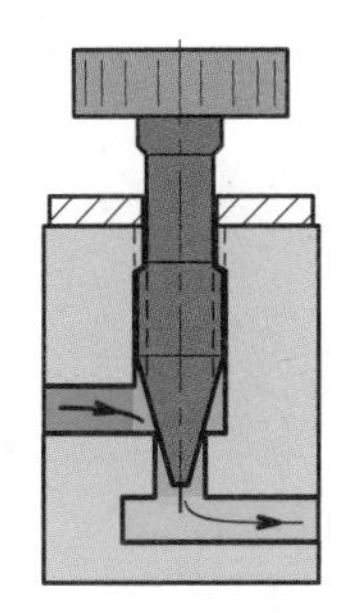

Fig. 7-8 Variable-orifice needle valve

Needle valves in a circuit

The circuit in fig. 7-9 consists of a 20 l/min *(5.3 gpm)* fixed displacement pump, relief valve, directional control valve, a flow control with variable orifice, and a cylinder with a piston area of 20 cm^2 *(3 in^2)*. With the relief valve set at 200 bar *(3000 psi)*, the pump attempts to push its flow through the orifice.

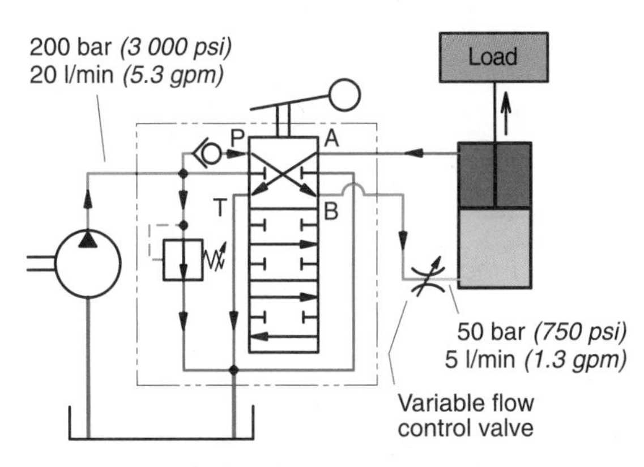

Fig. 7-9 Orifice restricting flow to a cylinder in a circuit

Because of the restricting orifice, only 5 l/min *(1.3 gpm)* passes through the orifice and out to the actuator; the pump pressure reaches the relief valve setting. The remaining fluid, 15 l/min *(4.0 gpm)* goes over the relief valve. The piston rod moves at the rate of 4 cm/s *(8 ft/min);* refer to Formula 7-2.

By turning the knob out, opening the needle valve orifice, more flow will pass through the valve and out to the cylinder before the relief valve setting is reached; at a flow of 15 l/min *(4 gpm),* the rod speed increases to 13 cm/s *(26 ft/min).*

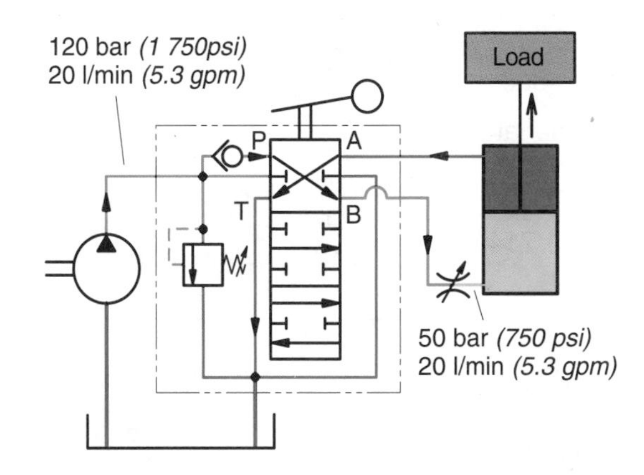

Fig. 7-10 Orifice size increased

By turning the knob in the opposite direction, thereby decreasing the size of the orifice, less flow will pass through the flow control valve before the relief setting is reached. Rod speed decreases since the cylinder receives less flow.

Formula 7-2 Cylinder rod speed is a function of flow and piston area

In metric terms:

$$V = \frac{Q}{A} \times 16.7$$

where: 'v' is rod speed in cm/s
'Q' flow in l/min
'A' piston area in cm^2.

In Imperial terms:

$$V = \frac{Q}{A} \times 19.25$$

where: 'v' is rod speed in *ft/min*
'Q' flow in *gpm*
'A' piston area in *in^2*

Pressure differential affects flow

When the size of the needle valve orifice in the above examples is changed, the pressure differential, ΔP, across the orifice (fig. 7-9) is still 150 bar *(2250 psi)*. Pressure ahead of the orifice is at the relief valve setting and pressure after the orifice is the workload pressure.

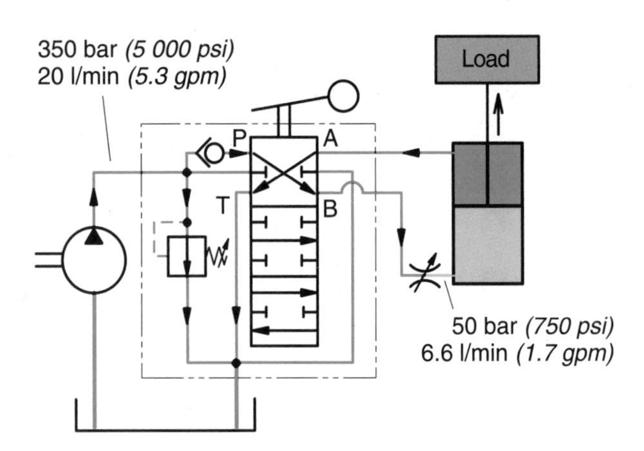

Fig. 7-11 Increased pressure increases flow through an orifice

Formula 7-3 Power loss over a relief valve or orifice

In metric terms:

$$P = \frac{Q \times \Delta P}{600}$$

where: 'P' is power [kW]
'Q' flow [l/min]
ΔP' pressure drop [bar]

In Imperial terms:

$$P = \frac{Q \times \Delta P}{1\,714}$$

where: 'P' is power [hp]
'Q' flow [gpm]
'ΔP' pressure drop [psi]

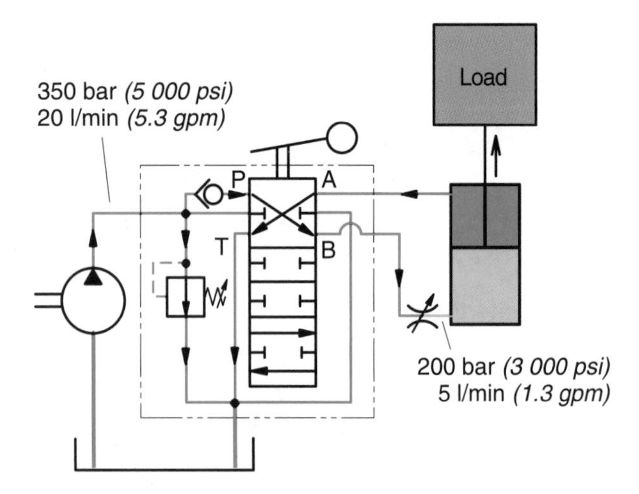

Fig. 7-12 Same circuit as in fig. 7-11 but with increased load pressure

If the flow control were to be opened sufficiently (fig. 7-10), ΔP would decrease, and pump pressure would decrease below the relief valve setting. The entire flow, 20 l/min *(5.3 gpm)* would then go to the cylinder. At a pump pressure of 120 bar *(1750 psi),* the ΔP over the orifice would be 70 bar *(1000 psi)* as load pressure is still 50 bar *(750 psi).*

The corresponding cylinder rod speed is 17 cm/s *(34 ft/min).* Flow through the orifice is a function of the ΔP over it. As previously stated, pressure in a hydraulic system equals potential energy. The greater the difference in pressure across an orifice, the more energy is available to drive flow through it.

Relief valve setting increased

In fig. 7-11, the relief valve setting has been increased to 350 bar *(5000 psi),* with the workload pressure and the setting of the needle valve remaining unchanged (compare with fig. 7-9).

This is now the pressure ahead of the needle valve and the ΔP over the orifice has increased to 300 bar *(4250 psi).* Therefore, the flow through it has increased to some 6.6 l/min *(1.7 gpm)* and the rod speed to 6 cm/s *(11 ft/min).*

It should be noted, that, in this example, 13.4 l/min *(3.6 gpm)* is being dumped over the pressure relief valve at a ΔP of 350 bar *(5000 psi)* which corresponds to an energy loss of 7.8 kW *(10.5 hp);* refer to formula 7-3.

At the same time, the useful power developed to lift the load is only 0.55 kW *(0.75 hp)* or 5% of the output power from the pump; 95% goes to waste. This can consequently be tolerated only temporarily. Otherwise, the hydraulic system will overheat.

Work load pressure increased

With the relief valve setting remaining at 350 bar *(5000 psi)* but workload pressure increased to 200 bar *(3000 psi)* because of a larger load (fig. 7-12), the flow through the orifice is again down to 5 l/min *(1.3 gpm)* and the rod speed to 4 cm/s *(8 ft/min);* compare with fig. 7-9.

The power loss over the relief valve is the same as before, 7.8 kW *(10.5 hp),* but the useful power has increased to 1.7 kW *(2.3 hp).* The reason for this is that the power loss over the orifice, 1.25 kW *(1.5 hp),* has decreased because of the lower ΔP, 150 bar *(2000 psi).*

Throttling check valve

When flow restriction is required in only one direction, the orifice is combined with a check valve (fig. 7-13). This type is in-line mounted and the throttling effect (upper illustration) can be changed by rotating the outer sleeve. The flow is 'free' (low ΔP) through the valve when the flow direction is reversed (lower illustration).

Pressure compensated flow control valves
As can be seen from the previous examples, any change in pressure upstream or downstream of a metering orifice affects the flow through the orifice, resulting in a change of actuator speed. These pressure changes must be minimized, or compensated for, before an orifice can meter fluid acceptably.

Nonadjustable constant flow valves

Needle valves are designated non-compensated flow control valves. They are good metering devices as long as the ΔP across the valve remains constant.

If more precise metering is desired, a pressure compensated flow control valve is required; it will compensate for pressure changes upstream or downstream of the orifice.

Pressure compensated, non-variable constant flow valves can be found in two main versions. Both supply a constant factory set flow in one direction, independent of pressure (when the input flow is higher than the regulated, downstream flow).

The simplest version is a straight, two-way restrictor valve (fig. 7-14), while the other, bypass valve (fig. 7-15), has a third port, through which the surplus oil is directed to tank. In both versions, the flow is free in the opposite direction (not shown).

Function

A spring-biased, precision-ground spool with throttling flow control slots and a sharp edged orifice (or orifices) moves inside the housing. The flow through the orifice results in a ΔP, which balances the spool against the spring force; the spool will always adjust itself to give

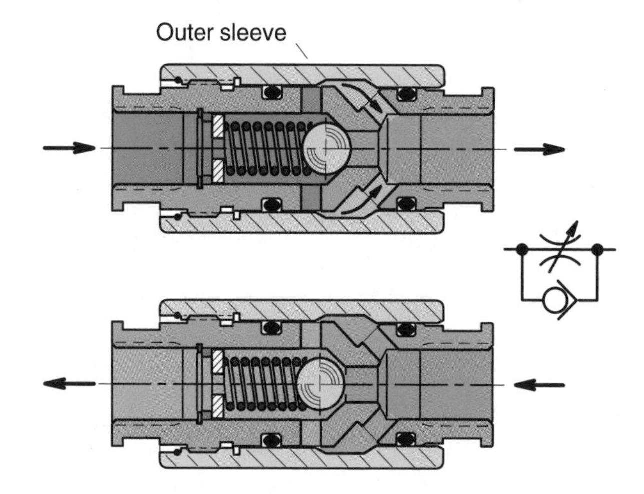

Fig. 7-13 Throttling check valve

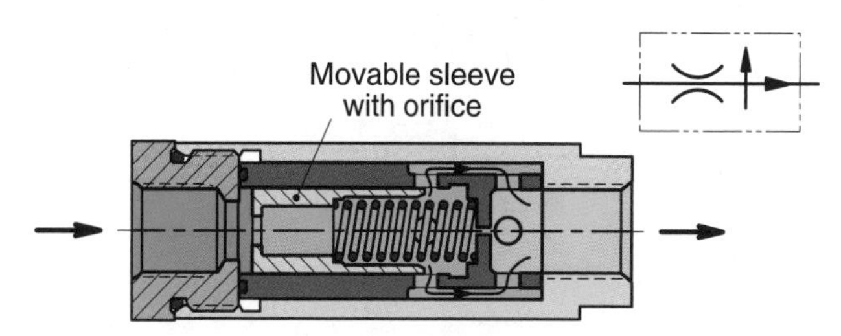

Fig. 7-14 Line mounted, two-way, pre-set, constant flow valve

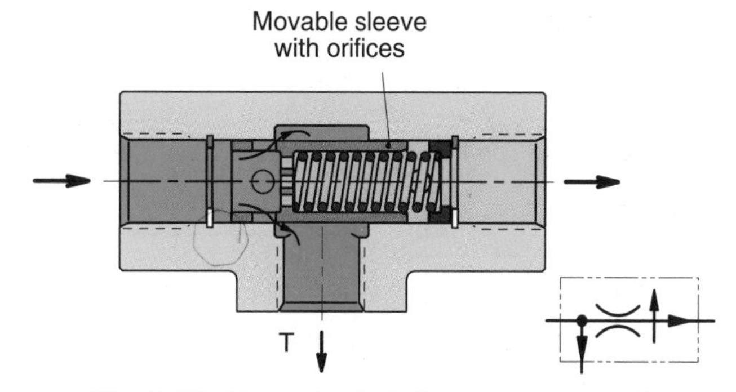

Fig. 7-15 Line mounted, three-way, pre-set, constant flow valve

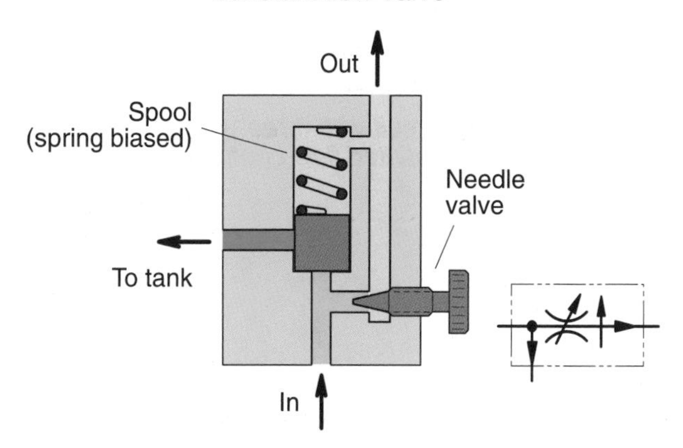

Fig. 7-16 Pressure compensated flow control valve

the set flow, independent of pressure variations in the fluid before and after the valve (within limits).

The spool in the three-way version will simultaneously open a bypass to tank for the surplus oil. The pressure ahead of the valve is always the same as the downstream load pressure plus the pressure created by the spring which is usually 7–15 bar *(100–200 psi).*

Adjustable pressure compensated flow control valves

The three-way, adjustable orifice, pressure compensated flow control valve, shown in fig. 7-16, consists of a valve body with inlet, outlet, secondary or tank port, a needle valve, a compensator spool, and a spool biasing spring.

Function

The compensator spool in the valve shown (fig. 7-17) develops a constant ΔP across the needle valve orifice by opening and closing the secondary passage to tank.

In the non-activated condition, the compensator spool is biased in the closed (non-passing) position by the bias spring. If the spring has a value of 7 bar *(100 psi)*, the ΔP between the 'in' and 'out' ports will be limited to that value.

When the valve is operating and there is no pressure in the outlet port, the pressure ahead of the needle valve attempts to rise to that of the valve setting. When the inlet port pressure reaches 7 bar *(100 psi),* the spool uncovers the secondary passage, thus limiting the inlet port pressure (ahead of the needle valve) to 7 bar *(100 psi).*

A constant pressure ahead of the needle valve orifice does not necessarily guarantee a constant flow rate. If the pressure in the outlet port (after the orifice) changes, the pressure differential across the orifice changes and, consequently, so does the flow.

To compensate for this situation, pressure after the needle valve orifice is added to the top of the piston (fig. 7-17) by means of a pilot passage. Two pressures now bias the spool: the spring pressure and the fluid pressure after the needle valve.

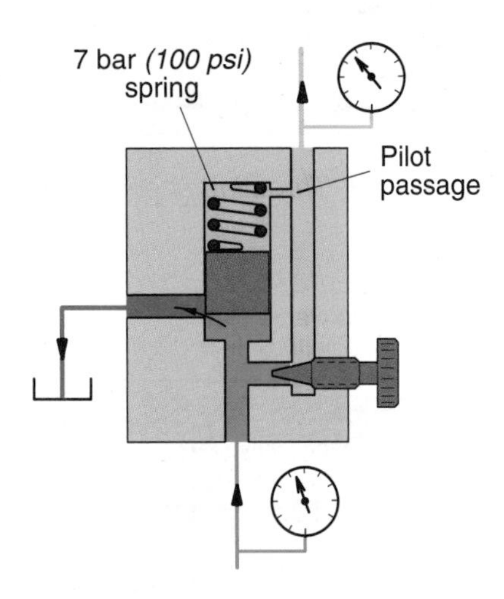

Fig. 7-17 Pressure differential in a flow control valve

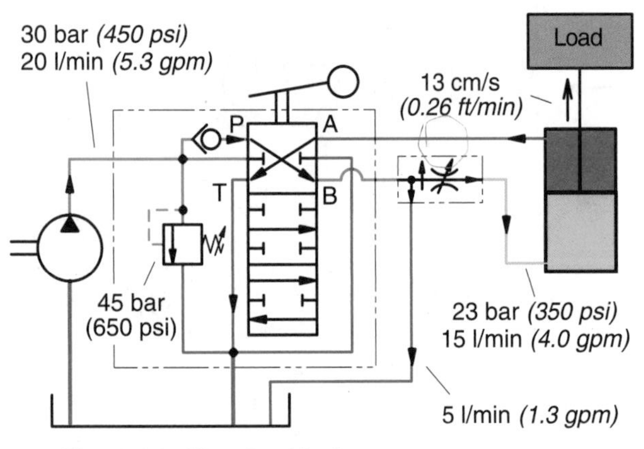

Fig. 7-18 Circuit with three-way, pressure compensated flow control valve

With a 7 bar *(100 psi)* spring, fluid pressure in the inlet port would always be limited to that value plus the pressure in the outlet port. As long as the system relief valve setting is high enough, the ΔP pressure over the needle valve will always be that of the spring. In this way, the same amount of pressure is available to develop a flow through the orifice, regardless of changes in pressure.

Three-way, pressure-compensated flow control valves in a circuit

In the illustrated circuit (fig. 7-18), the flow control valve is set for 15 l/min *(4 gpm)*. The relief valve setting is 45 bar *(650 psi)*. Workload pressure is 23 bar *(350 psi)*. The spring biasing the compensator spool has a value of 7 bar *(100 psi)*.

When the load is lifted (as shown) the workload pressure and the spring bias the compensator spool. The pump attempts to push its total flow of 20 l/min *(5.3 gpm)* through the needle valve orifice. When pressure ahead of the needle valve reaches 30 bar *(450 psi)*, the compensator spool uncovers the secondary passage.

Of this pressure, 23 bar *(350 psi)* is used to overcome the resistance of the load and 7 bar *(100 psi)* is used to develop a flow rate through the needle valve. The flow rate in this case is 15 l/min *(4.0 gpm)*.

The remaining 5 l/min *(1.3 gpm)* is bypassed back to the tank at a pressure which is 15 bar *(200 psi)* below the relief valve setting, thus reducing the generation of heat in the system.

Workload pressure and relief valve setting increased

If the workload pressure were increased to 30 bar *(450 psi)*, pressure ahead of the flow control orifice would still be limited to 37 bar *(550 psi)* due to the bias spring (fig. 7-19). Thereby a ΔP of 7 bar *(100 psi)* would still be available to develop the same flow rate.

Note that when a three-way, bypass-type compensated flow control valve is used in a circuit, excess flow may not be diverted over the relief valve. The operating, upstream pressure of the flow control is always the pressure from the bias spring plus the workload pressure.

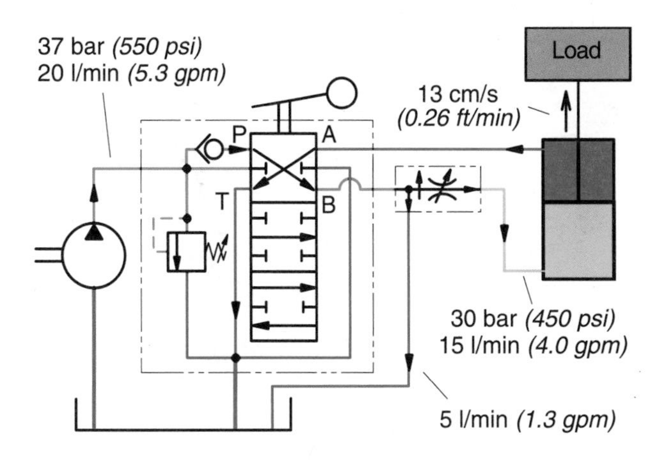

Fig. 7-19 Circuit with three-way pressure compensated flow control valve

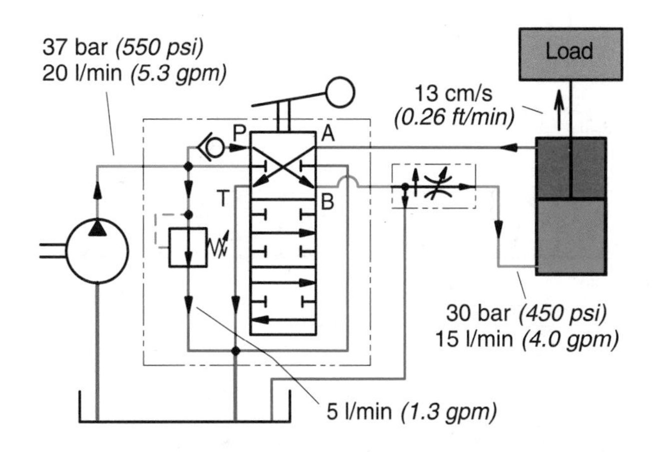

Fig. 7-20 Relief valve setting at load pressure; no flow control pressure compensation

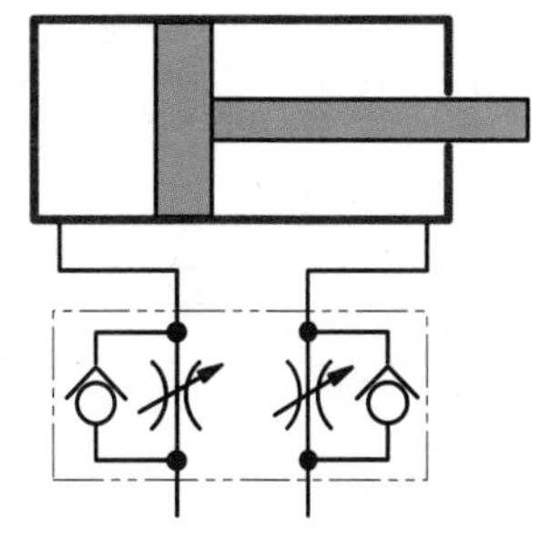

Fig. 7-21 Cylinder with meter-in flow controls

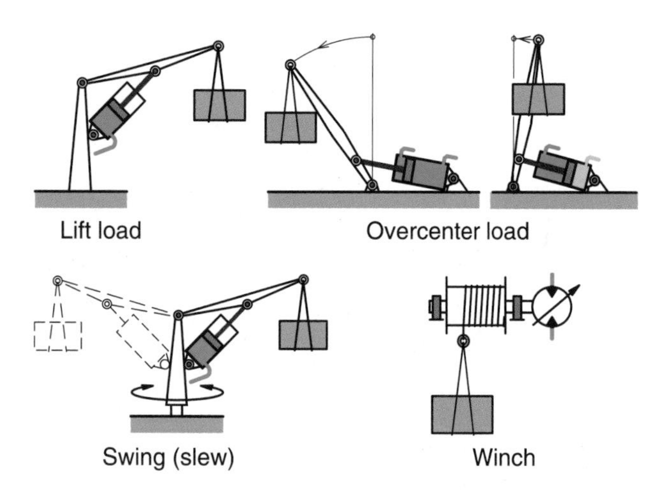

Fig. 7-22 Load examples

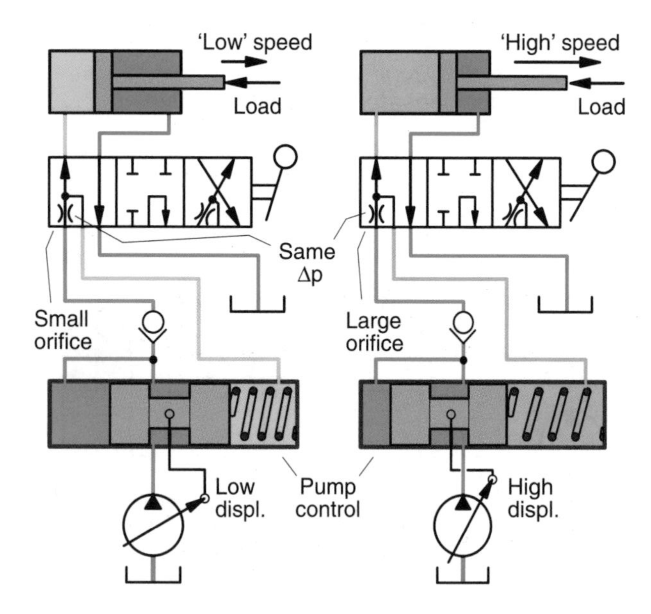

Fig. 7-23 Work function speed control in a load sensing system

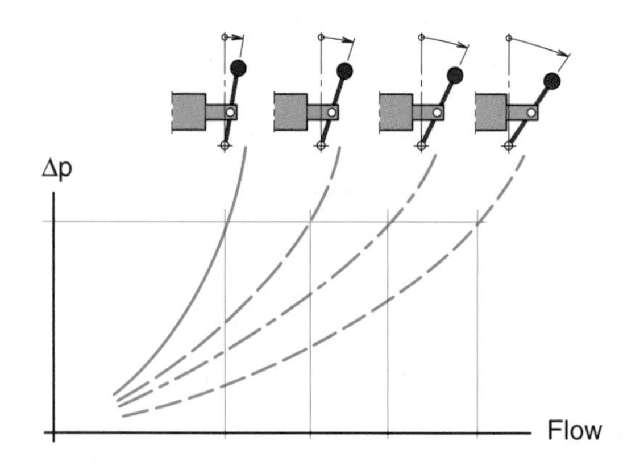

Fig. 7-24 Flow versus spool (handle) position

Since excess flow does not return to the tank over the relief valve, a higher relief valve setting has no effect on the operation of the flow control. However, if the relief valve were not set at least to the bias spring value plus the workload pressure, the compensator spool in the flow control would not operate. Consequently, flow through the flow control would not be pressure compensated, and excess flow would go over the relief instead (fig. 7-20).

Other flow control applications

A bypass flow control valve can only control flow to an actuator. This type of flow control valve is used as a priority flow divider in typical mobile systems such as in lift trucks.

A flow control valve used as a flow divider has a fixed orifice instead of an adjustable needle valve orifice. The priority circuit flow requirements are fulfilled first, then the secondary (tank) circuit receives the remaining flow.

Meter-in circuits

Until now, when the operation of a particular flow control valve was described in a circuit, the flow control was positioned in the circuit directly before the actuator whose speed it was controlling. In this arrangement, all the flow is metered (controlled) as it enters the actuator. This is termed a 'meter-in' circuit.

Fig. 7-21 shows a simple way of arranging meter-in control in both flow directions of a cylinder. This is achieved by using a twin throttle (flow)/check valve. The valve restricts the flow entering a cylinder port, and allows free flow, through the check valve, out of the cylinder. The meter-in circuit is used to control the speed of an actuator, which always works against a positive load. In other words, while the orifice is metering fluid to an actuator, the workload pressure has a continuous, positive value. An example is a load, which is being lifted, as in a the lift load illustration (fig. 7-22).

A meter-in circuit is the only kind in which a bypass type pressure compensated flow control valve, earlier described, can be used.

Fig. 7-23 shows the most common way to control the speed of a work function. In this figure, the variable restriction flow control is replaced by the main spool of a directional control valve.

The compensator spool is shown enlarged. By moving the hand lever, thereby varying the restriction, a determined output flow is produced for each lever position.

The diagram, fig. 7-24, illustrates flow as a function of the spool position. When the spool is pulled out further (to a new position), flow increases until the corresponding ΔP has been reached. The flow now remains the same, irrespective of work port pressure.

Meter-out circuit

In some applications, the work load changes direction and/or magnitude. Examples are when the load passes over the top center of a boom as shown in fig. 7-22 (overcenter load), or the workload pressure suddenly changes to zero (e.g. when a drill breaks through stock), or when the swing (slew) function of an excavator is driven by it's own load. In these cases, the load may run away if not controlled.

A flow control valve, placed at the outlet port of an actuator, controls the rate of flow exiting an extending actuator. This is a 'meter-out' circuit (fig. 7-25) and gives positive speed control to actuators used in drilling, sawing, slewing, boring and dumping operations.

A meter-out circuit is a very popular flow control circuit, especially in winch and swing motor applications (fig. 7-22). If both meter-in and meter-out control is required in an actuator port, two throttling/check valves (facing each other) are connected in series.

Lowering brake valve

The lowering brake valve is a special version of the meter-out control. It is often used when the lowering speed must be limited. A typical application is the lowering function of the main boom in a truck (lorry) crane; fig. 7-22, lift load.

This function is pressure compensated, providing a preset, maximum lowering speed which is independent of load pressure. In some valve designs that function is built into the main spool of the directional control valve (fig. 7-26). It consists of a drilled-through poppet which is spring biased; the required maximum flow is obtained by adding suitable shims behind the spring.

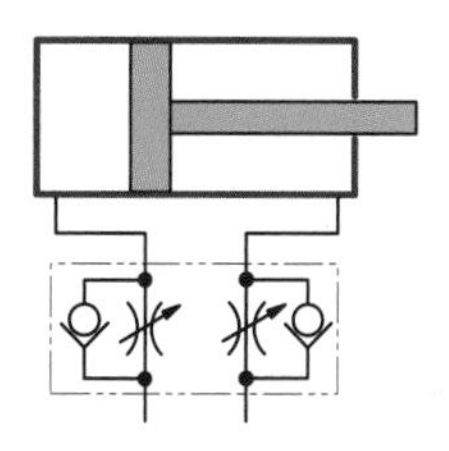

Fig. 7-25 Cylinder with meter-out flow controls

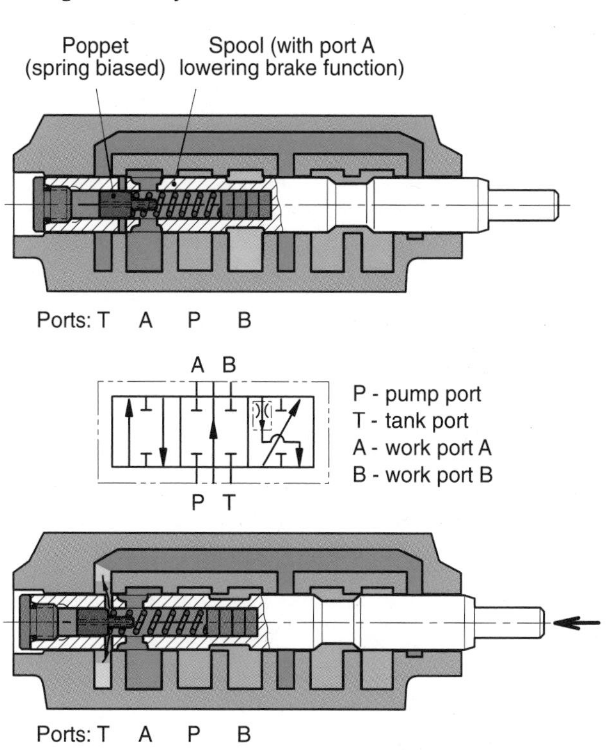

Fig. 7-26 Directional control valve with lowering brake valve built into the spool

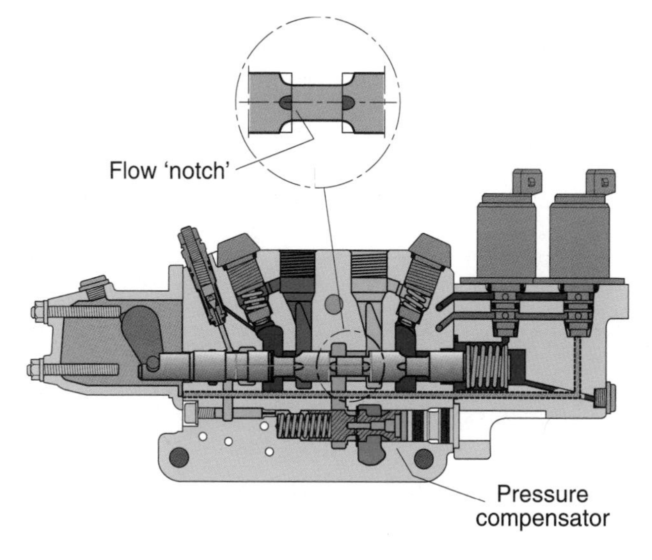

Fig. 7-27 Cutaway of a load sensing, directional control valve

When the spool is being pushed to the left (lower illustration), the poppet moves due to the spring force, and fluid starts to flow from actuator port 'A', through the drilled holes in the spool, to tank. At the same time, because of the restriction between the poppet and the holes, a reduced pressure is acting on the spring side of the poppet. Full port 'A' pressure is, however, communicated to the opposite side of the poppet through the drilling.

The three forces acting on the poppet will finally balance which in turn determines a particular flow through the lowering brake valve.

Flow control via flow notches on the spool of a directional control valve

Due to the needs to continuously vary the flow in a mobile hydraulic system, it is very convenient to utilize the directional control valve spool as a variable flow control.

This can be accomplished by stamping or machining control notches on the valve spool (fig. 7-27). By varying their design, a suitable characteristic for a particular application can be obtained.

The load sensing valve shown in fig. 7-27 is also equipped with a pressure compensator ahead of the main spool. The compensator (schematically shown in fig. 7-23) maintains a constant pressure difference and, therefore, makes the meter-in circuit independent of variations in load and pump pressures. The spool acts as a variable restrictor, providing a specific flow for the chosen spool position.

The main spool can also be designed to utilize flow forces, thereby achieving pressure compensation and, in turn, eliminating a separate compensator spool. This design is often used to compensate the meter-out side; for additional information, refer to chapter 8, Directional control valves.

Summary

Flow control valves and functions are extensively used in mobile hydraulics. The main purpose of a flow control is to provide a desirable speed of an actuator such as a hydraulic cylinder or motor. This is done by limiting the flow to (or from) the actuator by means of various fixed or adjustable, restricting orifices.

By limiting the flow, a pressure drop over the restricting orifice is created. This, in turn, means that the corresponding energy (flow times pressure drop) is converted to heat which is lost partly to the surrounding atmosphere, partly appearing as a temperature increase of the fluid.

If the fluid temperature is allowed to increase excessively, the viscosity of the fluid (refer to chapter 11, Fluids) as well as its life, will decrease drastically. This may cause the hydraulic system of a machine to malfunction due to component damage.

A solution to this problem is to use a variable displacement pump that provides fluid required only for the currently engaged functions. It is therefore most important, that fluid flow is restricted as little as possible and within a short period of time so as to limit the amount of heat being generated.

As has been shown, there is a variety of flow control valves, each adapted to a particular task in the hydraulic system. The simplest flow control consists of a plug with a drilled-through hole (orifice) which is very common in pump and motor displacement controls.

Two- and three-way, pressure compensated flow control valves maintain (with certain limitations) a set flow irrespective of any upstream or downstream variation in pressure.

'Meter-in' and 'meter-out' are important concepts, explained in connection with the control of flow to and from an actuator. The 'lowering brake valve' is an example of a meter-out function used in loading cranes.

In directional control valves, the metering (in both flow directions) is usually provided by cut-outs, called notches, in the valve spool. They are many times complemented by a pressure compensator which is located upstream of the spool.

In mobile hydraulics, the directional control valve, dealt with in chapter 8, is the most common valve type for controlling the flow to and from actuators such as hydraulic cylinders and motors.

Chapter 7 exercise
Flow control valves

Instructions: Answer the following questions or solve the following problems as required.

1. Assuming the cylinders to be the same size and the loads to be equal, how does the flow rate through the needle valve differ from situation A to situation B?
   a. flow remains exactly the same
   b. flow decreases
   c. flow increases

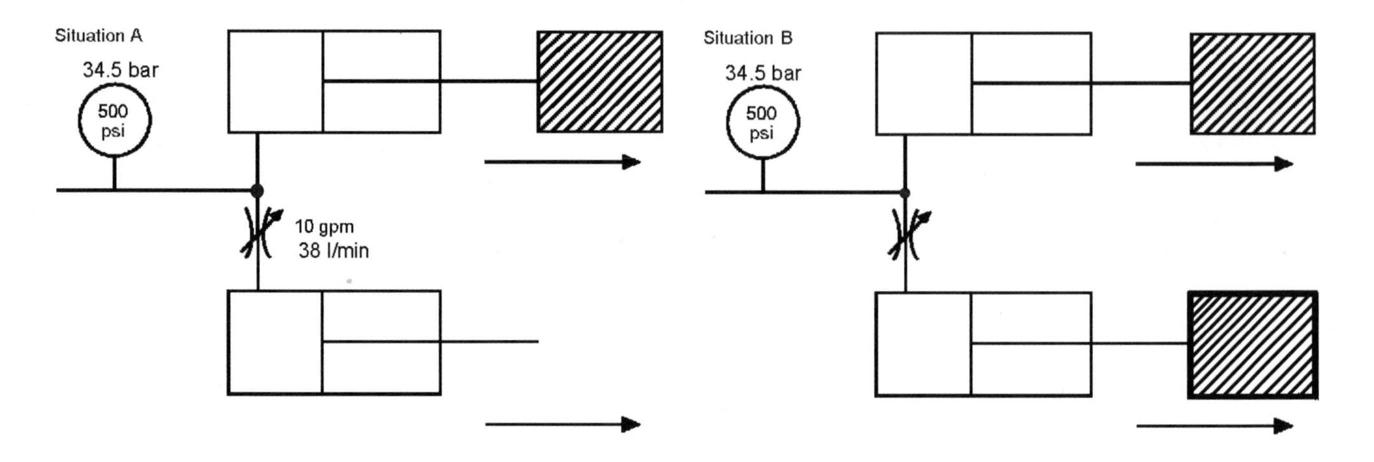

2. Assuming pipe friction to be zero, what do the gages read? (The spring acting on the spool in the pressure compensated flow control has a value of 35 bar *(100 psi).*

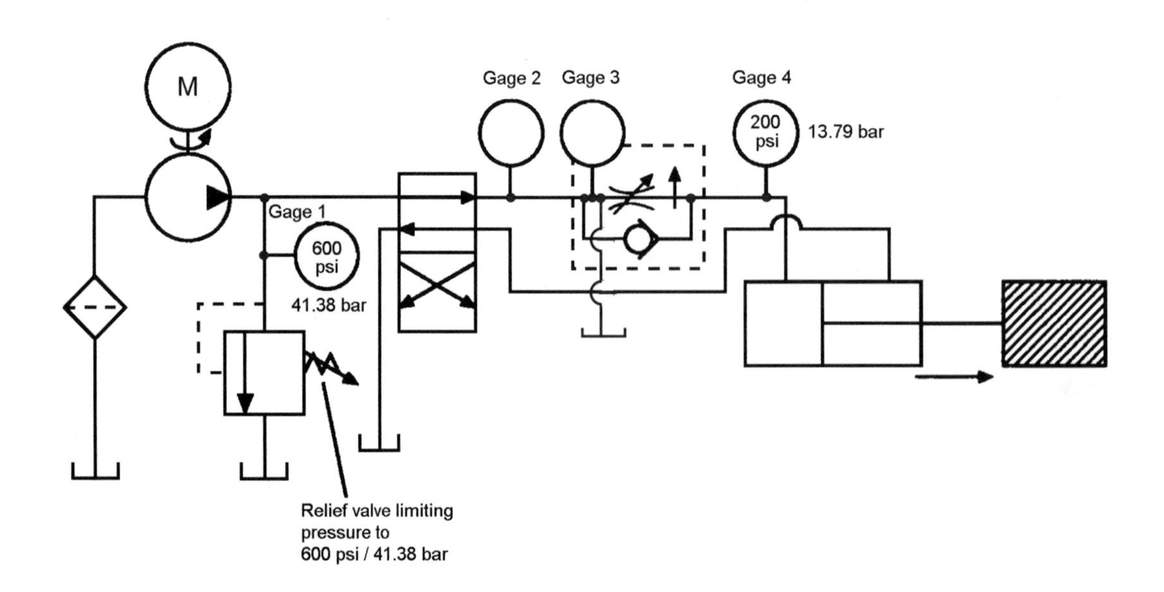

Chapter 7 exercise (cont'd.)
Flow control valves

3. Which of the following is an example of a variable orifice: _____
   a. gate valve
   b. globe valve
   c. needle valve
   d. none of the above

4. Pressure drop (ΔP) through an orifice (restriction) is proportional to the _____ of the flow through the orifice.
   a. square
   b. square root
   c. cube
   d. viscosity

5. Bypass flow controls should be used only in _____ circuits.
   a. meter-out
   b. breaking
   c. meter-in
   d. static

Chapter 8

Directional control valves

Directional control valves, DCV's, give a hydraulic system its operating characteristic. They control and regulate the pump flow to the selected function and in the right quantity (fig. 8-1).

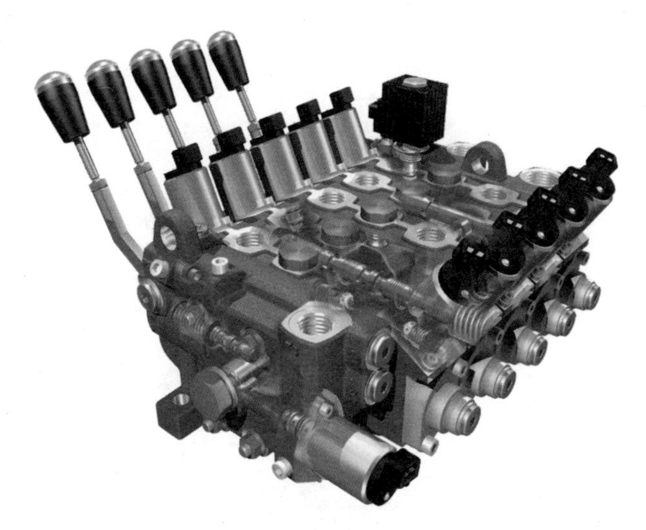

Fig. 8-1 Mobile directional control valve

The capacity, function and quality of the valve is a decisive factor in giving the mobile machine its desired operating characteristic. The valve can be combined with many integrated components, be remotely controlled or directly actuated.

The main pressure relief valve makes it possible to set the lifting capacity of the machine to a specified maximum value. Separate pressure relief and anti-cavitation valves in each actuator port prevent shock pressure from damaging the motor or cylinder.

The anti-cavitation valve provides makeup capability. When utilized, a lowering actuator movement can, without delay, be changed into a lifting movement. This applies also to an overcenter function. Furthermore, it eliminates the risk of cavitation and reduces the risk of seal damage.

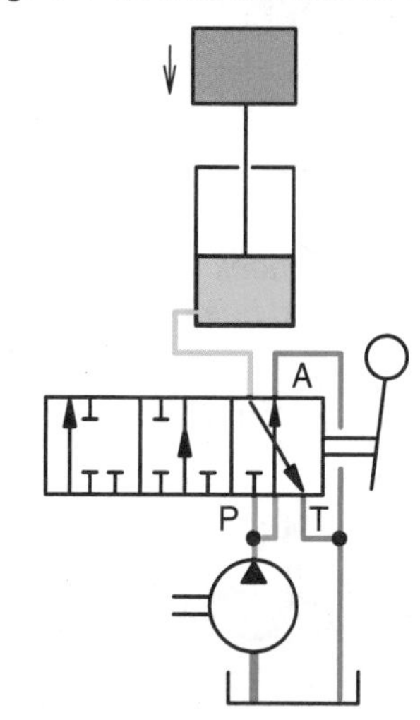

Fig. 8-2 Simple schematic with a single acting cylinder and a 5/3 control valve

DCV's are adaptable to various systems like constant flow (CF), constant pressure (CP) and load sensing (LS) as well as to a broad range of applications. Thus it is possible to design a hydraulic system that is the best compromise between economy, safety, function and operating characteristics. In multispool valves, internal flow passages can be connected in series and/or in parallel as well as externally between individual valve bodies.

Valves are normally named according to the number of ports and spool positions within the valve body, i.e. 5/3 which means three ports and three positions. In this example, the valve controls a single acting cylinder and opens and closes (blocks) five flow paths (fig. 8-2). Ports are named P (pump), T (tank or reservoir) and A & B (actuator ports).

To control a double acting cylinder (fig. 8-3), a 6/3 valve is required (four ports including actuator port B; three positions):

- Pos. 1: P-to-A and B-to-T
- Pos. 2: As shown in the center (neutral) position, ports A and B are closed, and port P is connected to T.

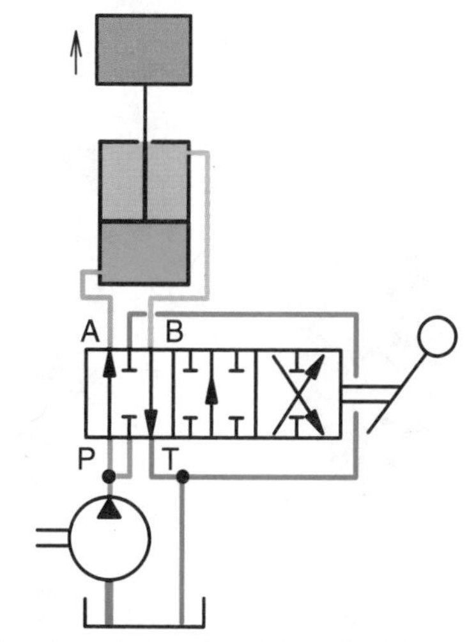

Fig. 8-3 Simple schematic with a double acting cylinder and a 6/3 control valve

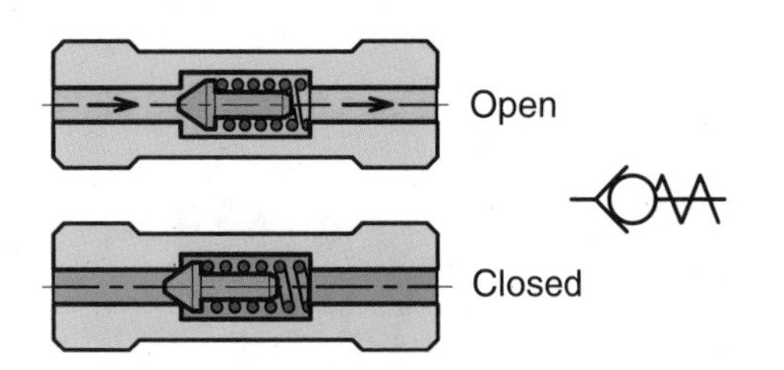

Fig. 8-4 Direct operated check valve (with symbol)

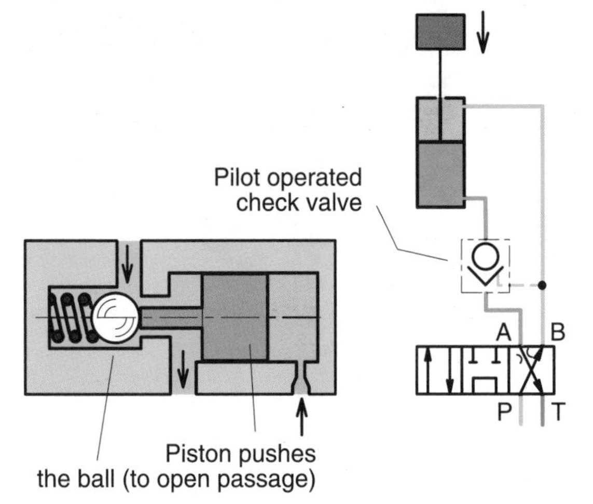

Fig. 8-5 Pilot operated check valve and simple cylinder circuit (load being lowered)

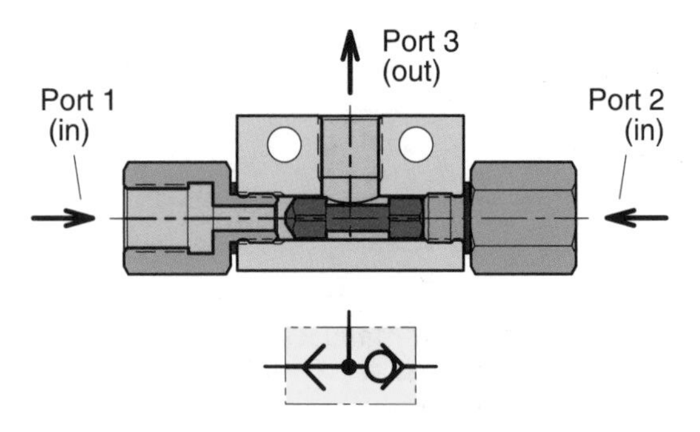

Fig. 8-6 Shuttle valve (with symbol)

Fig. 8-7 Cartridge valves (left: 3/2; right: 4/3)

- Pos. 3: P-to-B and A-to-T

As shown in the illustration, the valve controls the cylinder movement in either direction and locks the piston in any position as soon as the valve spool is returned to neutral.

The most common type of valve on a mobile machine is the spool type valve. Other valve designs are the poppet/seat valve and the combination of spool and seat valves.

Check valves

The simplest directional control valve (DCV), is the common check valve. It allows fluid to flow in one direction and blocks it in the other. It is either direct (fig. 8-4) or pilot operated (fig. 8-5).

A variant of the direct operated valve is the shuttle valve, often used in pilot and control signal lines (fig. 8-6). It should be noted that flow should always enter the valve as shown (through port 1 or 2). In the opposite direction, flow entering port 3 may exit the valve either through port 1 or 2.

If a check valve is preloaded by a spring, it is usually called a counterpressure or backpressure valve; it has a cracking pressure equal to the spring setting.

Cartridge valves

Directional control valves are available as cartridge valve assemblies for mounting in a manifold or screwed into a separate housing with ports for in-line mounting.

The cartridge valve can be controlled by a wet pin or armature solenoid with spring return to neutral. Manual override is normally available. Replacing the solenoid coil does not require valve disassembly.

For use in mobile applications, the solenoid voltage is 12 or 24 VDC. The valve function can be on-off or proportional. Single solenoid configurations are 2/2, 3/2 and 4/2; twin solenoids are 3/3 and 4/3 (fig. 8-7).

The 2/2 version is often designed as a poppet/seat valve with hydraulic internal pilot control for leakproof applications.

Customized functions can be obtained by combining different cartridge functions and machining the manifold accordingly. A modern DCV spool valve may contain several cartridge valves.

Seat/poppet valves

The seat or poppet valve differs from the spool valve mainly in that it is virtually leak free. This style valve is considered one of the most leak resistant found in a typical hydraulic system. It consists of many parts.

In order to achieve a normal 4/3 function, to operate a double acting cylinder, four seat valves are required to replace one spool in a DCV section. Due to the fact that there is no mechanical linkage between seat valves, certain limitations may arise when controlling several cylinders operating in parallel; the precise, simultaneous opening and closing of the poppets can be difficult to achieve.

On the other hand, this “freedom” can be an advantage in cases when independent functions are required on the A and B ports (fig. 8-8).

There are also valve designs that combine spool and seat/poppet valve cartridges. This type uses the spool as a meter-in element to both actuator ports, and two seat/poppet valves as meter-out elements in order to provide leak free load holding when controlling a double acting cylinder (fig. 8-9).

Spool valves

The spool type valve is the most common directional control valve found in mobile hydraulic systems. For this reason, we are concentrating on this valve type. Various valve functions will be examined and different hydraulic system types will be looked into.

The spool type DCV consists of a valve body with internal passages and ports, which are interconnected and synchronized by a movable part, the spool (fig. 8-10).

Mobile hydraulic directional control valves are of two main styles:

- the monoblock valve
- the stackable valve

Monoblock valves consist of a valve body for one or more spools, i.e. a 2-spool, 6-spool valve etc. Some design variations permit attachment of two or more monoblock valves together in order to

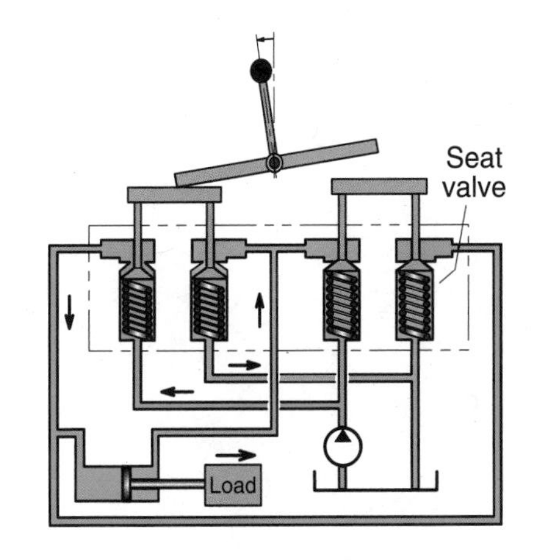

Fig. 8-8 Two seat valves engaged to extend the cylinder

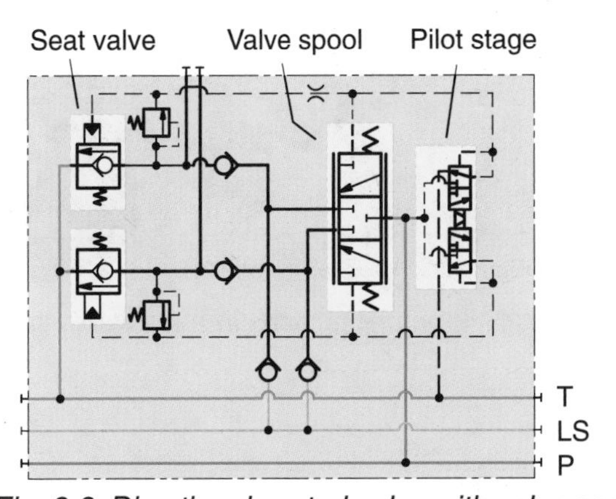

Fig. 8-9 Directional control valve with valve spool and seat valves

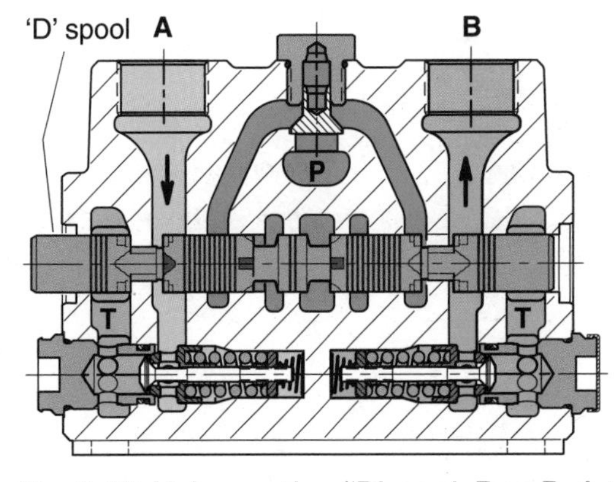

Fig. 8-10 Valve section ('D'spool; P-to-B; A-to-T)

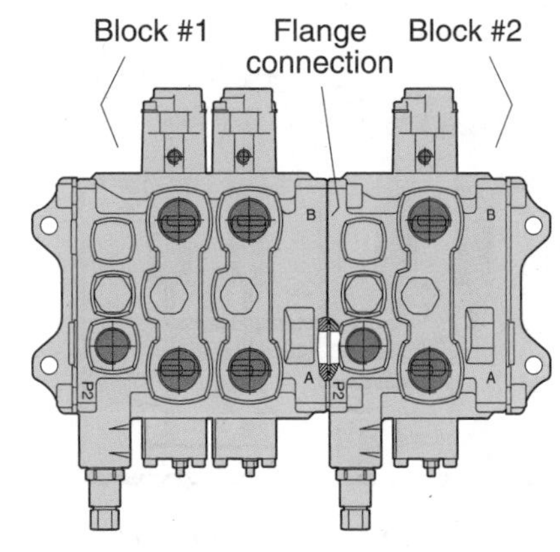

Fig. 8-11 Two monoblock valves flanged together

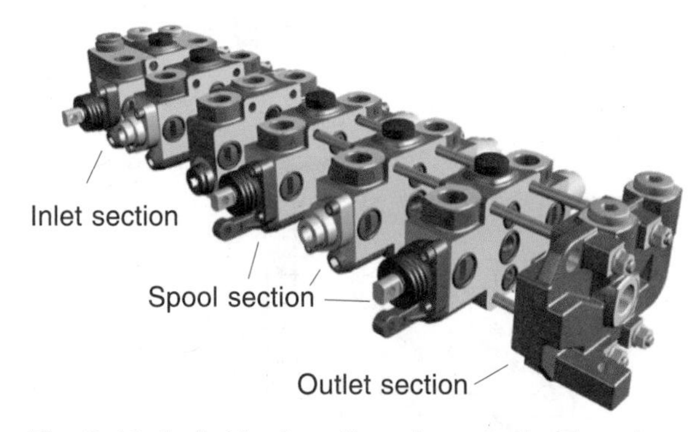

Fig. 8-12 Individual sections in a stackable valve

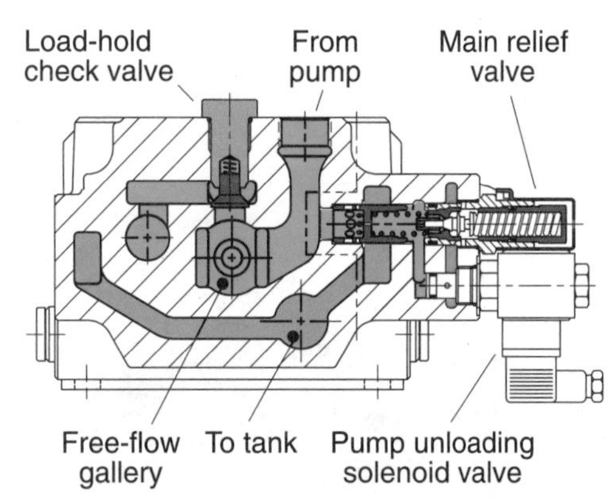

Fig. 8-13 CFO valve inlet with load-hold check valve

achieve specific configurations, i.e. multi-pump systems (fig. 8-11).

Stackable valves are normally designed for one or two spool sections and bolted together between an inlet and an outlet section (fig. 8-12).

Special manifolds can be installed between two sections of a stackable valve. These manifolds may include optional and/or customized functions, such as counterbalance valves, pressure controls, etc.

Monoblock valves are available with up to 6 sections, while stackable valves can go up to 12 sections. In comparison with stackable valves, monoblock valves have the following characteristics.

The monoblock valve is:

1) smaller and has a stiffer body (less risk of spool binding)
2) less prone to external leakage (fewer potential leakage paths)
3) more difficult to produce (casting, machining)
4) non-flexible in system design
5) more expensive to service, repair and replace

To obtain a stiffer body, some small, stackable valves are produced as two section blocks.

Valve inlet

Depending upon valve design, the inlet section (figs. 8-13 through 8-21) may contain:

- pump connection port (or ports)
- normally one tank connection port
- built-in main relief valve (pilot or direct operated)

The inlet section may also include an emergency stop, a common load-hold check valve, a special pilot copy function, a pump unloading function, connectors and/or accessories for external series or parallel connection to other valves and signal connections to pump compensators.

Depending upon the hydraulic system type and center conditions, additional accessories may be added.

Common load-hold check valve

A load-hold check valve is sometimes installed in the inlet of a CFO (constant flow, open center) valve. This prevents fluid in the actuator pressure port from flowing back into the free-flow passage (fig. 8-13).

The check valve is a substitute for individual load-hold check valves in each spool section and can be used if simultaneous operation of different sections does not occur.

Some valve designs incorporate load-hold checks in each valve actuator passage. This design will prevent the actuator from drifting due to leakage past the valve spool when the pressure in the pressure or free-flow passage is less than in the actuator ports. Also, this design will minimize actuator drift due to leakage past the valve spool to tank.

When valve blocks are connected in parallel in a single or multi-pump system, load-hold check valves are required in subsequent blocks (fig. 8-14; 'NC').

Bypass valve/main relief valve

In a closed center valve, a bypass/inlet compensator is installed in the valve inlet. It passes surplus fluid to the tank during operation as well as the entire pump flow when the system is inactive.

The bypass valve is preloaded with a spring which produces a pressure difference between the pump and load pressure. The highest load pressure is piloted (directed) into the spring chamber of the bypass valve through passages in the valve housing.

When the load pressure goes down to zero, the bypass valve directs all flow to tank at a pressure drop equal to the spring value; refer to figs. 8-15 and 8-16.

Some valve designs have an additional pump-unloading function that further reduces the pressure drop from 'P' to 'T' when no functions are used. An electrical design is shown in fig. 8-13, -17 and -18.

The bypass valve can also act as a pilot controlled main relief valve if provided with the pilot control components. If the pump pressure increases above the setting of the relief valve, the bypass valve opens and directs the entire pump. All hydraulic power can, consequently, be transformed to heat.

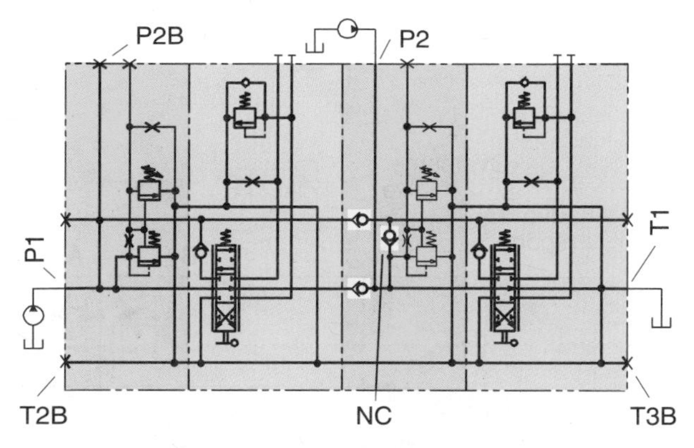

Fig. 8-14 CFO valves in a multi-pump operation; check valves allow any surplus flow from pump '1' to be used in valve '2'

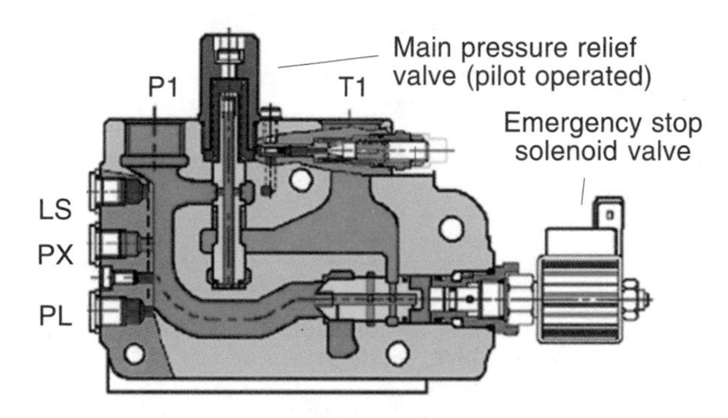

Fig. 8-15 CFC inlet section

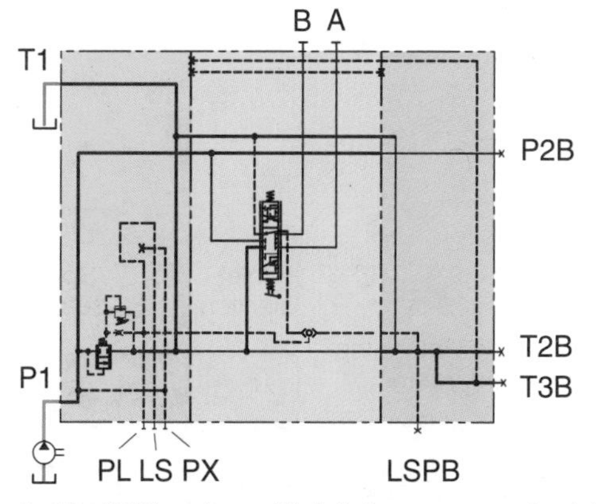

Fig. 8-16 CFC valve with inlet compensator/pilot operated main relief valve

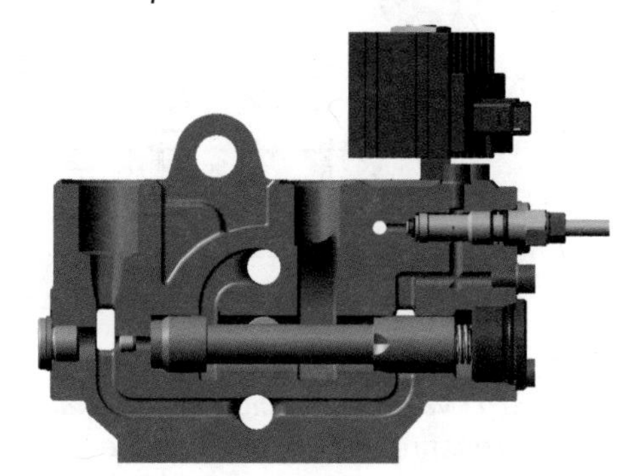

Fig. 8-17 CFC inlet section with combined pilot operated, main relief valve and electrohydraulic unloading valve

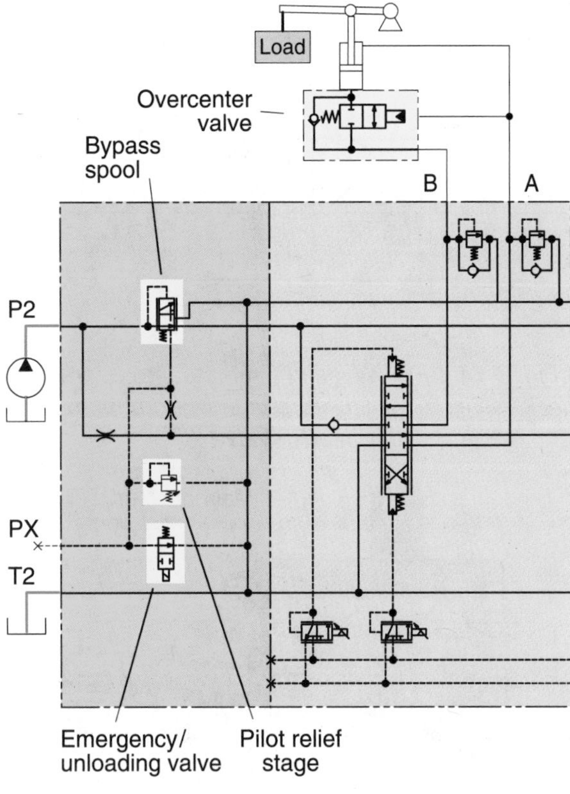

Fig. 8-18

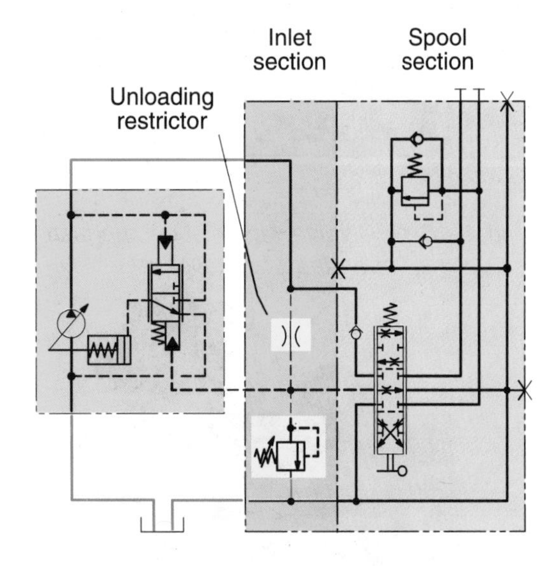

Fig. 8-19 CPU valve with unloading signal restrictor and pump pressure limiting valve in the inlet section

In a CFO valve, with an installed bypass spool, part of the oil can be directed to tank, and only a partial flow needs to go through the free-flow passage. This reduces the power loss in multispool valves when all spools are in the neutral position.

The pump unloading and emergency stop functions are activated by an electrically controlled on-off cartridge valve, which drains the control pressure from the end of the bypass spool, thereby opening a large passage from 'P' to 'T'. Together with an overcenter valve, this function enables the machine to be equipped with an emergency stop function (figs. 8-15 and 8-18).

Emergency stop function

According to the European Community's 'Machinery Directive', machines must be equipped with one or more emergency stop functions to enable actual or impending danger to be averted.

The Machinery Directive states: "The emergency STOP device must have clearly identifiable, clearly visible and quickly accessible controls. It must stop the dangerous process as quickly as possible without creating additional hazards."

Furthermore, the energy supply to the function must be cut off when the emergency stop device is activated. As shown, the electrically controlled, pump unloading function fulfills these requirements. Note that the solenoid must be activated before any work can be done; refer to figs. 8-15, -18 and -20.

Unloading signal orifice

The CPU (constant pressure, unloading) valve with an unloading function normally works with a variable displacement pump. The pump's compensator is striving to keep the pressure at a constant, preset value by increasing or decreasing the pump's output flow (fig. 8-19).

If the pump has a specific compensator control, it is possible to run the system with an unloading function when the system is nonactive. The unloading signal is taken from the valve inlet, and a small orifice lets some pump flow pass to tank through the valve's restricted 'open-center line', when all spools are in the neutral position.

This zero signal is also communicated to the valve inlet and the pump compensator, which regulates the pressure down to a 'standby level', which is preset by the pump's compensator, approx. 20 bar (*300 psi*).

When any valve spool is activated, the 'open-center' line is closed and the signal pressure is equal to the pump pressure, which then automatically increases from the standby setting to the maximum 'constant pressure' setting.

The maximum pressure setting is preset by the pump's compensator or by the relief valve in the inlet section as shown in fig. 8-19. This means that the pump operates at either of two pressure levels:

- maximum
- standby

If the pump reacts quickly to pressure changes, it can create pressure 'overshoots' or 'spikes" when a load is activated. For this reason, the use of a fast acting (direct operated) pressure relief valve, in the DCV's inlet, is highly recommended to reduce the pressure spikes (fig. 8-20).

Load sensing signal connector

In a load sensing system (LS), the valve inlet is very similar to that of the CFC or CPU valve. The pump is a variable displacement type with a standby and maximum pressure compensator.

The load sensing signal to the pump is taken from the valve inlet and directed to the pump's LS compensator. (In CFC valves the LS signal is directed to the end of the bypass spool).

Consequently, the valve inlet does not require a bypass function, because the pump compensator provides and maintains a constant pressure difference, ΔP, between the pump pressure and load sensing signal, by changing its flow output (fig. 8-20).

The ΔP is preset at the pump compensator in the same manner as for the bypass valve in the CFC valve.

The main difference between CPU and LS valves is that the CPU just sends an '"on-off" signal to the pump compensator, while the LS valve sends a signal that is uniformly variable with the load pressure.

The LS valve should have a fast acting pressure relief valve in the inlet for the same reason as the CPU system above.

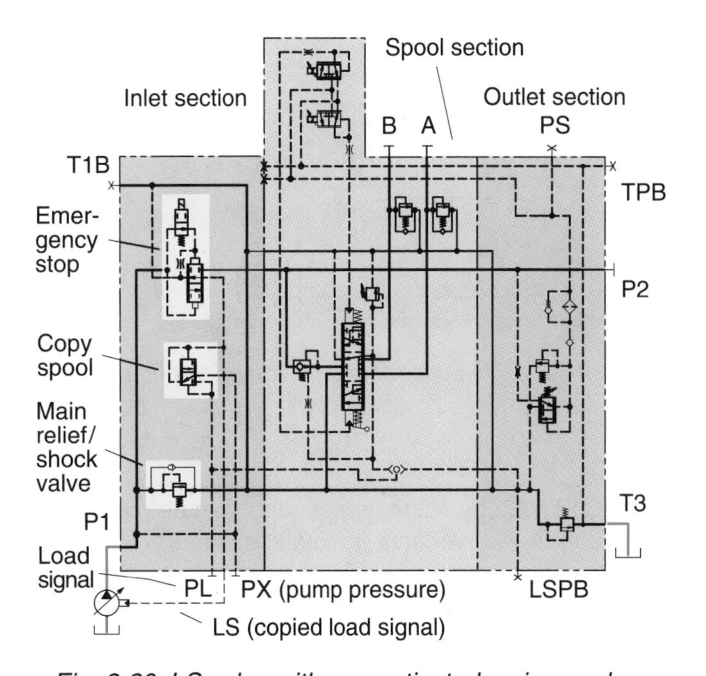

Fig. 8-20 LS valve with non-activated main spool (inlet section with direct acting main relief valve, copy spool for the load signal, an electrically controlled emergency stop valve; the LS signal is drained to tank)

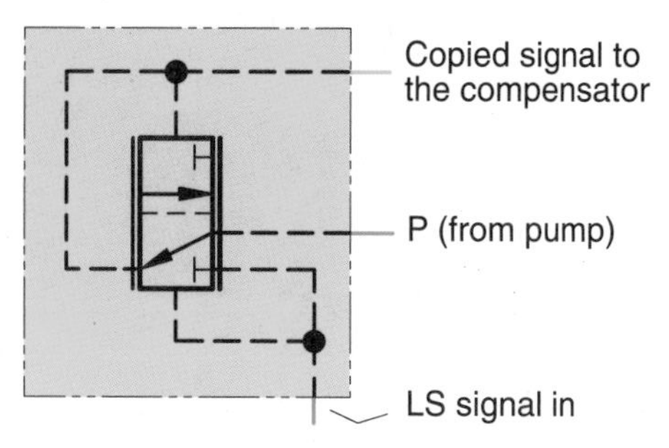

Fig. 8-21 The copy function

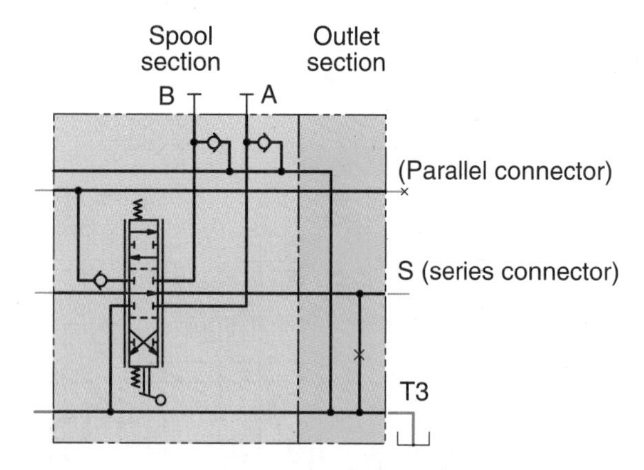

Fig. 8-22 CFO outlet section with series and parallel connectors

NOTE: Many of the components can be placed either in the inlet or in the outlet section.

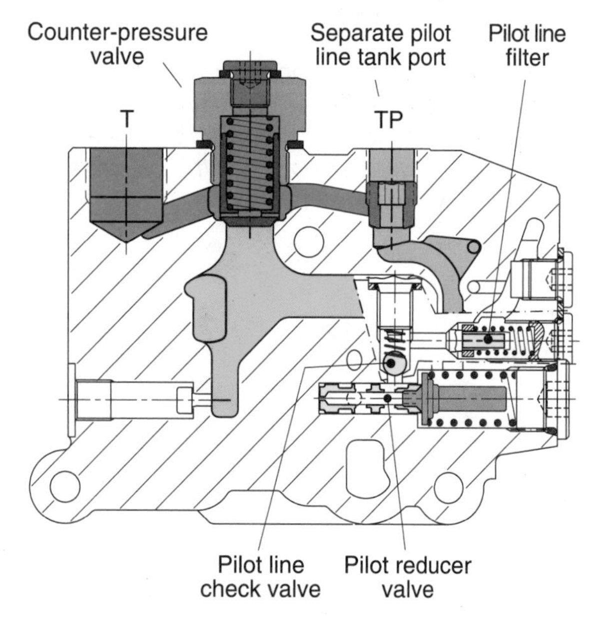

Fig. 8-23 CFC/LS valve outlet section

In the valve design shown in figs. 8-15 and -20, gauge ports are shown for load, pump, and LS signals as well as emergency shutdown and load sensing signal copying functions.

Copy function

This function eliminates any loss of pressurized fluid from the cylinder when an LS signal is sent to the pump compensator. This prevents disruptive 'micro sinking' of the load, at the beginning of the lifting operation.

This is because the signal fluid is taken from the pump instead of the load. The signal pressure is copied, and fluid from the pump is sent to the pump compensator; see figs. 8-20 and -21.

Valve outlet

The valve outlet may contain:

- one or two tank ports
- a pump port
- series (S) and parallel (P) connectors and accessories to subsequent valves
- counterpressure/backpressure valve and pressure reducing valve for the pilot supply (figs. 8-22, and -23).

Series connections

One of the tank ports is normally used for a 'power-beyond' connection. This high pressure carryover connection separates the free flow passage from the tank passage inside the valve. This function makes it possible for two or more valves to be operated in series. One pump is then able to supply both valves. To implement this function it may be necessary to modify the valve. This modification could be as simple as inserting flow diverter into the valves tank port. This connects the open-center line passage through all valves to tank via the last valve. This is done when a series connection of CFO valves is desired. It must be noted that the downstream valves receive only the "leftover" flow from the first valve. The downstream valves will receive full pump flow only when the spools in the first valve are in their neutral or center positions. Another note when connecting valves in this fashion is that each valve requires its own tank line (fig. 8-22). A power-beyond connection is installed when a series connection of a CFO valve is desired, connecting the open center line passage (S) through all valves to tank (fig. 8-22).

In a CPU system, the unloading signal line can be connected the same way from valve to valve and finally to tank.

LS valves normally have an outlet signal line connection for a series connection of the LS signal from subsequent valves; refer to port LSPB in fig. 8-20.

Counterpressure/backpressure valve

The counterpressure/backpressure valve installs in one of the tank ports. It increases the return gallery pressure inside the valve in order to improve anti-cavitation and makeup flow characteristics.

Increased gallery pressure is desirable especially in load sensing systems where the piston and rod end areas of a cylinder differ greatly, or during fast lowering movements when the pump is unloaded (fig. 8-23).

In an open-center valve with integrated pilot operated spools, the counterpressure/backpressure valve maintains the required, minimum pilot pressure.

Counterpressure/backpressure valves can be controlled with an internal or external pilot pressure. Alternatively, it has a fixed, bias spring setting.

In some valve designs, the counterpressure/backpressure valve is activated electrically; refer to fig. 8-24.

If the DCV has a built-in pilot pressure supply, it is recommended that the tank connection be separated from the main tank line, which is affected by the counterpressure/backpressure (refer to pos. TP, fig. 8-23).

Internal pilot pressure supply

Internal pilot pressure supply is a valve function, which can work both as a pressure reducing valve or a pressure relief valve in the pilot circuit.

It reduces the pump pressure to an appropriate level and supplies pilot pressure internally to electrohydraulic spool controls (figs. 8-23 and 8-24).

For safety reasons, this reducing valve is also equipped with a relief function that prevents

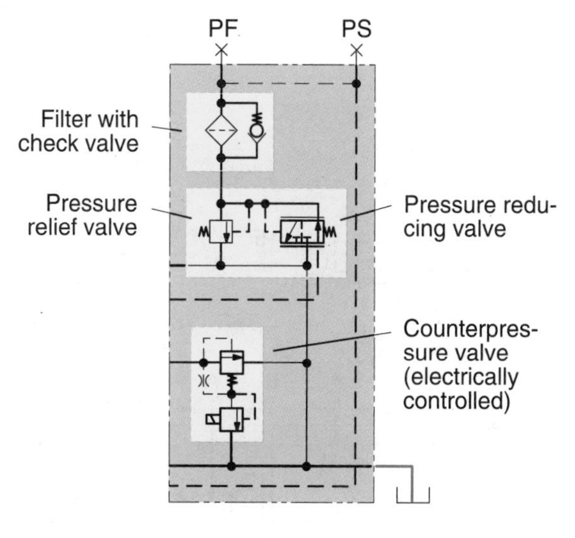

Fig. 8-24 Valve outlet with pilot pressure supply

maximum allowable, reduced pilot pressure to be exceeded.

The illustration also shows a coarse strainer with a bypass function in the internal pilot pressure supply circuit. The strainer protects the pilot circuit from contamination, especially during system startup.

The design shown includes connection possibility to an external filter for pilot pressure fluid, enabling the pilot circuit to be supplied with cleaner fluid, compared to the fluid in the working system (fig. 8-24, 'PF' and 'PS').

A check valve prevents the pilot fluid from leaking back into the pump line, thereby enabling the pressure in the pilot supply circuit to be maintained in the event of a temporary reduction in pump pressure, e.g. during a rapid cylinder lowering.

In order to secure pilot pressure during a reduction in pump pressure, an accumulator should be connected externally to port 'PS'. Pilot pressure can now be tapped off the PS port supplying e.g. hydraulic remote control joysticks or an external accumulator/brake system.

Note that many of the components can be placed either in the inlet or outlet section depending upon the valve manufacturer. As has been shown, these end sections may contain many functions and become very complex. However, the spool section of the mobile DCV is the business section of the entire valve. It is this section that directs and controls the hydraulic energy to the actuators.

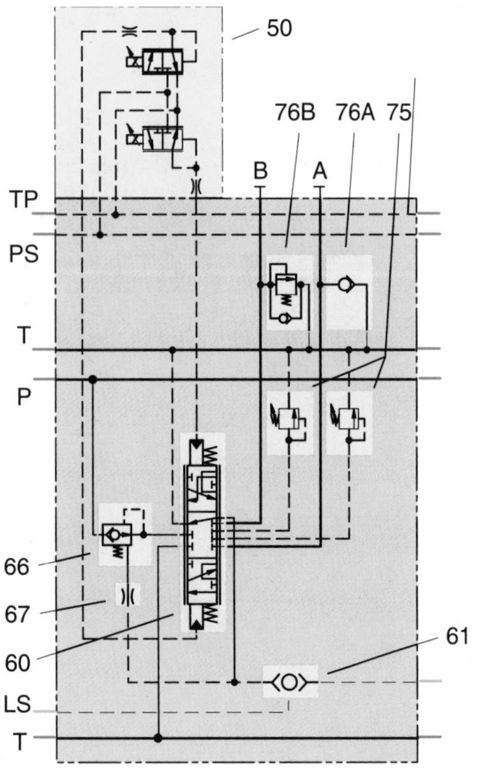

Fig. 8-25 LS valve spol section schematic

Valve section

Depending on the hydraulic system, valve type, and design, a valve section may contain the following components shown in figs. 8-25 and -26. The modern load-sensing valve illustrated contains:

- main spool (60), here remotely controlled with electrohydraulic actuators (50)
- shuttle valve (61) for transfer of LS signal to the valve inlet
- load holding check valve, here combined with pressure compensator (66) and LS signal dampening restrictor (67)
- pressure reducing valves maximizing the lifting force (75)

- shock valve (76B) combined with makeup check valve or with a separate makeup check valve (76A)
- feed and/or lowering brake valve maximizing lift or lowering function speed (not shown).

Fig. 8-26 shows an LS valve section where the spool (positioned almost to the right extreme) is remotely controlled through proportional, electro-hydraulic solenoid valves.

The spool can also be operated with the manual override. The spool stroke is limited by externally adjustable stroke limiters (72).

In the illustration, the left hand end cap is supplied with pilot pressure from the right hand cartridge proportional solenoid valve through drillings (passages) in the valve housing while the right hand end cap is drained to tank.

As shown, the spool, connects P-to-B and A-to-T. The load signal pressure from port B is fed to the pressure reducing valve as well as to the spring chamber of the pressure compensator through small drillings (passages) in the spool.

The actuator ports are protected by combined shock and anti-cavitation cartridges.

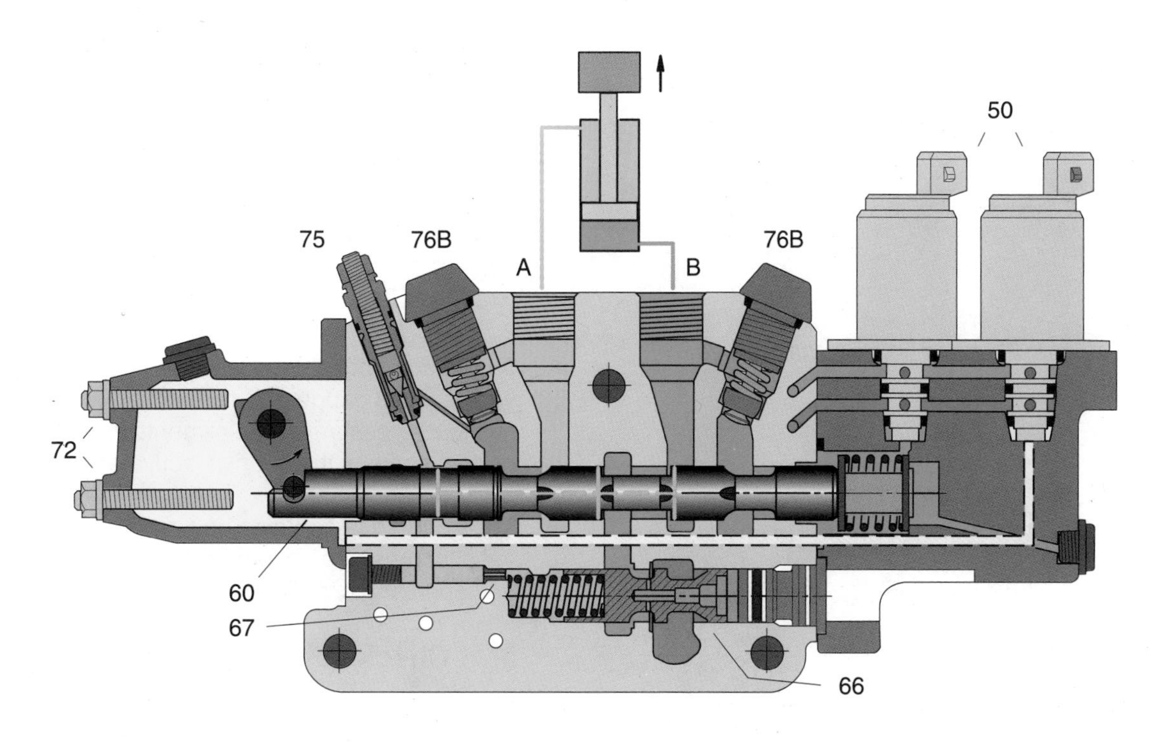

Fig. 8-26 LS valve spool section

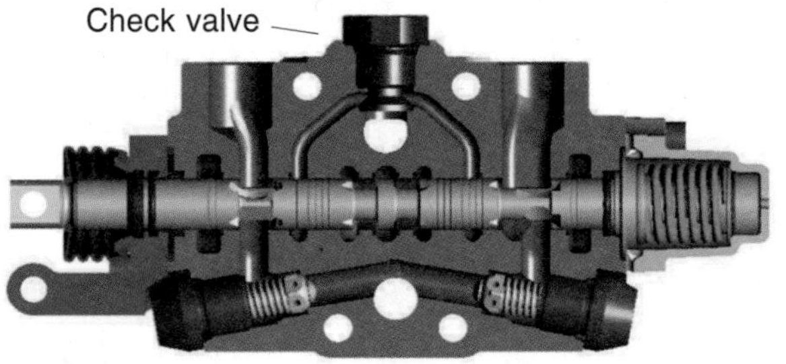

Check valve

Fig. 8-27 CFO valve section with load holding check valve

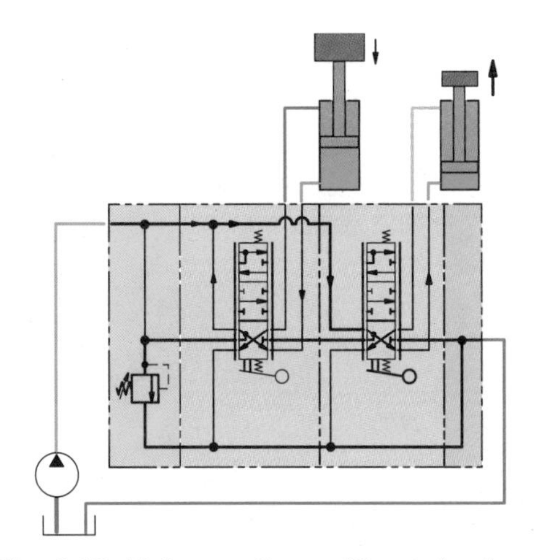

Fig. 8-28 Valve sections without check valves

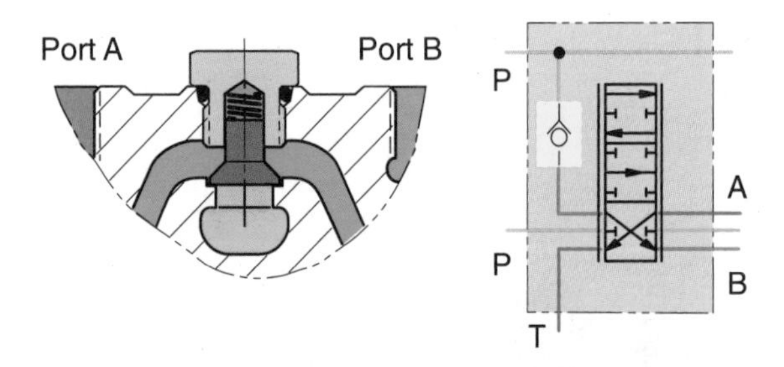

Fig. 8-29 Load holding check valve

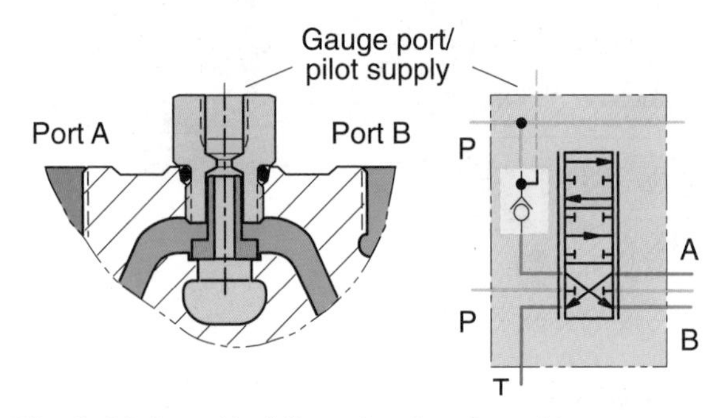

Fig. 8-30 Load holding check valve with gauge port

Load holding check valve

If the load pressure is higher than the pump pressure and the valve spool is moved to the 'lift' position, the movement can be quite the opposite if a load holding check valve is not installed in the supply passage. The check valve (fig. 8-27) prevents reverse flow when the pump pressure is not high enough to lift the load.

In a circuit with two lifting functions (but without check valve) being operated simultaneously with non-equal load pressures (fig. 8-28), the lighter load goes up while the heavier goes down, helping lift the lighter load. The cylinder with the light load increases its speed until the operator adjusts the position (throttling areas) of the individual spools until both loads go up.

If a load holding check valve as shown in fig. 8-29 were installed, the heavier load would be locked in place until the pump pressure had increased sufficiently to lift it.

Load hold check valves in CP and LS systems are of little use as long as the variable displacement pump operates within its regulating range. This is because the pump pressure is equal to, or higher than, the load pressure.

When the pump has reached its maximum displacement, however, the function changes dramatically (as the pump now acts as a fixed displacement pump). The pump adapts to the lowest load pressure in the system and the heavier load's fluid moves to the cylinder with the lightest load. For this reason, the use of load holding check valves should always be considered, even in CP and LS systems.

Check valve gauge port

A common option used with load holding check valves is the gauge port connection which simplifies load pressure measurements in the actuator port as soon as the spool is activated.

The gauge port can also be used to supply external pilot pressure as long as it doesn't disturb any operating cycles (fig. 8-30).

Check valve restrictor/orifice

The load holding check valve can also be combined with an adjustable restrictor (flow control orifice) as illustrated in fig. 8-31.

The valve throttles (reduces) the supply flow to the spool, thereby reducing the actuator speed. This is seen in constant pressure (CP) systems where valve spools are not pressure compensated, i.e. manually operated spools. This increases the operator's sense of control. However, when the lowest possible pressure drop in the supply line is reached, the cavity boring can be left empty (fig. 8-32).

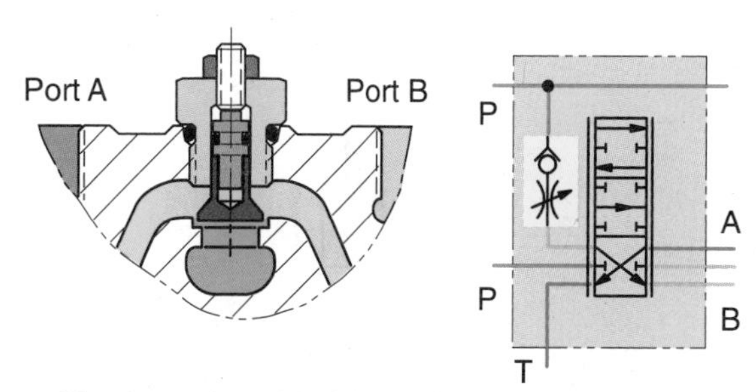

Fig. 8-31 Load holding check valve restrictor

Lift/feed brake

Sometimes a lift/feed brake device is installed in the same port utilized for the load holding check valve.

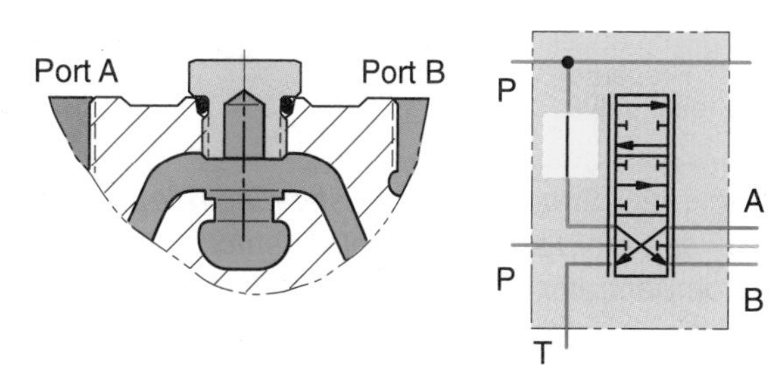

Fig. 8-32 Check valve port (plugged)

This cartridge valve (fig. 8-33) is a pressure compensated, constant flow valve. It insures that the maximum actuator speed is not influenced by variations in load pressure.

The feed brake is a special case of meter-in control for CFO valves. It is used as a 'safety' device when maximum actuator speed must be secured. An application for this device is the swing (slew) function on a truck (lorry) crane.

The design shown in fig. 8-33 consists of a plug/ sleeve, a drilled-through spool and a spring that is shimmed to the desired flow. The spring pushes the spool down, letting pump flow pass to the actuator via the valve spool.

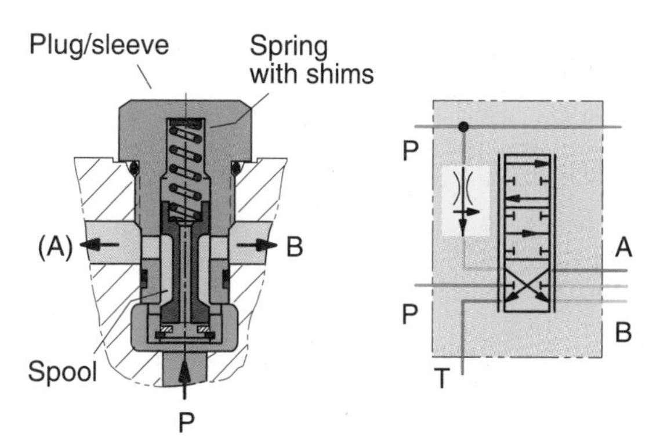

Fig. 8-33 Lift/feed brake

Load pressure acts on both sides of the spool through the drilling. A higher flow around the spool results in lower flow forces trying to close it. By compressing the spring, the flow forces are balanced and output flow to the actuator will be kept relatively constant.

NOTE: The pump pressure can easily reach its maximum level and direct excess flow to tank through the main pressure relief valve if this function is used alone.

Pressure reducing valve

In some DCV designs, a pressure reducing valve is installed in place of the load-holding check valve.

The purpose is to keep a certain, limited, pressure level on the secondary (actuator) side. This results in a maximized supply pressure to the actuator concerned (fig. 8-34).

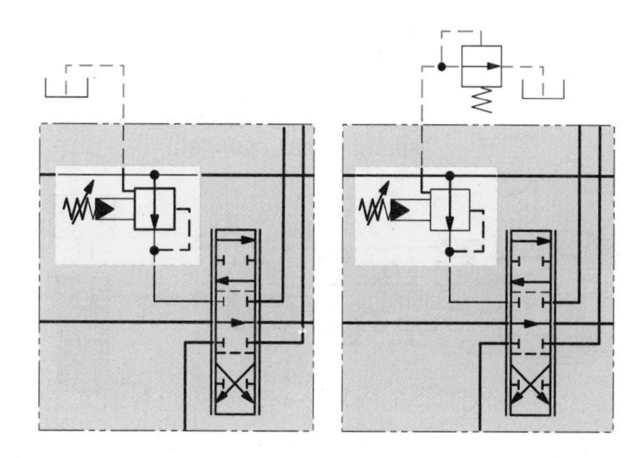
Fig. 8-34 Pressure reducing valve on the supply side

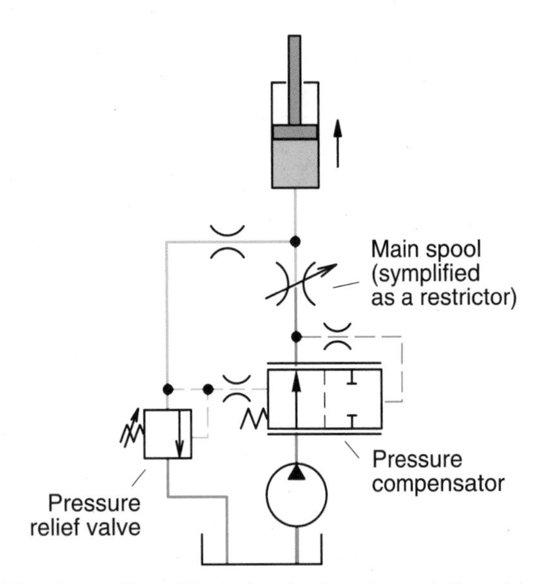

Fig. 8-35 Simplified circuit showing relationship between main spool (variable restrictor), pressure compensator and pressure reducing function.

As an example, the pressure reducing valve works as a safety device and prevents the crane on a vehicle from being overloaded by limiting the lift capacity.

NOTE: In CFO valves, the excess flow passes through the main pressure relief valve to tank if the entire pump capacity is not utilized. However, this is not the case in CP and LS systems with variable displacement pumps. It is important to combine the above function with load hold check valves described earlier.

The LS valve can also be equipped with 'individual pressure reducing functions' for the actuator ports. These are used for those functions that require a lower maximum pressure than the normal operating pressure in the system.

The LS signal relief valve (fig. 8-35 and '75' in figs. 8-25 and -26) is adjustable. It reduces the supply pressure, by limiting LS signal pressure behind the compensator, so the feed (supply) pressure in the section, in which it is installed, does not exceed the preset value.

The use of the reducing function limits the pressure by using a pilot flow of only 2 l/min *(0.5 gpm)* or less. Since the illustrated relief valve is of the two-way type, the pressure shocks that occur after it (in the actuator port) must be limited by using a port relief valve.

The port relief valve should be set slightly, about 10 bar *(150 psi)*, above the reducing valve setting. Note that, in this case, the valve section must be fitted with a pressure compensator.

Another alternative way of limiting the actuator port pressure is to use the 'port relief' alone. When the pressure setting is reached, the valve opens, dumping a large flow to tank. This, however, results in a huge power loss (refer to chapter 6, pressure valves).

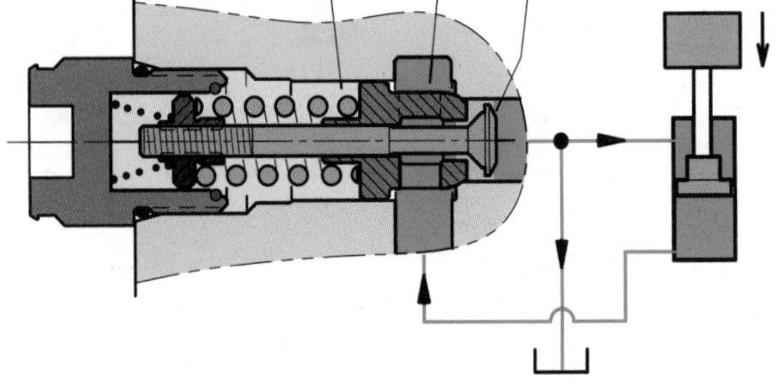

Fig. 8-36 Cartridge type port relief valve

Port relief and anti-cavitation valves

A cartridge style valve is normally used in actuator ports as a relief (shock) valve, and is combined with an anti-cavitation function (figs. 8-36 and 8-38, '2'). Its purpose is to protect the valve and actuator from pressure spikes and excessive pressure spikes in the system.

Very rapid opening (short response time), element stability, tightness (leak resistance) and good pressure characteristics are required of this valve.

The cartridge, as shown, is held against its housing seat by a light spring and the pressure in the actuator port. When the actuator port pressure increases to the setting of the shock valve, the valve poppet opens and fluid from the actuator passage flows to tank.

The diagram in fig. 8-37 shows typical performance curves for a relief valve cartridge rated at 70 l/min *(18 gpm)*.

The anti-cavitation (makeup) function facilitates fluid flow from the tank passage to the actuator port in the event of underpressure. The entire cartridge element moves back against the light (conical) spring and opens the passage from tank to the actuator port.

Depending upon the type of cartridge valve selected, the valve housing boring or cavity (as illustrated in fig. 8-36), can be used for various functions. This makes the mobile DCV very adaptable.

Additional actuator port accessories

As an alternative to the port relief valve, the actuator port can be fitted with a separate anti-cavitation valve. This will enable fluid to flow from the tank passage to the actuator port to prevent underpressure and cavitation (fig. 8-38, '3').

The diagram (fig. 8-39) shows pressure drop (ΔP) versus flow for a typical anti-cavitation check valve as measured between the tank connection and the actuator port.

This connection is sometimes blocked with a plug (fig. 8-38, '1'), eliminating both the shock valve and the anti-cavitation function.

A fourth alternative is to leave the tank-to-actuator passage open ('4'). This is used together with a type 'E' spool for a single acting actuator such as a plunger or ram cylinder which does not require a second actuator port; it can therefore be plugged (or be omitted).

The second type of a single acting cylinder has a vent port that can be connected to the valve actuator port shown in '4'. Thus, any leakage over the piston is automatically directed to tank, and protection against cylinder corrosion is achieved automatically as no air can enter the cylinder.

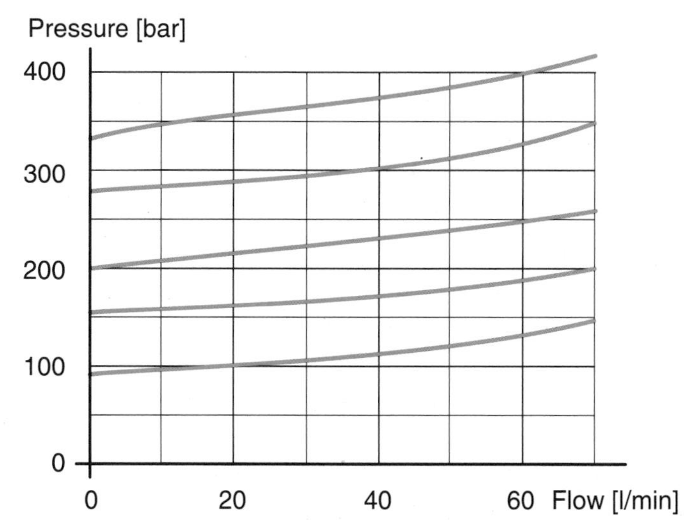

Fig. 8-37 Pressure-versus-flow of a cartridge valve at selected opening pressures

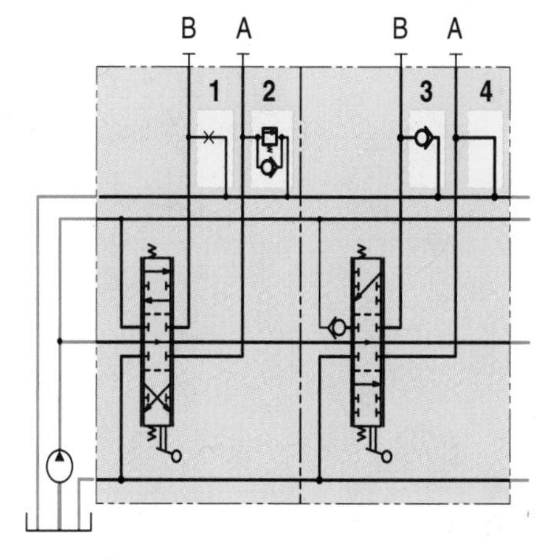

Fig. 8-38 Spool section schematic showing various actuator port functions

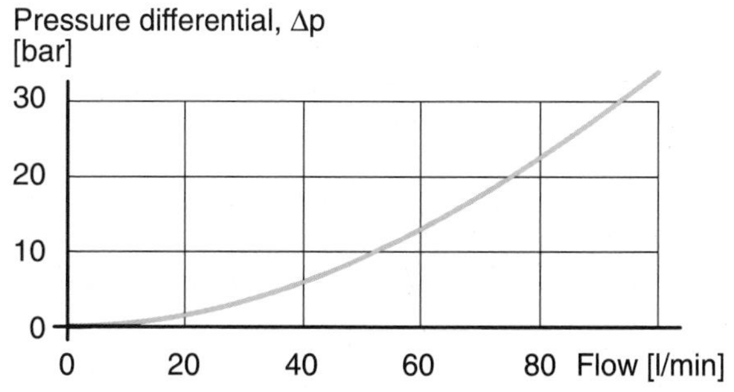

Fig. 8-39 Make-up flow characteristic of an anti-cavitation valve

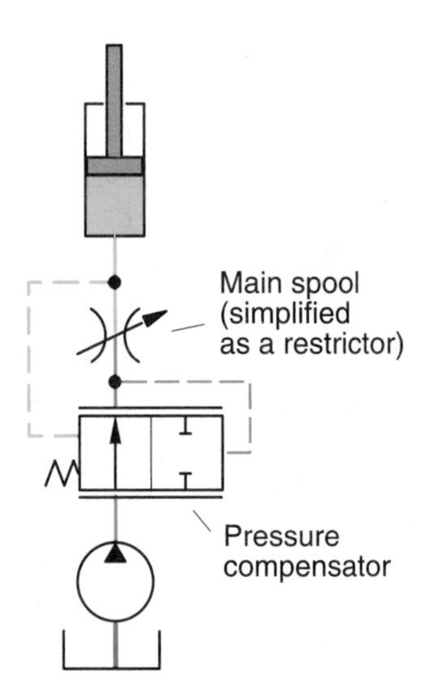

Fig. 8-40 Pressure compensator circuit (main spool shown as variable restrictor)

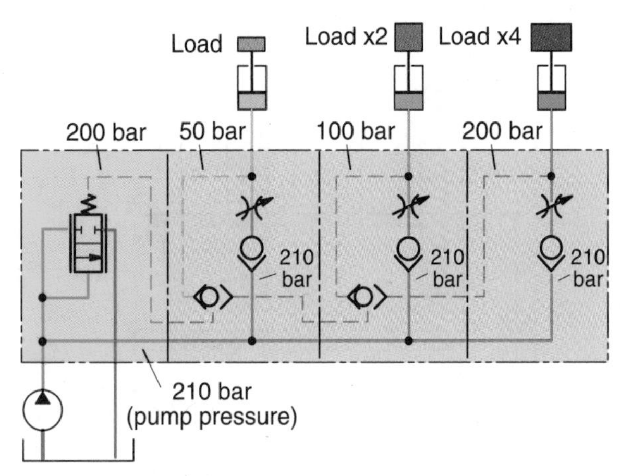

Fig. 8-41 Valve with meter-in restrictors

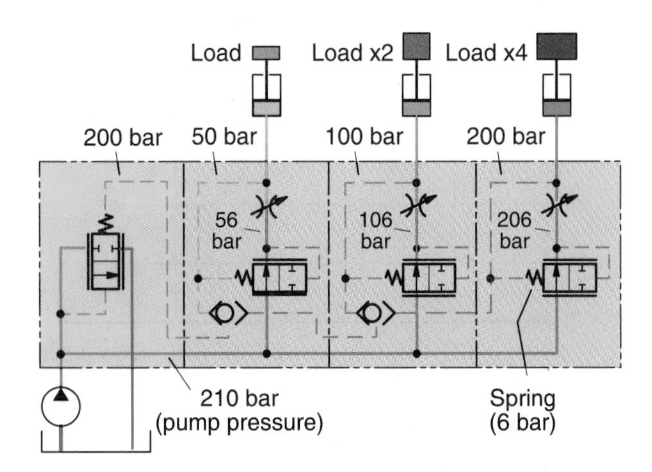

Fig. 8-42 Valve with individual pressure compensators

Pressure compensator

The primary purpose of a pressure compensator is to maintain a selected, constant flow to a function, regardless of system pressure variations. This is of particular value in lifting operations. In cases where several functions are operated simultaneously, each spool section should be equipped with an individual, integrated pressure compensator. Subject to the available pump capacity, the sections so equipped deliver the preselected, constant flow regardless of other simultaneously operated functions, and regardless of variations in load or supply pressure.

This type of integrated pressure compensator should have a fast response time and incorporate a built-in load-holding check valve; refer to figs. 8-26, '61', and 8-40.

As previously explained in chapters 6 and 7, the function of the pressure compensator is to keep a constant pressure difference, ΔP, across the main spool.

This results in a constant flow output which is a function of the opening (flow) area, determined by the position of the spool and by the compensator spring.

A weaker or stronger spring, or an adjustable setting device, can be utilized to meet the demands of a particular flow requirement. Another way is to vary the size of the main spool's opening (flow) area.

Fig. 8-41 shows a three-section valve in which spools are shown as 'meter-in' restrictors. Sections are equipped with supply check valves and the inlet with a common compensator. Pump pressure equals the highest load pressure plus the compensator spring setting, 200+10 bar *(3000+150 psi)*. The other loads are consequently not compensated.

Fig. 8-42 shows the valve equipped with individual compensators; all loads are individually compensated on the meter-in side; refer also to fig. 8-41.

Load-sensing signal system

The load pressure sensing signal system consists of a number of shuttle valves, which compare the load signals from different spool sections; the highest signal is connected to the inlet section of the valve (fig. 8-20).

In the inlet section, the signal is communicated either to the copying spool and the pump compensator, or directly to the pump compensator in the absence of a copying spool.

The system permits a certain amount of fluid flow in the load signal line to the pump without the signal level of the load being affected. This, in turn, permits a simpler system design with the added advantage of allowing external logic systems to be installed in the load sensing circuit.

Thanks to the drain flow in the pump's LS compensator, the fluid is always kept warm in the LS circuit, resulting in better operating functions and faster response during winter operations; refer to fig. 8-43.

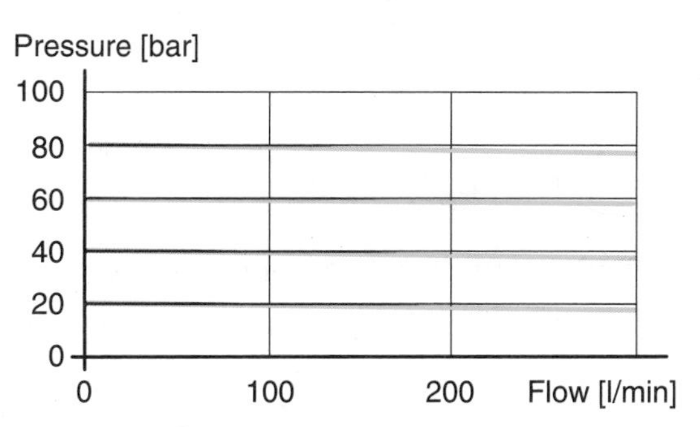

Fig. 8-43 Compensator characteristic at selected setting pressures

Spools

Common requirements of a mobile, directional control valve are:

- low spool leakage
- low lever forces
- good operating characteristics

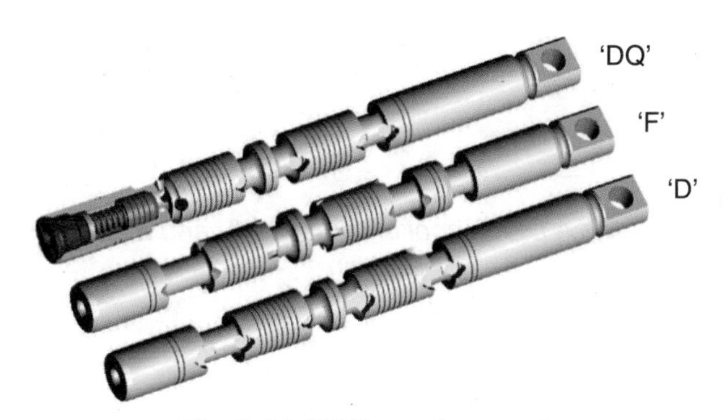

Fig. 8-44 DCV spool examples

The valve spool is the most important link between the lever being moved by the operator, and the corresponding movement of the controlled actuator.

The spool design must consequently be matched to the specific demands of each function. This is done by choosing suitable valve spool parameters.

A specific spool design is selected that matches the valve function in the hydraulic system, the required control, load characteristic and spool control.

Spool nomenclature

The standard spool symbol contains the number of spool positions within the valve, the function of the spool and the number of flow paths. The number of ports in one square equals the number of flow paths or ways.

In hydraulic schematics, connected pipes and tubes are usually shown in the center, at-rest, spool position (fig. 8-45).

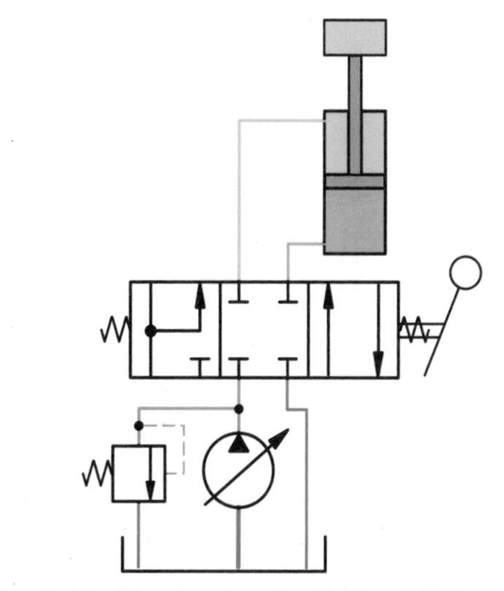

Fig. 8-45 Simple circuit with the DCV spool in the neutral position

Horizontal lines above and below the spool symbol means a "proportional function". Without the lines, the valve has an on-off function (fig. 8-46).

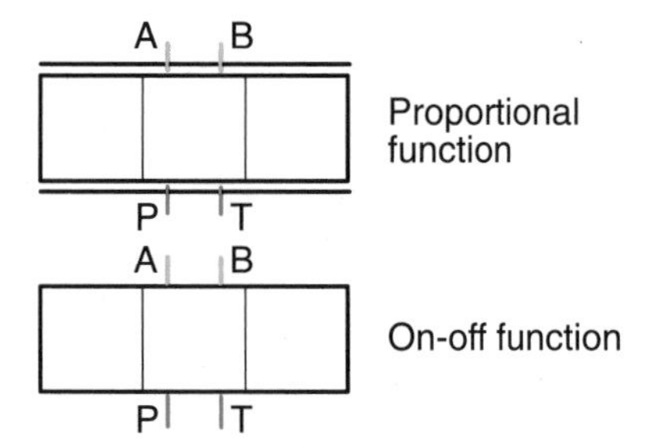

Fig. 8-46 Proportional and on-off spool symbols

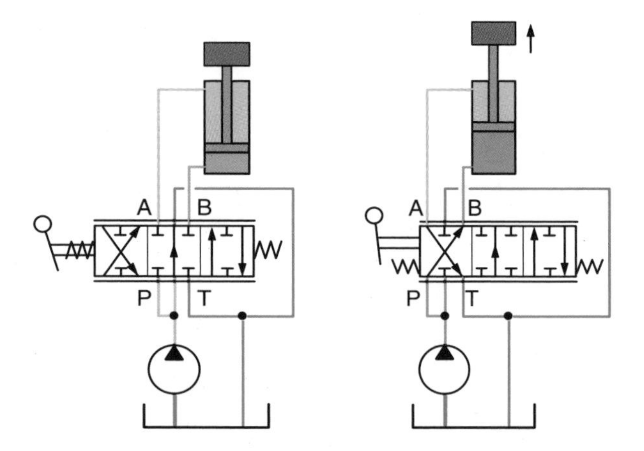

Fig. 8-47 A DCV spool in neutral (left) and a power position

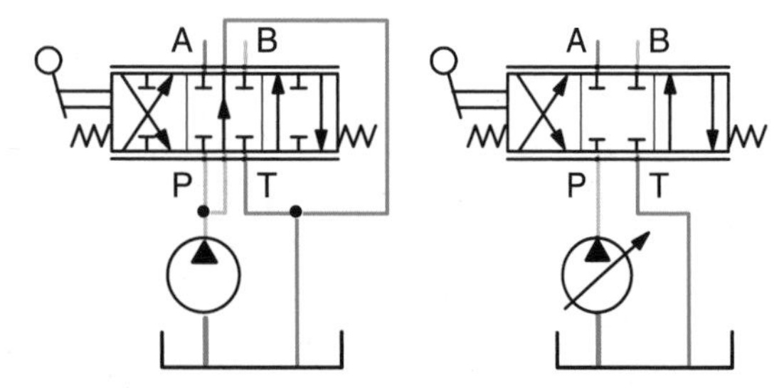

Fig. 8-48 Open (left) and closed center (simplified) circuits

Typical mobile directional control valves, DCV's, are used to control double-acting actuators such as cylinders and motors. In a CFO (constant flow, open circuit) function, a 6-way, three-position spool (designated '6/3') is required. The spool has three positions within the valve body:

- extreme left
- center
- extreme right

In the center position (fig. 8-47, left), pump flow is directed through the tank passage (gallery) in order to keep the pressure loss as low as possible, minimizing heat generation. Cylinder ports are blocked.

When the spool is moved to the right, the piston side of the cylinder is connected to pump and the rod side to tank; the cylinder extends. The load is lowered when the spool is moved in the opposite, left direction.

Note that hydraulic symbols in a schematic only shows the ***hydraulic*** function. In reality, this function can be achieved in many different ways.

For example, the same valve body is often used in both CFO and CP or CPU valves with different spools and can be shown as 4/3 or 6/3 valves in a schematic.

Spool center position

The spool in a directional control valve connects and disconnects various flow paths within the valve. Most mobile DVC spools have three positions.

The spool's extreme positions are related to the desired movement of the actuator. These positions are the 'power positions' of the valve which control the movement of the actuator. The center position is used to satisfy the needs of the system during idle.

Usually, two center conditions are used in mobile valves. The most common is the ***open center***. In neutral, flow passes around the spool either to the next spool or directly to tank (fig. 8-48, left illustration).

In the other, ***closed center*** condition, pressure is maintained in the valve inlet. The most common use of this center condition is in constant flow, close center valves (equipped with bypass valve) and in constant pressure and load sensing valves.

Pressure compensation

The pressure differential determined by the pump control, or by the bypass regulator in the valve inlet, results in a flow to the heaviest load that is always pressure compensated; flow through the actuator port is independent of variations in load pressure.

Functions can be pressure compensated in two ways:

1) By installing a separate compensator in each spool section which maintains a constant pressure drop over the spool orifice. This means that direct (manually) operated valves can be pressure compensated but the valve becomes more complex; compare figs. 8-26 and 8-52 (the function is explained in chapter 7, figs. 7-23 and -24. Fig. 8-49 shows a manually operated CFO valve without compensators.
2) By utilizing flow forces (so called *Bernoulli* forces) on a hydraulically operated, proportional valve spool. This provides for a simple and reliable function with pressure compensation of both the lifting and lowering movements.

The controlled flow rate remains constant for a specific control lever position, and is independent of variations in load or supply pressure; no additional components are required (fig. 8-50).

Spool leakage

Valve spools are normally hard-chromium plated, which ensures low friction and practically no wear (providing the oil is reasonably free from contamination). It is also important that the spool is centered in its bore to minimize leakage, friction and wear.

Therefore, spools are provided with 'balancing grooves', machined around the outside diameter of the spool which can slide inside the bore. The grooves create a 'hydrostatic bearing' and minimize friction (fig. 8-51). In addition, flow control grooves or cutouts are machined or stamped into the edges of the spool lands (fig. 8-52). The shape and size of these grooves determine the characteristic (flow-versus-spool position) of the valve.

Although small, the radial clearance between spool and bore causes internal leakage. To mini-

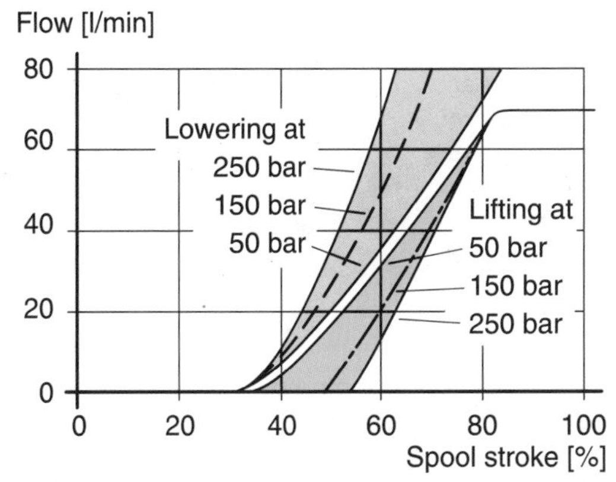

Fig. 8-49 Flow diagram for hand operated CFO valve

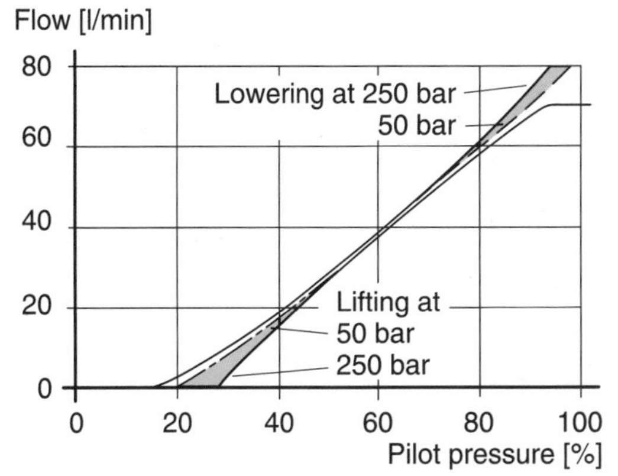

Fig. 8-50 Flow diagram for the CFO pilot operated valve (above)

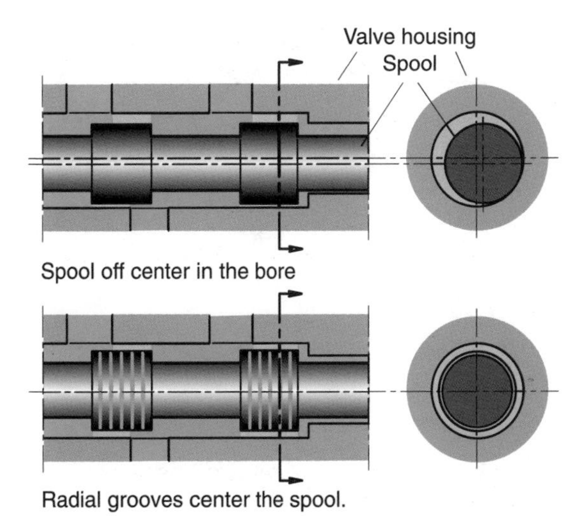

Fig. 8-51 Spool position in the valve housing bore

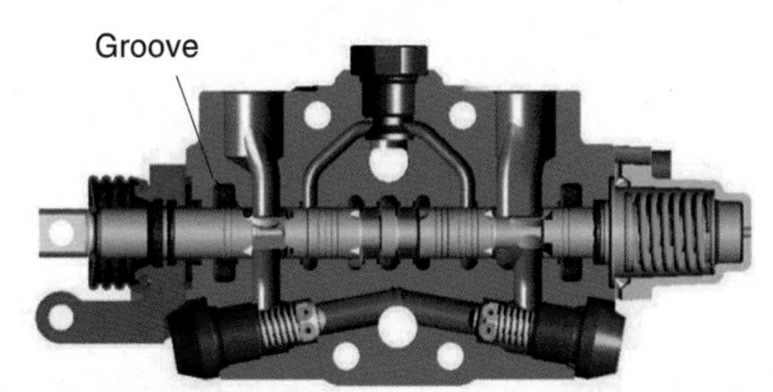

Fig. 8-52 Spool with grooves improve the operating characteristic

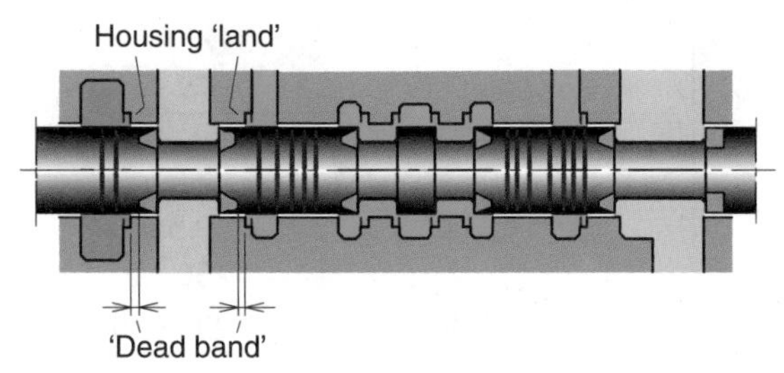

Fig. 8-53 Spool-housing overlap ("deadband")

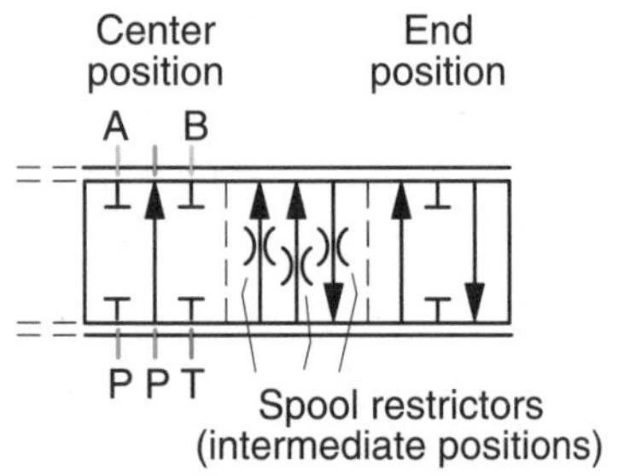

Fig. 8-54 Valve symbol - negative overlap

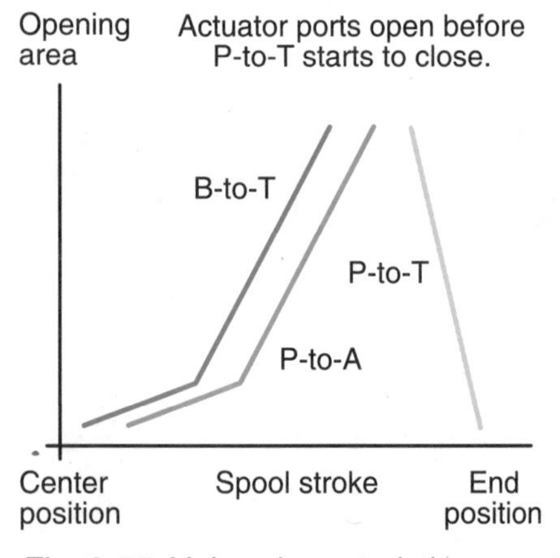

Fig. 8-55 Valve characteristic - negative overlap

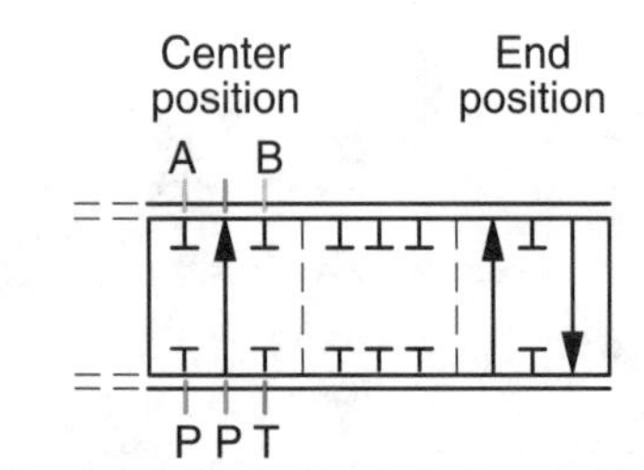

Fig. 8-56 Valve symbol - positive overlap

mize leakage in the neutral position, spools are usually designed with an axial "overlap". This overlap is often called ***deadband*** or ***dead zone*** (fig. 8-53).

The radial clearance between spool and valve housing is around 5 µm *(0.0002")*. The leakage over a spool land is determined by:

- clearance size
- pressure differential over the spool land
- viscosity of the fluid
- deadband length

NOTE: By machining the lands of the housing, the length of the deadband can be kept within a close tolerance. Consequently, max and min leakage also stays within a narrow range.

Spool overlap

The standard schematic symbol of a mobile valve shows only the spool function in the main positions.

It is, however, important to know what happens between these main positions, i.e. while the actuator is being controlled. This can be shown with additional squares that are set off with dashed lines.

In the CFO valve (with a double acting spool powering the back-and-forth movement of a cylinder), the three cooperating restrictors/orifices control the load. By varying the size of the restrictors/orifices, intermediate pressures can be obtained, thereby increasing control of the function (load).

If the spool has a ***negative overlap*** or ***underlap*** and the spool is moved to the left (to the right hand end position; figs. 8-54 and -55), the actuator port-to-tank (B-to-T) and pump-to-actuator port (P-to-A) opens before the pump-to-tank (P-to-T) closes. This results in a lower pressure buildup during operation (P-to-A and B-to-T); a low intermediate pressure means a smaller power loss.

In other words, all three restrictors are open simultaneously during part of the spool stroke and power consumption is optimized.

If the spool has a ***positive overlap*** (figs. 8-56 and -57), the pump-to-tank passage (P-to-T) closes before the actuator-to-tank (B-to-T) and pump-to-actuator (P-to-A) opens, which means that the pump provides maximum pressure, and that the system acts as a constant pressure system.

All ports are closed or blocked during the intermediate part of the spool stroke. With positive overlap, manoeuvering is optimized. The diagrams in figs. 8-55 and -57 show the difference in spool area between negative and positive overlap at a chosen spool position.

In actual valve designs, a number of compromises are found that positions the characteristic somewhere between the above two extremes - positive and negative overlap.

Zero overlap means that both pump-to-actuator (P-to-A) and actuator-to-tank (B-to-T) opens and pump to tank (P-to-T) closes at the same time; this condition is normally found in servo valves. More information on electrohydraulic valves can be found at the end of this chapter.

Spool configuration

Flow paths through a directional control valve are determined by the spool configuration. There are many spools to choose from and a suitable spool can be selected for a particular system type, flow, load condition and actuator area ratio. Spools are also available with varying degrees of pressure feedback.

D spool symbols for different system types are shown in fig. 8-58. In fig. 8-59, the most common spool types are presented with CFO symbols.

Spool type D

This is a double acting spool used to power an actuator in both directions. The actuator ports are typically blocked in the neutral position. The pump connection communicates with tank (CFO) or is blocked (CP, LS).

By shifting the spool to the left, pressure is connected to port A, and port B to tank. By shifting it to the right, pressure connects to B, and A to tank.

Spool type Dm

This is a double-acting spool with small drain or leakage passages (A-to-T and B-to-T) when in the neutral position. This minimizes pressure buildup in the actuator ports. This spool is used typically in combination with counterbalance valves.

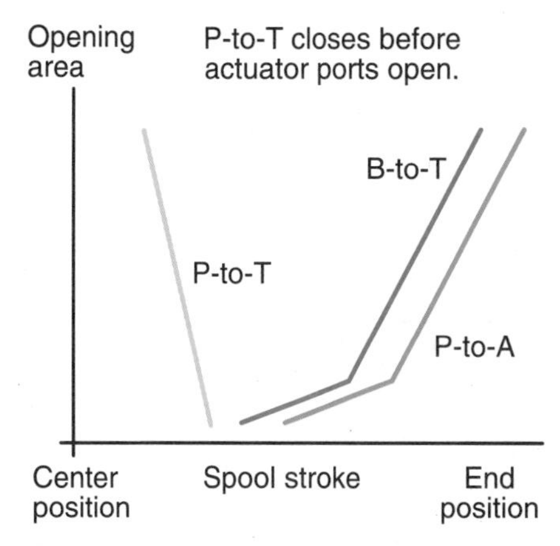

Fig. 8-57 Valve characteristic - positive overlap

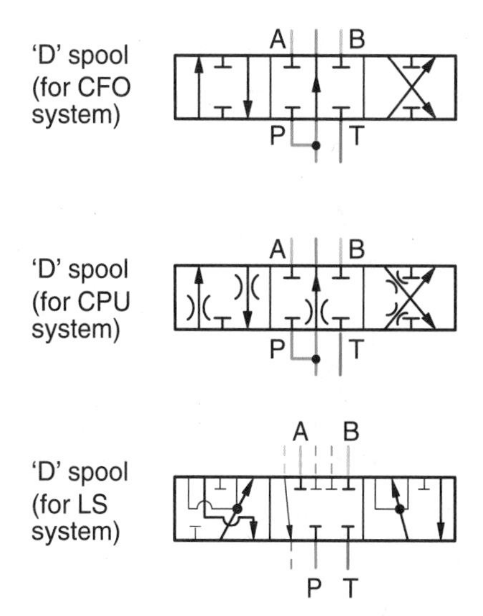

Fig. 8-58 Spool type 'D' symbols for CFO, CPU and LS systems

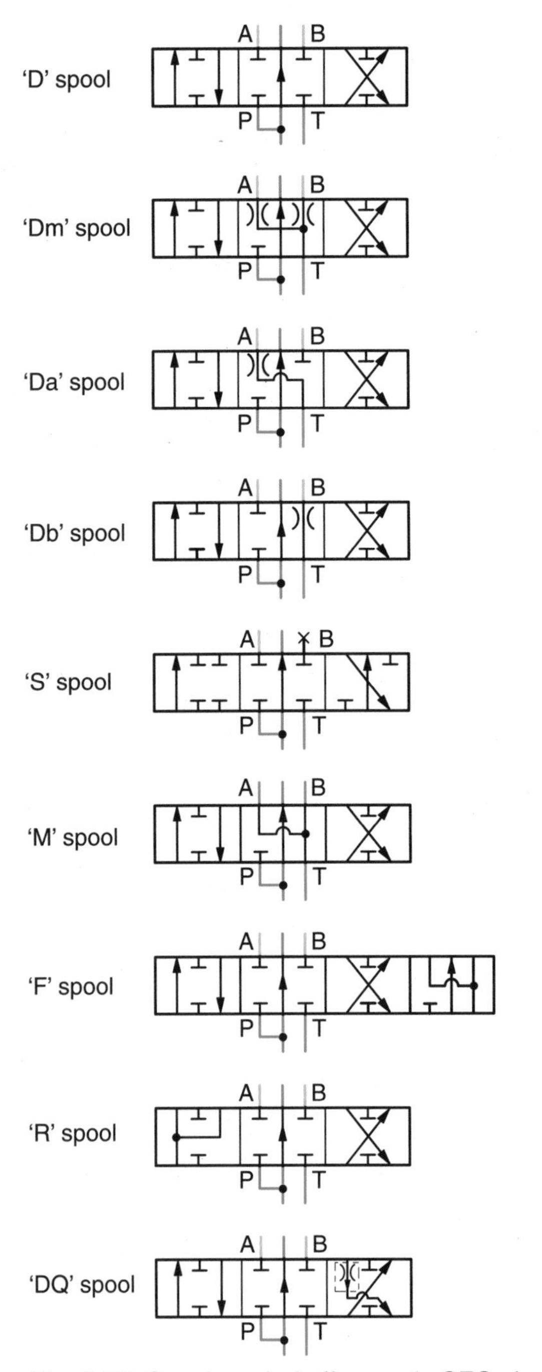

Fig. 8-59 Spool symbols (for use in CFO circuits)

Spool type Da

This double-acting spool has a drain or leakage passage A-to-T, which prevents pressure buildup in the A port when in the neutral position. The spool is used in combination with a counterbalance valve installed in the actuator line connected to port A.

Spool type Db

A double-acting spool with drain or leakage B-to-T preventing pressure buildup in the B port when in the neutral position. The spool is used in combination with a counterbalance valve in the actuator line connected to port B.

Spool type S

The two versions of this single acting spool allow either the A or the B port to be used to actuate a cylinder or unidirectional motor, while the other port is blocked.

The cylinder has either a built-in spring or a pushing load that retracts it. When lowering a load, retracting the cylinder, the pump-to-tank line is normally open to avoid unnecessary pressure buildup (CFO).

With an LS spool, the LS signal is not connected when lowering the load, thereby eliminating pressure buildup.

Spool type M

This is a motor spool, utilized to control hydraulic motors. Compared to the D spool, in the M spool, both motor ports are connected to tank in the center position.

When the spool is centered, an abrupt stop of the hydraulic motor is avoided and the risk of cavitation is reduced. Stopping a "living mass" or other high inertia load with e.g. a D spool could generate a very high pressure spike that might damage the hydraulic system.

In addition, the M spool is used in systems where pilot operated check valves or counterbalance valves are blocking the cylinder ports. In these cases, pressure buildup between the spool and the counterbalance valve is prevented, thereby eliminating the risk of check valve/counterbalance valve opening.

Spool type F

This spool is identical with the D spool in three of its positions. An additional fourth position is the ***float*** mode, similar to the mid-position of the M-spool.

In the float mode, both actuator ports are connected to tank, thereby allowing free flow in all directions. The pump port is connected to tank (CFO) or blocked (LS, CP). A spring loaded detent is generally employed to maintain this position.

As the name implies, the float position is used to allow the load to move freely. For instance, when a loader or scraper is working on a road, the bucket follows the contour of the road surface while the vehicle is driving forward.

Spool type R

This is a so-called regenerative spool. In neutral and one of the external positions, the R spool is exactly the same as the D spool.

In the opposite, external position (the regenerative position), the rod side of the cylinder provides a forced makeup flow to the piston side. This means that the cylinder's rod diameter area is the active, pressurized area. A higher speed with reduced force can thereby be obtained.

Common applications for the R spool are the telescopic outer boom on a crane and the ejection circuit on a scraper.

Spool type DQ

This is a double acting spool for a double-acting cylinder where the ports are blocked in the neutral position. An integrated flow limiting valve (lowering brake) limits the lowering speed; refer to chapter 7. Normally, the lowering brake valve is pressure compensated, preventing the maximum lowering speed from being influenced by variations in load pressure. The lowering brake function can be attached to all spools shown previously.

Spool nomenclature

The spools described above are examples of the most common types but many other spool functions are available. It is, in principle, possible to design a spool for a specific application. The nomenclature of a particular spool is, however, not standardized.

Spool selection

Spool selection is one of the most delicate matters for the hydraulic system designer. In a hydraulic system, the application, function and operator control alternatives must be considered. The following details some general guidelines.

When selecting spools for a CFO (constant flow, open-center) system, the pump flow is of main interest. The entire pump flow must pass through the valve's open-center line back to tank. Flow forces affect all spools in the valve and increase with increased pump flow as well as the corresponding pressure drop. When the spools are in neutral, they are in a so called ***hydraulic balance***.

The balance changes when the spool is being moved and ports open to the actuator. The spool is now affected by flow forces in both actuator ports as well as of those in the open-center line.

With a very large pump and/or actuator flow, there is always a risk of the spool being unable to return to neutral from the outer position by itself.

Therefore it is important **not** to exceed the flow limit stated by the valve manufacturer.

In a CP (constant pressure) system, the spool restrictors/orifice areas in the pump-to-actuator and actuator-to-tank valve passages control the function speed as long as there is a sufficiently high pump pressure available.

When selecting spools for a CP system, the function speed (the flow through the actuator port), is the first concern. Of second is the pressure difference, ΔP, between the pump and the load.

When estimating the actuator speed, the flow to the actuator or the spool opening area (A), the simple pressure drop formula (Formula 8-1) can be used for P-to-A and B-to-T calculations.

In an LS (load-sensing) system, the load pressure in the actuator port is used as a control signal to the compensator of the load-sensing pump, which reacts up to the required pressure level.

Formula 8-1 Pressure drop over a spool restriction

$$\Delta P = C \times \frac{q^2}{A^2}$$

where: 'ΔP' is pressure drop over the orifice (the restricting ring area between spool and housing)
'C' a constant (which depends upon the shape of the orifice and the specific gravity of the fluid)
'q' flow through the orifice
'A' orifice area.

If several loads are controlled simultaneously, the pump operates somewhat above the highest load pressure level. This load is consequently pressure compensated through the pump's compensator while the lighter, simultaneously operated loads, work in much the same way as in a CP system (if the valve sections are not provided with separate compensators).

When selecting spools for the LS system, there are additional requirements that should be satisfied:

1) In order to provide the best possible control for the operator as well as low system losses, the pressure drop (DP) relationship between meter-in (pump-to-actuator) and meter-out (actuator-to-tank) restrictions/ orifices, as shown in chapter 6, must be matched to the controlled function. The cylinder's piston-to-rod area ratio and the type of load affect this relationship.

2) The load can be either pushing or pulling, or be both (pushing and pulling).

   - A pure ***pushing*** load reacts against a meter-in (pump) restriction and doesn't need a meter-out restriction.

   - A pure ***pulling*** load acts against the meter-out (tank) restriction and doesn't need substantial meter-in restriction (fig. 8-60).

3) A double acting load requires a more careful investigation where the movements in both directions have to be analyzed separately.

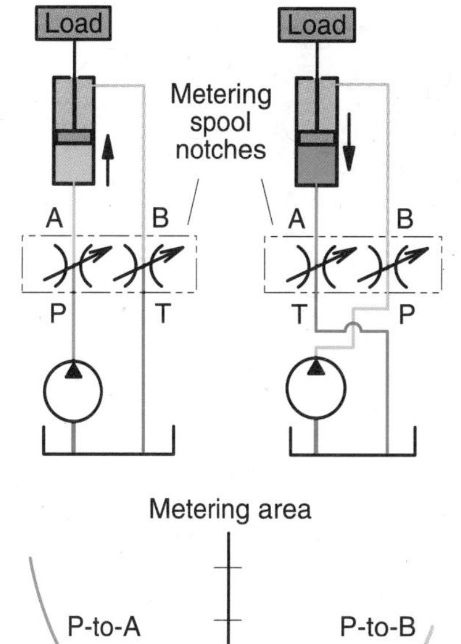

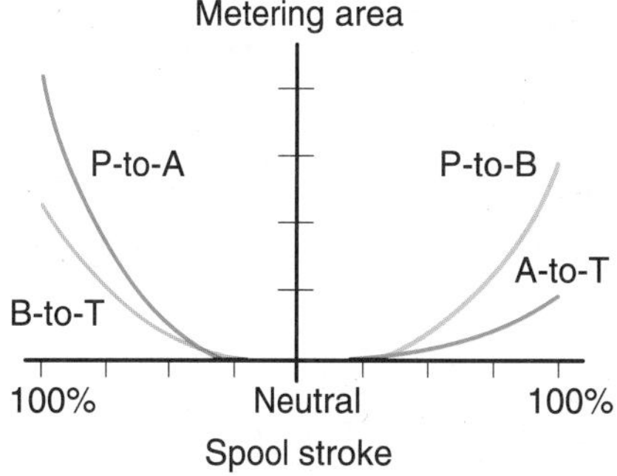

Fig. 8-60 Spool metering (orifice) areas as a function of spool position

Spool actuators

The spool in a directional control valve can be positioned anywhere between the left and right stroke limiters. The position is determined by a spool actuator which can be mechanical, manual, electric, hydraulic or pneumatic, or a combination of these methods (shown in fig. 8-61).

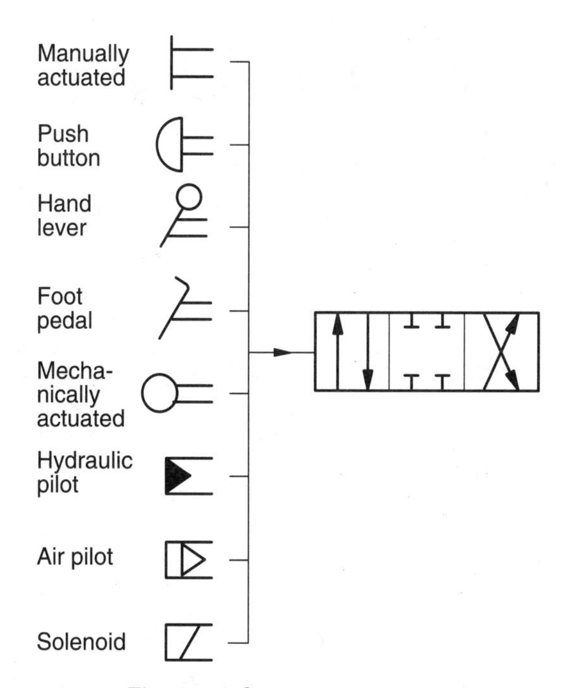

Fig. 8-61 Spool actuators

Mechanical actuators

A simple type of mechanical actuator is the plunger, equipped with a roller (fig. 8-62). The roller moves against a cam which is attached to a cylinder rod.

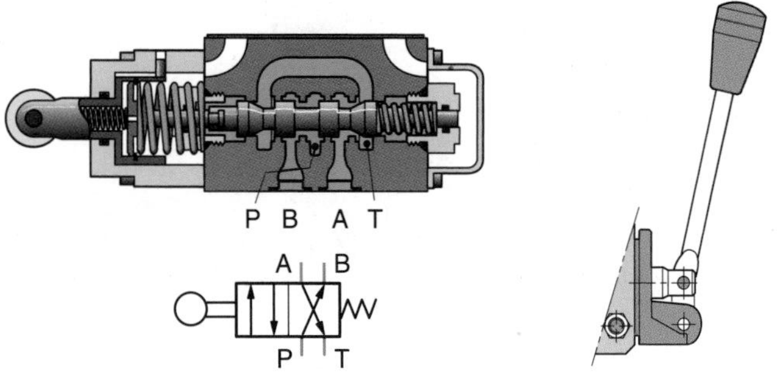

Fig. 8-62 Cam operated spool *Fig. 8-63 Hand lever*

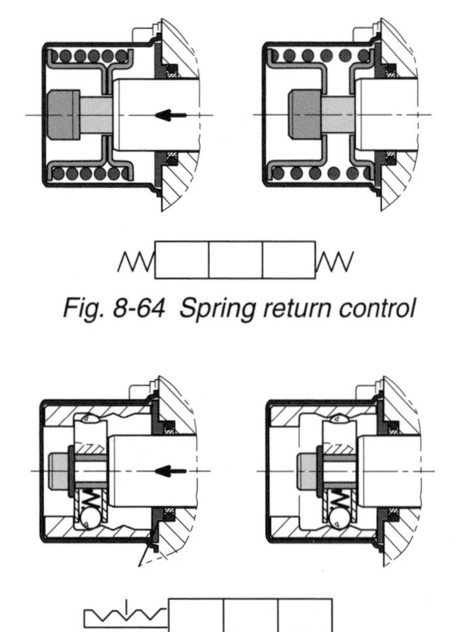

Fig. 8-64 Spring return control

Fig. 8-65 Three-position detent control

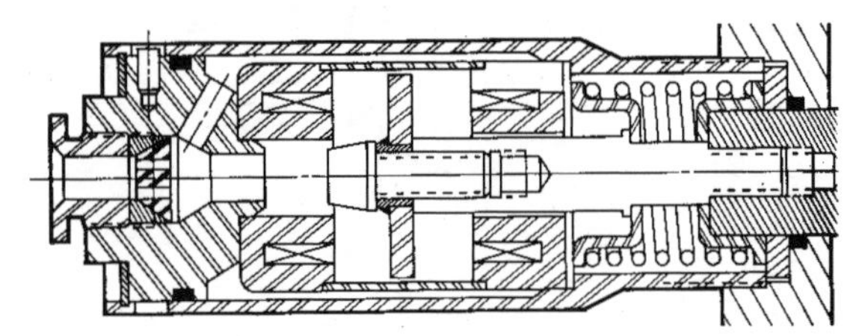

Fig. 8-65a Magnetic detent

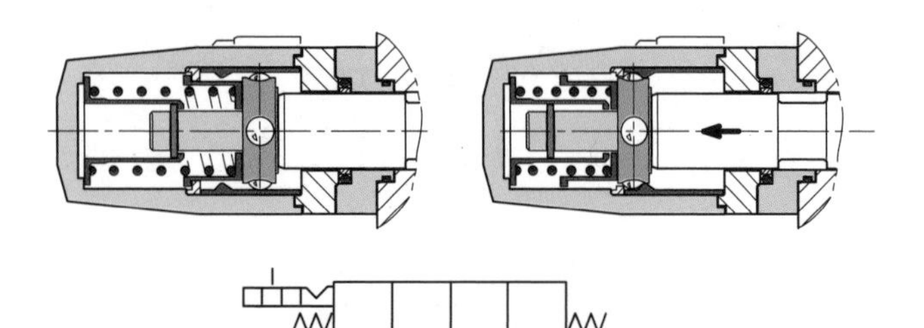

Fig. 8-66 Four-position 'float' control

Plungers are used on DCV's where the operation must be sequenced. This is a requirement when valve shifting must occur at the same time as the actuator reaches a specific position.

Manual actuators

A directional control valve, DCV, whose spools are moved by muscle power is known as a manually operated or manually actuated valve. Manual actuators include hand levers (fig. 8-63), push buttons and foot pedals.

The most common manual operating control is the hand lever, which the spool pushes or pulls into position. The control normally contains a spring, which returns the spool to the center (neutral) position as soon as the lever is released (fig. 8-64).

In a detented spool control (fig. 8-65), the detent keeps the spool in any of three chosen positions.

There is also the non-mechanical detent known as a magnetic detent (fig. 8-65a). In the magnetic detent, a "clapper" is attached to the spool end. The clapper moves between two electromagnetic coils, known as "pole-pieces." Therefore, no matter which direction the clapper is moved, it moves in the direction of one of the pole-pieces. When full stroke is reached, the clapper makes contact with a pole-piece which is constantly energized. It is then held in position as long as electrical current is supplied to the pole-piece coil.

A combination of a detented and a spring centered spool control is also available. For instance, a float ('F') spool must have a detent in the floating (fourth) position (6/4 spool). In the other three, normal positions, the spool is spring centered (fig. 8-66).

Pilot operated spools

DCV spools can be shifted with either pneumatic or hydraulic pressure. Pilot pressure is applied to the spool ends, or to a separate pilot piston connected to the spool.

Pneumatic pilot control

Movement of the spool can be achieved with a small *pneumatic cylinder* (fig. 8-67) together with a common spring centering device. The cylinder is controlled proportionally with a pneumatic joystick, or on-off through solenoid valves.

When one of the solenoids is activated (fig. 8-68), air pressure is supplied to one side of the cylinder and the spool moves to the corresponding end position. The pneumatic solenoid control is normally utilized for remote, on-off operation of the DCV.

The pneumatic spool control, when installed on one end of the valve, makes it possible to use a lever on the other spool end as seen in the illustration.

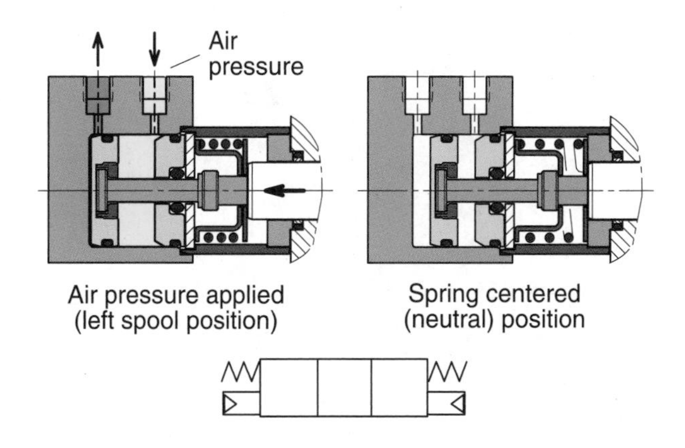

Fig. 8-67 Pneumatic pilot control

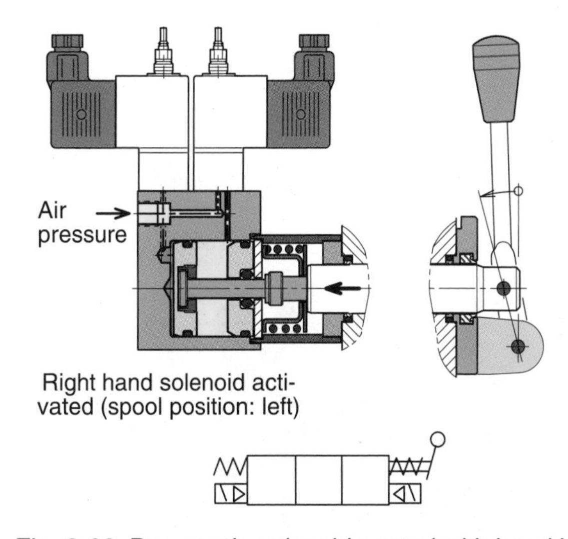

Fig. 8-68 Pneumatic solenoid control with hand lever

Hydraulic pilot control

In a hydraulic pilot control (fig. 8-69), pilot pressure is acting on one end of the spool, overcoming the force from the spring which (in this design) is located at the other end.

Since the compression of the spring develops a constant force-versus-deflection increase, the spool movement is directly proportional to the increase in pilot pressure. This, in turn, provides very good tracking characteristics and repeatability from the operator's control device.

The other advantage with this type of control is that the valve housing lacks o-rings, which means less hysteresis (friction) and increased control and resolution (minimum discernible change).

An additional advantage with the hydraulic pilot control is that the spring force (centering the spool) can be adjusted externally and that the spool stroke can be limited, without disassembling the control.

The hydraulic spool actuator is pilot controlled from a hydraulic joystick or similar device; refer to chapter 9, Remote control systems.

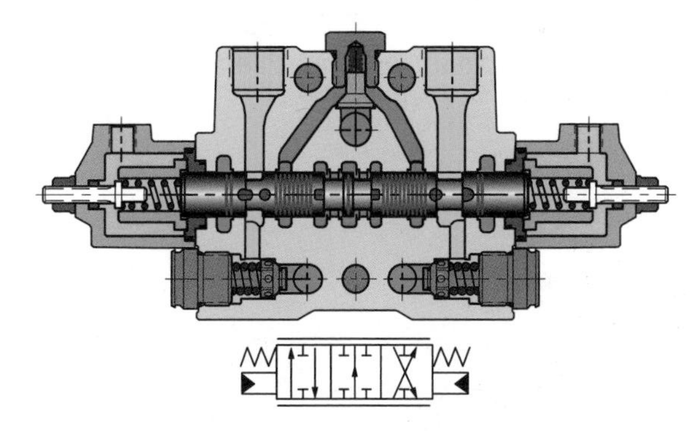

Fig. 8-69 Hydraulic, pilot operated control

Solenoid operated, on-off control

On directly actuated valves, the armature of the solenoid is pulled into the coil and pushes the spool against a spring force. The movement of the spool requires a large magnetic force which is the main reason why only small valves (up to nominal size NG 10 or $^3/_8$") are direct, solenoid operated (fig. 8-70). In mobile equipment, solenoids are usually 12 or 24 VDC.

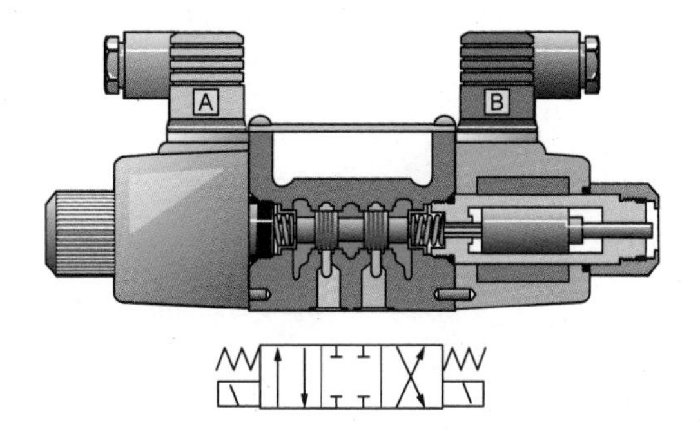

Fig. 8-70 Direct solenoid operated control

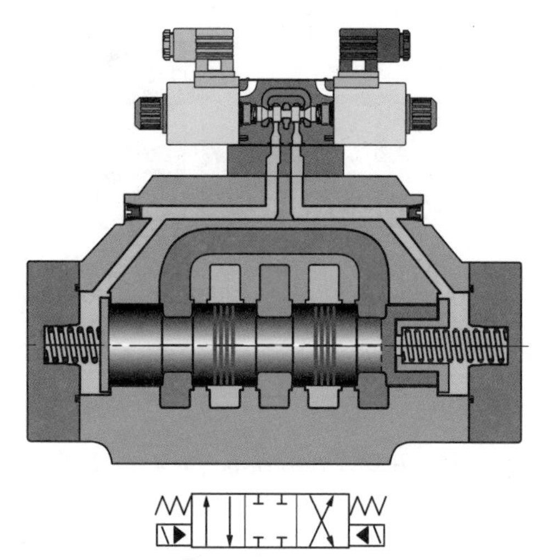

Fig. 8-71 *Electrohydraulic, pilot operated control*

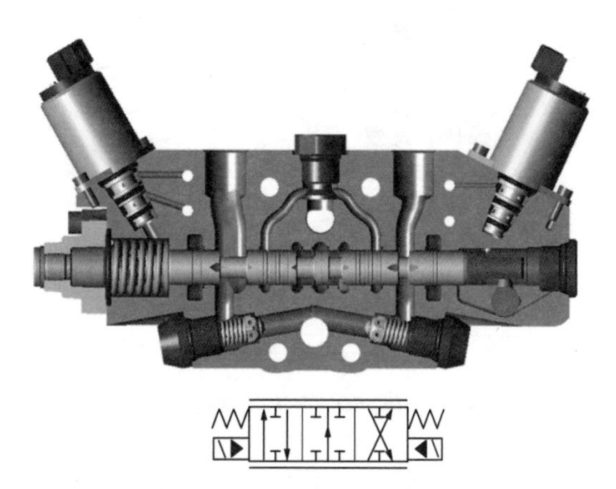

Fig. 8-72 *Proportional, pilot operated solenoid valve (mobile type)*

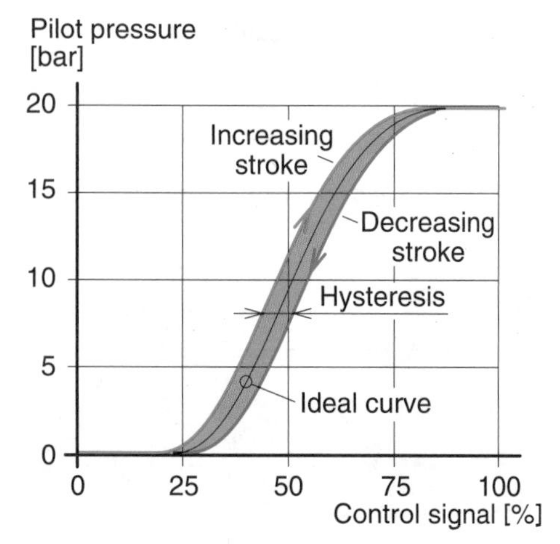

Fig. 8-73 *Pilot pressure versus control signal showing hysteresis*

Larger valves are pilot operated via a small solenoid valve controlling a pilot signal (fig. 8-71). When activated, the armature pushes on the pilot spool which moves to an extreme position, and the available pilot pressure enters the end cap of the main spool valve as shown.

Solenoids are designated *wet* or *dry* depending on whether air or oil surrounds the armature. The wet type is often used in mobile applications as the armature is more protected against wear, more damped and better cooled as compared to the dry armature.

Compared to the AC solenoid, the DC solenoid has high operating reliability and “soft switching”, and it doesn’t burn out easily if the spool jams.

Electrohydraulic, proportional pilot control

This style of spool control contains proportional solenoids, where the solenoid is the interface between the electronic and hydraulic control signals.

Proportional solenoids generate an output force, proportional to the input current; the force is constant over the length of the stroke. The force controls a pressure reducing valve that produces a corresponding hydraulic control pressure to the spool ends of the DCV.

In modern multispool valves, pilot pressure is supplied internally from the main pump via a pressure reducing valve and fed to the solenoids through a common internal pilot pressure supply channel/passage. The proportional solenoid then opens the pilot line to the spool end cap, increasing the pressure in proportion to the incoming, electric control signal (fig. 8-72).

The electric signal is often of the PWM (pulse-width-modulated) type, where a ‘dither’ is superimposed on the command signal in order to reduce valve hysteresis (illustrated in fig. 8-73).

A directional control valve equipped with electro-hydraulic remote control and adequate electronic control equipment (which reduces friction and the effects of varying temperatures), will perform close to that of a servo valve. Further improvement is achieved by installing a spool position sensor, connected to the control electronics, which monitors the position of the spool.

Another pilot stage design, the pulsar type pressure control valve, is a normally closed, spring-biased solenoid-actuated high-speed on-off valve. It works as a variable restrictor between the pilot pressure supply and the spool end cap (fig. 8-74). The cap has a fixed drain restrictor/orifice connected to tank.

The solenoid, when energized with a PWM signal, produces a control pressure that is directly proportional to the duty cycle. The wider the pulse width, the more oil flows into the pilot chamber and through the drain orifice, which, in turn, increases the pilot pressure correspondingly.

Fluid that exits the cartridge has to pass through a small restrictor/orifice, which results in a backpressure, proportional to the duty cycle as determined by the operator of the machine. This pressure then pushes on the main spool, biased by the centering spring on the opposite side.

A third design, shown in fig. 8-75, uses the main pump pressure which is internally fed to the pilot spool. When the solenoids are inactive, the pilot pressure line is closed and the spool end caps drain to tank. The main spool is spring centered.

When the solenoid is activated, the armature pushes the pilot spool to the left in proportion to the incoming signal. The pump pressure now passes to the left end of the main spool.

The right end of the spool is still connected to tank and the spool can now move to the right against its centering spring. Increased spring resistance results in higher pilot pressure on the left side. This pressure is also connected to the left side of the pilot spool, which starts to move to the right when the pressure force equals the solenoid force; this stops the movement of the main spool.

The travel speed can be regulated by restrictors/orifices, and maximum output flow by limiting the spool stroke.

The above valve types are normally controlled by electric or electronic joysticks.

Electrohydraulic directional flow control valves

Over the years an increasing need for higher response, stiffer systems, better flow characteris-

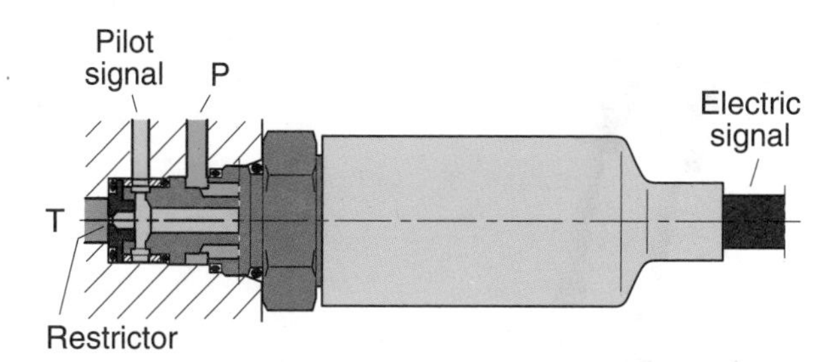

Fig. 8-74 Pulsar type on-off solenoid valve cartridge

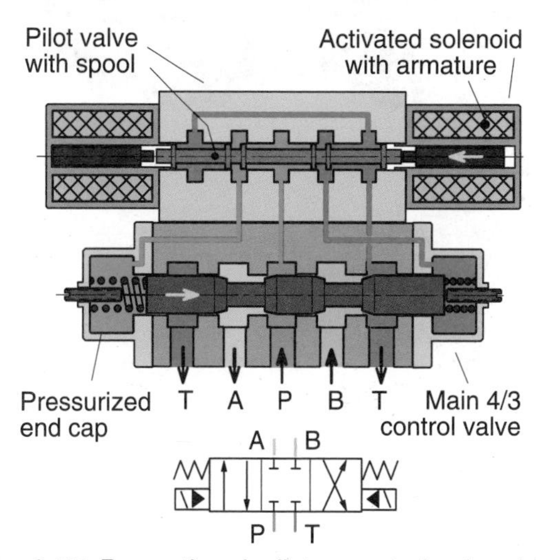

Fig. 8-75 Proportional, pilot operated solenoid valve (for manifold mounting)

tics and an efficient means of interfacing electronic control systems (joysticks, electronic controllers, etc.) with hydraulic control systems has arisen within the mobile hydraulic industry. The use of electrohydraulic directional flow control valves has met this need.

In general, there are two types of electrohydraulic valves; namely, proportional (fig. 8-76) and servo valves. Which type is used in a particular hydraulic system depends upon the sophistication and performance required by the system.

Fig. 8-76 Typical mobile style proportional valve

Proportional vs. servo

There are several areas that distinguish the electrohydraulic proportional valve from an electrohydraulic servo valve. These areas are in the overall response of the valve, the spool center condition, the hysteresis, repeatability, threshold of the valve, and the filtration requirements of the valves. As both valves (proportional and servo) are electrohydraulic the same terms are used to describe the construction, function and performance of these valves. Several terms are repeated in this section from previous sections, this has been done to better explain the valves and to reinforce their meaning.

Response

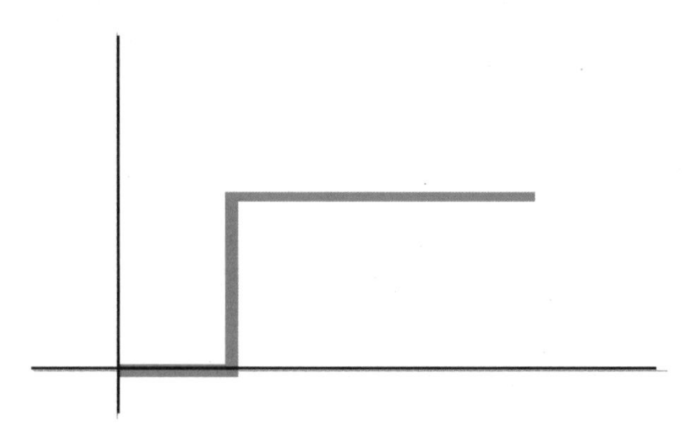

Fig. 8-77 Step response

The term response can have two different meanings. The first, “step response,” (fig. 8-77) is generally defined for proportional valves as the time required for the valve to achieve maximum rated flow (typically full spool shift) due to an electrical step input command signal. Depending on the size of the valve, “step response” times can range from 30 ms to 70 or 80 ms or more.

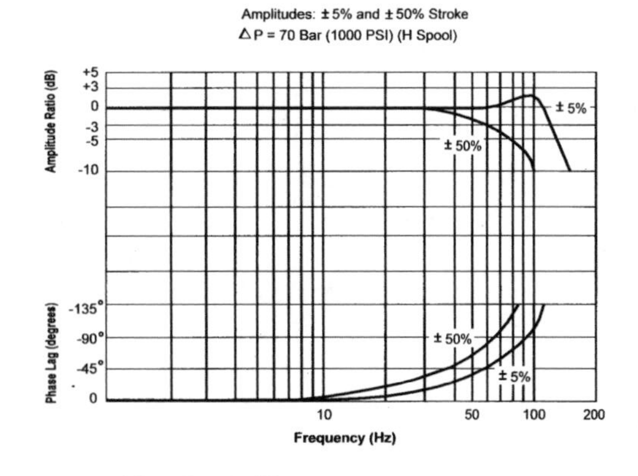

Fig. 8-78 Frequency response

The second “response” term is “frequency response” (fig. 8-78). This term is generally used when working with servo valves, though increasingly with proportional valves. Frequency response is expressed in hertz and is the frequency of a small sine wave command signal that causes the valve output amplitude to be -3db when compared with the valve output amplitude at a low frequency reference level (0.1Hz). The -3db point occurs when the valve output amplitude is 70.7% of the valves amplitude at the low frequency.

The “frequency response” for proportional valves vs. servo valves is 2-10 Hz vs. 10-300 Hz respectively.

Depending upon the system performance requirements, the spool center condition may be of more importance than its response characteristics.

Spool center condition

The spool center condition or crossover characteristics of the main valve spool are noticeably different between the proportional valve and servo valve especially when related to cost of valve, valve stability and operating characteristics.

Servo valves are "critically lapped" (fig. 8-79) by carefully matching both the width and the position of the spool lands to the flow ports within the valve body. In other words, the spool and valve body or sleeve are matched to produce a line-on-line or edge-on-edge contact by hand fitting (this adds greatly to their cost).

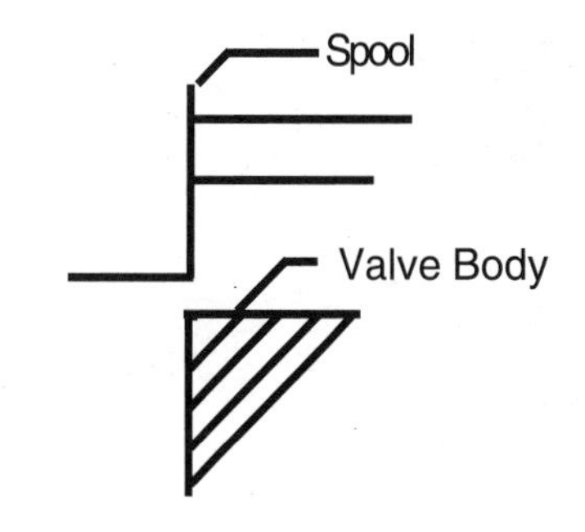

Fig. 8-79 Critically lapped spool

Proportional valves are designed so that the spools and valve bodies are interchangeable. This typically results in an "overlap" (fig, 8-80) of the spool and flow port on the order of 10-30% of the total spool stroke. This "overlap" creates a flow condition known as "deadband" or "deadzone." Though not a problem in circuits dealing with velocity control, "deadband" can cause instability and loss of resolution in a "closed loop" feedback positioning system.

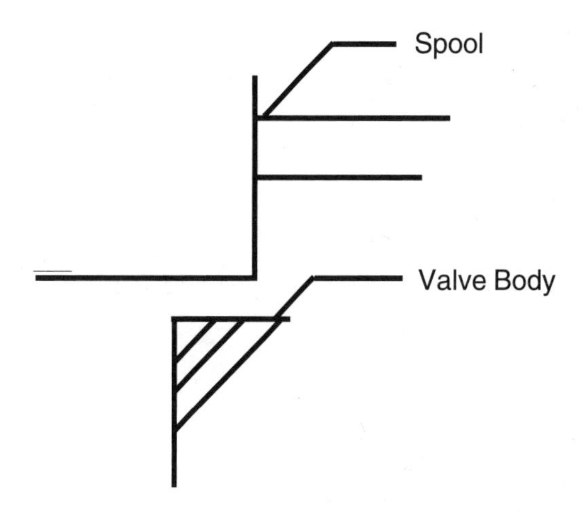

Fig. 8-80 Overlapped spool

Hysteresis, repeatability, threshold

Hysteresis is an accepted measurement of the difference in electrical command signal as percentage of rated electrical command for a given flow output level, when the flow output setting is alternately approached from above and below the desired flow output.

Repeatability is an indication of the ability of the valve to repeat a given flow when a given electrical command signal is repeatedly applied to the valve.

Threshold is the smallest discreet change in electrical command signal that will produce a corresponding change in the output flow.

"Open loop" proportional valves may exhibit levels as high as 10%. Typical levels for "closed loop" proportional and servo valves are 3% and less.

Detailed definitions to these electrohydraulic terms and others can be found in the Parker *Lexicon III -- Directory of Electrohydraulic Terms and Electrohydraulic Analogies.*"

Filtration requirements

Particulate contamination is the enemy of all hydraulic systems and especially servo valves. Because of their close tolerances, filtration requirements of 3 micrometres are specified. Proportional valves are a little more tolerant of contamination and require filtration of 10 micrometres.

NOTE: Some proportional valves use small servo valves as the pilot head, thus, requiring additional filtration for the flow of fluid being supplied to the pilot head.

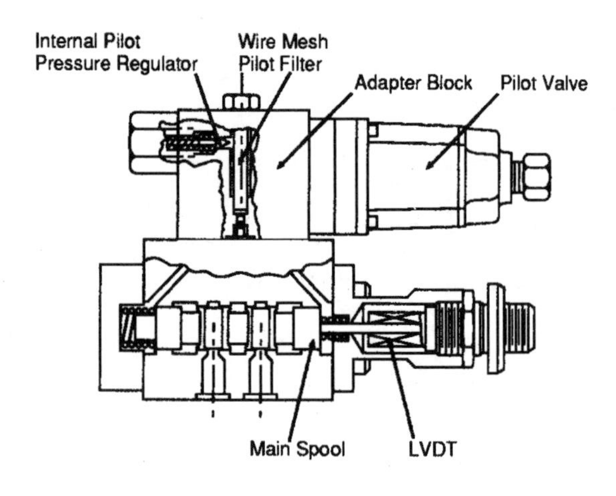

Fig. 8-81 Pilot operated proportional valve

Proportional valve construction

A typical pilot operated industrial proportional valve (fig. 8-81) consists of the torque motor pilot valve, adapter block, wire mesh pilot filter, internal pilot pressure regulator, main spool and body, and LVDT (Linear Variable Differential Transducer).

Another style of proportional valve uses proportional solenoids to operate the main valve spool direct with a positional transducer attached to the end of the valve spool to provide a feedback signal.

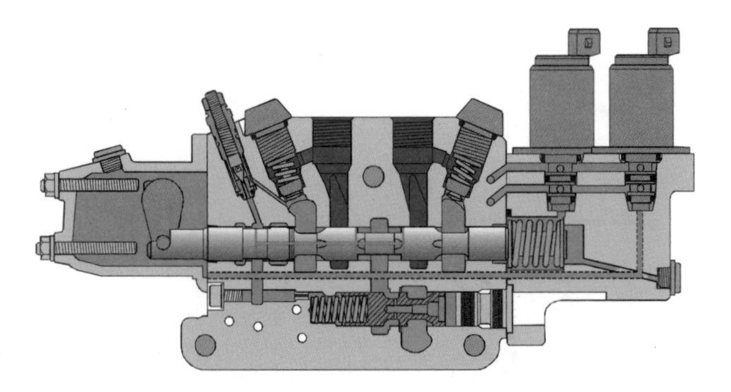

Fig. 8-82 Mobile style proportional valve

In the mobile style proportional valve (fig. 8-82) the spool position transducer is replaced with a spring (force feedback) or is eliminated all together. In this later case it is the operater that closes the loop and commends the spool position.

Still another type of proportional solenoid controlled valve design is used, that of direct operation.

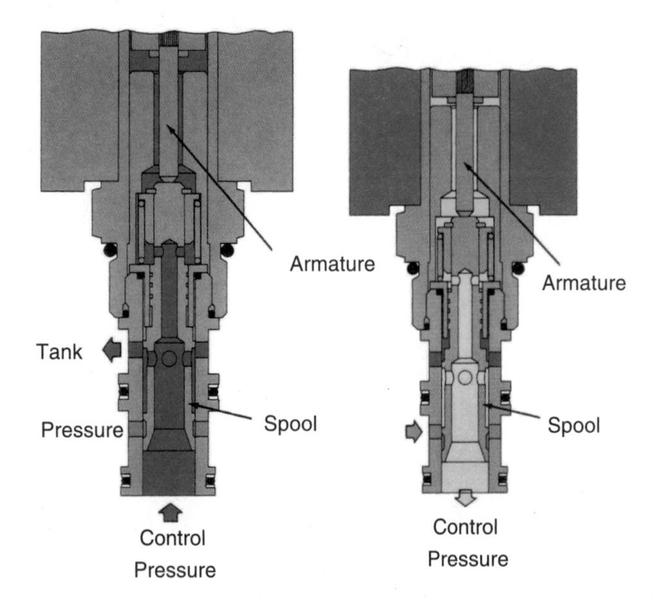

Fig. 8-83 Direct operated proportional valve

How a direct operated proportional solenoid directional valve works

The armature and spool are held in the neutral position by a centering spring (fig. 8-83). The pressure port is blocked by the spool lands. The control pressure port is connected to tank via the hollow spool. This proportional valve functions as a three way valve and could be used as the pilot valve of a pilot operated mobile proportional valve.

When the solenoid is energized the spool and armature are held down by the electromagnetic force of the solenoid compressing the centering spring. Fluid can now flow from the pressure port through the hollow spool to the control pressure port and out to the system or valve spool it is commanding. The amount of flow through the valve will be based on the amount of current flowing through the solenoid. Increasing or decreasing the solenoid current changes the flow, and resulting control pressure.

Some valve designs include a positional transducer or LVDT attached directly to the spool which measures the precise movement (position) of the spool and feeds this back to the electronics as a voltage signal.

Briefly in the electronics, the feedback signal and the solenoid input signal are compared generating what is called an "error." The electronics will then supply a voltage signal to the proportional solenoid moving the spool in either direction until the "error" equals zero. A detailed discussion of the control electronics is beyond the scope of this text.

At the same time, flow from the valve is increased or decreased.

How a pilot operated proportional solenoid controlled directional valve works

The main valve spool is held in the center condition by springs (fig. 8-84). Ports P, A, B and T are blocked by the land areas of the spool. With neither proportional solenoid pilot valve energized, the pilot supply flow, to the ends of the main spool, is blocked.

Energizing proportional solenoid "A" ports pilot control oil to the right spring cavity of the main spool (fig. 8-85). This converts the electrical signal into a pressure signal. This pressure signal forces the main valve spool to the left compressing the centering spring. At the same time pilot control oil is directed back to tank from the other end of the main spool. The metering slots on the main spool open progressively based upon the amount of pressure supplied by the pilot valve.

Just the reverse occurs when proportional solenoid "B" is energized. Pilot control oil is routed to the left end of the main spool via an internal pilot passage (fig. 8-86).

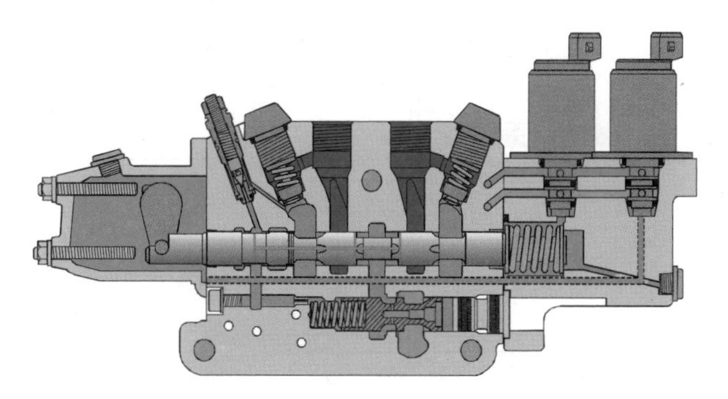

Fig. 8-84 Proportional valve in neutral

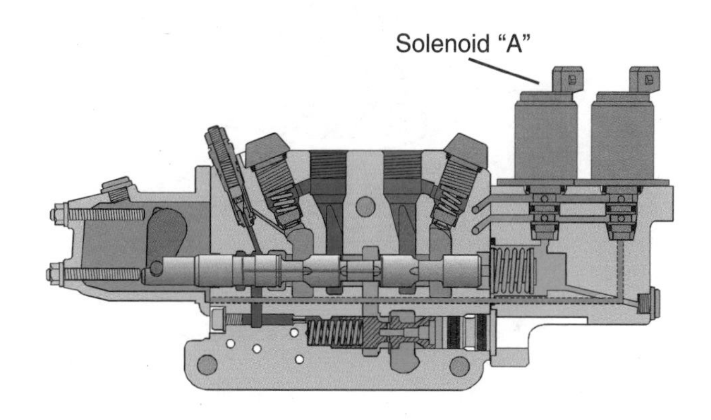

Fig. 8-85 Proportional valve activated

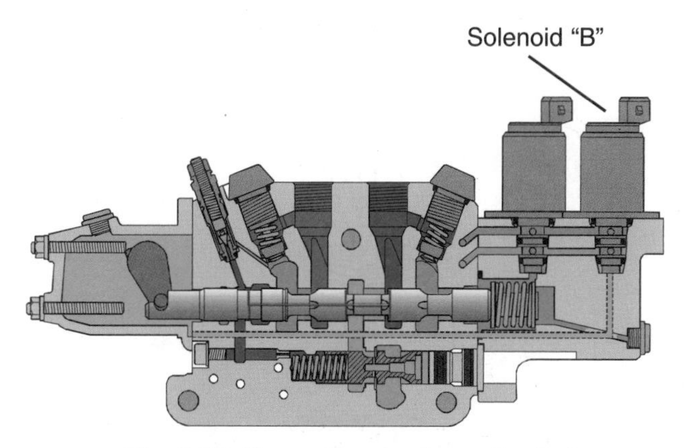

Fig. 8-86 Proportional valve activated

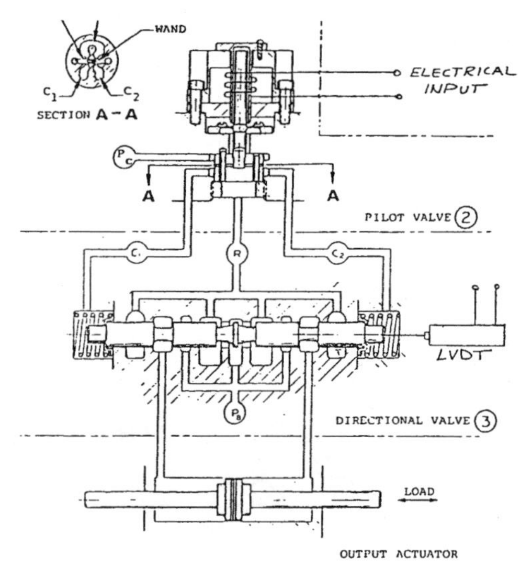

Fig. 8-87 Torque motor piloted proportional valve

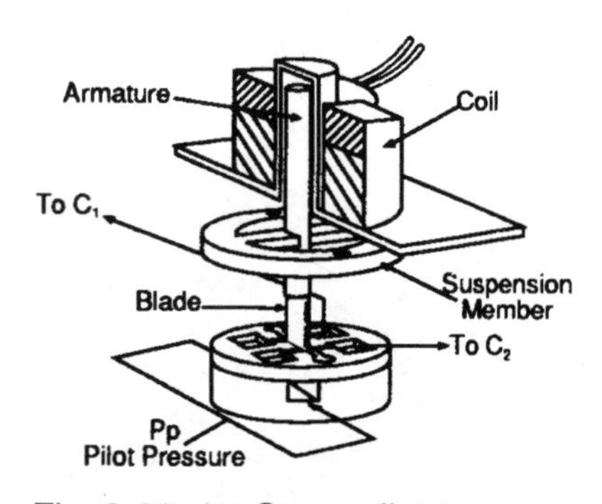

Fig. 8-88 1st Stage pilot-torque

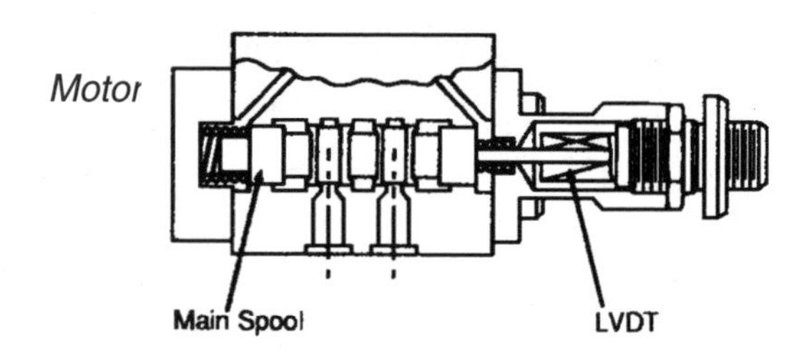

Fig. 8-89 Main stage

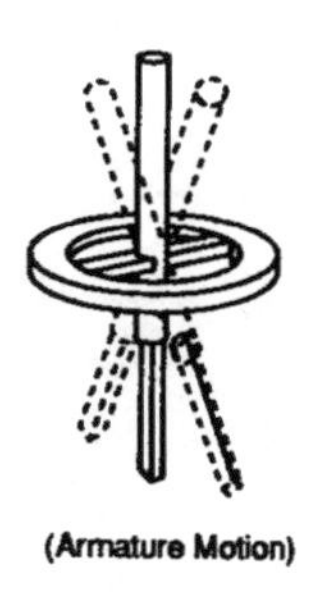

Fig. 8-90 Torque motor armature

If a positional transducer or LVDT is attached to the main spool it functions in the same way as previously discussed.

A pressure differential proportional directional valve (torque motor pilot)

A pressure differential type proportional directional valve is a two stage unit. The pilot valve sometimes called the 1st stage and the main spool or 2nd stage valve (fig. 8-87).

1st stage pilot valve consists of

The pilot valve basically consists of a coil, armature, suspension member, diverter plate, and blade (fig. 8-88). Some pilot valves incorporate a built-in relief valve and filter to limit pilot pressure and prevent contamination of the orifice(s) in the pilot stage.

The combination of magnet, coil and armature assembly is commonly called a torque motor.

2nd stage or main stage consists of

The 2nd stage or main valve consists of a spool, return springs, and possibly an LVDT (Linear Variable Differential Transformer). This stage is quite similar to a standard directional control valve (fig. 8-89).

How the pilot valve works

Current passing through the coil will create a north or south pole on top of the armature. The polarity of the current will determine the direction of movement of the armature between the poles of the magnet. As the armature is attracted to one of the magnet poles, it pivots on the suspension member and thus moves the diverter blade (fig. 8-90).

This movement will be proportional to the magnitude of the input command signal. The blade in the center position partially interrupts each jet in a manner to provide equal pressure in both receivers (C1 & C2). These receiver ports are connected, via internal passages, to opposite ends of the second stage or main spool.

When the blade moves from center, it will increase the interruption of the other jet stream. This will decrease the pressure in one receiver and increase the pressure in the other receiver (fig. 8-91). The resulting differential pressure between C1 and C2 will cause the second stage or main spool to shift proportionally.

How the main valve works

As was explained, the valve spool is moved back and forth by the differential pressure generated between C1 and C2 (fig. 8-92). However, these are not the only forces acting on the spool. Additional forces due to flow forces, dirt, friction and pressure loading can cause this type of proportional valve main spool to change position. To counteract these problems, the operator must change the input command signal, or an LVDT can be attached to the spool which generates an accurate electrical signal which is feedback to the electronics to indicate the spool position, this is known as position feedback.

The LVDT feedback is compared to the input command signal within the support electronics. If the two are not equal, the electronics increases or decreases the electrical power to the pilot coil, thereby adjusting the delta p across P1 to P2. This repositions the spool where the command inputs indicate it should be.

Force feedback

Another form of feedback incorporates force or pressure feedback (fig. 8-93). This style of feedback utilizes pressure on the end of the pilot spool, opposite the energized solenoid end, to drive the pilot spool back the center or neutral position. It is this pilot pressure that forces the spool back against the solenoid force created by the current passing through the solenoid.

The difference between these two forms of feedback (position and force) is that with position feedback the spool position is always being monitored by the position transducer. The position transducer tells the support electronics wether or not the valve spool is in the correct position. With force feedback the only thing known is that the pressure on the end of the pilot spool has increased to a level that forces the spool back against the solenoid force. Nothing is **known about the position of the spool.**

If more precise control is required then a servo valve is required. Typically these are mobile applications of industrial servo valves and depending on the application may be manufactured to higher standards.

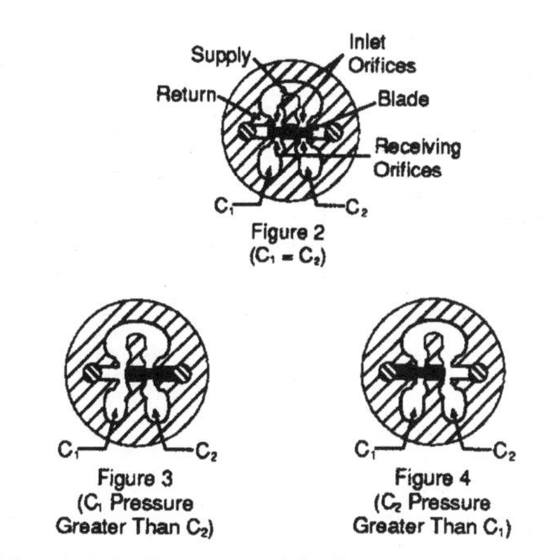

Fig. 8-91 Armature blade movement

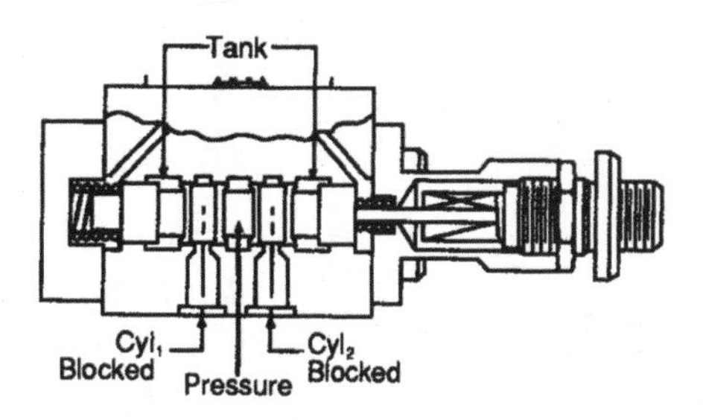

Fig. 8-92 Proportional valve main stage

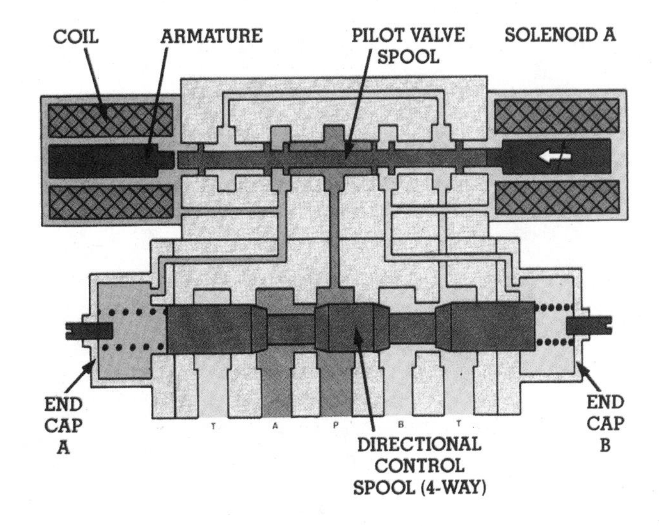

Fig. 8-93 Force feedback proportional valve

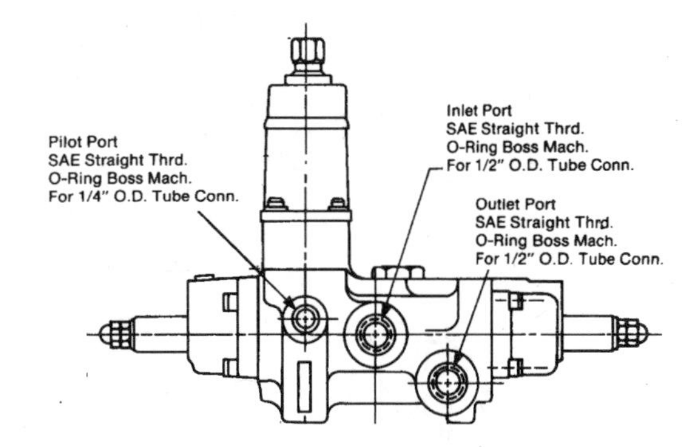

Fig. 8-94 Servo valve

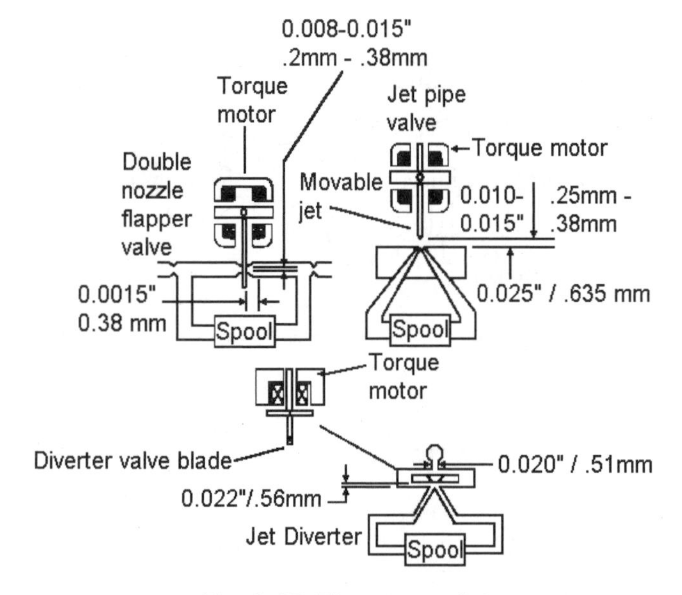

Fig. 8-95 First stage pilot

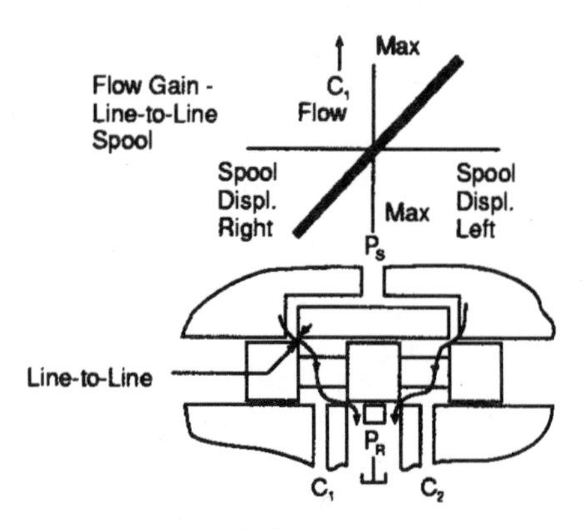

Fig. 8-96 Line-on-line servo spool

What a mobile servo valve consists of

A typical servo valve consists of a first stage (pilot valve) mounted onto the second stage (spool valve) with a mechanical feedback of spool position to the torque motor assembly (fig. 8-94).

Types of first stages

There are three common first stage designs, flapper nozzle, jet pipe and jet diverter (fig. 8-94). The minimum orifice for a flapper nozzle valve is the 0.38 mm *(0.0015")* clearance between the flapper and the 0.25mm/0.38mm (0.010/0.015") diameter nozzle (fig. 8-95).

A jet pipe valve has a typical nozzle diameter of 0.2mm/0.25mm (0.008/0.010") for the minimum orifice.

A jet diverter valve has a minimum orifice of 0.51mm(0.020").

It is not within the context of this textbook to suggest which first stage is best suited for an application, but it should be noted that the larger the orifice, the more contamination tolerant is the pilot valve.

Types of 2nd stage spool designs

The condition or matching of the second stage spool lands at the center position can and does vary depending upon the requirements of the system and/or tolerances during manufacturing.

The most common requirement dictates an edge condition of the spool lands to valve body ports to be as close to "line to line" as possible. Optional spool land matching includes "underlapped" and "overlapped" conditions. Each one of these creates unique flow characteristics within the valve.

Line-on-line condition

This "line-on-line" condition results in an ideal flow gain plot where the output flow to the cylinder ports is zero with the spool in the center position and increases immediately with spool travel (fig. 8-96).

Underlapped condition

An "underlapped" condition has more clearance between the spool land edges and metering notches or ports in the spool sleeve or valve body. This results in a higher leakage flow in the center spool position. Note that with this condition, the spool must travel through the underlap before proportional flow begins to the cylinder port (fig. 8-97).

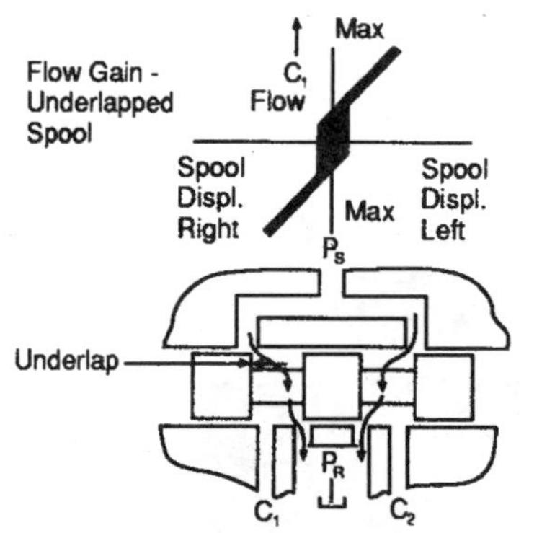

Fig. 8-97 Underlapped servo spool

Overlapped condition

An "overlapped" condition will reduce leakage flow to a minimum and no cylinder port flow will occur until the spool has traveled through the overlap (fig. 8-98). It should be noted that this configuration creates what is known as "deadband" or "deadzone" in the valve operation.

"Deadband" is a zone of valve movement in either direction from center where no actuator (cylinder or motor) response to input signal occurs.

NOTE: This condition is common in proportional valves previously discussed.

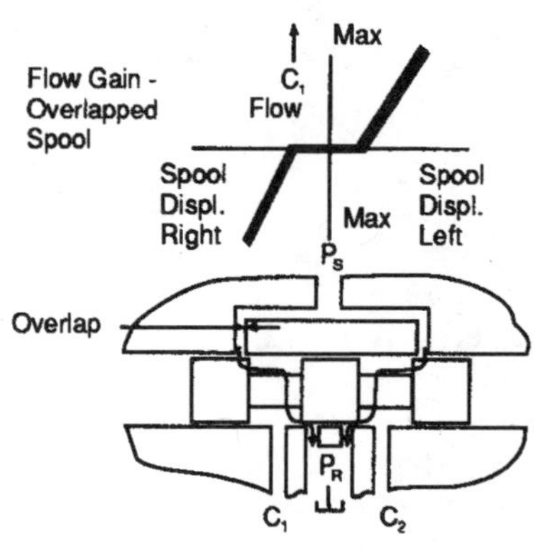

Fig. 8-98 Overlapped servo spool

How a servo valve works

A typical first stage section operation was explained in the proportional valve operating discussion and it is the same for servo valves.

The differential pressure created by the first stage is applied across the ends of the second stage spool and will cause it to move. In order to locate the spool into a position that is proportional to the electrical input command a feedback spring connects the first stage armature and second stage spool together.

This spring can be considered as a cantilever beam that is sized to provide a linear resisting force that is equal to the torque motor force for every spool position. When there is no electrical command, the diverter blade/armature assembly, the feedback spring, and the center position of the spool have a center line relationship (fig. 8-99).

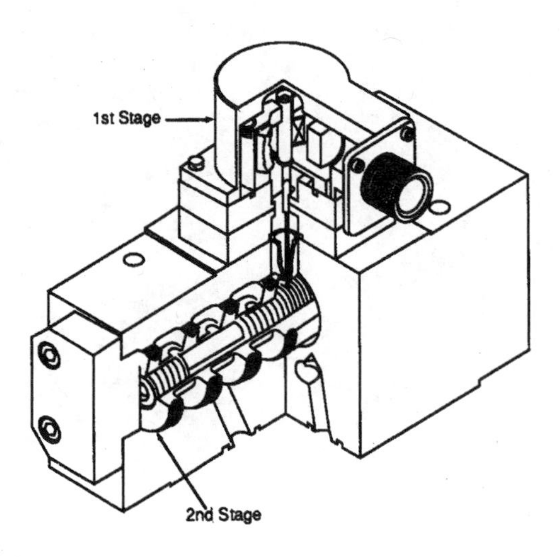

Fig. 8-99 Servo valve

Applying an input command signal to the torque motor causes the diverter blade/ armature assembly to move generating a differential pressure between the two pilot signals (PC1 and PC2). Again the spool is moved and at the same time deflecting the feedback spring developing a torque in opposition to the motor torque. When the spool has moved to the point where these two torques are equal, the diverter blade/armature assembly, the feedback spring and the spool are essentially recentered and spool movement stops at this new position related to input command signal.

As the input command signal is reduced to zero, the torques go to zero and the spool returns to its center position.

Detailed explanation of proportional and servo valves and their application in a circuit is beyond the scope of this text.

Wether it is a proportional or servo valve that is chosen for a specific system, both require additional support electronics.

Fig. 8-100 Electrohydraulic components

Support electronics for electrohydraulic systems

The required support electronics will take the form of "joystick" input devices, digital controllers, "black box" electronics, sensors and power supplies to mention just a few of these items (fig. 8-100).

In the typical mobile electrohydraulic application the input command is supplied by a "joystick." However, this device can not supply the required electrical signal (power) to directly drive the electrohydraulic valve, not even just the pilot stage.

Fig. 8-101

To drive a mobile proportional or servo valve with this level of input command requires that the signal be boosted or "amplified." This amplification is accomplished by support electronics contained on a driver or amplifier card or in the digital controller. It is this card or controller that not only amplifies the command signal, but also may contain such additional electronic circuits as, ramps, maximum, minimum and gain settings. Communication between the various elements of the electrohydraulic system (joystick, sensors, proportional or servo valve, digital controller, etc.) is achieved via a serial bus system such as CAN. These same parameters can be incorporated into more sophisticated systems.

Such sophisticated systems (fig. 8-101) could permit the control of several machine systems by one controller. The power of such a controller would allow the integration of additional options such as monitoring and faultfinding/troubleshooting routines. With onboard controllers continuous monitoring of machine functions by a graphic display mounted in the operator's cab is possible. This display presents information that permits the operator to adjust the system to meet the changing load and working conditions.

Some systems have the capability to communicate with a laptop computer permitting system diagnostics by trained personnel.

These various components can be combined into packaged systems (fig. 8-102) to meet specific application requirements. These packaged systems could include:

- application specific software
- digital and analog controllers
- on board calibration
- digital displays
- data logging
- Pulse Width Modulation (PWM) output

Fig. 8-102 System package

Each of these circuits, whether analog or digital, adds another level of control to the input signal as well as additional levels of circuit/system complexity.

Ramps control the rate of change of the command signal which in turn affects the actuator's acceleration and deceleration operating characteristics.

Maximum controls set electronic limits that prevent the valve spool exceeding. In other words, the valve spool will not move beyond a set point regardless of the input command signal.

Minimum settings, also known as "deadband eliminators," electronically minimize the dead zone within the valve. This is accomplished by prepositioning the valve spool. This requires slightly energizing the pilot stage of the valve.

Gain can be thought of as the sensitivity of the valve. Increasing the gain means that for a small input command signal change an even greater output change is seen. However, at some point to high a gain setting will result in the system going unstable. That is, the system will break into uncontrolled oscillation.

These electronic components require either an AC or a DC power source. Typically the required power is DC, direct current. This is most often supplied via the machines own DC electrical

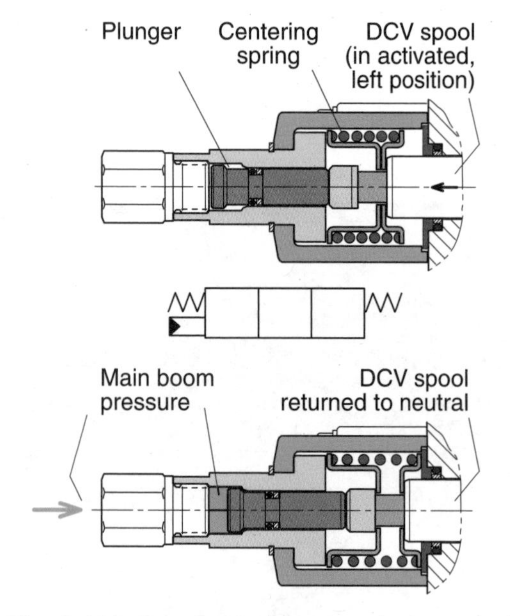

Fig. 8-103 Overload protection device on a DCV

system and may be 12, 24 or 48 volts DC, depending on the system. Though not common, AC, alternating current, is found on some applications and can be 120 or 220 volts. In the case of the AC the voltage is stepped down (lowered) and rectified (converted) to DC. This may be down by a separate power supply or it may be done on the driver card.

With the addition of electronics to the mobile hydraulic system the complexity of the system has increased but the ease of remote control has also increased.

Overload protection device

The overload protection device installs on the spool control which can be either manually or remotely operated.

Truck (lorry) cranes are required to have some kind of protective overload (POL) system in case the load is excessive. Fig. 8-103 shows a POL assembly installed on a manual, lever operated spool control.

The pilot operated plunger or piston forces the spool back to neutral as soon as the main boom pressure tends to exceed a predetermined level. The plunger force is strong enough to prevent the lever from being accidentally moved in the wrong, pressure increasing direction.

Pilot pressure is supplied by the main boom function via a sequence valve.

Parallel and series circuits Multiple-spool circuits

When comparing various multiple-spool circuits, some features are common to all:

- Each spool controls the movement of the actuator connected to the corresponding valve ports
- In a CFO valve, the bypass channel is connected in series with each spool. If any of the valve spools are actuated, the by-pass passage is blocked and consequently the pressure increases.

Parallel or series circuits

The way in which the pressure and tank passages of multiple spool valves are interconnected is the major difference between multiple spool circuits.

Multiple-spool circuits (pressure/tank) are:

- parallel - parallel
- series - parallel
- series - series

Parallel–parallel circuits

Each of the above circuits has a particular operating characteristic. In the parallel-parallel circuit, the pressure passages leading to the spools are in parallel, and the tank passages from each spool are also in parallel.

The arrangement allows both separate and/or simultaneous operation of the actuator (or actuators). If more than one spool is actuated at the same time, the pump flow will follow the path of least resistance.

This is the most common arrangement and can be achieved inside a valve block or between several, separate valve blocks. Therefore, in terms of function, the valve blocks act as one, big block. The function is the same as if the pump were connected to one large valve.

Fig. 8-104 shows two CFO (constant flow, open center) valve blocks connected in parallel with external piping.

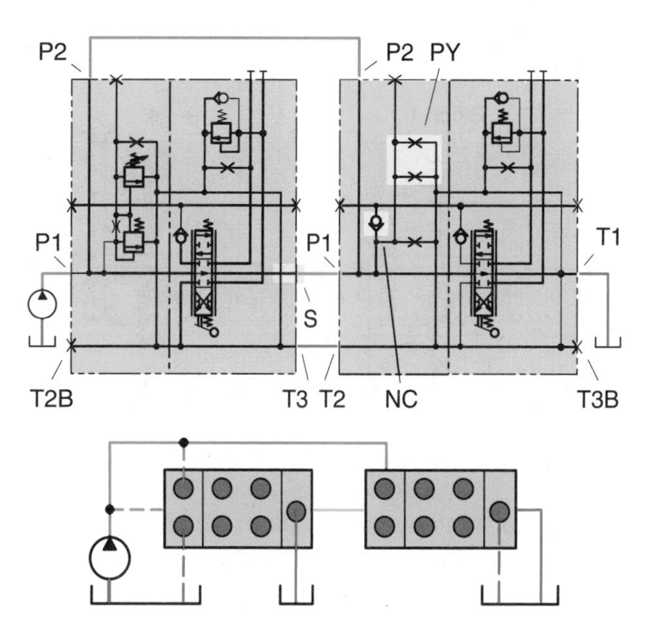

Fig. 8-104 Two CFO valve blocks connected in parallel through external piping

Fig. 8-105 illustrates two valve blocks connected in parallel in a CP (constant pressure) system.

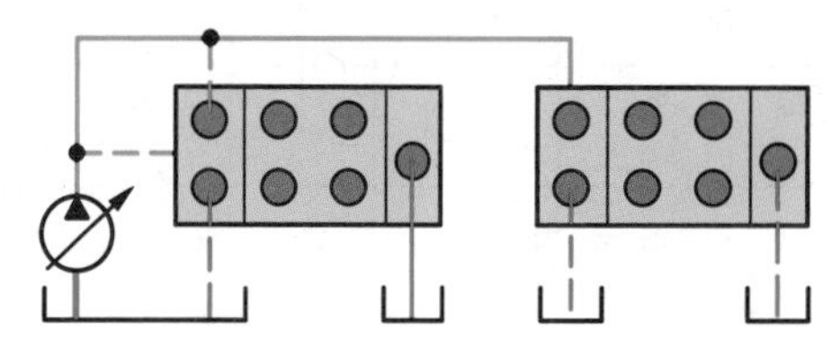

Fig. 8-105 CP system with two valve blocks in parallel

Fig. 8-106 shows parallel connection in an unloaded CPU (constant pressure) system; the unloading signal line goes through the open center passages of the valves.

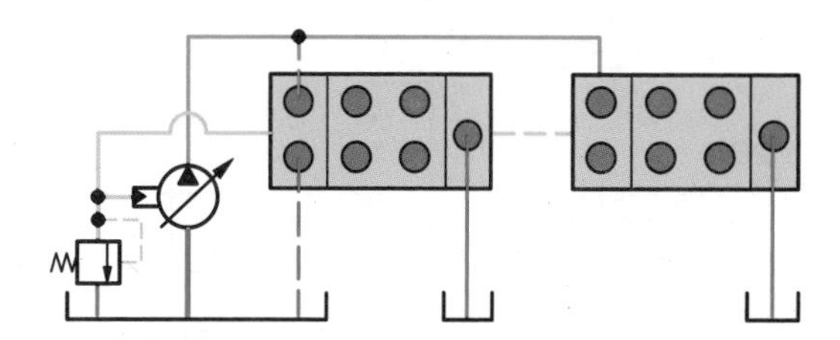

Fig. 8-106 Two valve blocks in parallel in a CPU system

Fig. 8-107, finally, illustrates parallel connection in an LS (load-sensing) system.

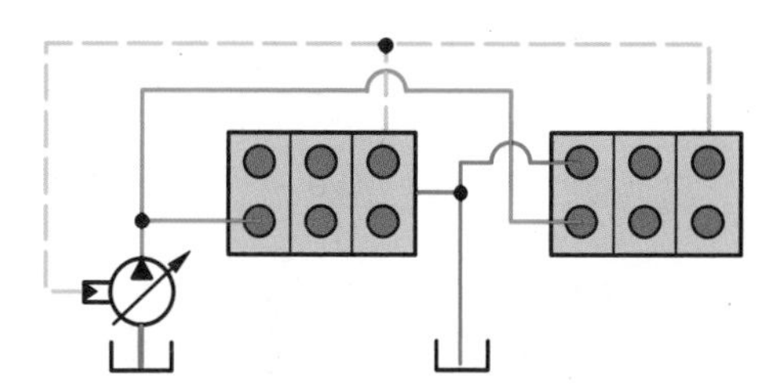

Fig. 8-107 Two valve blocks in parallel in an LS system

In a multipump system, each valve block is supplied from its own pump (fig. 8-108). The flow from pump 'P2' feeds the spool sections in block '2' only, where all sections are fed in parallel. The flow from pump 'P1' is used in the same way in block '1', but can also be used in block '2' (and in subsequent blocks) without limitation, since the pressure galleries/passages of the two blocks are connected. The illustration shows two valves externally connected.

Series–parallel circuits

In this circuit, the pressure passages are connected in series with the spools and the tank passages are in parallel from the spools.

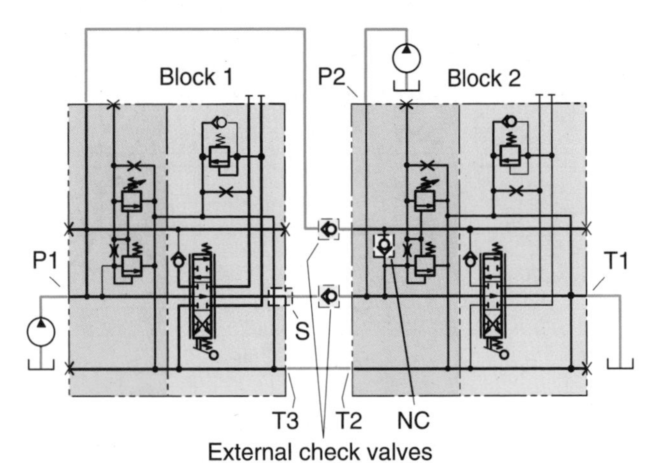

Fig. 8-108 Multi-pump operation with two CFO valve blocks connected in parallel

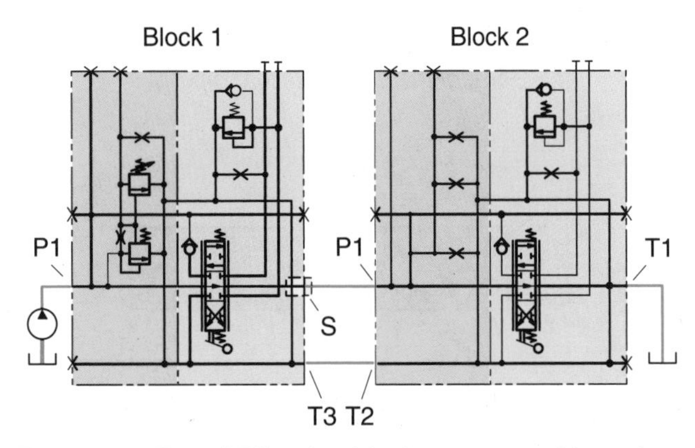

Fig. 8-109 Two CFO valve blocks connected in series

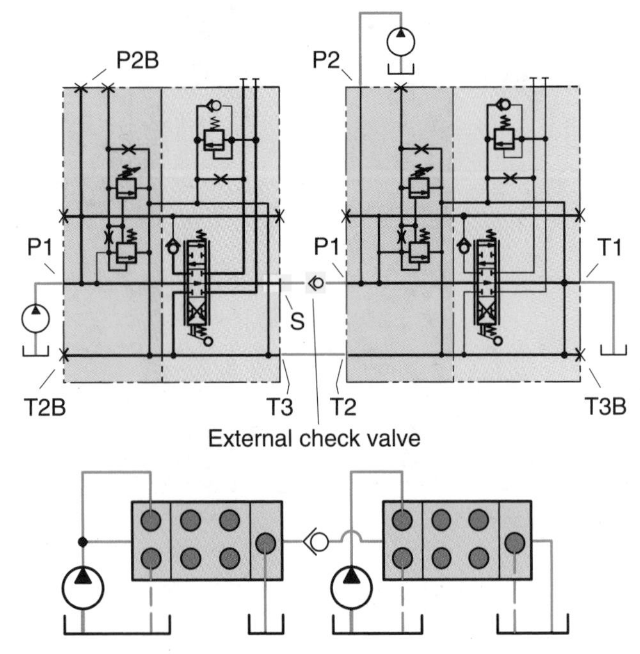

Fig. 8-110 Multi-pump CFO system; valves connected in series

In this circuitry actuators are operated in sequence. That is, the spool that is closest to the valve inlet takes priority of the pump flow. This can be done inside a multispool valve by blocking the parallel passage to the next spool or between valve sections in a stackable valve alternatively between separate valve blocks by using a high pressure carryover or power-beyond connection (designated 'S').

This power-beyond connector is used in CFO (constant flow, open center) valves and separates the 'open center' passage from the tank passage in the first valve, allowing pump pressure to pass to the next valve. This is possible if all spools in the first valve are in the neutral position (i.e. not actuated).

In series connected, separate valves each valve must have a separate return line to tank. If all valves within the circuit require the same pressure, only *one* main relief valve (fig. 8-109) is needed.

If the valves require different pressure levels, each valve must have its own relief valve, the first having the highest setting.

In a multipump system, an additional pump is connected to valve '2'. This valve receives the flow from pump '2' plus any remaining flow from valve '1'.

The flow from pump '2' feeds spool sections only in block '2', where all sections are fed in parallel. A check valve stops the flow from pump '2' from entering the preceding block.

Fig. 8-110 shows a schematic with a multipump CFO system (valves connected in series) and the same system in block diagram form.

Series-series Circuit

In this style valve, the pressure passages are in series to the valve spools and the tank passages are in series from the spools (fig. 8-111).

Just as with the series-parallel valve, the actuators connected to this valve are operated in a sequence. That is, all pump flow goes to the actuated/shift spool closest to the valve inlet.

Unlike previous valves, in the series-series valve, if downstream spools are actuated/shifted, their actuator(s) will operate. This is because the downstream actuator(s) is/are receiving the discharge flow from the upstream actuator(s). Also, single-acting cylinders can be retracted only if none of the downstream spools have been actuated/shifted.

It should be pointed out that with this type valve circuit when more than one spool is actuated/shifted, the actuator(s) loads are additive.

The next section will compare typical valve systems found on mobile equipment. Their efficiency and operational characteristics will also be discussed.

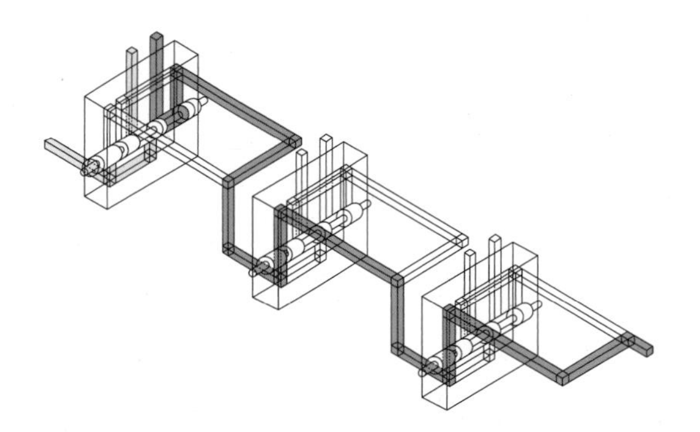

Fig. 8-111

Comparing standard mobile hydraulic systems

In the following, typical systems found in mobile equipment are compared, and the energy efficiency and control of each version will be discussed.

As explained in chapter 2, the field of hydraulics can be divided into two areas:

- hydrodynamics
- hydrostatics

Hydrodynamics is the study and use of liquids in motion, such as in the torque converter of an automobile or tractor transmission.

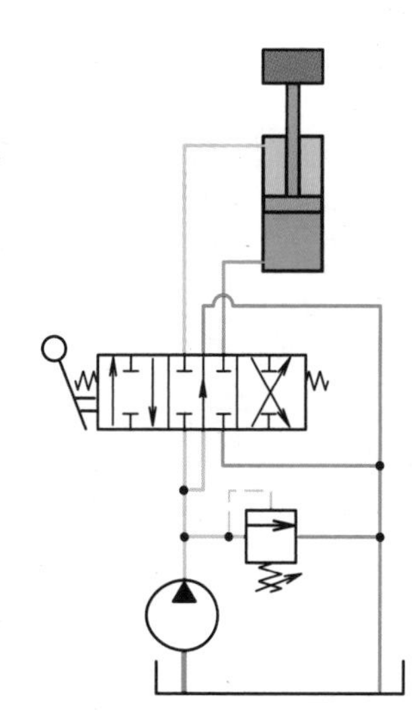

Fig. 8-112 Fixed displacement pump in an open center circuit

The hydraulic systems shown in the following belong to the other area, *hydrostatics.* In a hydrostatic device, energy is transmitted through a static force - pressure. Hydrostatic systems can further be divided into open loop ("open") or closed loop ("closed") systems.

In an open-center system, the main system flow circulates from the reservoir, through the system and back to the reservoir. The pump always has a defined inlet and outlet (low and high pressure) sides (fig. 8-112).

In a closed system, the main flow goes from the pump outlet, through the system and back to the pump inlet (fig. 8-113). Pump inlet and outlet ports change to suit the working direction.

A common example of a closed system is the *hydrostatic transmission.*

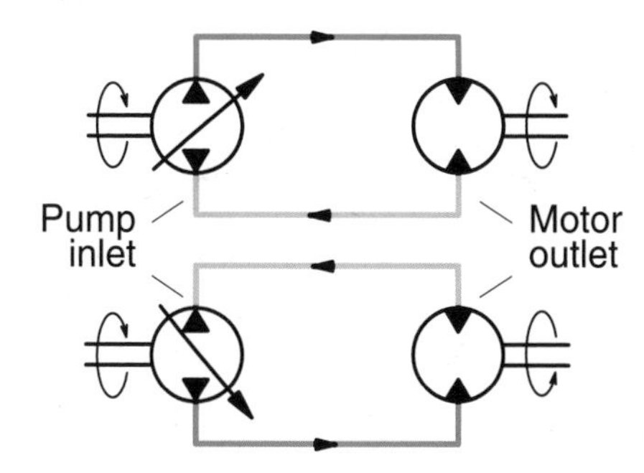

Fig. 8-113 Closed loop hydrostatic transmission

Three basic, open loop systems are covered here (hydrostatic transmissions were dealt with in chapter 4):

- constant flow (CF)
- constant pressure (CP)
- load sensing (LS)

CF constant flow system

In a constant flow system, the flow remains constant at a given pump speed while pressure flows according to demand.

The CFO (open center) system is a proven system for mobile machines. When compared to other systems, it contains less complicated components, and is relatively insensitive to contaminants in the hydraulic fluid.

Any fluid that is not directed to an actuator is directed back to tank through the free-flow gallery/passage in the valve. When several lifting functions are activated simultaneously, the heaviest load determines the pressure. Simultaneously operated functions should therefore have roughly the same pressure needs, or be divided into separate circuits to minimize cross-functional interference and provide good operating economy.

If the larger part of the pump capacity is utilized, the CFO system is very economical to operate. For this reason, it is important for the pump to be sized correctly.

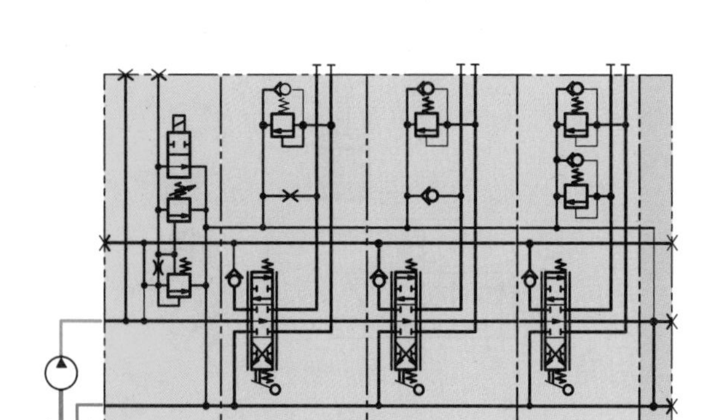

Fig. 8-114 Common CFO circuit

In a constant flow system, two different directional control valves can be identified:

- the open center (CFO) valve (fig. 8-114)
- the closed center (CFC) valve (fig. 8-115)

These valves exhibit completely different control characteristics.

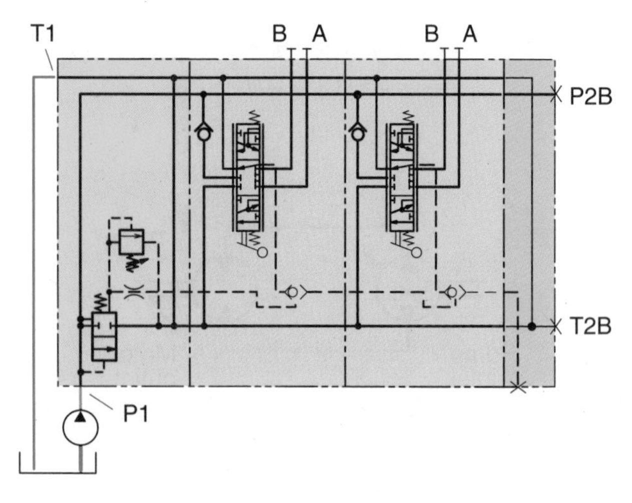

Fig. 8-115 Common CFC circuit

The CFC system differs from the CFO system primarily in that surplus, excess, hydraulic fluid is returned to tank through a separate bypass valve or inlet compensator. This means that the pressure difference, ΔP, between the pump and the heaviest load can be maintained at a constant value. The flow to the heaviest load is therefore independent of the load.

Operating characteristics

Manually operated CFO valves have no uniform relationship between lever stroke and the speed of the load or function.

The speed of the function is influenced by:

- pump flow
- size, direction and movement of the load
- other loads operated simultaneously
- valve spool position
- fluid temperature

The reason is that, when several functions are activated at the same time, fluid flows is redistributed so that the pressure drop in the flow paths becomes equal.

In a CFO valve, the spool can be seen as a restrictor/orifice bridge consisting of three restrictors/orifices for each controlled function (as illustrated in fig. 8-116):

- pump to tank (P-to-T)
- pump to actuator A (P-to-A)
- actuator B to tank (B-to-T).

The three restrictors/orifices are mechanically connected as they are part of the spool (in spool type DCV's). When the spool is moved, the opening area of the individual restrictors/orifices vary at the same time. (The corresponding spool symbol, with the three restrictors, is shown in fig. 8-117).

The ΔP over the restrictors/orifices (regulation losses) at no load is called the *intermediate pressure*, which can be manipulated by spool design either for better control (fig. 8-118, spool '1') or for better efficiency (spool '2').

The more the intermediate pressure is raised, the closer the system approaches a constant pressure system.

The use of application and function matched valve spools gives considerably better control characteristics when spools are operated simultaneously. In some cases, this can result in higher power losses during the "fine metering" stage.

In certain applications, however, this characteristic is taken advantage of, as it enables the operator to sense the weight of the load he is handling (called *force control*).

In some valve designs, such as in hydraulic, proportional pilot controlled CFO valves, the valve spool can be pressure compensated, which means that the regulated flow rate remains constant for a given lever position, regardless of pressure variations in the system (called *speed*

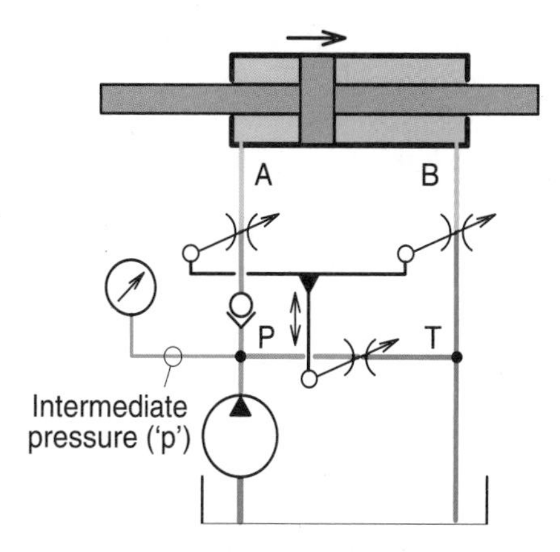

Fig. 8-116 Measuring intermediate pressure in a simplified CFO circuit

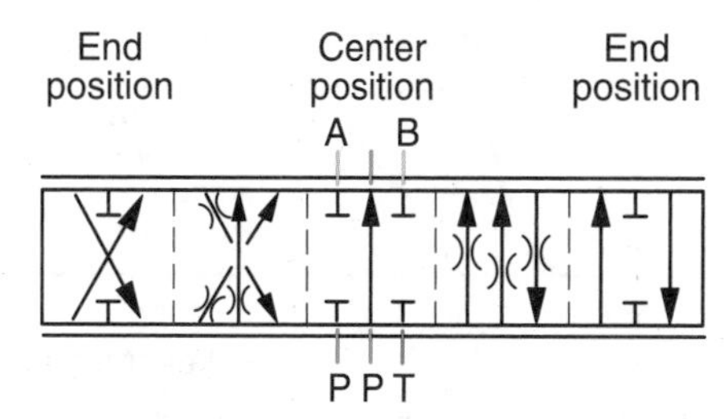

Fig. 8-117 Negative overlap spool symbol showing restrictors

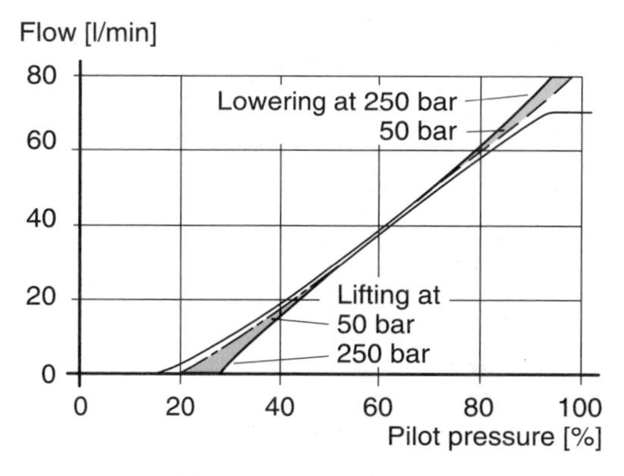

Fig. 8-118 Intermediate pressure 'p' versus spool position of two spool designs

control). It provides a larger control range which means higher resolution and less interference between simultaneously operated functions (compare figs. 8-49 and -50).

The operating characteristic of the CFC system is largely independent of load pressure. The spools can be pressure compensated, as above, and are consequently not influenced by other loads operated simultaneously. Alternatively, sections can be provided with separate pressure compensators.

The CFC system has the same operating characteristic as the load sensing system (shown below), but is not as efficient.

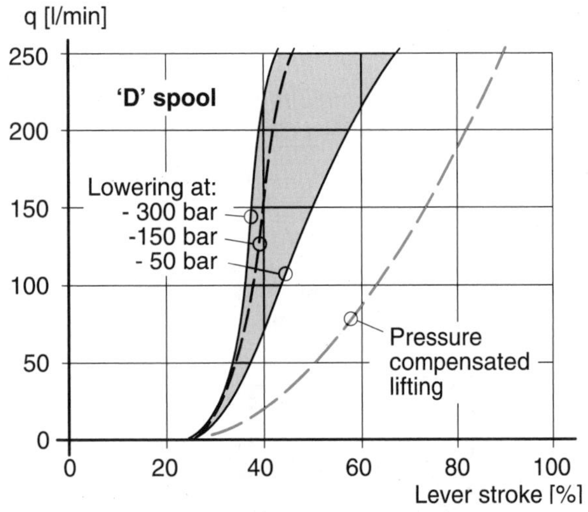

Applications

The CFO system is typically used in applications where basic components are utilized and the demand for interference-free operation of several simultaneously operated functions is not a requirement.

The design of such a system usually results in low-cost components and installation, while energy consumption, to a large extent, depends on the duty cycle of the machine and the sizing of the system. This design can, however, be very energy consuming and generate much heat.

The CFC system is used to advantage where high demands are made on the operating characteristic. Fig. 8-119 shows a flow-versus-lever position diagram of a hand operated valve. When in the lifting mode, flow is independent of load because of the inlet compensator. Lowering loads are not, however, load pressure compensated.

Fig. 8-119 Operating characteristic of a manually operated spool in a CFC circuit

The diagram in fig. 8-120 shows the same valve as above but, in this case, the spool is pilot operated. As can be seen, both lifting and lowering loads are pressure compensated due to the design of a spool that utilizes flow (Bernoulli) forces.

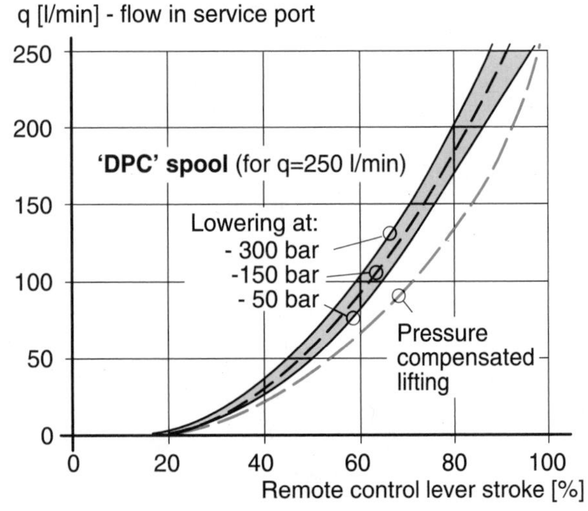

Fig. 8-120 Operating characteristic of a pilot operated spool in a CFC circuit

Forklift trucks, backhoe loaders, excavators, other earth moving machines, refuse vehicles as well as cranes are examples of suitable applications for constant flow (CFO, CFC) systems.

CP constant pressure system

In a CP system, the pressure is constant while the flow varies according to the demand of the system.

The constant pressure system is very simple in design, using uncomplicated components. The pump, generally of the variable displacement type, is controlled to keep the pressure constant. It should be sized to supply the total of the accumulated, maximum flows for simultaneously operated functions. The system is less sensitive to pressure drops as compared to a CFO system.

Constant pressure is available as long as the pump capacity is not exceeded. This type of pump, supplying a closed center directional control valve, is the main difference as compared to the CF system, which uses a fixed displacement pump and an open-center directional control valve; refer to fig. 8-121.

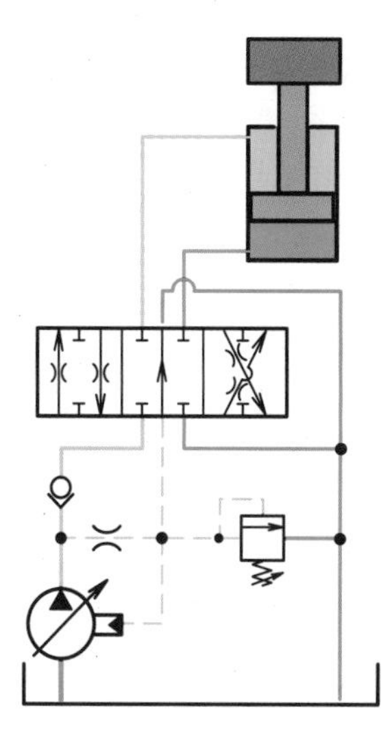

Fig. 8-121 Variable displacement pump in a CPU circuit

CPU constant pressure, unloaded system

In a simple, constant pressure system, maximum pressure is maintained by the pressure compensator of the pump.

The constant, high pressure causes spool leakage into the work (actuator) ports of the DCV when it is in the neutral, nonactive position. This can result in a creeping movement of the actuators. In order to minimize this and the corresponding, wasted energy, a version of the CP system, the CPU (fig. 8-121) can be utilized.

In the CPU system, an hydraulic signal from the DCV causes the pump to unload at the "at rest" position. As soon as a control spool is actuated, the pump provides maximum pressure.

The CPU is referred to as a constant pressure-unloaded system (schematic shown in fig. 8-122).

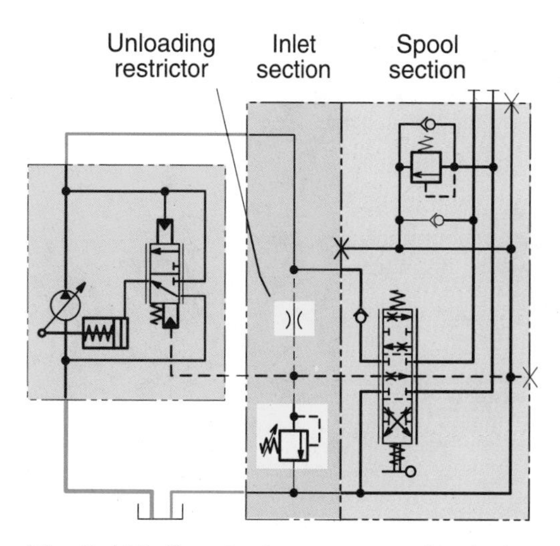

Fig. 8-122 Constant pressure, unloaded system

Another version of the CP system is one that operates at different pressure levels:

- lower pressure for functions that do not need high pump pressure
- higher pressure for heavily loaded actuators

Changeover is carried out by additional pilot control signals. An increase in the number of pressure levels involved may also increase the system's efficiency but will also increase the number of required components.

Operating characteristics

The CP system has excellent operating characteristics, and the various functions do not influence each other. Generally it has good anti-cavitation characteristics, which means that a crane's boom lowering movement can change to a lifting movement without any delay in passing the overcenter position.

The combination of spool area openings and the pressure difference between pump and load pressures results in the desired speed (flow) to the function.

The more the load pressure increases, the smaller the pressure difference which results in a slower function speed (smoother movements), which, in turn, means that the operator can utilize the entire lever stroke with improved resolution (excellent control characteristics) at the heaviest loads.

Valve designs, with hydraulic and/or electrohydraulic proportional remote controls, can incorporate pressure compensated spools, providing improved control characteristics (fig. 8-123).

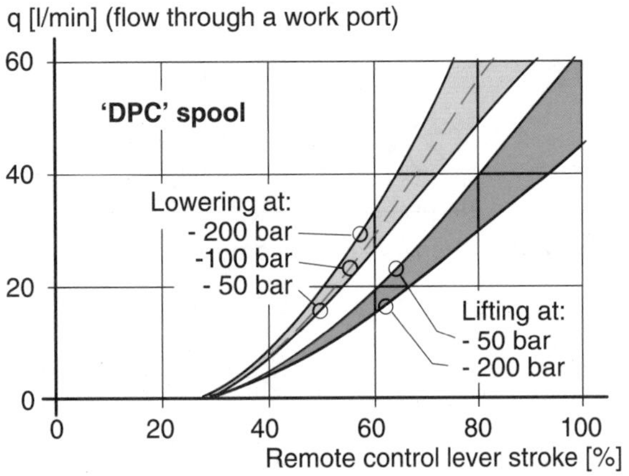

q [l/min] (flow through a work port)
'DPC' spool
Lowering at:
- 200 bar
-100 bar
- 50 bar
Lifting at:
- 50 bar
- 200 bar
Remote control lever stroke [%]

Fig. 8-123 Operating characteristic of a DPC spool in a remotely controlled, CP system

If the flow requirement of the system exceeds the capacity of the pump, the pressure level cannot be maintained (the heaviest function stops moving) and the interference-free simultaneous operation of several functions deteriorates.

Applications

CP systems are used when large power levels, high control accuracy and interference-free, simultaneous operation of several functions is a requirement.

CP systems are used in advanced forestry machines, mobile work platforms, and rock drilling equipment. The system is uncomplicated in design but may have to be matched to a specific application in order to provide the best combination of control, operating characteristic and economy.

LS load sensing system

In a load sensing system, both pressure and flow are matched, as closely as possible, to the load requirement.

The directional control valve regulates pump displacement via a load pressure signal, providing a constant ΔP between the pump outlet and the load sensing signal; the pump pressure adjusts to the heaviest load.

The CFC system is sometimes referred to as load sensing. From the viewpoint of control technology however, a load sensing system is more advanced than both the CFC and CPU systems. Load sensing makes greater demands on the system designer.

The combination of good operating characteristics and high efficiency require the spools of the directional control valve to be matched to the actual load profile and actuator size (fig. 8-124).

As in constant flow systems, simultaneously operated functions should have roughly the same pressure demands. If not, they should be divided into separate, smaller circuits so that optimal efficiency can be obtained.

Load sensing signal systems can be constructed in different ways. In certain valves, the signal is taken from a logic, shuttle valve system, and in others through check valves (fig. 8-125).

In more advanced designs, a special load sensing signal system (including a copy spool) allows the signal fluid to circulate in the load sensing signal line without any fluid being taken from the work or actuator port. In this design, the pump reacts faster to command changes, especially at lower temperatures. It also eliminates the small load "dip" when starting a lifting function.

Pump compensators or controls can also differ. Certain compensators (regulators) consume no fluid from the load sensing signal line. Others have a bleed orifice to the tank line, resulting in only a small energy loss. In yet another version, fluid feeds into the signal line, a flow that must be drained in a suitable way as shown in fig. 8-126.

Operating characteristics

The LS system has excellent operating characteristics, particularly in connection with pressure compensated valves. Efficiency and operating economy are generally better than those of other system versions.

With respect to the operating characteristic, the pressure compensated LS system can be compared with the CP system. Lowering movements in the LS system can be carried out without the

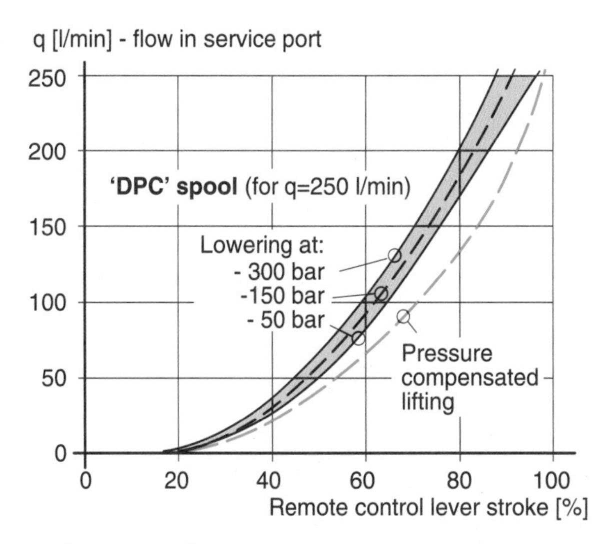

Fig. 8-124 Operating characteristic of a pilot operated spool in an LS circuit

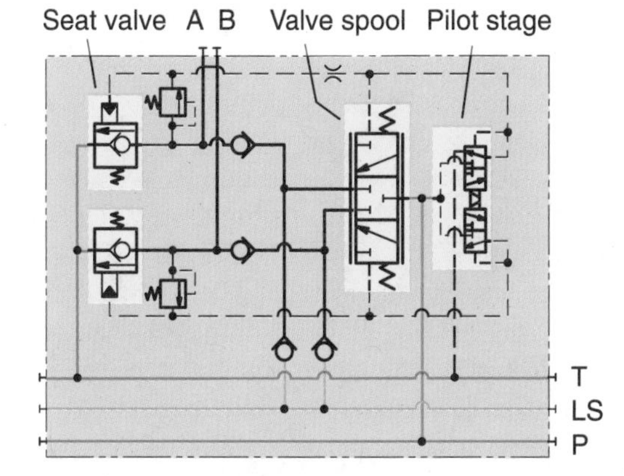

Fig. 8-125 Directional control valve with LS signal check valves

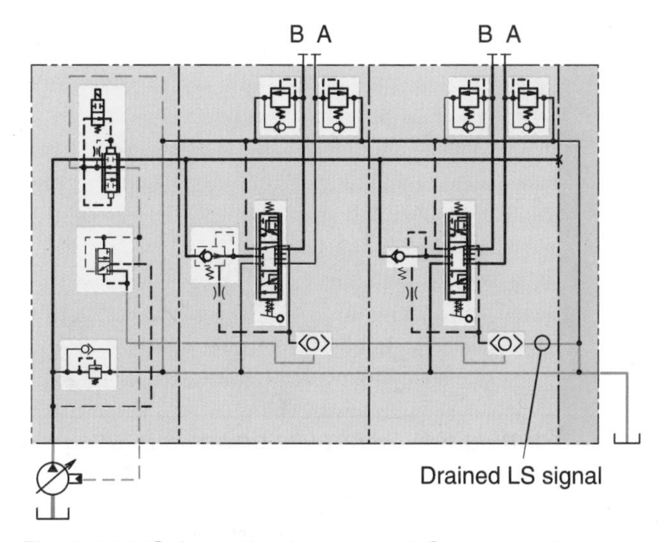

Fig. 8-126 Schematic showing an LS system where the LS signal is copied to the pump control or, alternatively, drained to tank (when neither spool is activated).

pump being activated (no LS signal). If the cylinder must be refilled during this process, great demands are made on the anti-cavitation (makeup) characteristic of the valve.

LS valves should, therefore, be fitted with built-in anti-cavitation valves and sometimes also with a counterpressure/backpressure valve in the tank line in order to improve the makeup capabilities and to meet the above requirements.

Applications

The load sensing system is suitable for use on equipment where one phase of the working cycle requires a large flow under moderate pressure, while another phase requires a small flow at high pressure.

The capability of the system to adjust to both pressure and flow results in less power consumption and permits the use of a smaller power source (prime mover).

For maximum power control, pumps with torque or horsepower controls are often used in LS systems. (For more detail refer to chapter 3.)

The LS system, with pressure compensated functions, is suitable where simultaneous and interference-free operation of several work functions is required.

LS applications are found in mining equipment, combine harvesters, street sweeping machines, backhoe loaders, front-end loaders, excavators, cranes and forestry equipment.

Power loss

A simple power consumption comparison of three system types is shown in figs. 8-127 through -129.

The comparison is made when operating:

A) one function with low pressure and high flow
B) one function with high pressure and low flow
C) three simultaneously operated functions with small differences in pressure and flow

The negative part of the histogram should be seen as load lowering which always is a loss, at least until methods for power regeneration have been developed.

Formula 8-2 Hydraulic power is a function of pressure and flow

$P = p \times q$ where: 'P' is power
'p' pressure
'q' flow.

Power loss in each case can easily be calculated by using Formula 8-2. The red area in the diagrams represents lost power.

CF constant flow system

Case **A**: With a fixed displacement pump, the entire flow is used to operate a function. The only losses are from regulation and component friction (fig. 8-127, 'A').

Case **B**: A high pressure function is running with a small flow. The pump adjusts to the load requirement and must deliver the surplus flow either through the throttled, open-center line (CFO) or through a bypass valve (CFC).

Alternatively, the surplus flow passes over the system relief valve in the directional control valve. This results in huge losses and high heat generation ('B').

Case **C**: The total losses in this area depends on the total, used flow. The higher the unused part of the flow, the higher the losses ('C').

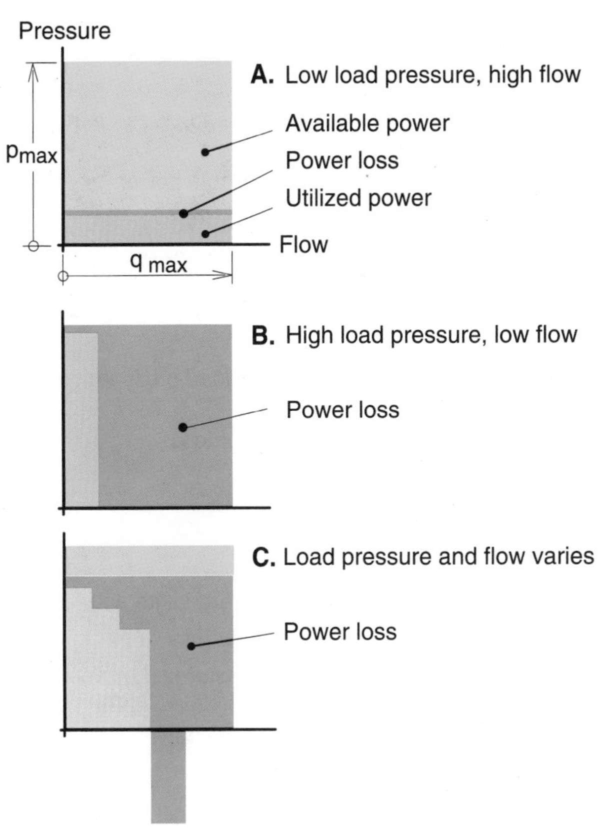

Fig. 8-127 Power loss in a CF system

CP constant pressure systems

In this example, a low pressure function with high flow is operating. The variable displacement pump is set for a much higher pressure level, thus causing high power losses (fig. 8-128, 'A').

A high, near constant pressure function is running. The pump adjusts the flow to the system requirements, and power losses are very small ('B').

Pressure requirements are lower than those of the pump pressure setting, yielding some losses. Depending upon the lowering function, lowering loads also cause losses. The cylinder must be refilled on the low pressure side by fluid that is taken from the high pressure side and partially from the low pressure side via anti-cavitation valves as makeup flow ('C').

LS load sensing systems

In fig. 8-128, 'A', a low pressure, high flow function is running without other losses than that from the pump control. This function is opposite to the one above, but it has the same, small losses for the same reason. The pump adapts to flow and pressure requirements ('B').

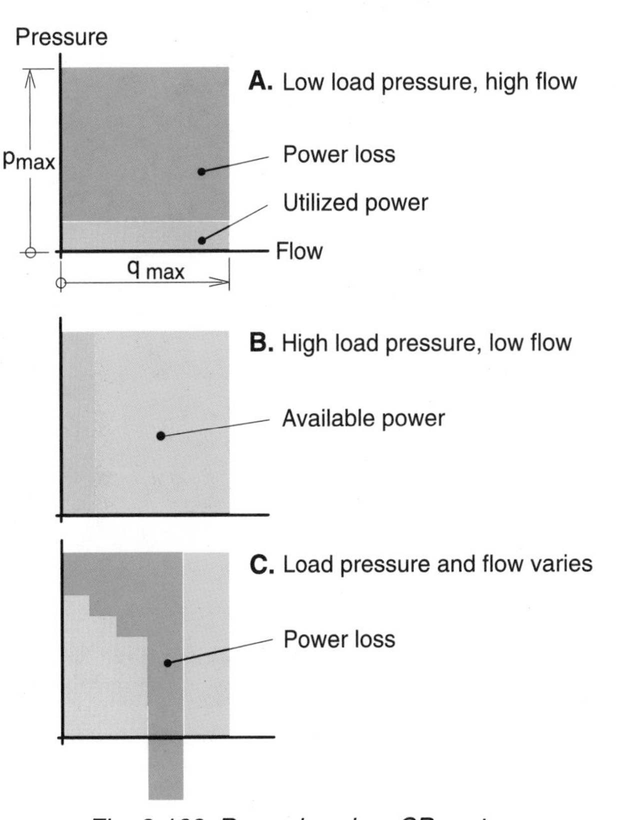

Fig. 8-128 Power loss in a CP system

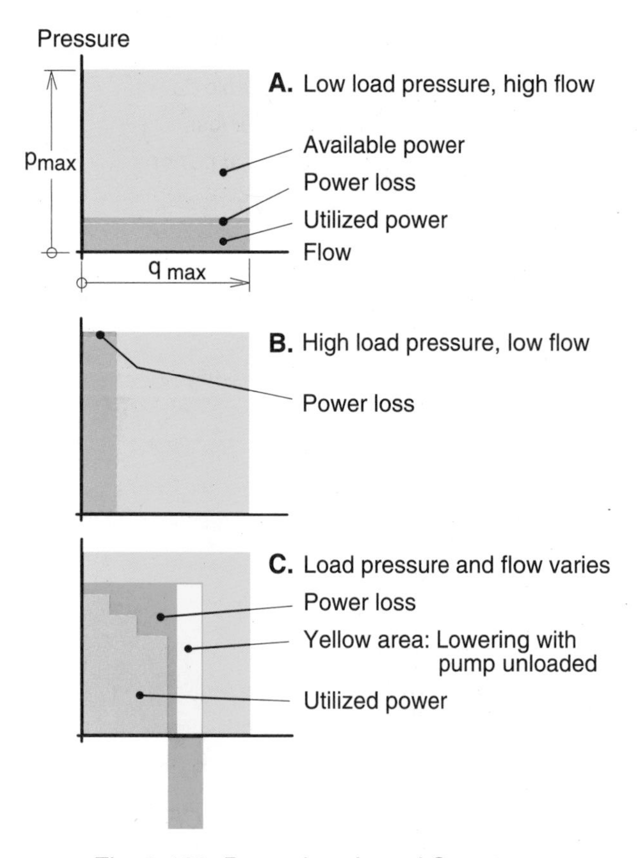

Fig. 8-129 Power loss in an LS system

Here it is easier to see the advantages of an LS system. Pressure and flow are adjusted to the requirement of the system. Losses are caused by the difference between load and compensator pressures ('C').

When in the lowering mode, the makeup flow to the low pressure side of the cylinder is taken via the anti-cavitation valves from the tank line, i.e. no flow is taken from the pump circuit.

This is possible if the DCV is equipped with separate pressure compensators in the valve sections.

Chapter 8 Exercise

Directions: Match the descriptive statement with the component being described.

1. If a check valve is preloaded by a spring, it is usually called a(n) _____ valve.

2. It is the purpose of the _____ to protect the DCV and actuator from pressure spikes and excessive pressure spikes in the system.

3. The monoblock valve has a(n) _____ body.

4. This is used to power an actuator in both directions and typically the actuator ports are blocked in the neutral position. _____

5. This function makes it possible for two or more valves to be operated in series. One pump can supply both valves. _____

6. This type of electronic signal is commonly used to control proportional solenoids. _____

7. According to the European Community's *Machinery Directive*, machines must be equipped with one or more of these functions. _____

8. It is the primary purpose of the _____ to maintain a selected, constant flow to a function, regardless of system pressure variations.

9. This function eliminates any loss of pressurized fluid from the cylinder when an LS signal is sent to the pump compensator. _____

10. In a(n) _____ system, pressure and flow are matched as closely as possible to the load.

11. This style of feedback utilizes pressure on the end of the pilot spool to drive the pilot spool back to the center position. _____

12. The _____ contains a "clapper" and two "pole-pieces."

13. In a _____ system the flow remains constant while pressure floats according to demand.

14. This electrohydraulic term has two meanings: step and _____.

15. Mobile hydraulic DCVs are of two main styles: monoblock and _____.

a. backpressure
b. stackable
c. stiffer
d. emergency stop
e. power beyond
f. copy
g. relief & anti-cavitation
h. pressure compensator
i. spool type D
j. magnetic detent
k. PWM
l. frequency
m. pressure
n. CF
o. LS

Chapter 9

Remote controls in mobile hydraulic systems

Introduction

In this chapter, we will concentrate on remote control of mobile directional control valves used to operate hydraulic actuators, such as cylinders and motors (fig. 9-1).

Fig. 9-1

Direct control, which is what remote control supplements or replaces, will be reviewed first. Direct control is the oldest, simplest and most common method of shifting spools within directional control valves. This is accomplished by means of a mechanical lever that is mounted directly to the end of the valve spool (fig. 9-2).

Characteristics of direct control

The operator must be in direct contact with the directional valve.

- When more than one function is activated at the same time, it can be difficult for the operator to reach all levers.
- The lever forces can be high, owing to the spring, flow and frictional forces.
- Heat, smell and noise caused by the oil flowing through the system can be uncomfortable for the operator.
- In the event of oil leakage, the operator's place of work is quickly polluted.
- There is very little to go wrong in direct control mechanisms.

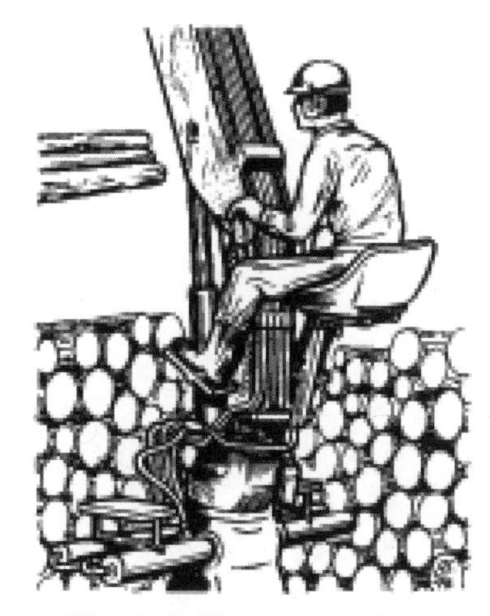

Fig. 9-2 Direct control
Reprinted from The Off-Road Vehicle Volume 1, courtesy of CPPA

Over the past few decades, direct control has been increasingly replaced by remote control. In remote-controlled valves, the spool is shifted by a pneumatic, hydraulic or electric pilot. The pilots are controlled by a remote device. There is no mechanical linkage between the remote control device and the spool (fig. 9-3).

Characteristics of remote control

- operator can control the machine from a safe distance.
- improved hydraulic performance, such as higher resolution, repeatability and smoother, more gentle operation.
- easy to operate more than one function at the same time.

Fig. 9-3 Remote control
Reprinted from The Off-Road Vehicle Volume 1, courtesy of CPPA

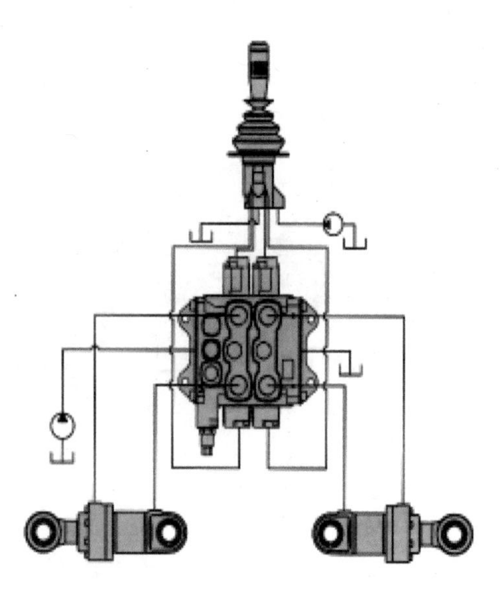

Fig. 9-4 Main components in remote control of DCVs are spool actuator and the remote control device.

- lower lever forces, which reduce the risk of strain injury to the operator.
- minimal noise, smell and heat from the hydraulics at the operator's station
- more safety considerations can be met with the aid of remote control systems.
- choice of control location

The two main components in the remote control of directional control valves (fig. 9-4) are the spool actuator and the remote control device. The spool actuator is the component that actually shifts, or moves, the spool. It is mounted directly on the directional control valve. The remote control device is the component that controls the spool actuator. It is mounted at the operator's control station.

While actual spool-shifting principles have changed little in recent years, the revolution in electronics and microprocessor technology has affected the design of remote control devices. This gives today's control systems engineer almost infinite opportunities to downsize, integrate and refine remote control systems.

Before touching on electronics and integrated control systems, let us look at the different ways in which remote control can be achieved.

Different means of remote control

Remote control can be achieved in different ways, depending on the type of control device. The following arrangements are common:

1. A remote switch sends an electric ON/OFF signal to an ON/OFF solenoid that shifts the spool directly.
2. A remote switch sends an electric ON/OFF signal to an ON/OFF solenoid that opens/closes pneumatic or hydraulic signal pressure to the spool actuator.
3. A variable pneumatic remote control lever sends a proportional pneumatic control signal pressure to the spool actuator.
4. A variable hydraulic remote control lever sends a proportional hydraulic control signal pressure to the spool actuator.
5. A variable electric remote control lever sends a proportional electric signal to a proportional solenoid, which opens/closes proportionally a hydraulic control signal pressure to the spool actuator.

Spool actuators

Two main types of spool actuators are used, depending on the sealing arrangement at the spool ends in the directional control valve.

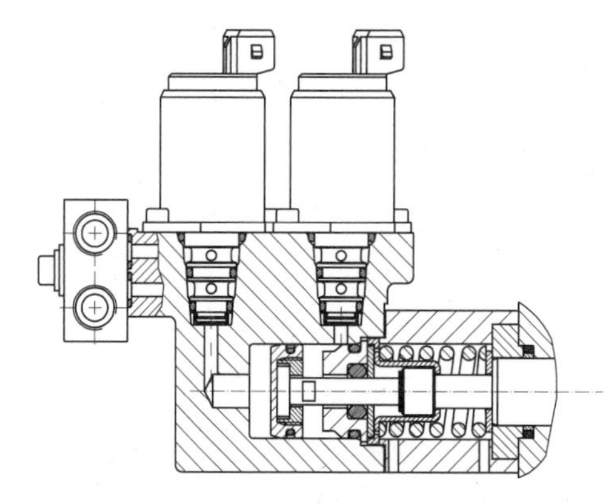

Fig. 9-5 Open spool actuator

Open spool actuators

If the spools are sealed with o-rings to prevent external oil leakage, the spool actuator is called an open spool actuator. When o-rings are used to provide a seal around the spool, an increase in friction is seen when the spool is shifted. Open spool actuators can be controlled either directly by means of a lever, or by remote control. Direct control is, however, more common (fig. 9-5).

Closed spool actuators

With closed spool actuators, there are no o-rings around the spool and this results in a certain amount of leakage into the spool cap. This leakage drains to the valve's tank gallery or passage. Since friction is low, thanks to the absence of o-rings, there is minimal hysteresis. With closed spool actuators, the shifting of the spool is by means of hydraulic remote control.

This enables the spool to be pressure compensated by exploiting the flow forces (fig. 9-6).

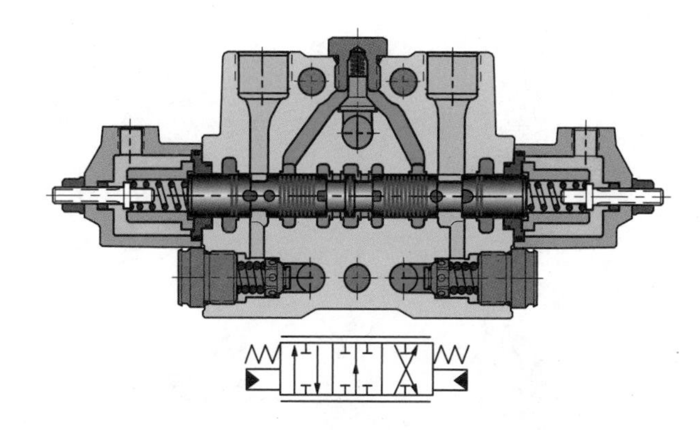

Fig. 9-6 Closed spool actuator

Pilot pressure

A reduced and stable hydraulic or pneumatic pilot pressure (also called the control pressure) is needed when hydraulic or pneumatic spool actuators and remote control devices are used. The pilot pressure is always considerably lower than the pressure in the main system.

Hydraulic pilot pressure can be obtained from a separate pump, or from the main hydraulic system via a pressure reducing valve. Pneumatic pilot pressure is obtained from an external source.

Hydraulic remote control

The control device consists of a remote control lever or pedal that sends a proportional hydraulic signal pressure directly to the spool actuator. The value of the pressure depends on the stroke of the lever or depression of the pedal. The spool actuator consists of caps that are mounted to, and sealed against, the directional valve. The caps contain connections for the pilot signal

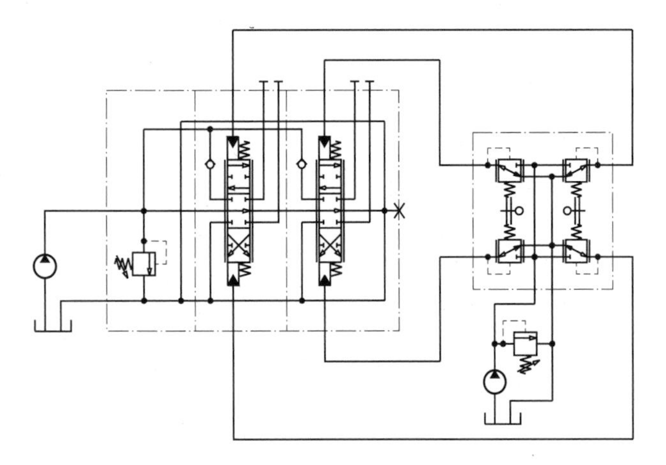

Fig. 9-7 Circuit diagram showing two-section PCL4 with two linear levers controlling one hydraulic directional valve containing two spool sections

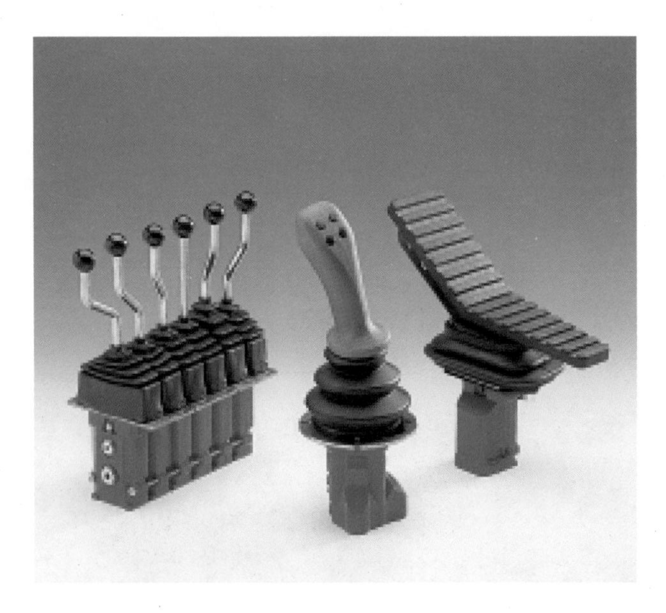

Fig. 9-8

pressure. The pilot pressure inside the cap shifts the spool against a force exerted by a spring on the opposite end of the spool. The spool is shifted proportionally with the value of the pressure and any influence exerted by the flow forces. See fig. 9-7.

Proportional hydraulic lever unit

In hydraulic terms, the proportional hydraulic lever unit is a three-way pressure reducing valve. It is supplied with constant pressure by an external source, and is also connected to the tank line. It also has connections to the spool actuator it controls.

The proportional hydraulic lever unit is designed to convert the physical position of its lever (i.e. the lever stroke) into a hydraulic pressure signal. It functions as a pressure reducing valve. The oil flows from the supply side to the spool actuator.

The lever can also reduce the hydraulic pressure to the spool actuator, thanks to its three-way function. Oil then flows from the spool actuator to tank (fig. 9-8).

Function

The main component of the proportional hydraulic lever unit is a spool. A spring acts on the upper end of the spool. The pressure output to the spool actuator acts on the lower end of the spool.

There is a transverse hole through the spool, which is connected with the lower end of the spool. In the housing, around the spool, are two holes that connect with the hole in the spool. One of the holes in the housing is located above the hole in the spool and the other below it. The upper hole in the housing is connected to tank and the lower one to the supply pressure.

When the spool is shifted downward, the lower hole in the housing is connected with the hole in the spool. When the spool is shifted upward, or is in neutral, the upper hole in the housing is connected with the hole in the spool. This is the tank connection.

When the lever or pedal is actuated, a guide pin compresses the preloaded spring package above the spool. The spring package transmits a force to the spool, causing it to move downward. The transverse hole in the spool connects with the lower hole in the housing, whereupon the pilot pressure is transmitted (via the hole in the spool) to the underside of the spool, and on to the spool actuator. When the pressure on the underside of the spool becomes equal to the force exerted by the spring package, i.e. when a pressure balance is struck, the spool moves back to its initial position.

The pressure under the spool therefore becomes proportional to the force of the spring package, which is representative of the physical position of the lever (its stroke). When the lever or pedal is released back to neutral, the spring force above the spool is reduced. The pressure on the underside of the spool therefore forces it upward, whereupon the hole in the spool is connected with the upper hole in the housing, i.e. with the tank. When the forces are in equilibrium, the spool has reached its neutral position. If the pressure under the spool continues to rise, the spool continues to move upward, opening the connection with tank more and more, thereby acting as a pressure relief valve (fig. 9-9).

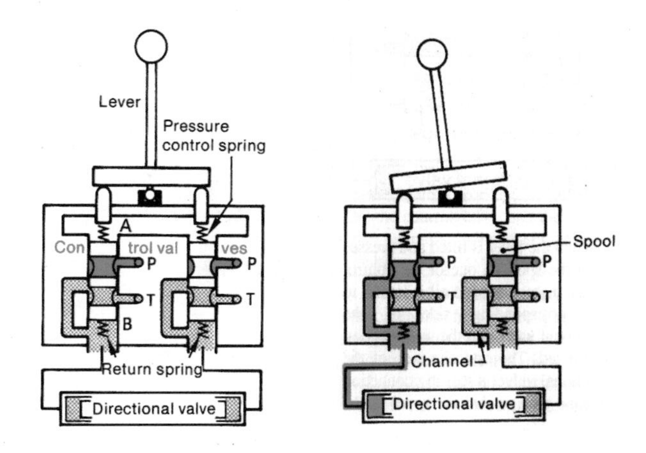

Fig. 9-9

Pneumatic remote control Electro-pneumatic ON/OFF control

An electro-pneumatic ON/OFF control device consists of a switch that closes or opens the electrical circuit between the battery and the solenoid valve in the spool actuator. The spool actuator is equipped with two ON/OFF solenoids located on either side of a piston. The solenoids open/close the air supply to either side of the piston. The air pressure acts on the piston, which in turn shifts the spool in the directional control valve to one end position or the other. A centering spring continuously tries to keep the spool in neutral (fig. 9-10).

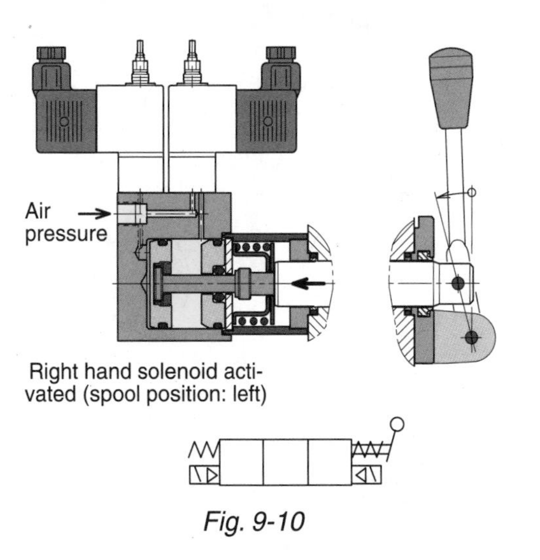

Fig. 9-10

Proportional pneumatic lever unit

The proportional pneumatic lever unit sends a pneumatic signal that is proportional to the stroke of the lever. The signal is sent directly to the spool actuator.

In terms of function, there is no difference between a proportional hydraulic and a proportional pneumatic lever unit, just that the latter uses air instead of oil as the pressure signal medium. Friction is greater in the pneumatic version as compared to the hydraulic version, with the result that hysteresis is greater (fig 9-11).

Fig. 9-11

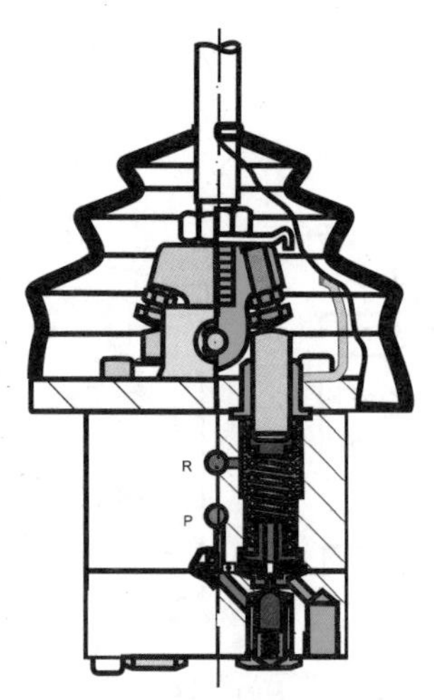

Fig. 9-12 Thumb switch to control auxiliary on/off functions

Fig. 9-13 Electrohydraulic remote control systems

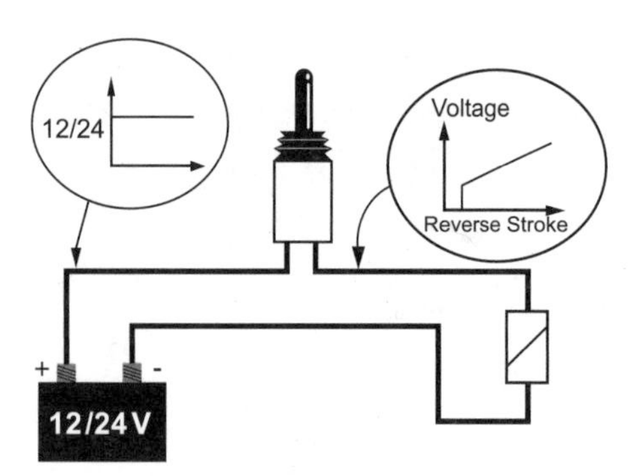

Fig. 9-14

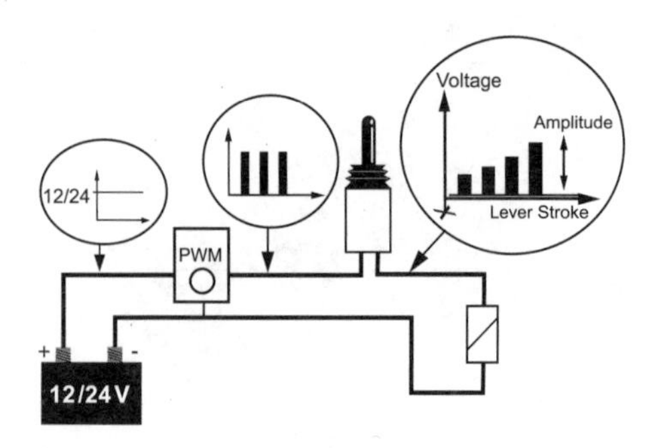

Fig. 9-15

Both hydraulic and pneumatic control devices can be equipped with a variety of lever options. The simplest of these consists of a handle with ball at the end. Handles can be of different lengths, angles and balls of different sizes and shapes to improve ergonomics.

Some proportional levers also incorporate thumb switches to control auxiliary ON/OFF functions (fig. 9-12).

Electrohydraulic remote control

There are a wide range of electrohydraulic remote control systems available today, with varying degrees of complexity. The biggest advantage they provide is a good working environment for the operator. The systems are quiet, emit little heat, and are easy to install (fig. 9-13).

Moreover, they enable the machine manufacturer to design the cab to meet the needs and performance requirements of the operator and machine. Microprocessor-based systems make it possible to change operating parameters of the system easily, they enable the performance of the machine to be modified and different operators' settings can be saved.

Direct control by coordinate-lever units (joysticks)

A coordinate-lever unit (joystick), fig. 9-14, is based on a simple variable-power resistor, or potentiometer, which works as an attenuator. The joystick is connected directly to the solenoid. This solution can be used in systems where low cost is of more importance than high performance. Performance could easily be improved by using a signal from a pulse-width-modulator (PWM) signal.

Indirect control by coordinate-lever units (joysticks)

This system would typically include the following components:

- linear and/or multiple-axis coordinate levers (joysticks) and pedals, i.e. the remote control devices (fig 9-15)
- solenoids to control the pilot valve that direct oil to the spool actuators. (The solenoid is the interface between the electronics and the hydraulics in a system.)
- amplifiers that boost the electrical signals to the solenoids.

A more advanced design, fig. 9-16, would have a separate amplifier to convert the electric signals from the joystick into regulated signals to the valve solenoids. In this design a voltage is applied across the command potentiometer and, depending upon the direction of actuation, will result in an output voltage within the applied voltage range.

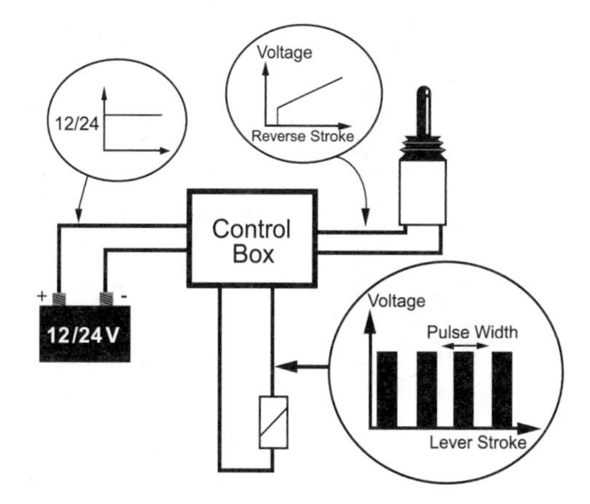

Fig. 9-16

Dual mode operation

Some control levers have independent potentiometers for each major direction or axis. As an example, in some lever designs, three-proportional, may contain 6 potentiometers. Such designs operate in a dual mode. The advantages of this design are higher resolution and ease of meeting high safety requirements, e.g. by letting zero to one volt be the neutral (deadband) range. The limitation of this design is the one connection for each potentiometer. This makes the amplifier and wiring harness more expensive, because there will be two inputs for each function and double the amount of wires in the harness (fig. 9-17).

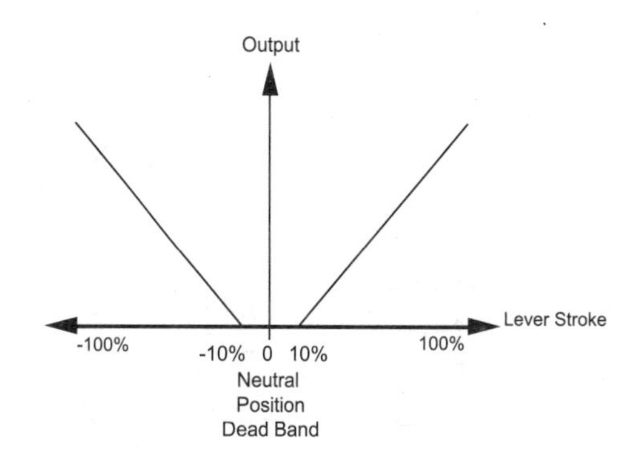

Fig. 9-17

This limitation has forced manufacturers to use a single mode design with one potentiometer (fig. 9-18).

Single mode operation

In the single mode design, the lever works as a voltage divider, where the neutral position is half of the applied supply voltage (fig. 9-19). This means that a lever with a 5 volt potential will have a neutral position at 2,5 volts (inactive), and zero and 5 volts will mean full speed in the respective directions. This could be very dangerous if a poor connector or short circuit to ground or supply causes the commend signal go to either zero or 5 volt. Therefore, the levers usually contain built-in fault detection. This can be accomplished with end resistors built into the lever for fault detection purposes (fig. 9-20). Their purpose is to prevent the output voltage from reaching the ground or supply voltage (permissible range, e.g. 10-90 % of the applied voltage).

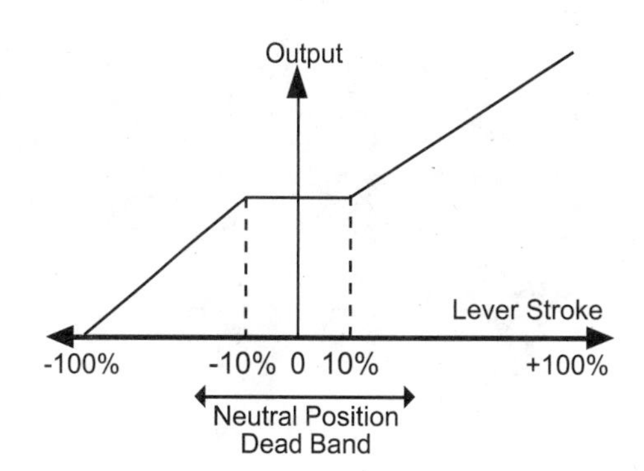

Fig. 9-18

Contactless levers

A limitation of potentiometer-based levers is their service life. After prolonged use, the resistor track wears and the lever/pot must be repaired or replaced. For this reason, the use of

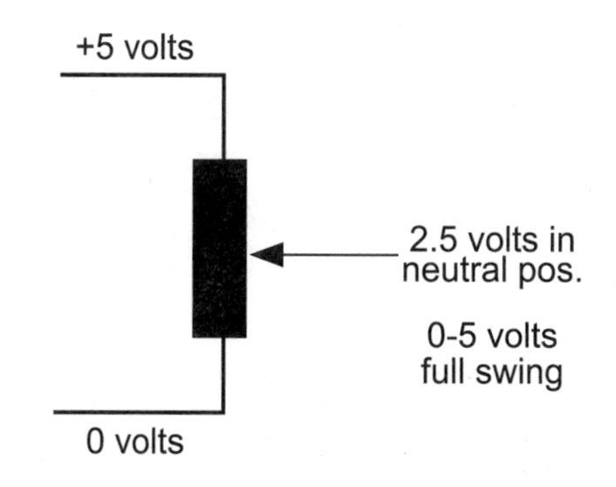

Fig. 9-19

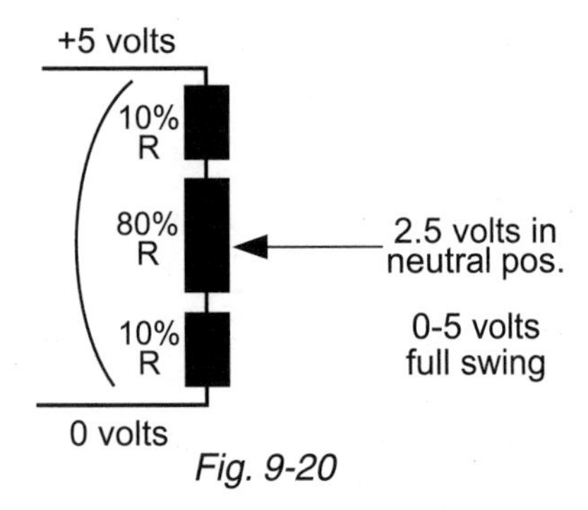

Fig. 9-20

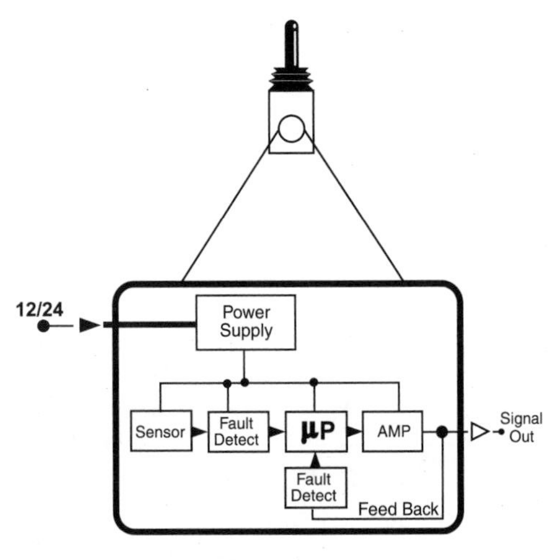

Fig. 9-21

Fig. 9-22 Mini-joysticks

contactless sensors, such as coils or Hall-effect sensors, is growing.

A Hall-effect sensor detects changes in a magnetic field and converts them into electric signals. Lever units with Hall-effect sensors typically have built-in electronics, including a fault detection circuit (fig. 9-21). Contactless sensors are being used more and more in modern machines, especially those with digital electronic remote-control systems.

Cab design ergonomics
Facts about ergonomics

Recently, a leading ergonomic study was done by The Job Study Department of the Swedish Forestry Institute. They studied the relationship between position and design of maneuvering levers and certain types of occupational injuries representative of machine operators.

This study is condensed as follows: It is possible to build machines with ergonomic principles in mind and without detrimental consequences to effectiveness. In fact, performance will most probably be enhanced.

The static strain on the vulnerable Trapezius muscles can be eased considerably by using armrest mounted, fingertip actuated controls (mini-joysticks). A poll made by the Institute reveals that more than 50 % of all machine operators suffer from pain in the neck, shoulders and back (fig. 9-22).

These pains are caused mainly by static strain on the Trapezius muscles. At long lever strokes, the operator is forced to hold his/her elevated above the armrest, which results in static strain in arms and back. Precision maneuvering with long lever strokes increases the muscular tension even more.

These problems can be resolved or diminished by designing armrests and operating control units so that the arms can actually rest upon the armrest. Also precision control becomes much easier using fingertip actuation instead of hand operated levers. Mini-joysticks make it possible for the operator to maneuver the machine with “fingertip” control, thus minimizing strain in the wrists, arms and shoulders. This has reduced substantially the number of strain-related injuries.

The levers can be optimized for individual applications and can include several proportional functions, as well as on-off functions. Furthermore they can include mechanical and/or electrical detents, which locks the lever in a certain position. They can have 'friction-brakes' that allows the lever to be positioned optionally e.g. giving constant speed for some travelling or rotating function (fig 9-23).

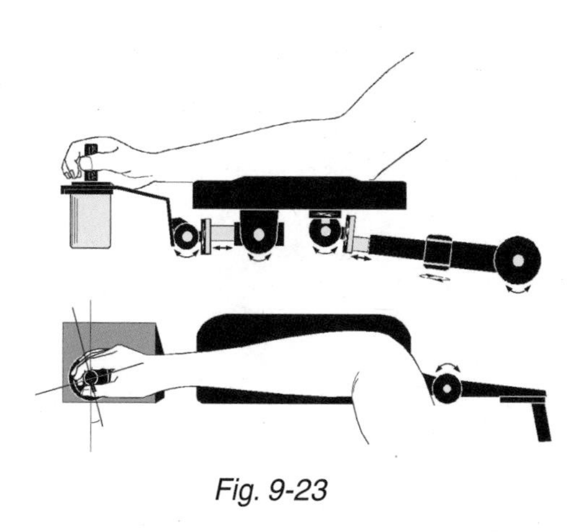

Fig. 9-23

The mini control unit provides relief for machine operators who suffer from pain in arms and shoulders, but equally important, the armrest mounted controls minimize the chances that such problems will arise.

Federal Forestry Work and the Institution for Forestry Technique at the Agricultural University (SLU) in Sweden made a follow up examination of 151 forestry machine operators that had switched over and used mini-levers for more than 6 months. The investigation showed a very positive development, not only in regards to problems in shoulders and neck, but also in arms, hands and fingers.

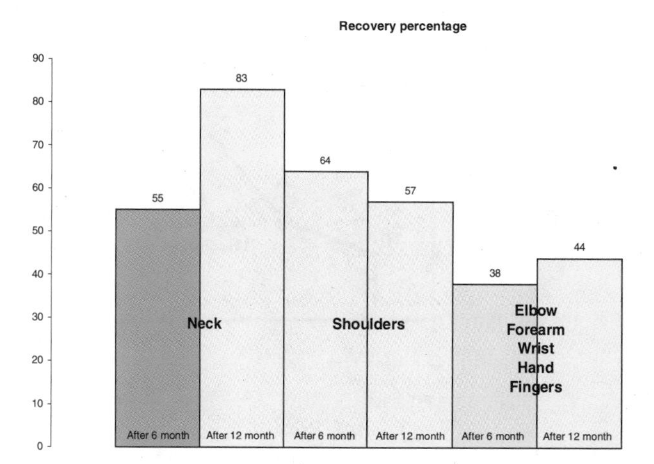

Fig. 9-24

The majority also found the machine smoother and easier to control with mini-levers. The graph shows how many of the operators had recovered from their injuries thanks to the mini-levers, see fig. 9-24.

Current feedback amplifiers

Hydraulic pressure through a proportional pilot valve is directly proportional to the electric current through its solenoid. However, the solenoid resistance changes with temperature. The higher the temperature, the higher the resistance (with a positive temperature coefficient) the lower the current through the solenoid and the lower the output pressure.

To compensate for this, most drive amplifiers are designed with current feedback circuitry. This means that, when the solenoid temperature rises (and resistance with it), the amplifier increases the applied voltage across the solenoid to maintain the same current. The current feedback circuitry thus compensates for losses in the wiring harness and/or electrical connectors, as well as for variations in the supply voltage (fig. 9-25).

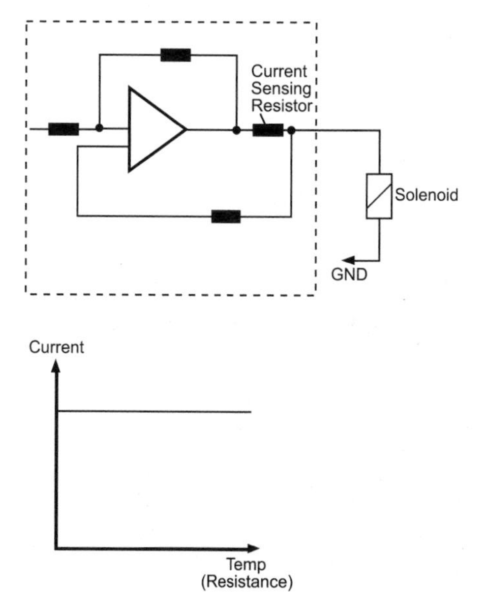

Fig. 9-25

Start and final current

The start current (sometimes called the threshold current) is adjusted so that it barely initiates the movement of the hydraulic function, while the final current determines the maximum speed of the function. It is important that the current level be adjusted accurately, so that maximum resolution is achieved (fig. 9-26).

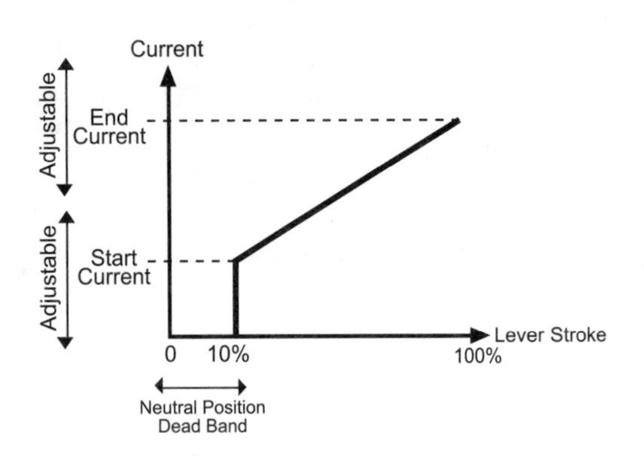

Fig. 9-26

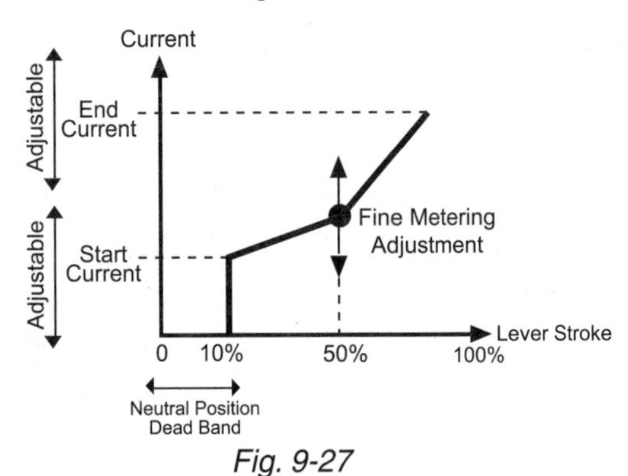

Fig. 9-27

Some systems may have features such as "fine metering control." Here, resolution is greater at the beginning of the lever stroke (fig. 9-27).

Hysteresis

Hysteresis in both the spool actuator and the main valve spool can be reduced by sending a "ripple current" signal to the solenoids. This is usually called "dither." Dither can be superimposed on the pulse-width-modulated (PWM) signal, or can be a sine wave on top of a DC signal. The pulses must be a certain frequency, usually between 50 and 200 Hz. Though some in the industrial environment may be as high as 40,000Hz (fig. 9-28).

.

Too low a dither frequency may be felt as small vibrations, whereas a too high frequency would result in increased hysteresis (bad resolution). The right frequency is the optimum between these two extremes.

Ramps

To obtain "smooth or gentle" control characteristics, most operators prefer ramp times at the start and at the stop of a function. Ramps also increase the service life of the system, both hydraulically and mechanically. A ramp can be either linear or nonlinear. Ramps are usually adjustable electronically and may be stored in memory in a digital remote control system. This

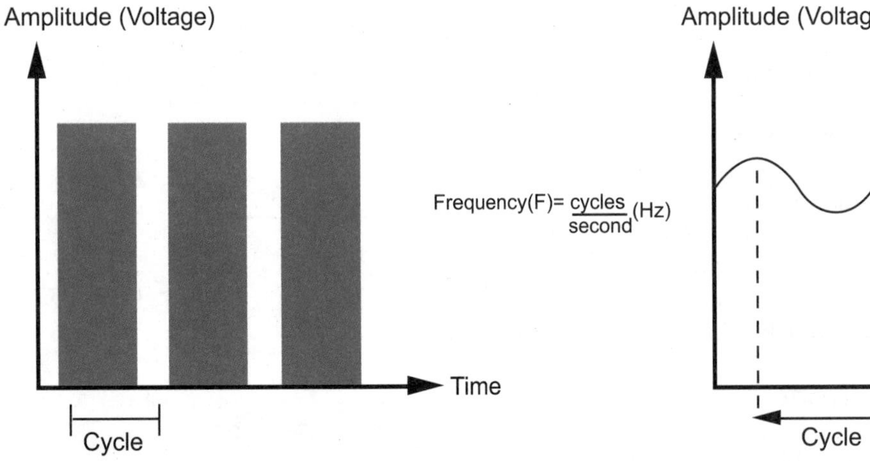

Fig. 9-28

makes it possible to have different settings stored for different operators and different work situations. See fig. 9-29.

Proportional solenoids

A proportional solenoid is the interface between the hydraulics and the electronics in an electro-hydraulic remote control system. Solenoids can be powered by either AC (alternating current) or DC (direct current), which is the most common type for mobile applications.

When using DC for proportional solenoids, the signal is often superimposed with a dither signal to diminish the hysteresis in the hydraulic components controlled.

Proportional solenoids are often used to control spools or poppets in directional control valves. They can be used to control many types of hydraulic valve function, e.g. pressure reduction, pressure relief, sequence opening, flow direction, etc. (fig. 9-30).

When used to control on-off valves, the solenoid normally receives a 2-level signal only, either 0 or maximum. On-off solenoids controlling pilot valves of the poppet type can be made to simulate a proportional hydraulic function. One example of this type solenoid develops a pulsating signal of varying duty cycle. See also chapter 8 Spool actuators.

Communication

Electronic remote control, whether analog or digital, must be connected electrically with all associated units. The wiring harness in mobile applications is of vital importance. All too often the electronics get blamed when poor connectors or broken wires are to blame. In the following sections some differences between analogue and digital communication are highlighted, and also what may cause problems.

Analog transmission

There are two basic ways of communicating between electronic units in a remote control system. The most common is by means of an analog signal. The analog signal is a constantly varying signal, which can be measured in volts or amperes, and is very easy to troubleshoot. A standard multimeter instrument is enough in most cases (fig. 9-31).

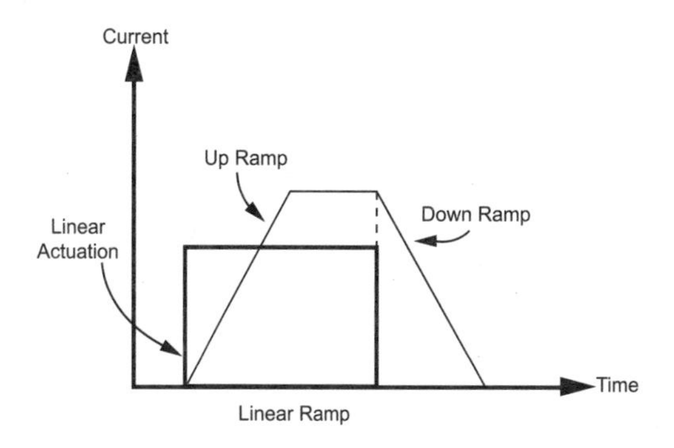

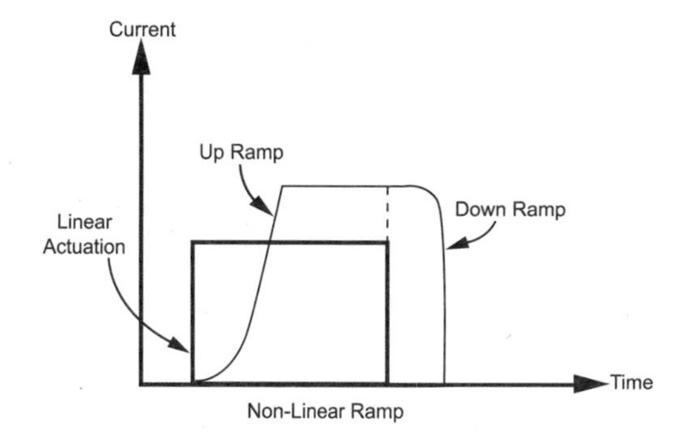

Fig. 9-29

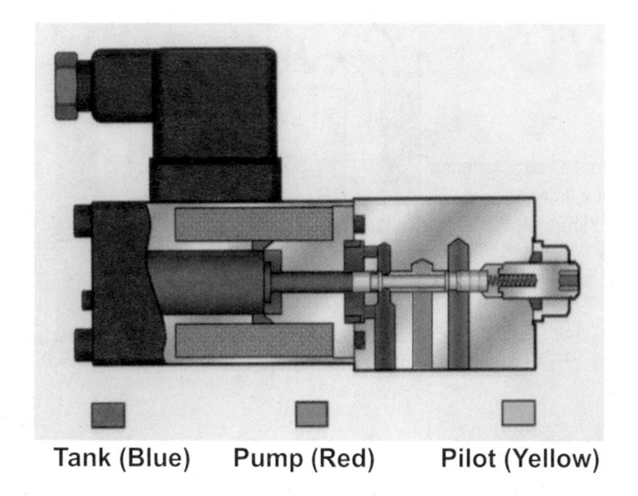

Fig. 9-30 Proportional solenoid

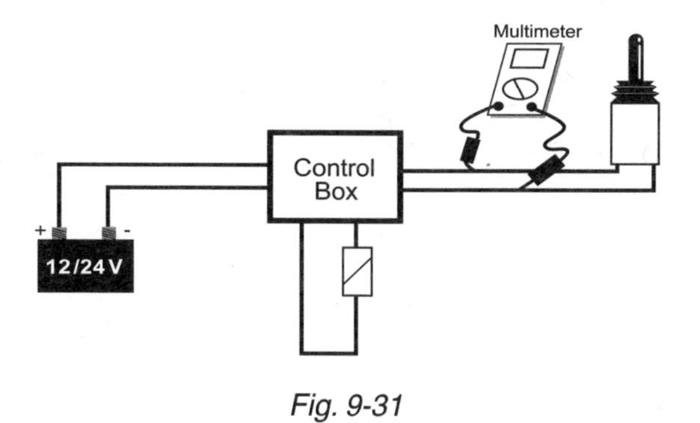

Fig. 9-31

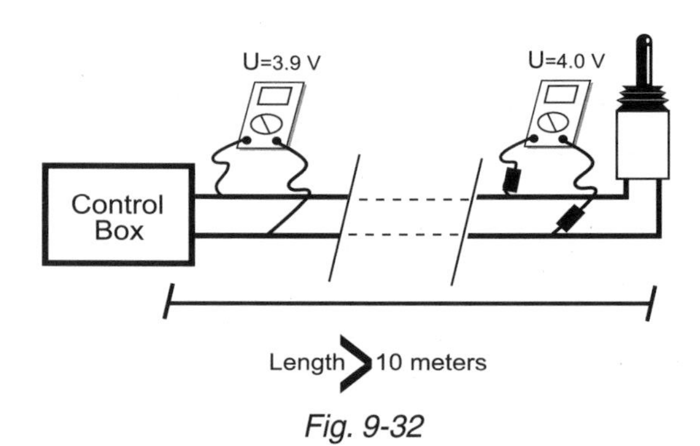

Fig. 9-32

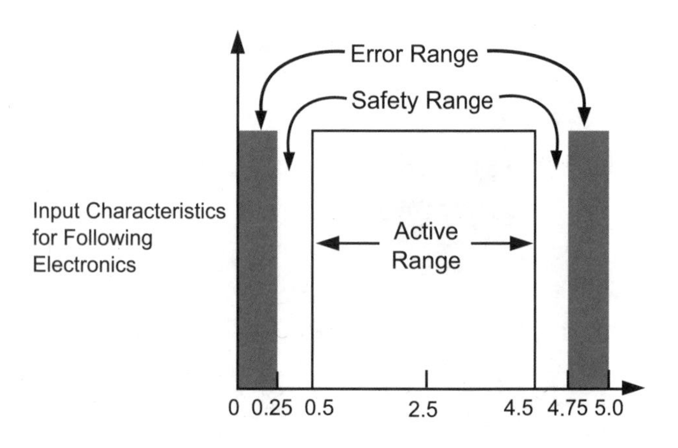

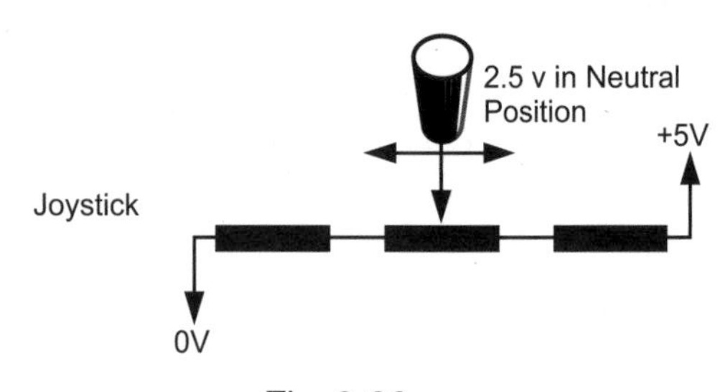

Fig. 9-33

Many different manufacturers offer levers and controls for analog systems. Such components would work in almost any application. This enables a variety of products to be connected together, provided they work within the same signal range.

A joystick working as a voltage divider will produce an output signal that is proportional to the lever stroke. As mentioned in the lever section, if there is a problem in the wiring harness due to poor connections or excessively long cables between the joystick and the amplifier, problems will arise. A signal reading of 4.0 volts at the joystick may read only 3.9 volts at the amplifier. The actuated function would then move slower than expected. In some analogue systems, this could be a problem. See fig. 9-32.

For example, a system using potentiometer-based levers with a 5-volt potential will have 2.5 volts in the neutral position. Anything else than this value would be hazardous if the system cannot detect 0 and 5 volts as a fault. Many systems therefore have a security level at 0 volts and at supply (usually 5 volts) (fig. 9-33).

Digital transmission

Digital communication between components (or nodes) doesn't have the same problems that analog has. In digital transmission, messages are sent over a "serial bus", which may consist of just 2 twisted wires. Each component/node has its own specific address and "listens" only for messages being sent to that address. If the receiver is unable to read the message, it shuts down its output and forces the system into a predefined secure mode. This usually means that the machine can only work in a specific programmed way. Exactly how the machine is going to work in practice must be determined/examined by a failure mode and effect analysis (FMEA) in order to ensure a safe solution.

The digital system is not as easy to troubleshoot as the analog, and usually it requires a PC with special software for that. However, some systems have "built-in intelligence" that informs the operator, via a display or indicator lamp, about what has gone wrong within the system (fig. 9-34).

Communication buses

There are different types of transmission buses. The most common is the RS232 standard serial communication bus in PCs. The bus in a mobile application is usually a 2-wire twisted communication between all nodes/components. The messages are sent over the bus (wire) at a certain speed (kbit/s) and each message has a certain length (measured in bytes). There is a protocol for how a node should read and interpret a message on the bus. The bus needs to be terminated at both ends to minimize reflection and disturbance from other electrical sources.

What seems to becoming a standard in mobile applications is the CAN (controller area network)-bus. Even though there are lot of protocols, only a few have been standardized, e.g. J1939 and, to some extent, CAN Kingdom. Detailed information on "bus" systems is beyond the scope of this text. Additional information can be found in textbooks specific to this topic.

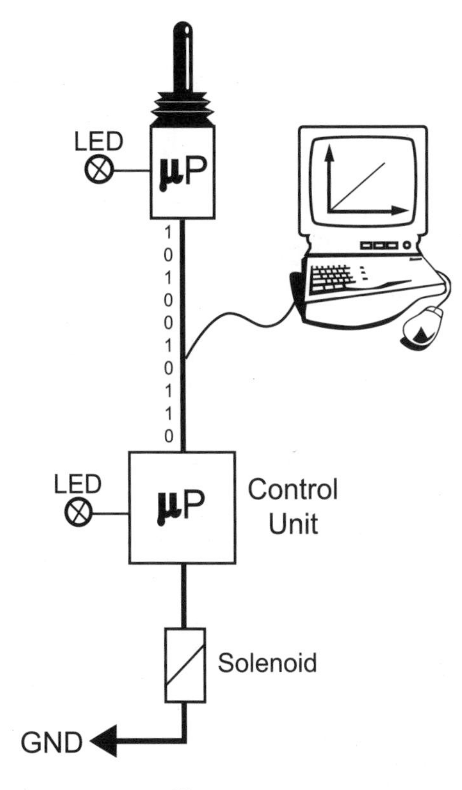

Fig. 9-34

Machine concept

The trend toward electronic control of hydraulic systems in mobile applications has been in progress for some time. The computer era has played a big roll in the advancement towards so-called "smart systems".

Functions

Fig. 9-35 shows typical functions that can be controlled/supervised in a machine concept.

Fig. 9-35 Functions that can be controlled/ supervised

Sensors

A "smart system" needs a lot of information about the machine performance. This is reported by strategically located sensors, which give feedback to the control system. Such sensors might, for instance, give information about:

- the position of hydraulic cylinders, determined by linear or angular transducers
- pressures in the hydraulic system, transmission and gearbox.
- temperatures in the hydraulic system, engine, cooling water, etc.
- the speeds of the engine, gearbox, hydraulic motors / pumps etc. in rev/min determined by frequency transducers.
- levelling or positioning related to the ground, by means of an inclinometer.

Fig. 9-36

Sensors can be based on analogue components or on analogue components with digital converters.

Some sophisticated sensors have a built-in processor, which can be connected to the communication-bus and become a node in the system (fig. 9-36).

Modular design

Modern control systems used today are based on a flexible module principle and usually include a display and a master control unit, together with specific software. Depending on the complexity of the system, it can be extended by using a serial bus, usually CAN-bus, to communicate with expansion units. An expansion unit usually has sufficient inputs and outputs to control an entire function, such as the working hydraulics on a crane.

Master control unit

The master control unit is the central unit in the system, where the specific software is downloaded and stored. Settings, parameters, graphics and texts, to which only authorized personnel have access, are also stored in this unit. See also section 'Communication with the system' below.

Interactive display

The interactive display is one way of communicating with the system. The other is usually by means of a PC, which enables extended reading of the system data and can also be used to execute adjustments in the system.

The primary function of a display is to enable the operator and service personnel to communicate with the vehicle system via a unit that is always available on the machine.

The control unit presents continuously updated vehicle information, such as vehicle and engine speed, pressures and temperatures. It can replace many conventional instruments in most machines.

Application specific information about the loading condition, margin to overload and other critical monitoring data can be presented graphically. If required, the operator's attention can be attracted by means of alarm functions, or an automatic slowing down or stopping of specific functions or the machine.

The display can also receive information from the operator during the trimming (tuning) of individual functions. Several settings can be saved for specific work situations or for different operators when the machine is working more than one shift a day.

During troubleshooting and servicing, the use of alarms and the facility to read the status and actual values of all inputs and outputs in the system enables faster identification of probable errors. See fig. 9-37.

Fig. 9-37

Expansion units

Expansion units are used in more advanced systems, where several inputs and outputs are needed. Typical examples are in systems controlling forestry harvesters, excavators, backhoe digger/loaders and mobile cranes.

In such cases, the system is expanded by means of expansion units connected via the serial bus, depending on the type of signals that need to be added. Sometimes it can be more cost efficient to add an expansion unit to reduce the amount of cables, since the CAN-bus system usually requires only 2 wires for the bus and 2 for the power supply.

Other CAN-bus based components such as joysticks can also be used to reduce the number of cables.

Communication with the system

During the start-up and evaluation of a "smart" machine, a PC-based Windows program is usually used as an aid to communicate with the system. The program handles configuration and measurement functions, the hardware and software update status, system authorization, and display texts in selectable languages and information about the specific machine.

During configuration, the "channel values" for all inputs and outputs are set. Examples of this might include the minimum and maximum values of current outputs or scaling factors and the logical operators for the inputs. "Parameter setting" for machinery functions such as engine regulation, gear shifting and monitoring and limitation functions for pressure and temperature are also part of the configuration (fig. 9-38).

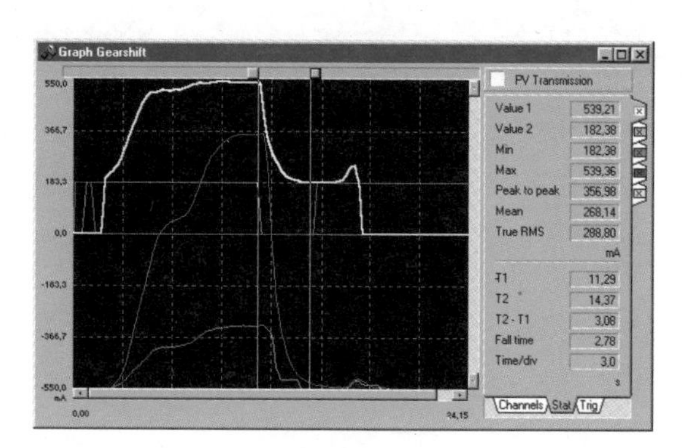

Fig. 9-38

Typical functions - engine control

The purpose of application software is to control the interaction between the various subsystems of the machine. Since this is frequently a question of how the torque and speed of the engine can be used in different operating situations, almost all machine control systems include some kind of electric or electronic control of the engine.

By controlling the injection fuel pump and by sensing the engine speed, it's possible to measure the power loaded on the engine. The power draw can then be adjusted by overriding the controls according to a predefined order of priority between simultaneously operated functions. Correct "power regulation" and "priority allocation" of a high performance machine results in reduced fuel consumption and noise level, sometimes it is also possible to reduce the horsepower or engine size.

Transmission control

The information received from strategically located sensors can be used to control the transmission in a vehicle. Both hydrostatic and hydrodynamic systems with powershift gearboxes can be controlled. The system can maintain the traction force in every work situation without stalling the machine.

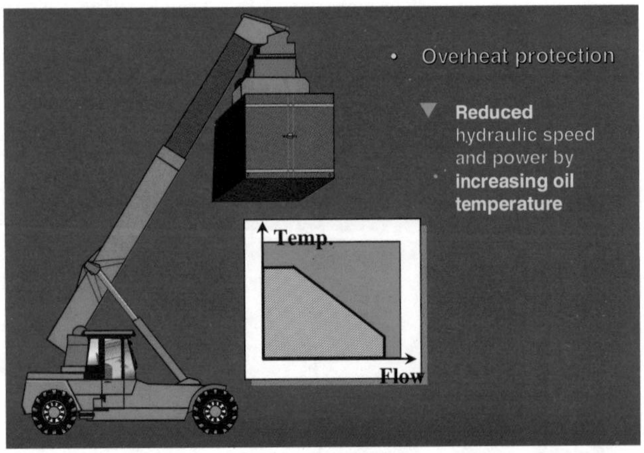

Temperature control

Temperature monitoring in a gearbox, for instance, can be used to inform the operator about high temperature. If the temperature should continue to rise, the speed of the vehicle can be reduced linearly until the temperature falls back to normal values. If the temperature rises above the maximum permissible value, the function can be permitted to continue operating, but at reduced output, or be shut down completely (fig. 9-39).

Fig. 9-39

Crane control

To achieve the highest possible productivity from a crane, individual settings for each hydraulic function, the start and final currents, must be possible. Ideally, it should be possible to set function-specific ramp and resolution values, and to store them for individual machine operators or particular work situations (applications).

With the aid of positional feedback sensors, it is possible to limit cylinder stroke and reduce the speed of the load when cylinder approaches an end position. The latter precaution greatly reduces the risk of metal fatigue and prolongs the service life of the cylinder/machine (fig. 9-40).

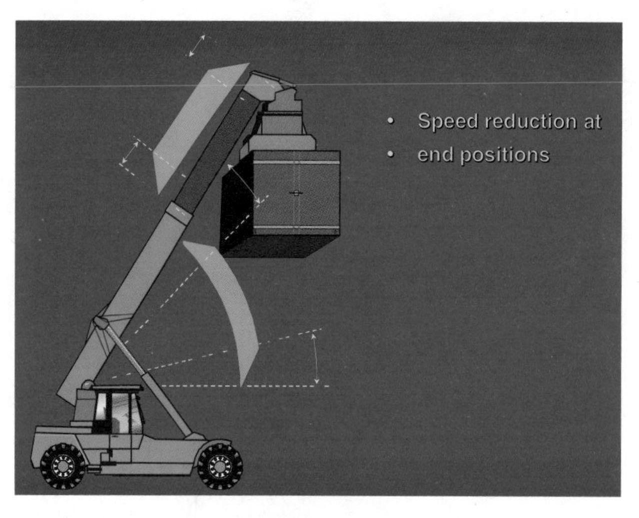

Fig. 9-40

Other functions

Other functions such as overload protection and maintaining the speed or position of an actuator, as well as envelope control, can be achieved easily. A great number of functions can be controlled with a "smart system." The possibilities are limited only by the developer's imagination.

Insulation from the environment

Water or moisture can cause short circuits and other problems in electrical connectors. Strong electric or magnetic fields can cause the system to activate. Both of these risks can be avoided today by correctly designed components and professional assembly. Manufacturers are able to seal electronic systems to prevent water reaching vital components, and can employ electronic filtering and shielding to improve the system's resistance to electromagnetic interference (EMI).

Some manufacturers now offer products that are protected and certified in accordance with various standards (fig. 9-41).

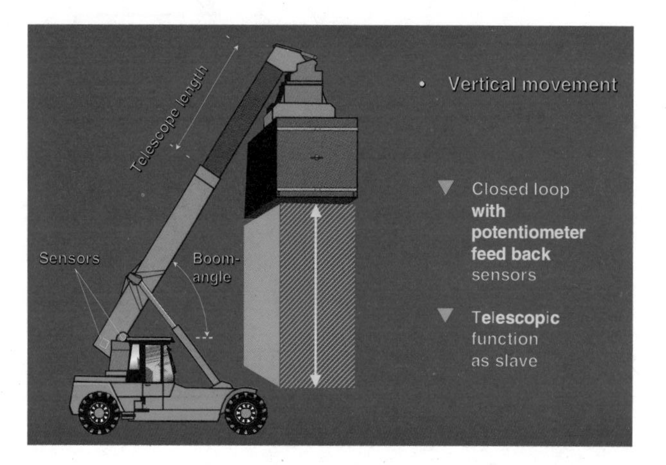

Fig. 9-41

Chapter 9 exercise
Exercise 1

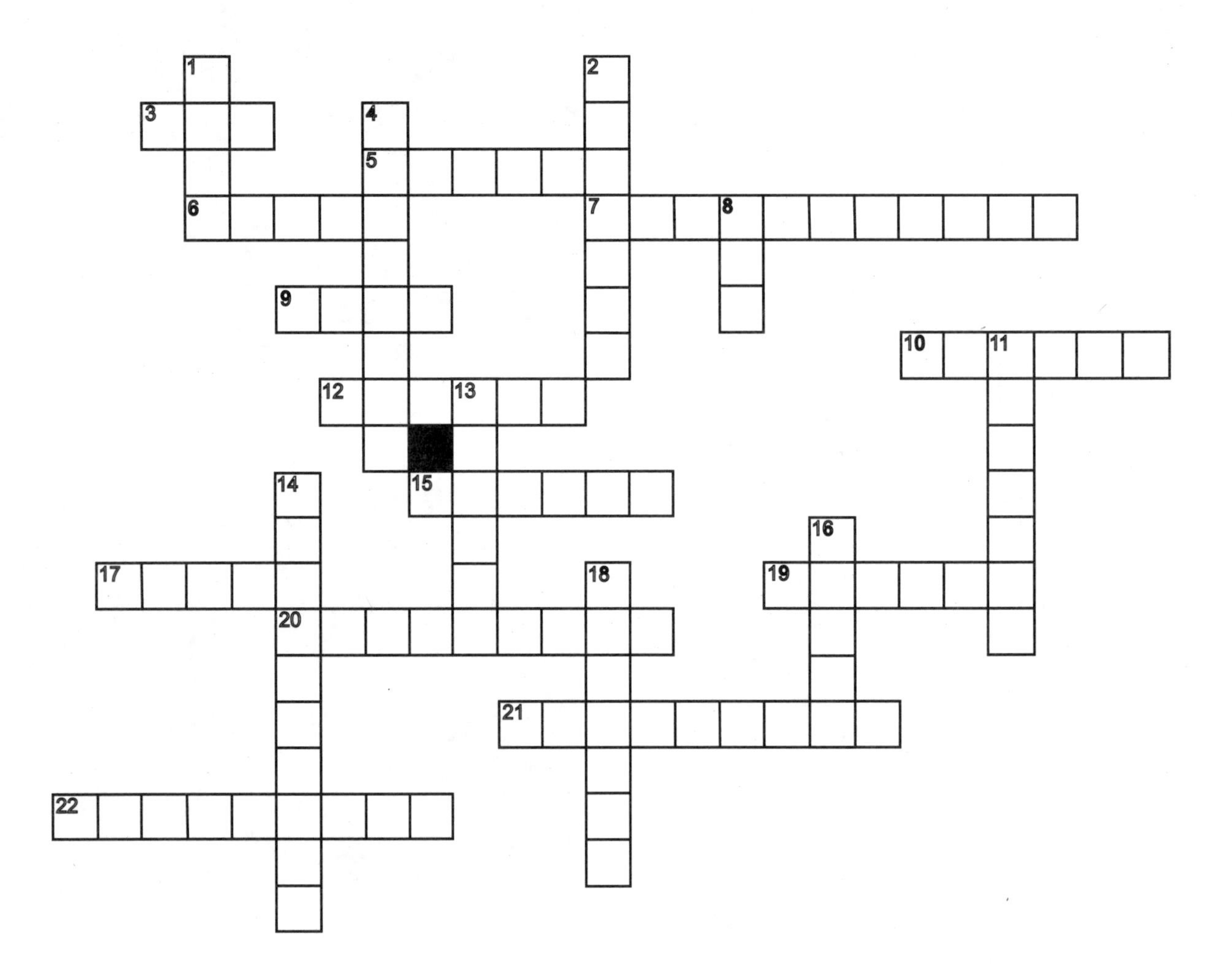

Across

3. Serial bus standard
5. A constantly varying signal
6. Basic device used to shift a valve spool
7. Type of display to communicate with the system
9. Mode of operation with high resolution
10. Central unit in a modular system design
12. One of two main spool actuators
15. A typical function monitored in a smart system
17. Devices that provide smooth control
19. Another term for ripple current characteristics
20. Static strain on these muscles can cause pain
21. Another term for start current
22. Units used in advanced systems to expand them

Down

1. Contactless sensor
2. Another form of electronic communication
4. Lever that sends a proportional control signal pressure to a spool actuator
8. Electronic filtering and shielding used to minimize this problem
11. Smart systems require these
13. Mode operation where lever works as a voltage divider
14. Can be reduced with ripple current
16. Current that determines maximum speed at a function
18. Feedback used to compensate for resistance change due to temperature change

Chapter 10

Fluids - Introduction

The fluid found in hydraulic systems performs two very important functions:

- to lubricate the system components
- to transmit the hydraulic energy throughout the system to the point of work

This chapter on hydraulic fluids is divided into the following three broad fluid categories:

1. petroleum base hydraulic fluids (page 1)
2. fire resistant fluids (page 19)
3. biodegradable fluids (page 29)

Petroleum base hydraulic fluids

A common fluid for a hydraulic system consists of paraffinic and naphthalenic petroleum oils which are blended for characteristics that make it suitable for use in a hydraulic system (fig. 10-1).

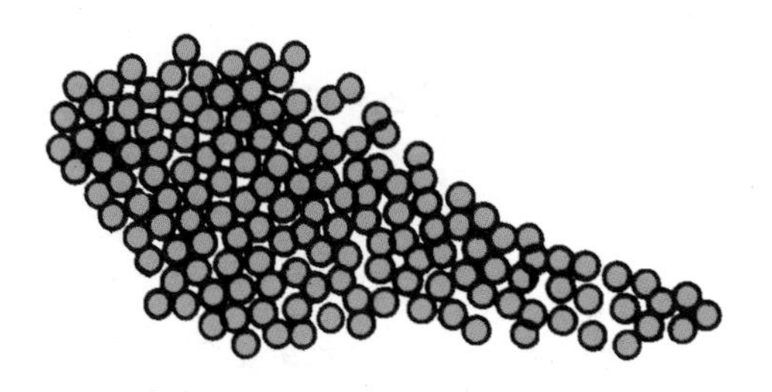

Fig. 10-1 Molecules in a petroleum oil

As was pointed out previously, hydraulic fluid is the substance used for transmitting energy from pump to actuator in a hydraulic system (fig. 10-2). The intent of this section is to concentrate on some characteristics of petroleum base fluid.

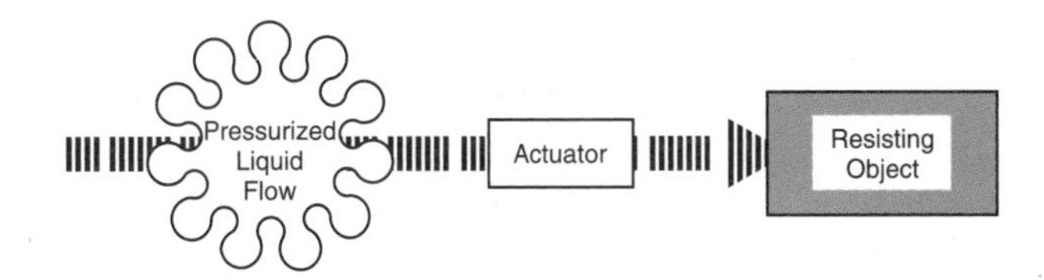

Fig. 10-2 Energy transmission in a hydraulic system

Besides acting as a medium for energy transmission, the second most important function of a petroleum base fluid is to act as a lubricant. The lesson begins by describing lubrication.

Lubrication

Lubrication is the process of reducing friction between moving surfaces which are in contact.

Lubrication is a very important function of hydraulic fluid. Without lubrication, friction would cause system components to wear excessively and excessive heat to be generated.

Friction

Friction is a force which can stop or retard the motion of a moving object. Assume that one surface of a clean, dry steel block is at rest on a similar surface. Any attempt to slide the block across the contacting surface would be resisted by a frictional force. Friction occurs because of surface roughness and welding of minute metal surfaces.

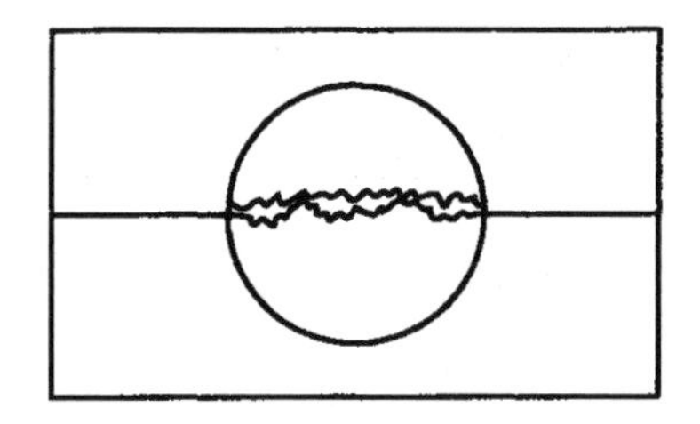

Fig. 10-3 Close-up of two smooth surfaces

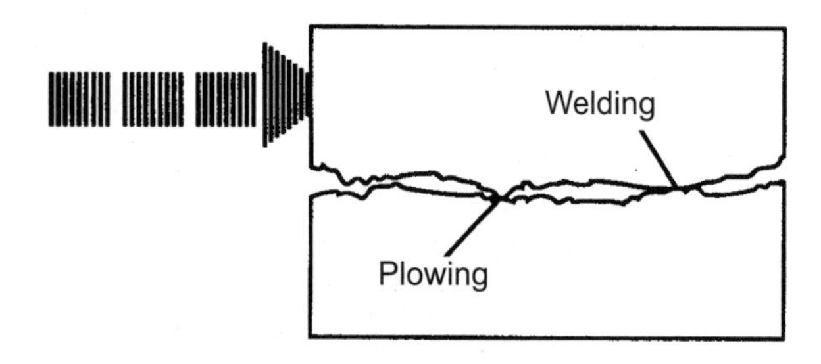

Fig. 10-4 Friction between two metal surfaces

Fig. 10-5 The lack of a lubricating film increases friction

If an apparently smooth surface of a typical component were magnified, it would appear to be quite irregular; even the best machining methods cannot eliminate these irregularities completely (fig. 10-3).

As surfaces are rubbed together, material is plowed, ripped, and worn away at a considerable rate. The rougher the surface and the greater the sliding force, the more friction will be developed.

Friction can also be related to the infinitely small welds which commonly occur between contacting metal. As a force is applied to mating surfaces, high points of a metal are deformed until they acquire a large enough base to support a force. This action tends to bond the material at the contacting points (fig. 10-4).

In moving the surfaces across one another, these tiny bonds are ruptured, and this action contributes to friction.

Previously, it was shown that a liquid consisted of continuously-moving molecules which could take the shape of its container. Also, it was learned that a liquid had a resistance to flow known as viscosity. In the following section, how a petroleum oil's capability of adhering to a surface and viscosity contribute to develop a lubricating film will be examined.

Fluid film

Interaction between metal surfaces can be greatly reduced by introducing a lubricating film between them. Not having a lubricating film between moving parts is similar to rowing a boat on land (fig. 10-5).

Any liquid will form a lubricating film, but some do a better job than others. Water, for example, was the first hydraulic fluid, but it was a poor industrial lubricant because its fluid film is not durable. Petroleum oil is a good lubricant because it forms a durable fluid film.

Lubricity

Lubricity refers to a liquid's ability to form a durable fluid film between contacting surfaces. This ability is directly related to:

a. a fluid's natural film thickness
b. a fluid's tendency to adhere to a surface

Petroleum oil has good lubricity. If at room temperature, a petroleum base hydraulic oil were poured on a steel plate, it would appear to wet, or adhere to the surface with a substantial fluid film (fig. 10-6).

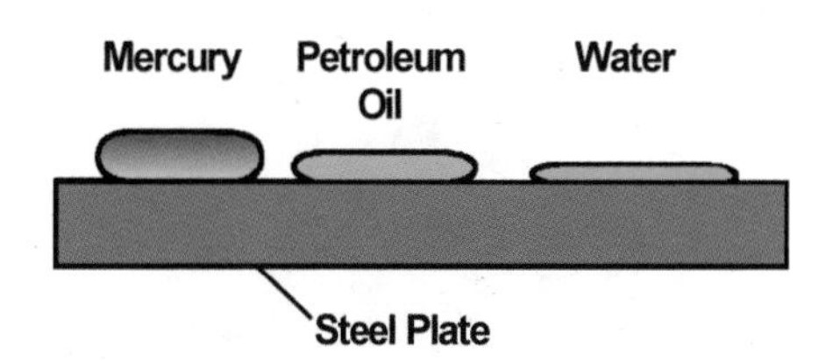

Fig. 10-6 Petroleum oil and water wet the surface, but mercury does not.

If water were likewise poured on the unprotected metal, it would appear to wet the surface, but its fluid film would be thin and therefore easily penetrated. For this reason, water has poor lubricity.

If the same procedure were followed with mercury, a thick fluid film would form, but it would show relatively little tendency to adhere to the steel.

As a matter of fact, the mercury could be broken up into little balls or beads. Even though mercury does form a thick fluid film, it also has poor lubricity because it does not tend to stick to the steel (ferrous) surface.

A liquid with good lubricity adheres to a surface and also develops a substantial fluid film. Of the fluids used in a hydraulic system, petroleum oil has been found to exhibit the best lubricity.

Viscosity affects a system

Up to this time, petroleum base hydraulic fluid has been required to preform two important functions:

1. to act as a medium for energy transmission
2. to lubricate internal moving parts of a system

Both of these functions, and consequently hydraulic systems in general, are influenced by fluid viscosity which is probably one of the most significant characteristics of a petroleum base fluid. In the following section, viscosity will be defined.

It will be shown in what manner it is measured, and then determined how viscosity affects heat generation, lubricity, hydrodynamic lubrication, and clearance flow. The starting point for this review is at the molecular level.

Liquid molecules

As with all liquids, petroleum base hydraulic fluid is made up of molecules which are attracted to one another (fig. 10-7). This attraction is much greater than the molecules of a gas, but is less

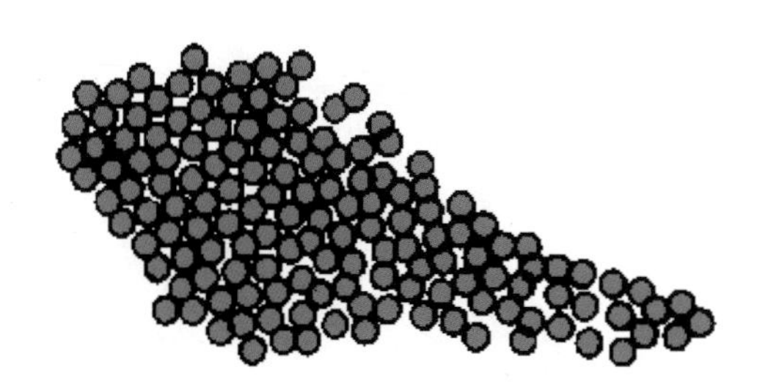

Fig. 10-7 Molecules in a liquid

Fig. 10-8 Liquid molecules are free to move

strong than the molecules of a solid, which are in a relatively fixed position.

Liquid molecules are free to slide past each other, and, as a matter of fact, they are continuously moving (fig. 10-8).

Viscosity

Viscosity is the **resistance** of a liquid's molecules to flow or slide past each other; it is sort of an internal friction. An example of a high viscosity liquid is honey or molasses; water or cooking oil is an example of a low viscosity liquid.

Viscosity affected by temperature

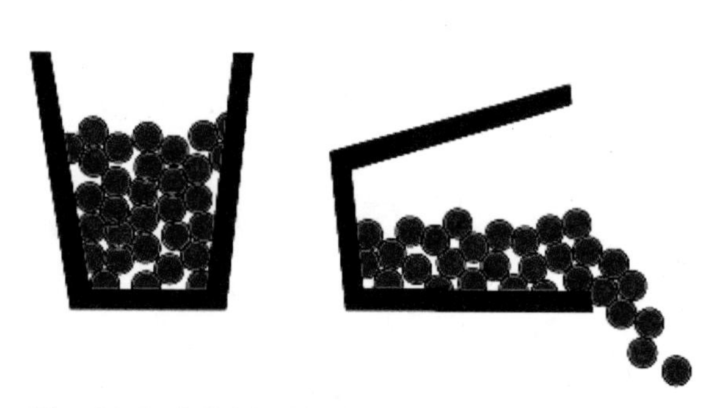

Fig. 10-9 A fluid with high viscosity doesn't pour easily.

As was indicated, a liquid is made up of molecules which are attracted to each other and continuously moving. Some experts feel that the more slowly molecules move, the greater are attractive forces resulting in an increased resistant to flow.

A bottle of molasses taken from a refrigerator consists of very slowly moving molecules which have large attractive forces; cold molasses has a high resistance to flow, a high viscosity. Trying to pour this liquid through a funnel would be a time consuming task (fig. 10-9).

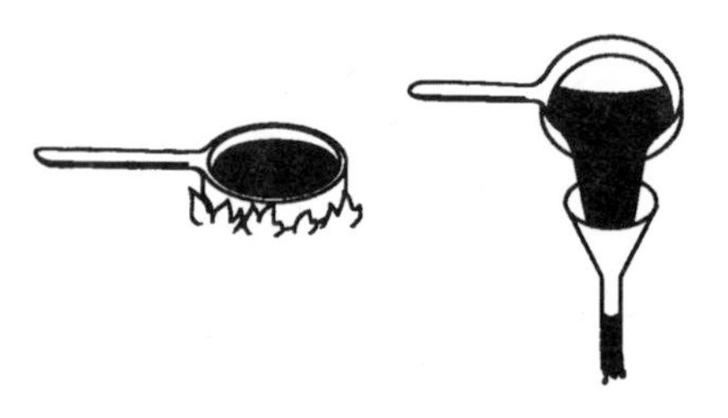

Fig. 10-10 Heating reduces the viscosity

Heating the molasses in a sauce pan adds energy to the molecules; molecular speed increases reducing attraction between molecules.

With less of a resistance to flow, a reduced viscosity, heated molasses can more easily flow through the funnel (fig. 10-10).

Generally, as temperature increases, viscosity of a liquid decreases.

Centistoke

In the SI system, the measure of liquid viscosity is the centistoke, abbreviated 'cSt'.

Saybolt universal second

In the Imperial system the common measure of liquid viscosity is the Saybolt Universal Second, abbreviated 'SUS' or 'SSU'.

NOTE: There is no direct relationship between the Saybolt Universal Second (SUS) and the centistoke (cSt).

The Saybolt Universal Second is named in honor of George M. Saybolt who, in 1919, offered to the United States Bureau of Standards his Saybolt Viscosimeter - a device which measured viscosity.

The use of the Saybolt Viscosimeter consisted of filling the apparatus with a liquid and heating the liquid to a specified temperature. With the liquid heated, a cork was pulled from the chamber bottom at the same time a stopwatch was started. The liquid then drained through an opening of a specific size until 60 milliliter (about 2 fluid ounces) was in a flask (fig. 10-11).

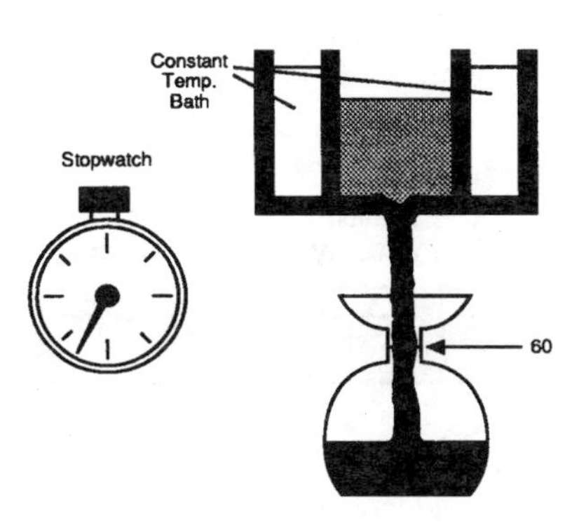

Fig. 10-11 Saybolt viscosimeter device

The stopwatch timed how long the liquid took to fill the flask. The result was a measure of viscosity in Saybolt Universal Seconds.

If an oil heated to a temperature of 38°C *(100°F)* took 143 seconds to fill the flask, its viscosity would be *143 SUS* at 38°C *(100°F)*, which corresponds to approximately 31 cSt.

If the same oil took 82 seconds to fill the flask when heated to 54°C (130°F), its viscosity would be 82 SUS at 54°C (130°F) or approx. 18 cSt.

Viscosity is always associated with a temperature. Yet, it is common to hear someone say he is using '150 SUS' or '32 cSt' oil. The 38°C (100°F) temperature is assumed.

Viscosity affected by pressure

Viscosity is affected by system pressure. As pressure in a system increases, viscosity also increases. This is pointed out by the illustrated graph (fig. 10-12).

The graph shows that for a common industrial hydraulic oil, viscosity increases 40% as pressure increases from zero 200 bar *(3000 psi)*.

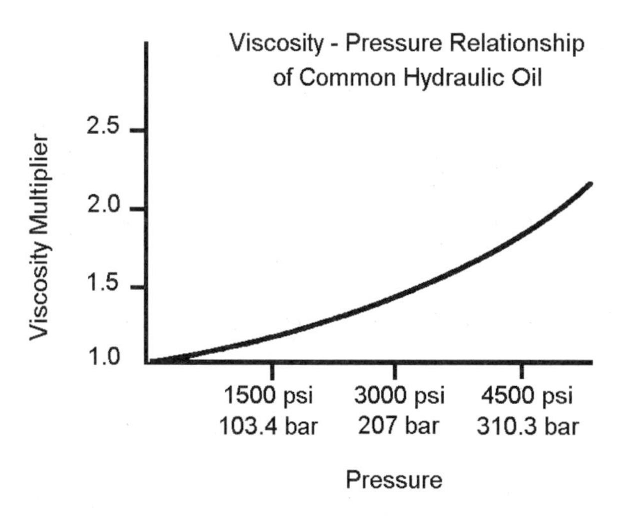

Fig. 10-12 Viscosity increases with pressure

Viscosity affects heat generation

Viscosity of a petroleum base hydraulic fluid affects heat generation.

A high viscosity liquid of 100 cSt *(500 SUS)*, having more internal resistance to flow, will cause more heat to be generated in a system than a low viscosity liquid of 32 cSt *(150 SUS)*. In many hydraulic systems, viscosity of an oil is 32-55 cSt *(150-250 SUS)* at 38°C *(100°F)*.

Component manufacturers will supply fluid viscosity range recommendations.

Viscosity affects lubricity

Since it is a resistance, viscosity may not at first appear to be a desirable characteristic. But, viscosity is a very important and desirable fluid characteristic since it affects lubricity.

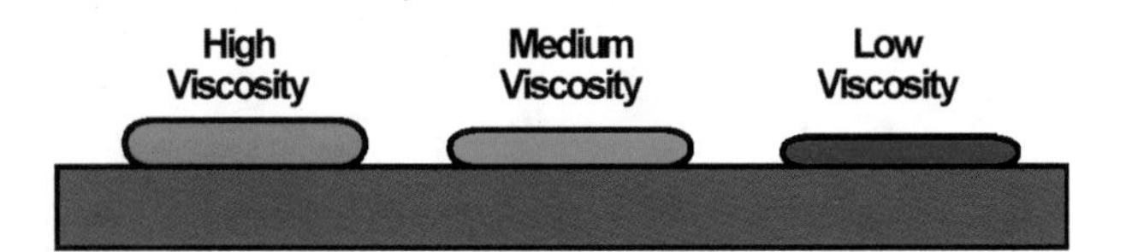

Fig. 10-13 Lubricating film thickness increases with viscosity.

It was illustrated that lubricity was dependent on an oil adhering to a metal surface and developing a substantial fluid film; viscosity affects fluid film (fig. 10-13). The higher the viscosity, the thicker will be the fluid film. Of course, the fluid must be capable of readily flowing so determination of an appropriate viscosity for a system is a compromise between its ability to form a fluid film and its ability to flow.

Viscosity affects hydrodynamic lubrication

The ability to form a durable fluid film is an important characteristic of petroleum base fluid. This ability is referred to as lubricity.

It may be felt that a fluid film would be difficult to maintain between moving parts, since any rapid movement would tend to scrape a surface clean. However, once parts begin to move, liquid viscosity does not usually allow this to happen.

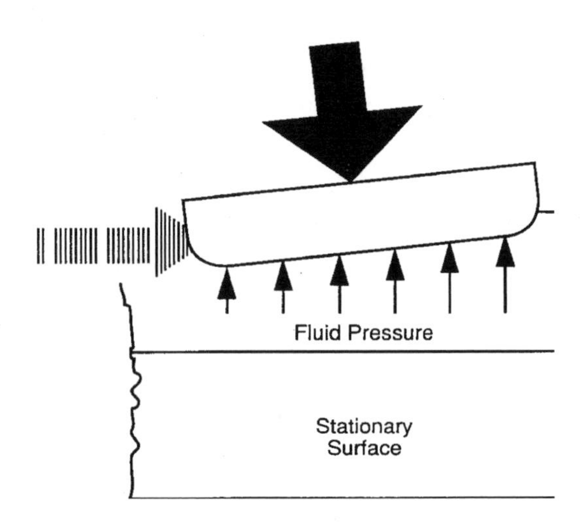

Fig. 10-14 Hydrodynamic lubrication

A metal block immersed in oil and at rest on a stationary metal surface is separated by the oil's fluid film. As a force moves the block, the leading edge rises because the oil resists getting out of its way due to the viscosity of the fluid (fig. 10-14). This action forms a fluid wedge under the block which floats it along like a boat planing on water. As long as pressure on the moving block does not become excessive, the fluid wedge would ward off normal attempts at penetration. This is known as **hydrodynamic lubrication**.

A low viscosity liquid like water would get out of the way too easily under low speed and high load conditions. The wedge would not fully form, resulting in a fluid film which could more easily be penetrated.

When system components are moving, they are lubricated by this hydrodynamic process. However, at start-up or when excessive pressure pushes a moving part through a hydrodynamic wedge, a liquid's ability to form a durable fluid film (lubricity) becomes very important.

Viscosity affects clearance flow

Another important effect of viscosity is its ability to help reduce leakage between clearances of close fitting moving parts.

Most components in a hydraulic system do not have a zero-leakage seal between internal moving parts. Frequently, the seal is metal-to-metal through which a small portion of fluid is continuously flowing and lubricating.

Metal-to-metal seals are found in:

- piston pumps between piston and piston bore
- gear pumps between gear and housing
- vane pumps between vane tip and cam ring
- cylinders between piston ring and cylinder bore
- control valves between spool and valve body.

To achieve the best seal possible, clearances between moving parts are kept to a minimum. However, clearance size is not so small that a fluid cannot pass through and lubricate. Size of a clearance is frequently a compromise between sealing, lubrication and leakage flow (fig. 10-15).

Clearances between internal moving parts of a hydraulic component are in effect orifices; they continually meter lubricative-leakage flow. Just as any orifice, flow through a component clearance is affected by fluid viscosity.

With a viscous fluid, leakage and therefore lubricative flow through a clearance will be reduced. On the other hand, a fluid which is not viscous enough, or too thin, means that excessive fluid passes through component metal-to-metal clearances. This results in less flow passing into the system and an unnecessary buildup of heat in the system (fig. 10-16).

Metal-to-metal clearances between component moving parts can be considered built-in fixed restrictions which are continually bleeding off flow. Too much leakage, then, can be harmful to the system. However, if too little flow passes through the clearance, the component may not be sufficiently lubricated. The system may become erratic and undependable as a result. A

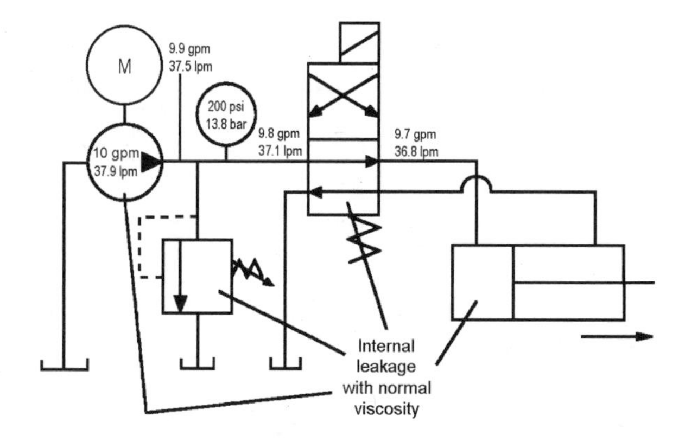

Fig. 10-15 Internal leakage with nominal viscosity

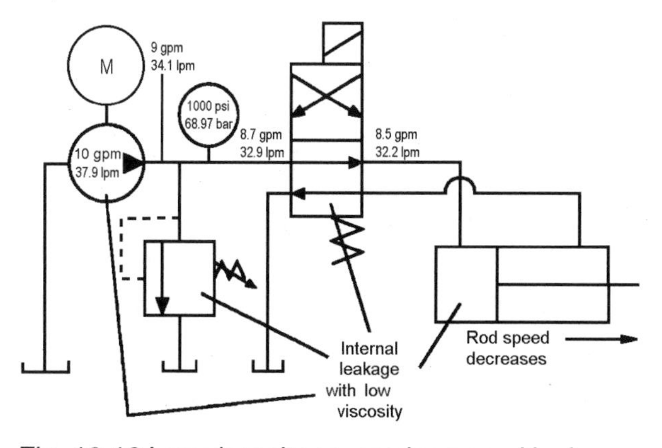

Fig. 10-16 Low viscosity causes increased leakage

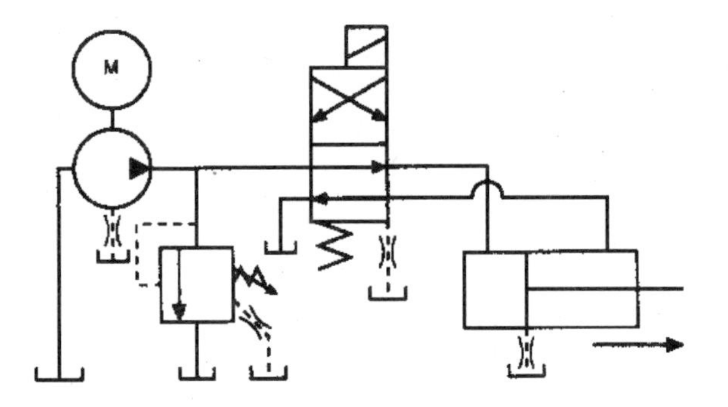

Fig. 10-17 'Correct' viscosity prevents component seizing and 'acceptable' leakage.

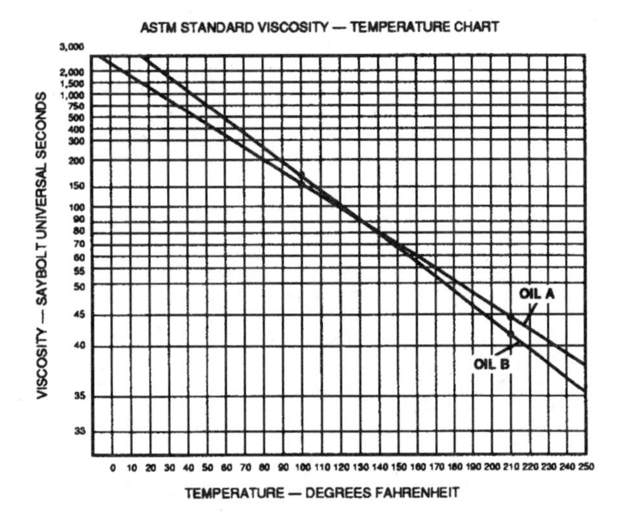

Fig. 10-18 Graph showing viscosity of two liquids.

happy medium can be found with a fluid of the appropriate viscosity (fig. 10-17).

Viscosity index

Since the viscosity of hydraulic oil does change with temperature and since viscosity is an important factor in a hydraulic system, systems which are not, or cannot be, maintained at a constant temperature (this is the case many times in mobile hydraulic systems) need an oil whose viscosity remains relatively stable over a given temperature range.

An oil's viscosity index illustrates how viscosity is affected with changes in temperature. This relationship can be depicted by a straight line using ASTM (American Society for Testing and Materials) standard viscosity-temperature charts for liquid petroleum products (fig. 10-18). When the viscosity of an oil at two temperatures is plotted on the paper, viscosity of the oil at any temperature can be determined by drawing a straight line connecting and running through the two points. (This can be done with any liquid petroleum product which does not have added chemicals that affect its natural viscosity-temperature relationship.)

The viscosities of two oils are plotted on the chart (fig. 10-18). The oil with the more horizontal line, 'A', has a higher viscosity index than oil 'B'. In these examples, oil 'A' has a viscosity of 33 cSt *(153 SUS)* at 38°C *(100°F)* and 9.5 cSt *(44 SUS)* at 100°C *(210°F)*. Oil 'B' has a viscosity of 36 cSt *(165 SUS)* at 38°C *(100°F)* and 9 cSt *(42 SUS)* at 100°C *(210°F).*

Oil 'A' (with the more horizontal line) has a higher viscosity index and its viscosity will change less with changes in temperature. When the term viscosity index, 'VI', was first adapted, a certain oil showing a very rapid change in viscosity with respect to temperature was assigned a viscosity index of 0, while the best oil type showing a small change with temperature was arbitrarily given a viscosity index of 100. Therefore, oils at that time all had VIs between 0 and 100. With up-to-date refining practices, oils can now have VIs above 100.

In a modern hydraulic system, an oil's viscosity index is generally required to be 90 or above. However, viscosity index means little for systems with a relatively constant operating temperature.

Oil operating range

Petroleum oil is an excellent lubricant for a mobile hydraulic system, but not at all viscosities. If viscosity of an oil were too low, its fluid film would be like water and consequently too thin. If an oil's viscosity were too high, insufficient amounts of fluid would flow into bearings and component clearances.

For this reason, manufacturers of rotating equipment (pumps, motors) which are especially dependent on proper bearing lubrication, specify the viscosity range at which their components are to be operated. When these components are sufficiently lubricated, it usually means the rest of the system is lubricated as well.

If components have a required viscosity range, then this information, along with the temperature range of the system, indicates the use of a specific oil.

For example, a particular mobile system at its operating temperature requires a minimum/maximum viscosity of 15-55 cSt *(80-250 SUS)*. If the operating temperature range were 25-60°C *(75-140°F)*, hydraulic fluid 'Y' would be used (fig. 10-19). If the temperature range were 45-80°C *(115-175°F)*, hydraulic fluid 'Z' would be used. Mobile hydraulic systems have run with temperatures in excess of 95°C *(200°F)*.

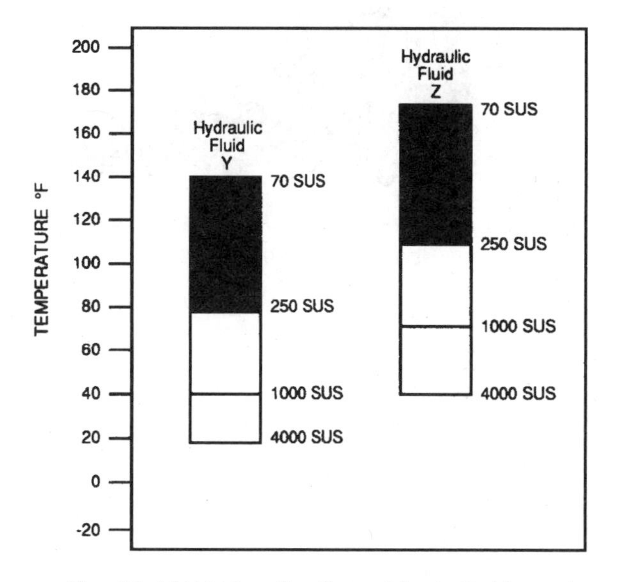

Fig. 10-19 Hydraulic oil must be suitable to the operating temperature range.

Since temperatures can become quite low in mobile equipment environments, an oil can become extremely viscous. To ensure that pumping mechanisms will fill, pump manufacturers also specify the maximum viscosity allowable at start-up.

In general, these viscosities are 200 cSt *(1000 SUS)*, 500 cSt *(2000 SUS)* and 1600 cSt *(7500 SUS)* for piston, vane and gear equipment, respectively.

Pour point

ASTM graph paper does not point this out, but at extremely low temperatures petroleum oil does not flow. At low temperatures, wax structures begin to form in hydraulic fluids containing any paraffinic base crude. These wax formations hinder and may even stop flow. Pour point (fig. 10-20) of a hydraulic fluid is the lowest temperature at which it will pour under an ASTM laboratory test. In an actual system, if the maximum viscosity start-up specification is adhered to, the

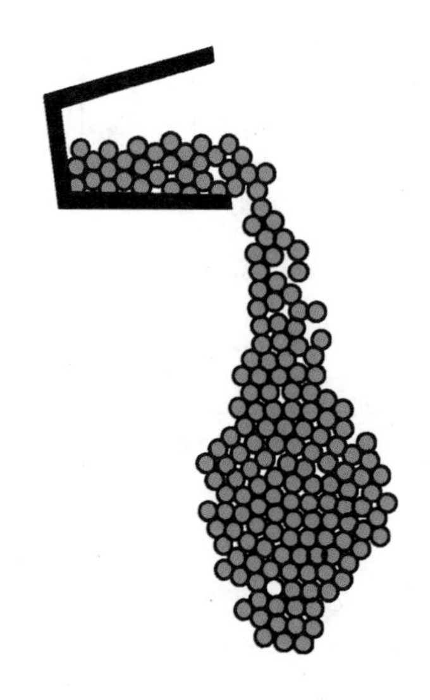

Fig. 10-20 Pour point - the lowest temperature at which the fluid can still be poured.

pour point of a fluid is generally not considered. But, when a hydraulic system has the possibility of operating under extremely low temperature conditions, the pour point of an oil should be at least -20°C *(20°F)* below the lowest expected temperature.

Pour points for various oils are indicated in the manufacturer's data sheet of the specific oil.

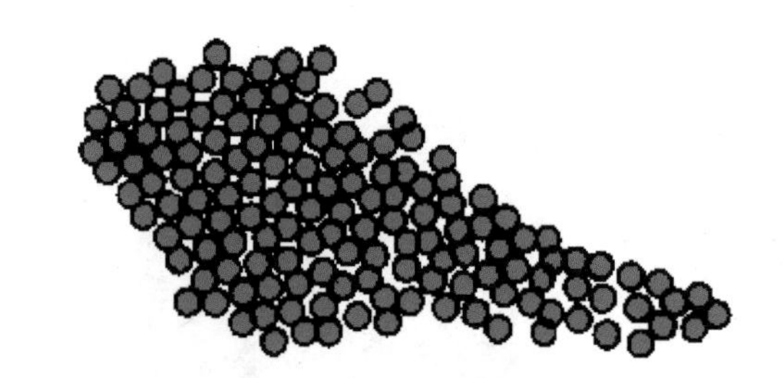

Fig. 10-21 Hydraulic fluid without additives may cause problems.

Oil problems and additives

In the day-to-day operation of a system, as petroleum base hydraulic fluid (fig. 10-21) performs its function, certain problems can arise which may affect both fluid and system. Some of these problems are high pressure lubrication, oil oxidation, and oil contamination with water, air bubbles, and dirt. To limit some of these problems, hydraulic fluids are equipped with chemical additives.

The problems encountered with hydraulic fluids and the usual additives used to limit the problems, are dealt with below. It should be realized, however, that chemical additives cannot solve every fluid problem, and that an oil cannot contain every available additive making it a super oil. Many fluid additives are not compatible with each other and therefore would give an unfavorable reaction if mixed.

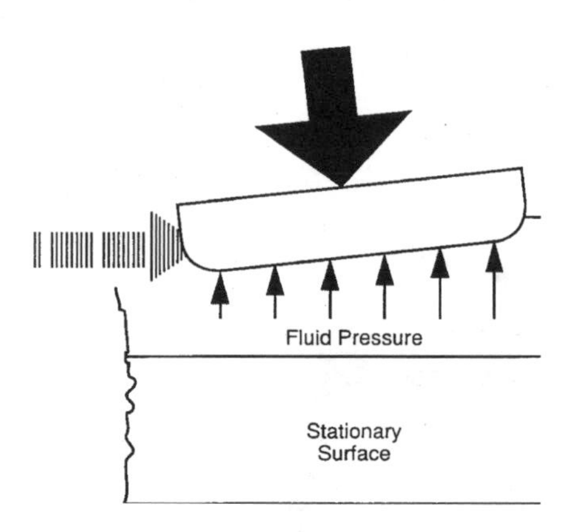

Fig. 10-22 At high pressures and low viscosities the lubricating film can break down.

High pressure lubrication

A good quality petroleum base hydraulic fluid is not a good enough lubricant for some systems. As pressures climb, the hydrodynamic fluid wedge (fig. 10-22) between moving parts has more of a tendency to break down. This means lubrication is more dependent upon a fluid's inherent lubricity.

To aid in lubricity or boundary lubrication at high pressures, hydraulic fluids are equipped with chemical additives. These additives become even more important as mobile hydraulic system pressures begin to exceed the 200-350 bar *(3000-5000 psi)* range. Oil additives help limit the resulting damage that would be caused by the tremendous forces that are generated between moving parts in mobile hydraulic systems operating in this range.

Antiwear additives

Antiwear, 'AW', or wear resistant, 'WR', additives can be divided into three types. One type, sometimes called an oiliness or lubricity agent, is a chemical made up of molecules that attach themselves vertically like blades of grass to metal surfaces (fig. 10-23). This creates a chemical film which acts as a solid when an attempt is made at penetration.

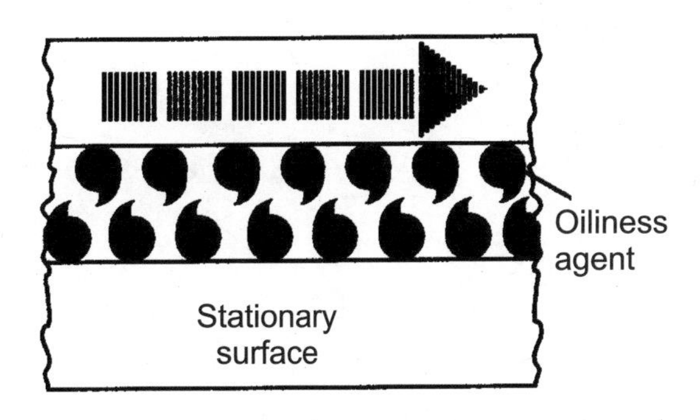

Fig. 10-23 An antiwear additive can prevent the lubricating film from breaking down.

The additive molecules support the load, allowing a moving part to slip by. But this film is not very durable, tending to break down at high temperatures.

Another type of antiwear additive chemically combines with a metal surface to form a protective film. This film forms as low frictional heat is generated between contacting points of moving surfaces. They serve to smooth out or polish surfaces so that friction is reduced.

Another antiwear agent, known as an extreme pressure, 'EP', additive, forms a film on a metal surface as high frictional heat is generated. In a high pressure system, as mechanical interaction between surfaces becomes excessive, heat becomes excessive and the surfaces attempt to weld together. The extreme pressure additive comes out of solution at this point, keeping the surfaces apart.

All three types of antiwear additives (fig. 10-24) are not found in the same fluid and are not used in the same applications. When oiliness agents are used, they are generally found in relatively low pressure systems, below 70 bar *(1000 psi)*.

WR - Wear resistant additive
AW - Anti-wear additive
EP - Extreme pressure additive

Fig. 10-24 Antiwear fluid additives

When extreme pressure additives are found in a hydraulic system, the system will probably be operating above 200 bar *(3000) psi*, or the same fluid that is used to lubricate gears and machine ways is also used in the hydraulic system.

A very common antiwear additive is the one which operates in the medium pressure range, 70-200 bar *(1000-3000 psi)*.

Check for high pressure lubrication

The check for a fluid's ability to give high pressure lubrication is the title of the oil or a manufacturer's catalog sheet. For example, with a Gulf Oil Co. fluid titled 'Harmony 48AW', the AW stands for antiwear. Or, with a Sun Oil Co. fluid titled 'Sunvis 816 WR', the WR indicates wear resistant.

Many refiners do not indicate the antiwear additive in an oil title. Consequently, the refiner's catalog or data sheet for a particular fluid must be referred to.

If excessive component wear has been a problem and the system's hydraulic fluid does not contain antiwear additives, switching to an oil with an antiwear additive will probably reduce the problem. This assumes, however, that the component wear was not the result of fluid contamination.

Oil oxidation

Oxidation is a process by which material chemically combines with oxygen; this is a common occurrence.

If you have ever taken a bite out of an apple, you know that the pulp quickly turns brown as it is exposed to air (fig. 10-25). The same process happens when a car's fender is scraped down to bare metal; the exposed metal reacts with oxygen in the air and rusts. Many things on earth, including oil, oxidize in this manner.

Fig. 10-25 Oxidation is caused by exposure to air

Oxidation of hydraulic fluid can be pinned down to basically two system locations - reservoir and pump outlet. In both cases, oil reacts with oxygen but in different ways and the oxidation products are not the same.

In a reservoir, the free surface of the oil reacts with oxygen in the air (fig. 10-26). The product of this reaction includes **weak acids** and **soaps**. Acids weaken and pit component surfaces; soaps coat surfaces and can plug pressure-sensing orifices and lubrication paths.

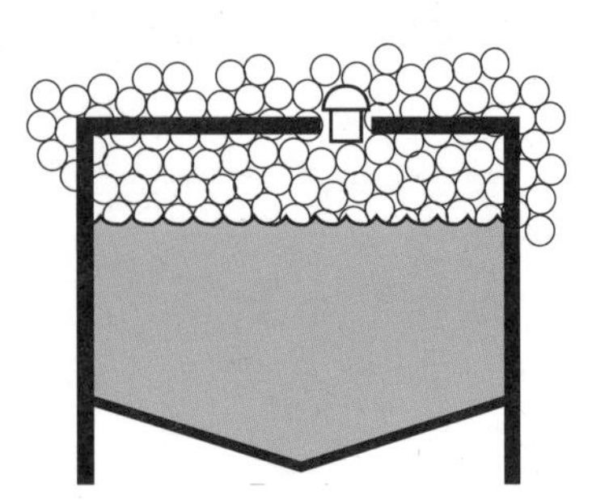

Fig. 10-26 Without an antioxidant, the fluid may foam.

Heat is a major contributor to oil oxidation in a reservoir. As a rule, oil oxidizes twice as fast as normal for every 10-11°C *(18-20°F)* rise in temperature above an average reservoir temperature of 55°C *(130°F)*.

The higher the oil temperature, in a mobile hydraulic system, the more often it, and system filters, should be changed. The cost of new oil as well as the disposal cost of the old oil and filters is the result of over - temperature operation. Reservoir oil also oxidizes more readily in the presence of **iron** and **copper** particles and **water** droplets.

Besides the reservoir, another location where oil oxidation occurs is at the pump outlet (fig. 10-27).

If air bubbles are present in a pump suction line as a result of an air leak in the suction line or returning fluid velocity churning up the reservoir, they suddenly collapse upon being exposed to high pressure at the pump outlet.

This action generates a high temperature which, according to some calculations, can rise to 1150°C *(2100°F)* when the bubble collapses from 200 to 0 bar *(3000 to 0 psi)*. The high temperature fries the oil, forming resinous products, and causes the oil to acquire a characteristic burnt odor (fig. 10-28).

As high-temperature oxidation at pump outlet occurs, resinous materials are formed, but dissolve in the oil. When a hot surface (pump rotor, relief valve spool) is encountered, resins come out of solution, forming a varnish or lacquer coating on the hot surface; this may cause moving parts to stick (fig. 10-29).

Resinous material can also form sludge which combines with dirt and floats around the system plugging small openings in valves and filters, and interferes with heat transfer to reservoir walls. Strong evidence exists that collapsing air bubbles at pump outlet is a major influence in rapid oil degradation.

Check for oxidized oil

A check for oxidized oil is performed by comparing a sample of the questionable fluid with a sample of new fluid out of a drum. With both fluids at the same temperature, fresh, new fluid will have a definite "body" and will tend to stick to one's fingers as it is poured over them. And, if thumb and forefinger are rubbed together, the fluid will feel slippery (fig. 10-30).

Oxidized fluid feels very much like water. As it is poured over one's fingers, oxidized oil runs off just as water. It exhibits little "body" and small tendency to adhere. Oil which has been oxidized by the high temperature collapse of air bubbles will also have a characteristic pungent odor (fig. 10-28).

If any fluid sample exhibits the characteristics of oxidized oil, its condition is questionable. In this case, the fluid should be sent to a lab for further

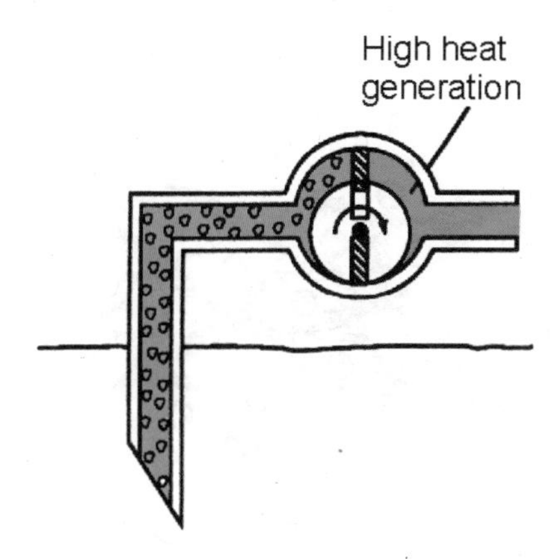

Fig. 10-27 High temperatures at the pump outlet caused by air in the fluid

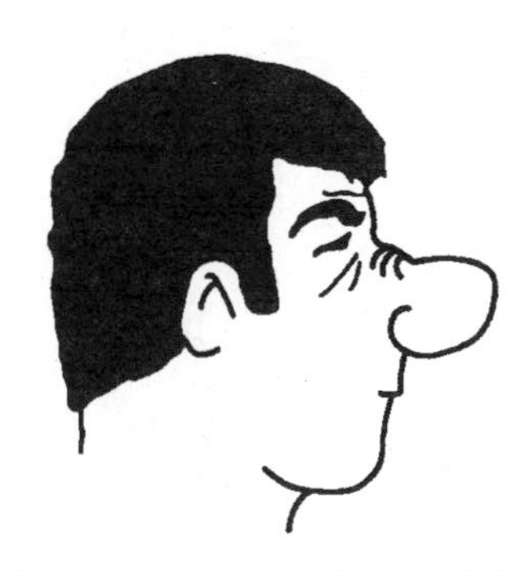

Fig. 10-28 You can easily smell 'fried oil'

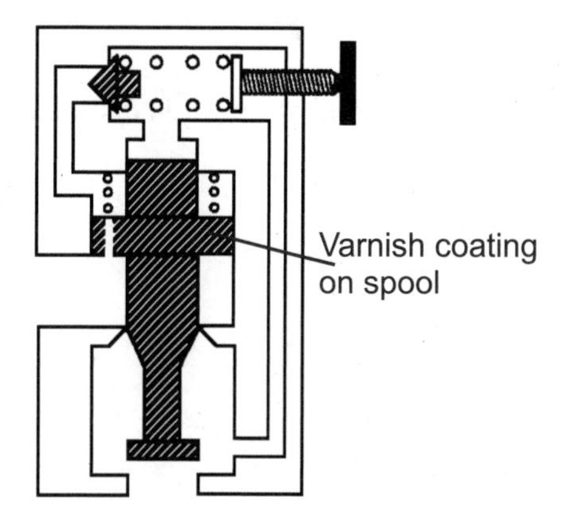

Fig. 10-29 Varnish on a part indicates hot fluid.

Fig. 10-30 New fluid has 'body'.

analysis. If this is impractical, the system should be drained, flushed and refilled with fresh fluid.

Water in hydraulic oil

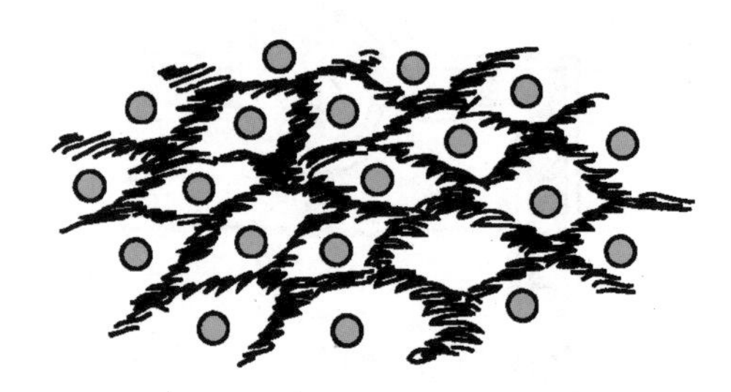

Fig. 10-31 Water droplets carried by the hydraulic fluid

All hydraulic oil contains water in varying amounts. We know from experience that water and oil do not mix (except for water soluble oils). Attempts to mix large amounts of water and oil will result in water settling out at the bottom of a tank. In small quantities, however, water is broken into small droplets which are carried throughout the system by an oil (fig. 10-31).

If an oil contains acidic and resinous products of oxidation, it has an increased tendency to take on water.

Check for water in hydraulic oil

A check for water in hydraulic oil is performed by comparing a sample of the questionable fluid with a sample of new fluid.

Fig. 10-32 Checking for water in hydraulic oil.

Holding a beaker or glass of fresh oil up to a light (fig. 10-32), you will notice that it looks crystal - it sparkles a little. If a fluid sample contains 0.5% water, it will appear dull or smoky. If the sample contains 1% water, it will look milky.

An additional means of checking for water in oil is heating a fluid sample which appears milky or smoky. If the sample clears after a time, the oil probably contained water.

If an oil contains a small percentage of water (less than 0.5%), it is usually not discarded unless the system is critical. For small percentages of water special water absorbing hydraulic filters are available. While in the fluid, water will hasten the oxidation process and reduce lubricity.

After time, water will evaporate, but its products of oxidation and chemical and mineral contamination will stay behind to cause further harm.

If an oil contains water in large quantities, much of it will eventually settle out. Centrifuging can be used to separate water and oil if time is important.

Rusting and corrosion

In the context of a hydraulic system, **corrosion** refers to a deterioration of a component surface due to a chemical attack by acidic products of oil oxidation. **Rusting** refers to the process of a ferrous surface oxidizing due to the presence of water in oil.

The process of corrosion dissolves metal and washes it away, reducing the metal part size and weight. On the other hand, rusting adds materials to a ferrous surface, increasing its size and weight. Since the efficiency of precision components is affected when their parts are either too large or too small, rusting and corrosion cannot be tolerated in a hydraulic system.

Rust and oxidation inhibitors

Rusting of ferrous component surfaces can be expected in a hydraulic system even if water is present in minute quantities; oil in its natural state does not provide adequate rust protection.

Since it is impossible in actual practice to keep water out of a hydraulic system, hydraulic fluids are generally equipped with a rust inhibitor which coats metal surfaces with a chemical film.

Oxidation due to the interaction of air and fluid in a system's reservoir generates a chain of products which eventually attack metal surfaces and cause further fluid oxidation to occur. An oxidation inhibitor is a chemical which interferes with the oxidation chain.

The high-temperature oxidation which occurs as air bubbles collapse at pump outlet, cannot be reduced by a chemical. This form of fluid oxidation can be eliminated by removing air bubbles from the fluid stream to pump inlet.

Rust and oxidation inhibitors should be basic additives for most hydraulic systems. Hydraulic fluids equipped with these additives are sometimes referred to as R & O oils; the high grade is R & O turbine quality. Lower quality turbine oil is still suitable for many hydraulic applications and is designated "R & O less-than-turbine quality".

R&O - Rust and oxidation inhibitors

Foaming

As oil returns to a reservoir, it should release any entrained air bubbles which have been acquired in the system. In some systems where leaks are prevalent and/or returning oil is churned up as it enters a reservoir, foaming of the oil can occur (fig. 10-33).

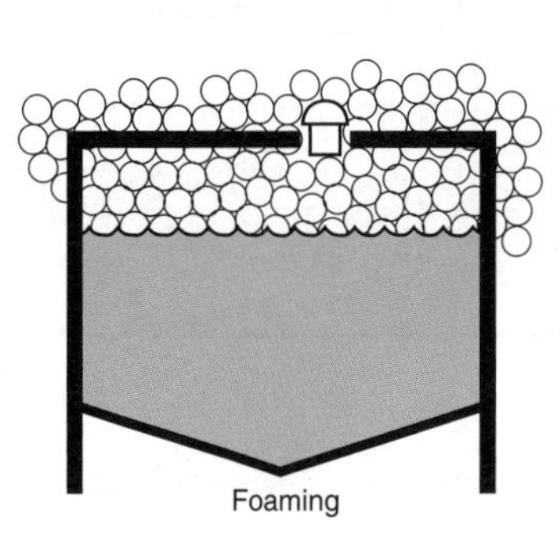

Fig. 10-33 Oil foaming.

As a result, entrained air is pumped throughout the system, causing spongy, erratic operation, rapid oil oxidation and noise. In more severe cases, oil foam could bubble out of a reservoir, creating a housekeeping problem.

Probably the best solution for alleviating foaming oil is to fix any system leaks and redesign the return part of the system with baffles or larger return lines which reduce fluid velocity. Sometimes, because of economics, convenience, or a lack of training, chemicals are used to solve the problem.

Anti-foam additives

In an attempt to discourage oil foaming, hydraulic fluids can be equipped with anti-foam additives. In some cases, these additives work by combining small air bubbles into large bubbles (fig. 10-34) which rise to a fluid surface and burst.

In other cases, these additives function by interfering with air release which action reduces foaming, but increases the amount of air bubbles in the system. If an anti-foam chemical is desired in an oil, care should be taken that the agent selected does allow air to escape.

Fig. 10-34 Anti-foam additive forms large air bubbles.

Check for foaming

A check for foaming oil is performed by taking a fluid sample. By draining or drawing fluid from a system's reservoir, it is possible to tell by sight whether air bubbles are present in the fluid. The sample should be taken as close as possible to the pump inlet line so that a representative sample of what is getting into the system can be taken.

Another indication that air bubbles are present in a system is noise. As air bubbles are swallowed by a pump, a high-pitched, erratic noise is emitted (figure 10-35). In some cases, a pump will periodically emit a loud bang as if someone were exploding firecrackers inside the pump housing.

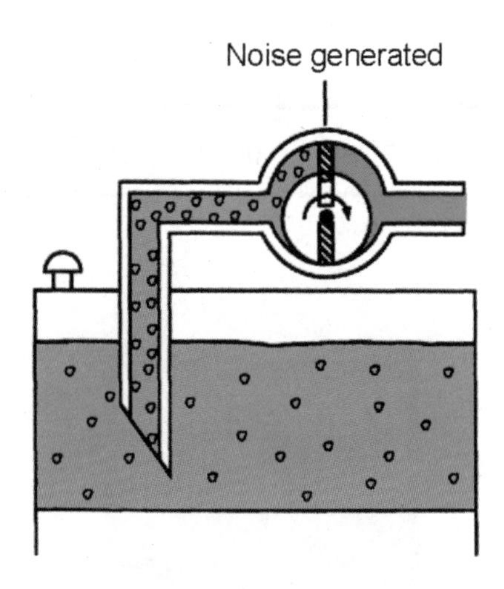

Fig. 10-35 Air bubbles can cause pump noise.

An additional indication of air bubbles is spongy system operation. This is evidenced by erratic actuator movements and erratic gage readings as a system is operating.

Dirt in the oil

The biggest problem with hydraulic oil in service is that it can easily become contaminated. The source of contamination can be water or air, but more frequently it is dirt.

Dirt in a hydraulic fluid can plug sensing orifices (fig. 10-36), cause moving parts to stick and wear excessively, and act as a catalyst to oxidize oil.

Dirt is an insoluble material in an oil which has several sources for contaminating oil. Dirt can be built into a system due to manufacturing, storing, and handling practices of system components and their assembly into a hydraulic system. Dirt can be generated within a system as a result of internal moving parts, flexing of component housings, and rust formation on reservoir walls (fig. 10-37).

Dirt can also be added to a system as a result of servicing failed system components (fig. 10-38), not servicing reservoir breathers, and cylinder rods pulling in dirt as they retract. There is a continuous influx of dirt into hydraulic fluid.

At present, there is no chemical additive which either keeps dirt out of, or removes dirt from, hydraulic fluids. Keeping dirt out of a system is the function of good system design and maintenance practices. Removing dirt from a fluid is the responsibility of filters and maintenance personnel.

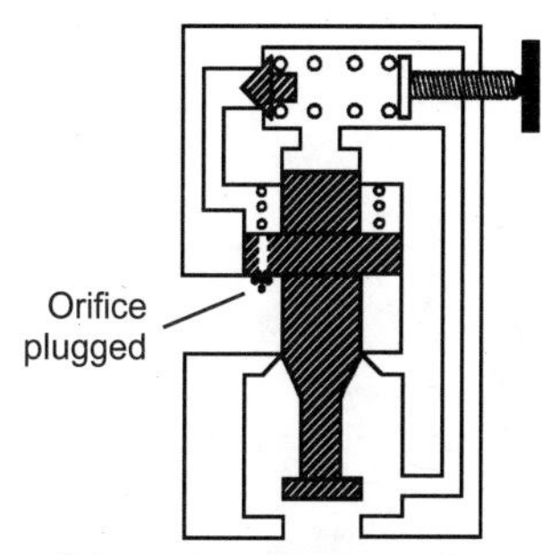

Fig. 10-36 Dirt can cause a valve spool to stick.

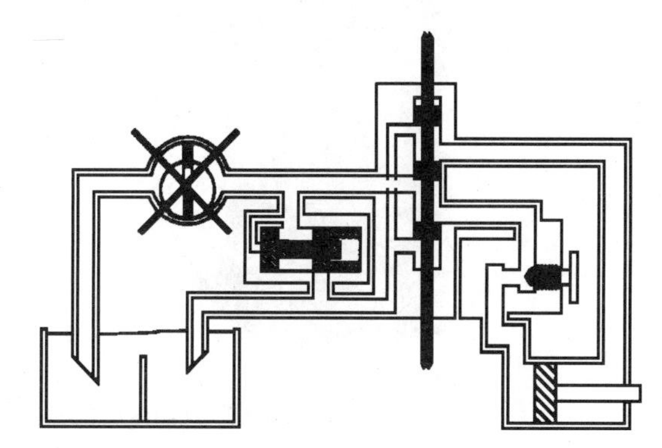

Fig. 10-37 Dirt can be generated in many parts of a hydraulic system.

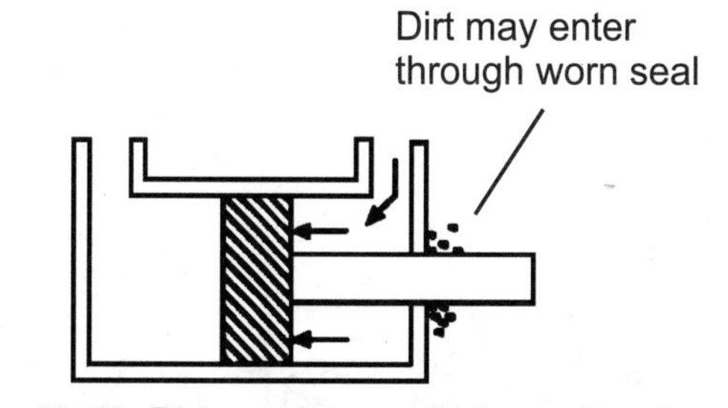

Fig. 10-38 Dirt passing a cylinder rod seal

Check for dirt in the oil

Trying to determine the dirt level of a fluid with the unaided eye is many times impossible. Holding a glass or beaker of hydraulic oil up to a light and inspecting for dirt is an inaccurate means of determining dirt contamination. Many harmful dirt particles for a hydraulic system are not normally visible. Determination of dirt contamination is best performed in a lab.

A check for dirt contamination in a hydraulic fluid is performed by checking indicators of a system's filters (fig. 10-39). Assuming that the filter element is appropriate for the system and that the indicator is functioning properly, the filter indicator will give an idea if the fluid is clean enough for the system.

With an indication of 'needs cleaning', the filter element should be serviced. If the indicator shows a bypassing condition, fluid is probably not clean enough, and the filter should be serviced at once.

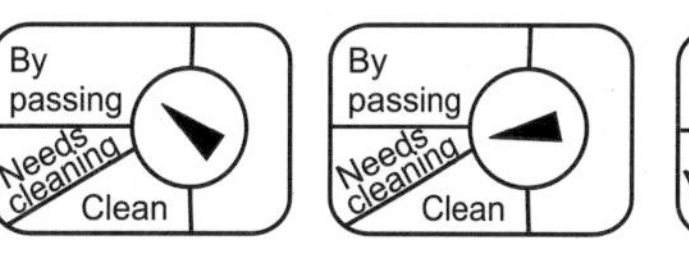

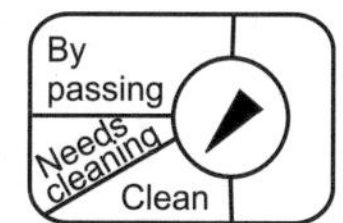

Fig. 10-39 Filter indicator readings

Hydraulic oil maintenance considerations

As has been pointed out, hydraulic oil has several functions in a system and it contains additives which aid it in performing these functions. Hydraulic oil is something special and it should be given special handling during storage, transfer to machine reservoir, and while operating in a system.

Keeping a fluid in top condition as it is stored, is a major consideration. Oil which becomes contaminated as it sits in a drum, is not only wasteful, but results in a false sense of security as oil supplies become depleted.

As a general rule, oil drums should be stored in a clean, dry place. If drums are stored outside, they should be stacked on their sides so that rain water does not collect on drum covers and leak past the seals into the oil (fig. 10-40).

*Fig. 10-40 Oil drums should **not** be stored upright.*

Transferring oil from barrel to reservoir (fig. 10-41) is another important consideration. Before the drum plugs are removed, the drum cover should be wiped clean. This procedure should also be followed for any apparatus or tools which will be used in the process such as hoses, pumps, funnels, reservoir filler hole, and the operator's hands.

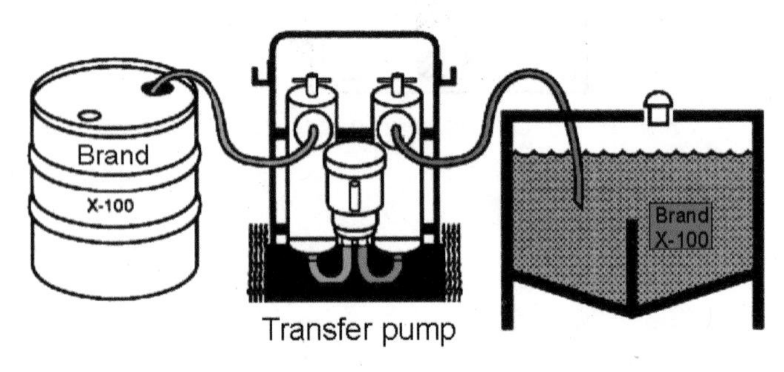

Fig. 10-41 Hydraulic oil being cleaned when transferred to a reservoir

Before the oil is actually placed in the reservoir, check to see that the barrel contains the correct fluid by brand name and viscosity. All hydraulic fluids do not contain the same additives. Mixing additives is not recommended unless authorized by the oil manufacturer.

Once an oil is in a system, it should be monitored and maintained at regular intervals. Maintenance of the oil includes filling a reservoir when its minimum oil level has been reached (with fluid the same as or compatible with the fluid in the reservoir), fixing leaks, and servicing filters.

Servicing filter elements is very helpful in keeping a fluid in top condition. Dirt can be very harmful to a fluid because it acts as a catalyst for oil decomposition. This is especially true if the dirt particles are ferrous, lead, or copper.

Filters usually remove a great percentage of dirt from a fluid stream. They do not remove dirt from the system, however; this is a maintenance function. Consequently, if filters are not maintained and cleaned when indicated, uncaught dirt not only passes to downstream components affecting their operation, but stored dirt on the filter element remains in the system contributing to oil decomposition (fig. 10-42).

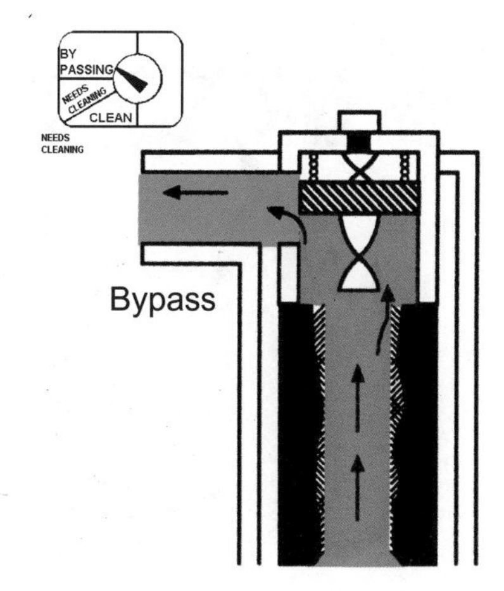

Fig. 10-42 Oil filters must be maintained

Cleaning wire mesh filter elements

When servicing a filter with a wire mesh element, the element may be cleaned (fig. 10-43).

Wire mesh filter elements can be cleaned in several ways. With relatively coarse elements, no one way is by itself better than another. The degree to which an element becomes clean depends on the care and effort used in the cleaning process, not to the specific cleaning method.

A common way of cleaning reusable elements is washing in a clean solvent or a hot soap-water-ammonia solution and blowing off the element with clean air. A soft bristle brush (new paint brush) is helpful in scrubbing the element. At no time should a wire brush or any abrasive material be used.

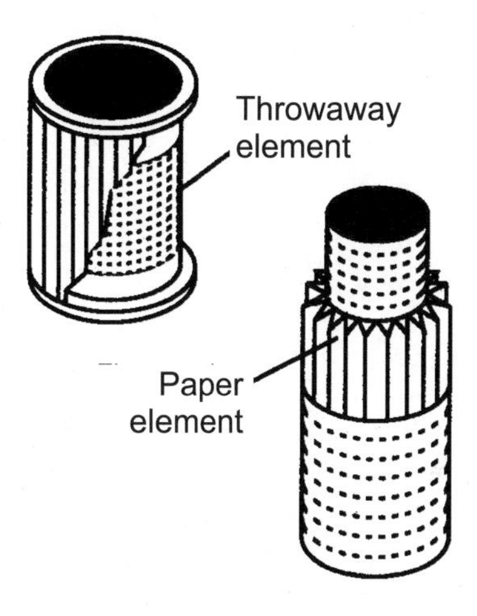

Fig. 10-43 Throw-away and reusable filter

To check the cleanliness of the element after cleaning, hold it up to a light (fig. 10-44). Any gray or dark areas indicate that the element must be recleaned.

Ultrasonic cleaning is a more expensive, but a more convenient way to clean elements. Dirty filter elements can be placed inside the ultrasonic device for a time and be removed clean and ready for reuse.

Wire cloth elements with ratings of 40 µm or less need ultrasonic cleaning to effectively restore element life.

Fig. 10-44 Check reusable filter element after cleaning.

Fire resistant hydraulic fluids

Since petroleum base hydraulic fluid is an excellent lubricant, mobile systems which use it as their energy transmission medium can look forward to years of dependable life.

But, in some systems or applications, such as underground mining, petroleum oil has a major disadvantage. Oil under pressure may spray (atomize) at a leak point and in the presence of an ignition source will ignite. This has been the source of many mobile equipment fires, both surface and underground.

Normally, petroleum oil in a system is not a high fire hazard. Petroleum oil is nonvolatile at room temperature and is capable of extinguishing a small flame like that of a match. However, a high pressure line with a pinhole leak spraying an oil mist into the air is a ***combustible mixture*** which can be easily ignited by an open flame (fig. 10-45). A leak of this nature can be considered a fuel nozzle.

Fig. 10-45 Oil mist in air is combustible.

In fire-hazardous environments where undisturbed operation and operator safety are of primary concern, and the ambient conditions are such that an ignition source may be present, fire resistant fluids are employed. These fluids are used with the knowledge that operating expenses will increase because the fluid is more expensive than petroleum oil and component life may decrease.

The intent of this section is to identify common types of fire resistant fluids for a hydraulic system, to see some problems of fire resistant fluids in service, and to indicate some maintenance considerations with respect to fire resistant hydraulic fluid.

Fire resistance determined

Fire resistant fluids are **not** fireproof. They are just as their name implies - **resistant** to fire. If fire resistant fluids are heated to a high enough temperature, they will burn.

Fire resistance of a particular fluid is determined by three test specifications (fig. 10-46):

- flash point
- fire point
- auto ignition temperature

Fig. 10-46 Fire resistance check-up

In describing the three tests below, petroleum base hydraulic fluid is used as the test fluid.

Flash point

Flash point of a fluid is the temperature to which a fluid must be heated to give off sufficient vapor to ignite when a test flame is applied.

As a petroleum oil or any liquid is heated, vapor is given off; in other words, liquid evaporates. With a petroleum oil heated to 175-230°C (*350-450°F),* enough vapor is given off from the oil's surface to ignite when a flame is applied. However, once the flame is removed, oil vapor ceases to burn.

Fire point

Fire point of a fluid is the temperature to which it must be heated to burn continuously (fig. 10-47) after a test flame has been removed.

When a petroleum base hydraulic oil is heated above that temperature, enough vapor is given off from the oil's surface to ignite when a flame is applied and to remain lit after the flame is removed.

Fig. 10-47 If enough heated, hydraulic oil will burn.

Auto ignition temperature

Auto ignition temperature of a fluid is the temperature at which it ignites without an external flame or spark (fig. 10-48).

Heating a petroleum base hydraulic oil between 260-370°C *(500-700°F)* will result in the fluid bursting into flame. This occurs without a flame present.

Fig. 10-48 Hydraulic oil will burst into flame at the auto-ignition temperature.

Types of fire resistant fluid

Hydraulic fluids which are classified as fire resistant have higher flash, fire, and auto ignition temperatures than petroleum base oil. These fluids can be divided into two types - water base and synthetic.

Water base fluid

Water was the fluid used in the first hydraulic systems. Water had several disadvantages as far as lubrication was concerned, but it did not burn.

When the need arose for a fire resistant hydraulic fluid, the initial action was to turn once again to water. However, since a certain amount of lubrication was demanded, oil was emulsified with the water.

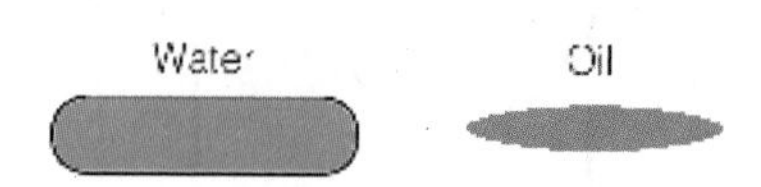

Fig. 10-49 In an emulsion, water and oil are not mixed.

Water-oil emulsion

A water base fire resistant fluid consisting of water and oil is not a mixture - oil and water do not mix (fig. 10-49).

Since this is the case, oil is broken down into extremely small droplets, usually by a chemical emulsifier (fig. 10-50).

Oil droplets are carried around by the water, increasing its lubricating qualities. If the fluid is ex-

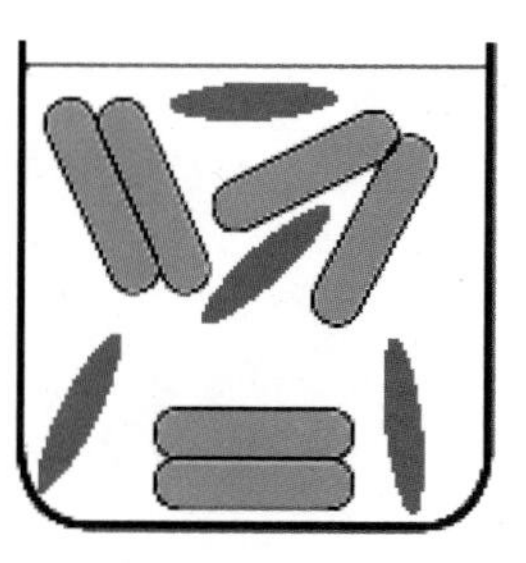

Fig. 10-50 An emulsifier breaks down the oil into small droplets carried by the water.

posed to fire, the water turns to steam, extinguishing the flame.

A two-phase, water-oil fluid is known as an emulsion. At the time this fluid was popular, a normal ratio of water to oil in an emulsion of this type was 60% water and 40% oil. Water was the dominant fluid and carrier of oil droplets.

Soluble oil fluid

Fire resistant hydraulic fluids which are predominantly water, are not normally found in present day hydraulic systems except where large amounts of fluid are lost due to leakage.

In these systems, reduced component life is sacrificed for an economical fire resistant fluid. The fluid is relatively inexpensive because of the high percentage of water - at least 90%.

A water base hydraulic fluid made up of water emulsified with 1-10% oil, is an 'oil-in-water emulsion' commonly referred to as soluble oil fluid (fig. 10-51). Anyone remarking that he is using 5% soluble oil in his system, is indicating that his fluid is made up of 95% water and 5% oil or chemical concentration.

Fig. 10-51 Soluble oil fluid with 1 to 10 % oil

Invert emulsion

A common water-oil emulsion of a modern hydraulic system is a creamy, white liquid made up of 60% oil and 40% water (fig. 10-52). As compared to a previous emulsion (60% water and 40% oil), the ratio of this emulsion is turned around or inverted.

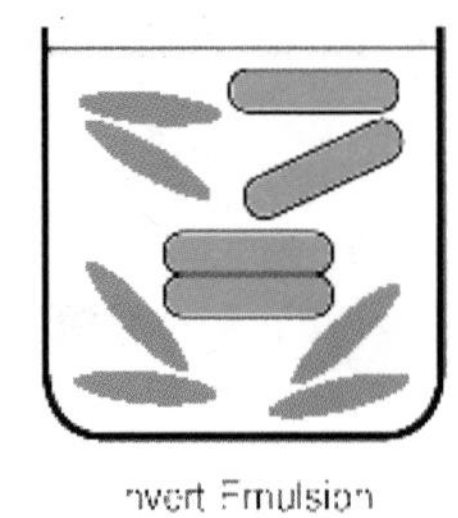

Fig. 10-52 Invert 60/40 emulsion - 60 % water and 40 % oil

Since oil is the dominant liquid and carrier of water droplets, invert emulsions have increased lubricating characteristics with a slight decrease in fire resistance.

Viscosity of water-oil emulsions

Viscosity of a water-oil emulsion is an important characteristic just as with petroleum hydraulic fluid. Since a soluble oil fluid contains a minimum of 90% water, its viscosity is basically that of water. Consequently, these fluids are rather poor lubricants.

An invert emulsion, on the other hand, normally consists of 60% oil. However, this does not mean that its fluid viscosity will be that of its base oil.

Viscosity of an invert emulsion operating in a typical hydraulic system will have a higher viscosity than a normal petroleum fluid for that system (fig. 10-53).

For example, a system operating with an invert emulsion may have a viscosity of 80 cSt *(375 SUS)* at 38°C *(100°F)*, whereas a petroleum oil would have a viscosity of 32 cSt *(150 SUS)* at 38°C *(100°F)*.

Because of the shearing action between the two fluid phases as it moves through the pump and system, invert emulsions exhibit a decrease in viscosity. To ensure that system components are properly lubricated, an invert emulsion with a higher than normal viscosity is used.

NOTE: ASTM graph paper does not properly depict the viscosity-temperature relationship of any invert emulsion or of fire resistant fluids in general.

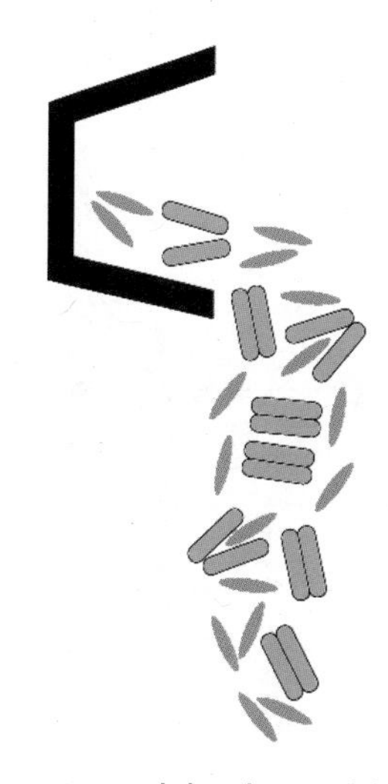

Fig. 10-53 Invert emulsion has a higher viscosity than normal hydraulic oil.

Problems with an invert emulsion

With a water base fire resistant fluid in a machine reservoir, certain problems can arise. Two problems specific to an invert emulsion are phase separation (fig. 10-54) and bacteria formation.

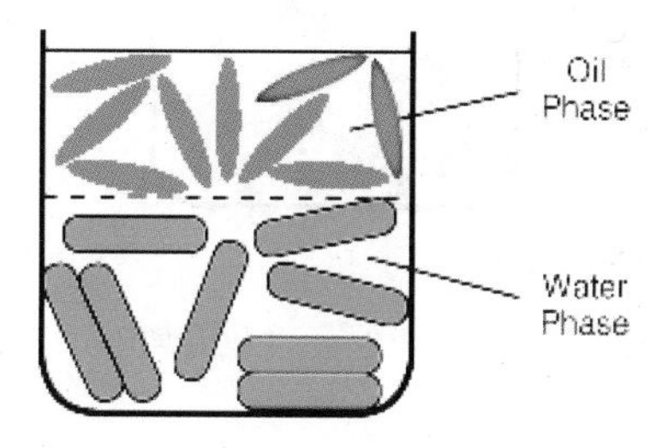

Fig. 10-54 Water and oil can separate in an invert emulsion.

Phase separation

Invert emulsion fluids are not designed to be operated at low temperatures. At 0°C *(32°F)*, ice slivers begin to form; at approximately -23.3°C *(10°F)*, the fluid freezes. Also, freezing and thawing of an invert emulsion cause the two phases to separate.

At the freezing point of water, 0°C *(32°F)*, some water droplets carried by the oil free themselves from the emulsion, forming ice crystals. As the system operates and temperature increases, ice crystals melt, but do not necessarily emulsify again. In this condition, the fluid has more tendency to rust system components and adversely affect lubrication.

Repeated freezing and thawing of an invert emulsion could cause water and oil phases to separate to a large degree. In this condition, it would be very difficult, if not impossible, to get the two liquids back together. As a result, fire resistance could be a serious problem.

Check for phase separation

A check for phase separation is performed by inspection. With the fluid in the reservoir mixed, it is difficult to determine whether oil and water phases have separated.

Fig. 10-55 Free water in an oil will settle to the bottom.

By draining off a fluid sample into a jar and allowing the fluid to rest for a period of time, any free water will settle to the bottom of the jar (fig. 10-55).

If phase separation is felt to be severe, the fluid representative should be contacted; he may recommend that the fluid be changed.

Bacteria formation

In some situations, under the proper temperature conditions, an invert emulsion can support the growth of bacteria (fig. 10-56).

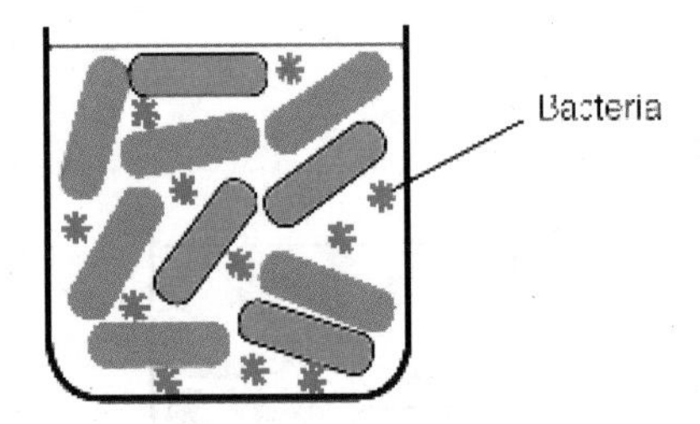

Fig. 10-56 Bacteria can plug a filter element.

Bacteria in large quantities can plug pressure sensing orifices of pressure control valves and pressure compensated flow controls. Bacteria can also plug filter elements. All these actions result in an undependable, nonproductive system.

Many invert emulsions are equipped with a bactericide additive to avoid this situation.

Check for bacteria formation

A check for bacteria formation is performed by sight and smell. If bacteria is present in an invert emulsion, inlet filters will appear to be coated by mucous or slime, and the bacteria will give off a very offensive odor (fig. 10-57).

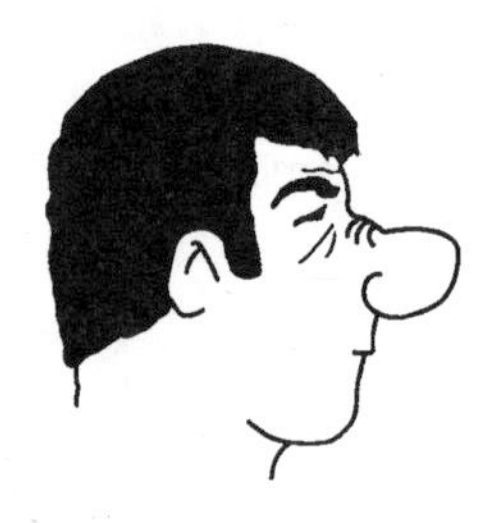

Fig. 10-57 You can smell the odor from bacteria in an emulsion.

If bacteria is present in an invert emulsion, it will most probably have to be changed.

Water glycol

Water glycol is another type of water base fire resistant fluid; it consists of water and a glycol which has a chemical structure very similar to automotive antifreeze.

Water glycol is many times dyed red or pink, and normally consists of 60% glycol and 40% water (fig. 10-58) along with a chemical thickener to increase its viscosity. The glycol actually mixes with the water.

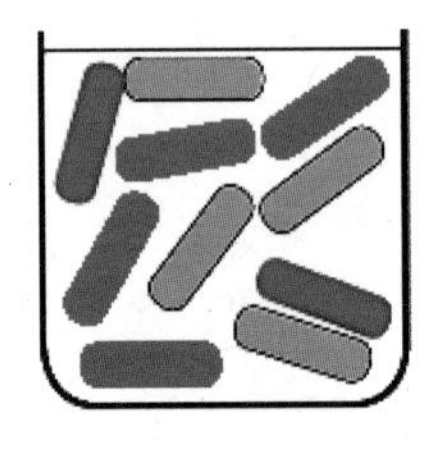

Fig. 10-58 Water glycol - normally 60% glycol and 40% water

The fluid is homogenous and not two-phase like an emulsion; that is, seen through a microscope, the fluid will not appear as separate droplets of water and glycol.

Water glycol fire resistant fluid works well at low temperatures.

Comparing invert emulsion with water glycol

When comparing an invert emulsion with a water glycol fluid (fig. 10-59), we find that:

a. It is more difficult to keep an emulsion stable than to maintain a water glycol solution
b. Stable invert emulsions have more lubricity
c. Invert emulsions are less expensive
d. Water glycol fluids are more fire resistant
e. Water glycol fluids operate better at low temperatures.

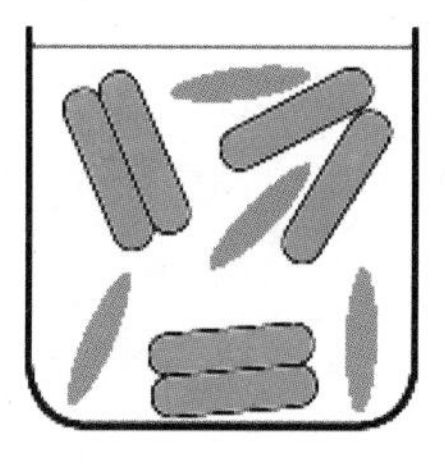

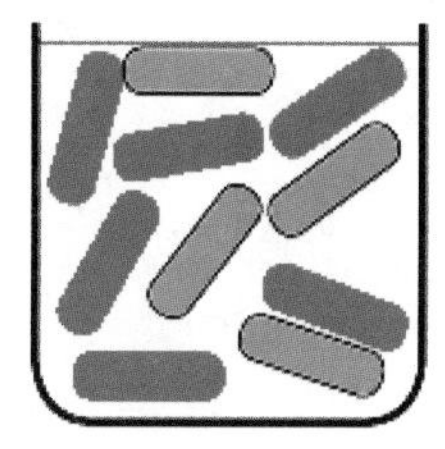

Fig. 10-59 Invert emulsion and water glycol don't have the same adavantages.

Problems with water base fluids

With a water base fire resistant fluid in a hydraulic reservoir, certain problems can arise. Some of these problems are reduced component life and water evaporation.

Lubricity of water base fluids

Since water base hydraulic fluids contain a significant percentage of water for fire resistance, they have an inherent disadvantage. With respect to petroleum oil, these fluids have reduced lubricity.

Lubricity and oiliness agents are added to the fluids, but reduced component life is a realistic expectation when these fluids are used. Because of this handicap, water base fire resistant fluids are not normally used in systems which operate above 125 bar *(1800 psi)*. Some component manufacturers will limit component operating pressures to 70 bar *(1000 psi)* when operating in a high water content system.

Of soluble oils, invert emulsions and water glycol fluids, stable invert emulsions have the best lubricity, followed by water glycol and soluble oil fluids, respectively (Table 10-1).

Table 10-1 Relative lubricating effectiveness

Lubricant	Lubricant derating factor
Petroleum oil	1.0
Invert emulsion	2.0
Water glycol	2.6

Water evaporation

Many fluid manufacturers recommend that water base fluids be operated at a maximum tempera-

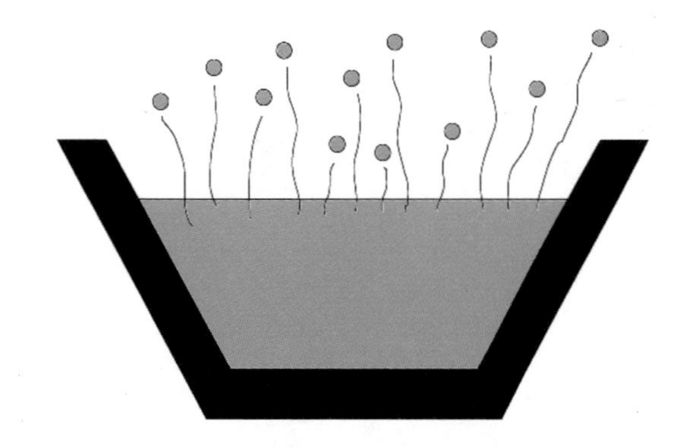

Fig. 10-60 Water evaporates excessively above 60°C (140°F).

ture of 60°C *(140°F)* with 50°C *(120°F)* being more desirable. Above 60°C *(140°F),* excessive water evaporation may occur (fig. 10-60). Not exceeding a fluid operating temperature of 50°C *(120°F)* to 60°C *(140°F)* in the typical mobile environment can be most challenging.

As water evaporates from a water base fluid, some undesirable things can happen. Water vapor escaping from the fluid can condense on unprotected ferrous parts causing rust formation. After a time, rust scale flakes off the unprotected metal surface, becoming a source of dirt for the entire system.

Water base fluids are generally equipped with rust inhibitors, but any unprotected metal surface which is not bathed by the liquid is subject to the attack from escaping water vapor.

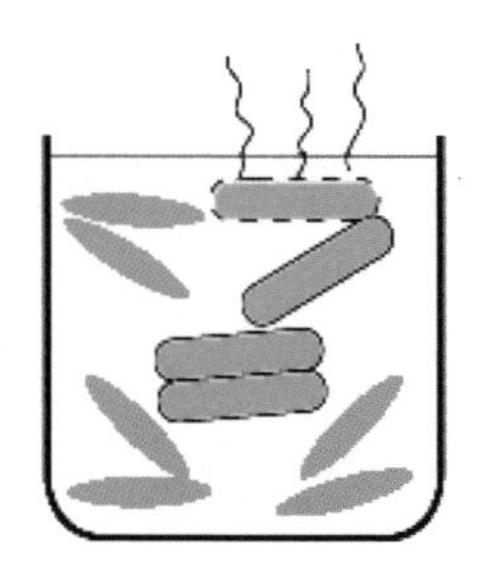

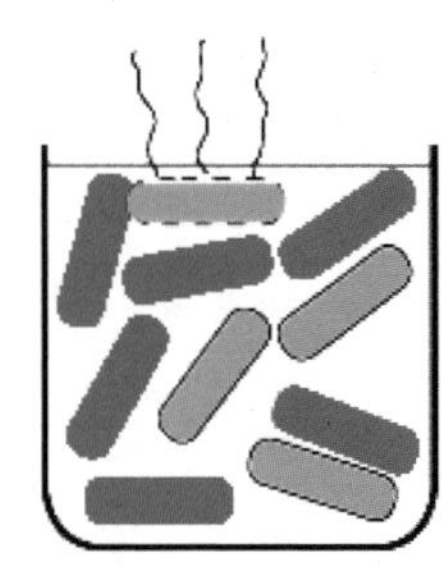

Fig. 10-61 Loss of water means less fire resistance.

Water evaporation affects a fluid's fire resistance. Since the percentage of water determines the fire resistance of a water base fluid, water loss due to evaporation results in the liquid being less resistant to burn (fig. 10-61).

Loss of water from an invert emulsion or water glycol also affects fluid viscosity. Water loss in a water glycol fluid increases fluid viscosity.

In an invert emulsion, water loss causes viscosity to decrease or may even result in the fluid becoming unstable.

To ensure maximum fire resistance and the proper viscosity, water content of water base fire resistant fluids should be monitored by lab analysis at regular intervals.

Synthetic fire resistant fluid

Synthetic fire resistant fluids (fig. 10-62) are man-made liquids which are praised for their resistance to burning while performing close to petroleum oil with respect to lubrication. The most common type of synthetic fire resistant fluid is *phosphate ester.*

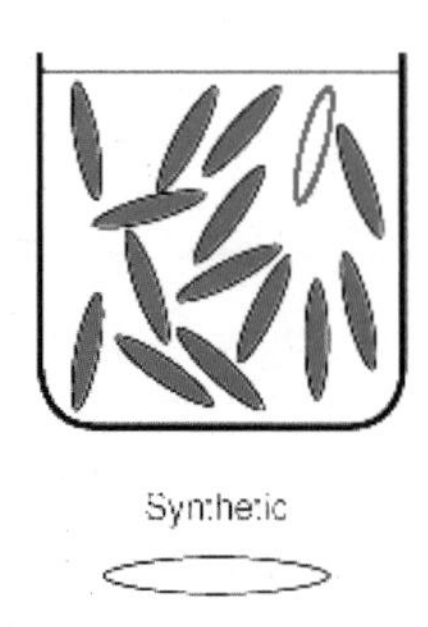

Fig. 10-62 Synthetic fire resistant fluids don't burn easily.

NOTE: Synthetic fire resistant fluids should not be confused with synthetic fluids such as silicones, silicate esters, dibasic acid esters, polyglycol ether compounds and polyol. These fluids have characteristics which are desirable for specific applications, but they are not normally considered fire resistant.

Phosphate ester fluids operate well at high pressure, provide excellent fire resistance but are very expensive. In high pressure systems where fire resistance is demanded, and the price of phosphate ester is prohibitive, a blend of phosphate ester and petroleum oil can be used (fig. 10-63).

This fluid has the lubricity the systems demands, but gives a fire resistance less than a phosphate ester fluid. Such blended fluids should be purchased from the fluid manufacturer. Blending fluids and mixing fluids are **not** the same.

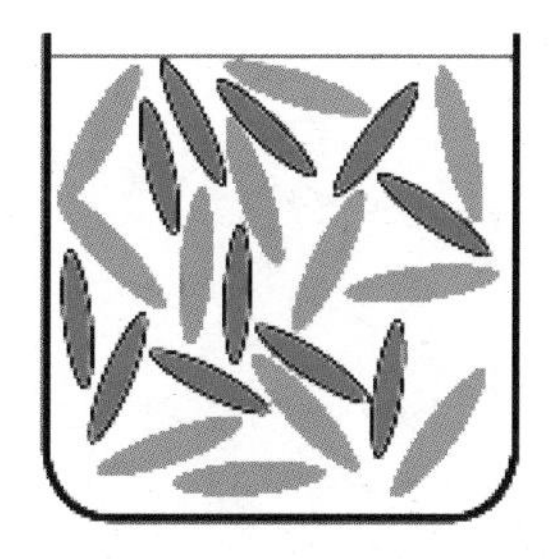

Fig. 10-63 A phosphate ester–petroleum oil blend can be a good compromise.

Comparing water base with synthetic fire resistant fluid

When comparing a water base fluid with a synthetic fire resistant fluid (fig. 10-64), we will find that:

a. Synthetic fluids exhibit more lubricity and can operate at higher pressures
b. Synthetic fluids are more expensive
c. Synthetic fluids are more fire resistant
d. Flash point, fire point and auto ignition temperature for a phosphate ester fire resistant fluid are approximately 235°C *(455°F)*, 350°C *(665°F)*, and 620°C *(1150°F)* respectively.

Fire resistance for water base fluid is not indicated by flash and fire temperature points as long as water is present in the fluid.

Auto ignition temperature for water glycol is approximately 600°C *(1100°F)* and for an invert emulsion around 440°C *(825°F)*.

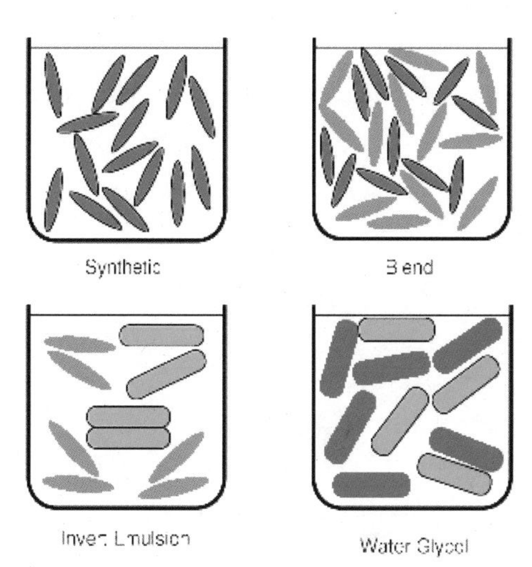

Fig. 10-64 Synthetic fluids have important advantages compared to water base fluids.

Problems with fire resistant fluids

Using a fire resistant fluid in a hydraulic system can result in certain problems (fig. 10-65). Some of these problems are compatibility with seals and protective coatings, foaming and air retention, and dirt retention.

Fig. 10-65 Fire resistant fluids have characteristics that must be addressed.

Compatibility with fire resistant fluids

A common material for sealing petroleum oil is nitrile rubber ('Buna N'). This material is also compatible with an invert emulsion as well as a water glycol. These types of seals would require change, if certain synthetic fluids, such as phosphate ester, were used.

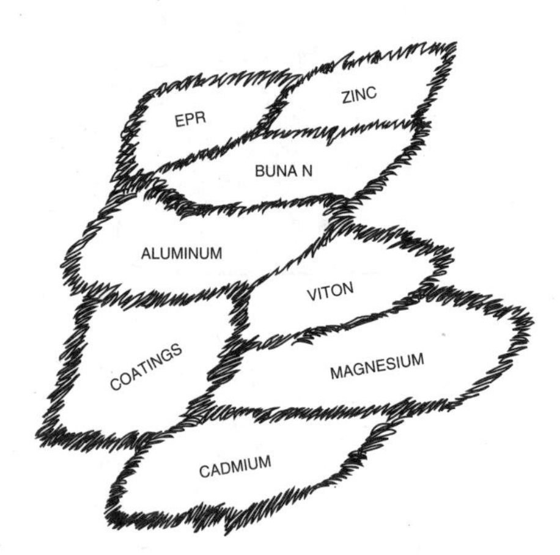

Fig. 10-66 Water glycol fluids can form residues with certain metals, which may be the reason for plugged filters and stuck valve spools.

When switching from petroleum oil to a water base fire resistant fluid, some problems could occur with protective coatings. If a reservoir interior is protected with petroleum compatible paints and varnishes, a water base fluid may dissolve the coating.

Water glycol fluids and some chemical concentrates are not compatible with some metals. They may attack zinc, cadmium, magnesium, and certain alloys of aluminum (fig. 10-66), generating gummy residues which plug orifices and filters and cause valve spools to stick.

It is recommended that parts which are alloyed or plated with these metals not be used with water glycol. Examples of such parts might be galvanized pipe, and zinc or cadmium plated strainers, fittings and reservoir accessories.

The common nitrile rubber seal material used for dynamic sealing of a petroleum base fluid is not acceptable for a phosphate ester or phosphate ester blend fluid. These fluids require 'Viton' (EPR), or any other suitable material.

Synthetic fire resistant fluids tend to dissolve petroleum compatible paints and varnishes; these fluids, however, do not attack common metals found in a hydraulic system.

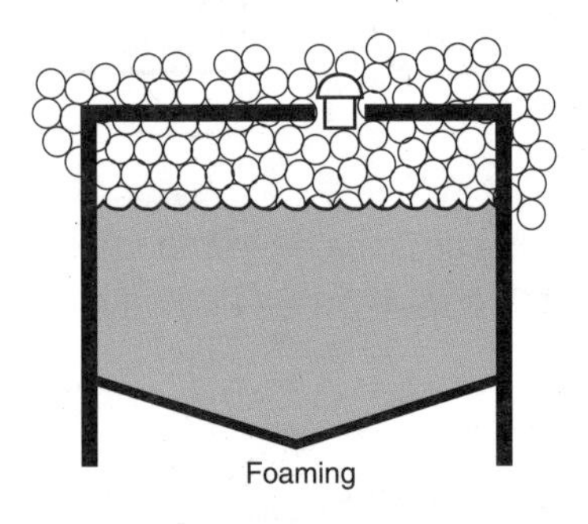

Fig. 10-67 Foaming must be prevented.

Foaming and air retention with fire resistant fluids

Water base and synthetic fire resistant fluids have more of a tendency to retain air and to foam compared to petroleum oil. After returning to the reservoir, fire resistant fluids require more time in a reservoir to give up any accumulated air bubbles (fig. 10-67).

Consequently, systems using fire resistant fluids should have larger reservoirs than comparable systems using petroleum oil.

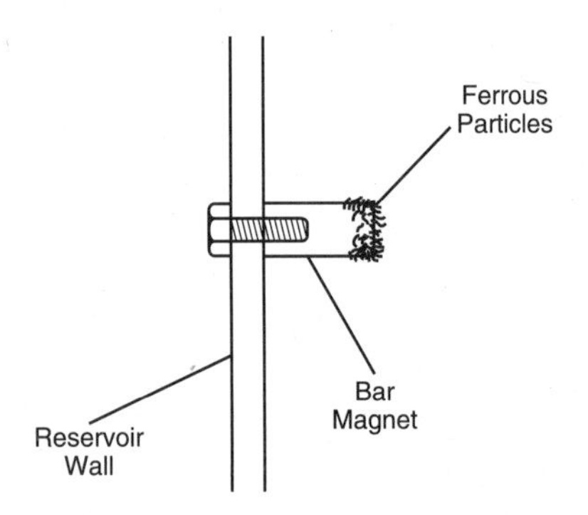

Fig. 10-68 A magnet accumulates steel particles.

Dirt retention with fire resistant fluids

As it returns to a reservoir, fire resistant fluids have more of a tendency to retain dirt in suspension compared to petroleum oil. A fluid is supposed to allow any reasonably sized dirt to settle to reservoir bottom, but a fire resistant fluid tends to hold the dirt. When a fire resistant fluid is used in a system, good fluid filtration should be a prime consideration. And, the use of magnets should not be overlooked (fig. 10-68).

Maintenance considerations

Maintenance considerations of fire resistant fluids with regard to storage are basically the same as for petroleum oil. Store barrels on their sides (fig. 10-69) so that water does not collect on barrel tops and leak into the fluid.

Fig. 10-69 Drums with fire resistant fluids should be stored on their sides.

Invert emulsion fluids have additional storage requirements. These fluids can have their stability affected by repeated freezing and thawing. Care should be taken then to ensure that the fluid does not freeze.

Transferring oil from barrel to reservoir is another important consideration. Before the drum plugs are removed, the drum cover should be cleaned. This procedure should also be followed for any apparatus or tools which will be used in the process such as hoses, pumps, funnels, reservoir filler holes, and the operator's hands.

A check should be made to see that the barrel contains the correct fluid by brand name and viscosity. And, if a pump is used to transfer fire resistant fluid, care should be taken that the pump is not filled with a different fluid and that pump materials and connector assemblies are compatible with the fluid.

With the fire resistant fluid in the reservoir, it should be maintained at regular intervals. Maintenance of the fluid includes filling a reservoir before its minimum oil level has been reached, fixing leaks and servicing filters.

Water base hydraulic fluid should be regularly checked (fig. 10-70) for its water content since its concentration must be kept within a narrow range as it affects viscosity and fire resistance.

Fig. 10-70 Check water base fluids regularly.

Adding water to an invert emulsion is not normally recommended because of the critical mixing process which is demanded.

Adding water to a water glycol solution is common but it is not a simple matter of running a hose to the reservoir from the nearest tap.

Makeup water should be free of mineral deposits which could otherwise contaminate a system. Distilled steam condensate or deionized water are suitable for use in a water glycol solution. The amount of water to be added is determined after analysis of a fluid sample by a lab.

Scheme	Country
ASTM D 6046-96 standard [2]	USA
ISO/CD 15380 draft standard [3]	International
Umweltbundesamt (UBA) RAL-UZ 79 Blue-Angel eco-label [4]	Germany
UBA WGK Water hazard classification [5]	Germany
VAMIL regulation [6]	Netherlands
City of Gothenburg 'Clean lubricants' project [7]	Sweden
SS 15 54 34 standard [8]	Sweden

Table 10-2

Fluid	ISO Code
Mineral oils	HH, HL, HLP, HM, HV, HLPD
Fire resistant	HFA, HFAE, HFAS, HFB, HFC, HFD
Environmentally friendly	HPG, HTG, HE

Table 10-3

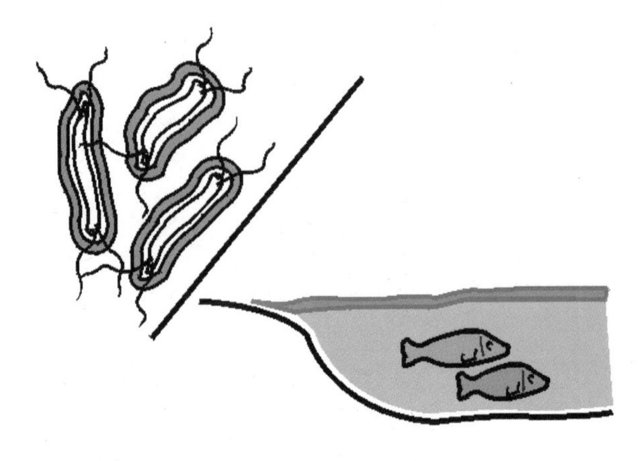

Figure 10-71

Biodegradable Hydraulic Fluids

Because of the growing concern for the environment, mobile equipment users and manufacturers have expressed concern over such potential problems as oil leakage, oil spills and the disposal of spent hydraulic oil and filters. In addition, recent governmental legislation has helped drive the operating costs of a mobile hydraulic system almost to the breaking point. If there is an accident with a piece of mobile equipment and a spill of hydraulic oil (petroleum oil) should result, the mandatory clean up of the environment would be most costly. However, an alternative to traditional mineral base oil and associated problems is gaining interest and support. The alternative is a biodegradable oil, an environmentally friendly or "green" oil. Many European countries, such as Germany (VDMA 24568, WGK, VDMA 24569) and the Scandinavian countries (SS155434, VAMIL) have enacted tough bio-standards. These standards, as well as ISO [ISO/CD 15380 (draft)] standards, (Table 10-2) have been in effect for some time. Not only have countries enacted such legislation, but several mobile equipment manufacturers (Caterpillar) have established "bio-fluid" performance standards relative to their equipment. In addition, ISO has assigned a code to describe each type of hydraulic fluid (Table 10-3).

Just as with petroleum fluids, the concern over the performance characteristics of the biodegradable fluids (i.e.. viscosity, temperature range and stability, metal and elastomer compatibility, lubricating ability, oxidation stability, air release, foaming, water removal, etc.) are of great importance to the mobile hydraulic system designer and maintenance personnel. However, regardless of the base stock employed the fluid must meet "Biodegradable Standards."

To be classified "biodegradable" a fluid must meet two conditions. The first is, depending upon the standard, that 70-80% of the base fluid will break down (biodegrade) by microorganisms (bacteria) within a given length of time. The second requirement is that the fluid must pass a toxicity test. Basically this means that the fluid does not contaminate the ground water and does not harm living organisms.

Biodegradable fluids can be broadly classified into two groups: **non-mineral oils** and **water**. The non-mineral oils would include the following types: vegetable oils, esters (natural & synthetic), polyglycols and mixtures. It is these non-mineral oils that will be examined first (fig. 10-71).

Vegetable Based Oils

Early attempts at developing a biodegradable hydraulic fluid began with the addition of a chemical additive package (which could be potentially toxic), found in mineral oil, to a vegetable oil base stock (fig. 10-72). The reason for this approach was one of cost and the feeling that this biodegradable fluid, being more environmentally friendly, would simply degrade (dissolve) in a very short time leaving behind only the chemical additives. The clean up of these additives would be far less costly than with the entire mineral oil spill. However, progress has been made in the development of additives which are relatively less toxic.

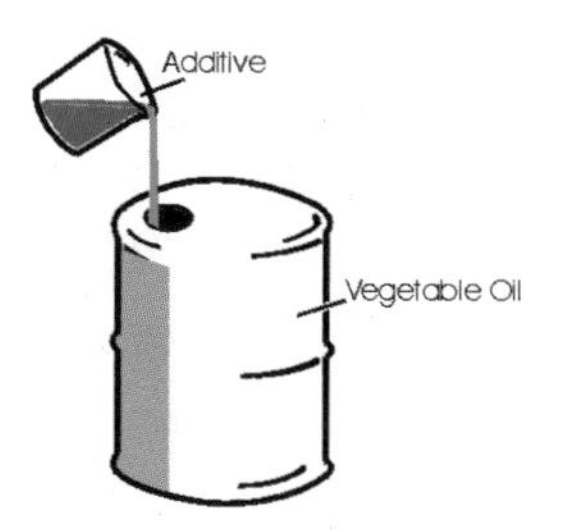

Figure 10-72

The sources for this vegetable based oils included: *canola oil, sunflower oil, soybean oil, coconut oil, palm kernel oil and rapeseed oil* (fig. 10-73). These base fluids are known as triglycerides. Several of these renewable resources (rapeseed) have seen application in Europe with others (soybean) being investigated in the United States. In addition, these fluids form a group with the ISO designation HETG or HTG.

Figure 10-73 Palm, soybean and sunflower

Unfortunately this group of fluids did not exhibit the hoped for performance characteristics of mineral oil, even though vegetable oils do exhibit several characteristics that are better than those found in mineral oils. These are low volatility, high flash points, viscosity index and lubricity. However, research is continuing to develop new "bio" additives (antioxidants, rust preventers, antiwear, etc.) and is even employing genetic engineering to improve plant stock and resulting oil performance characteristics. Significant progress has been made in this area.

But even with all this effort, many of these fluids exhibit poor thermal and oxidation stability. Poor oxidation stability can lead to the generation of sludge. The maximum upper temperature range for these fluids may be as low as 60°C *(140°F)*. With mobile hydraulic system temperatures typically in the 94°C *(201°F)* range or higher and with the possibility of long idle periods at subzero temperatures potentially gelling the fluid, for the time being, the application of vegetable oil based fluids may be limited (fig. 10-74). As new additives are developed, the pour point of the fluid may be lowered to a point that will permit the use

Figure 10-74 Mobile equipment works in temperature extremes.

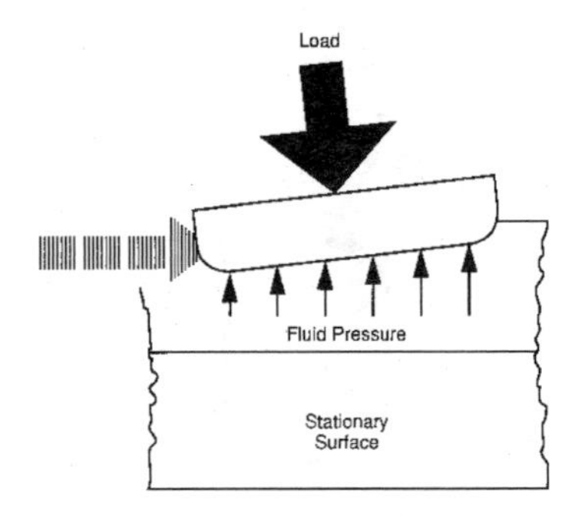

Figure 10-75

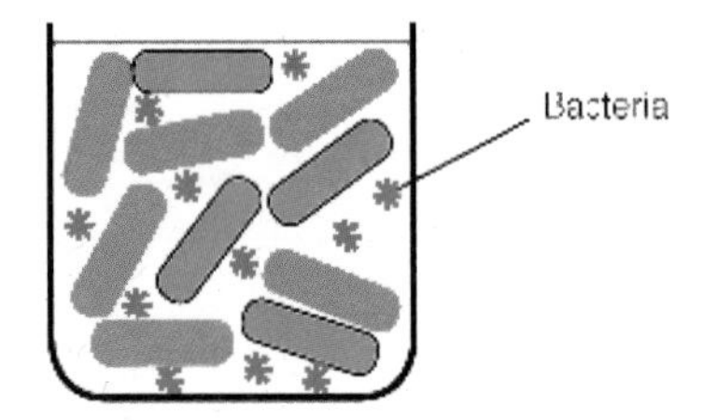

Figure 10-76 Bacteria can be present when there is water in the system.

of this type fluid. Until such time reservoir heaters may be required. As the mobile hydraulic system reservoir is reduced in size to save weight and space, the system oil temperature will increase, correspondingly increasing the importance of the fluid's oxidation stability. Also, mobile hydraulic system pressures are presently in the 207-345 bar/ 2.1-3.5 MPa *(3000-5000 psi)* range, with 413 bar/4.1 MPa *(6000 psi)* system pressures common in the near future. As system pressures increase it will be a challenge for the vegetable based hydraulic fluid to maintain its load-carrying capability (fig. 10-75).

It would seem that this type of fluid, along with the typical fluid temperatures and the possibility of water (in some amount) being in the system, would present an ideal environment for bacteria growth (fig. 10-76). However, to date this has not been encountered in any substantial amount.

With increasing demands being made on the system fluid, other fluid alternatives are becoming necessary. Fluids known as "synthetics" show the possibility of meeting these challenges.

Ester Based Fluids

To date the type of fluid that shows the most promise as a biodegradable replacement for mineral oil is the "esters." ISO classifies these fluids as "HEES" or "HE." Esters can be divided into natural and synthetic.

The "natural" ester fluids are derived from renewable resources, such as rapeseed and are also known as "oleates." However, they exhibit the same poor oxidation stability that was discussed previously. But with proper base stock selection and additive formulation, these fluids have operated successfully at temperatures of 90°C *(194°F)*. These fluids also biodegrade better then the "synthetic" ester fluids.

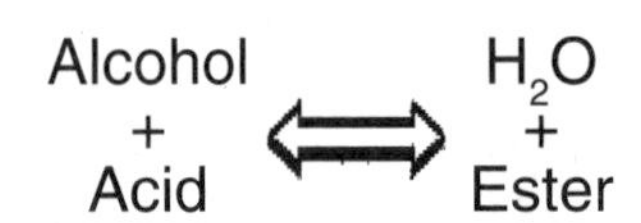

An example of a "synthetic" ester is a petrochemical product known as "adipate." It is this fluid that is capable of withstanding fluid temperatures above 60°C *(140°F)*. All fluids will react with oxygen, and esters are no exception. The byproduct of this reaction is the formation of sludge and a subsequent buildup of acidity in the system.

This acidity problem can be made worse if water is present in the system. An ester is the result of the reaction between an acid and alcohol. This reaction is reversible in the presence of water and the proper temperature, thus increasing the acidity of the system and resultant component damage. The resultant contamination, due to wear, can act as a catalyst for this process and hasten the reaction. The rate of this reaction will **double** for every 10°C *(22°F)* increase in temperature. An ester based fluid's water solubility increases with temperature, from 1500 ppm (0.15%) at 20°C *(68°F)* to 3000 ppm (0.3%) at 70°C *(158°F)*. Bio-fluids can also exhibit greatly differing saturation levels with some as low as 500 ppm (0.05%) to others as high as 2000 ppm (0.2%). Some tests have shown that as little as 1000 ppm (0.1%) water in the system can start this "hydrolysis" reaction (6000 ppm-10,000 ppm (0.6-1.0%) water is typically found in mobile equipment hydraulic oil).

Water can enter the system from a number of sources: (fig. 10-77) rain, refilling of the reservoir, contaminated new fluid, moisture in the air, "hosing off" of the equipment as well as the equipment working under water and splashing due to the environment. There are several methods that can be used to minimize the water content of the fluid. The first is to properly store the fluid before it is placed into the piece of mobile equipment. The container of fluid must remain sealed and stored in a weather protected area such as a building (fig. 10-78). For large amounts of free water, a centrifuge can be used, but this technique will reduce the water content to only the saturation level of the fluid. Further reduction in water content can be achieved by utilizing water removal equipment (fig. 10-79) that operates on a vacuum principle. Water removal filters are available as well as reservoir filter/breather caps equipped with a chemical drier. This style cap will remove much of the moisture from the air as it enters and exits the reservoir due to the change in fluid level within the reservoir. In addition, a modified reservoir design or sealed reservoir may be necessary. With the trend toward reducing the size of the reservoir, an ester's lower ability to handle heat and high viscosity index, the system fluid temperature will be higher than normal.

These bio-fluids have another consideration; that is, they will react with certain metals in the system and with certain additives (zinc) that may remain if the system previously contained mineral based oil. The elevated temperature (100°-300°C/*212°-572°F*) (fig. 10-80) that can result due to the compression of an air bubble could reduce the effectiveness of a zinc based antiwear addi-

Figure 10-77 Water can enter a mobile hydraulic system from various sources.
Reprinted with Permission SAE Technical Paper 981497 © 1998 Society of Automotive Engineers, Inc.

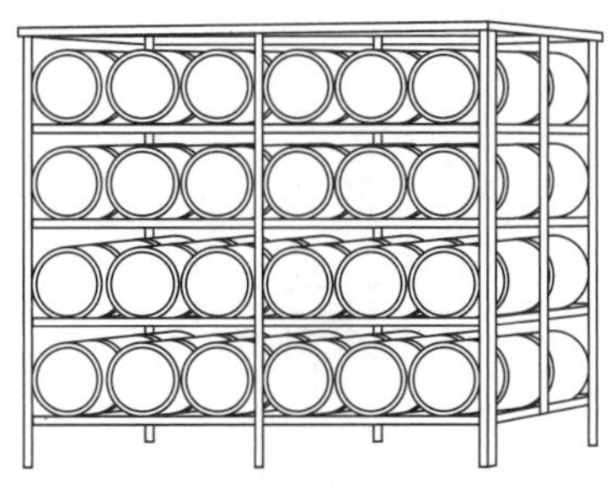

Figure 10-78 Proper fluid storage in important in minimizing water content of fluids.

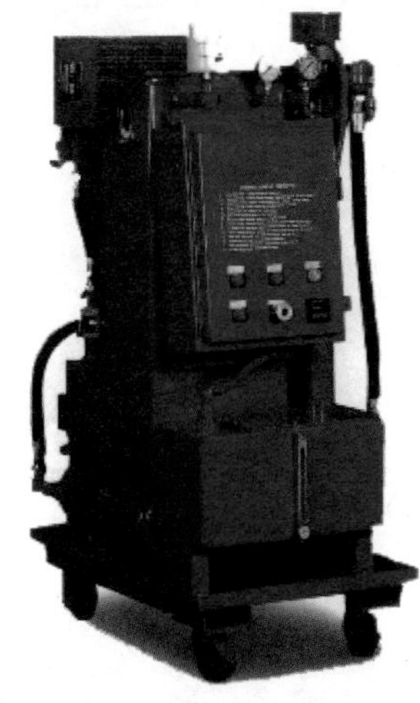

Fig. 10-79 A coalescer is another way of speeding up the reprocessing of oil contaminated with water.

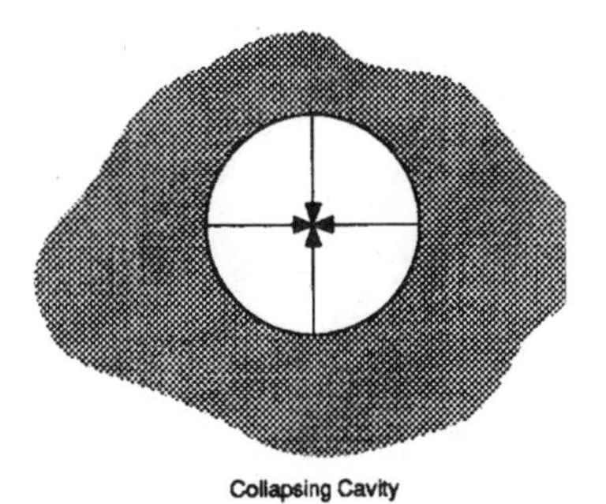

Figure 10-80

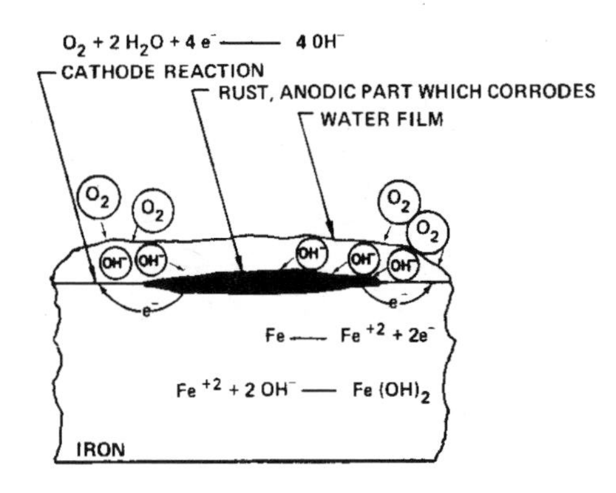

Fig. 10-81 Contaminants react chemically with the fluid and system components. (Courtesy E. C. Fitch, Fluid Contamination Control, FES, Inc.)

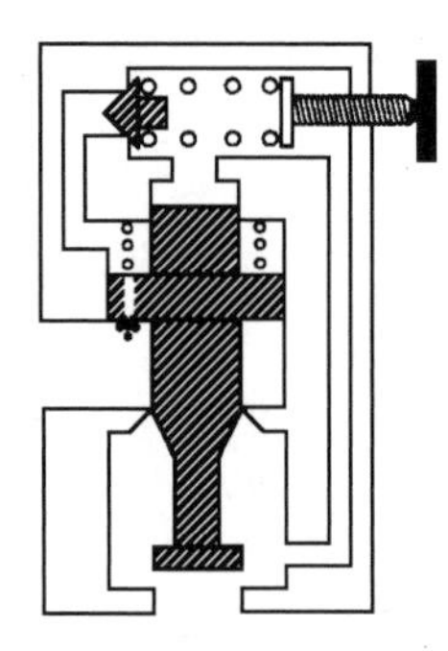

Figure 10-82

tive by as much as 56%. The elevated temperature would also act as a catalyst and increase the oxidation rate. The chemical reaction with metals could include alloys of brass and bronze which are two metals commonly used for bearings, pump port plates and piston slippers or shoes. In some cases iron has been found to be more reactive than copper (fig. 10-81).

Still another bio-fluid consideration is the compatibility of the fluid with the elastomer material found in the system seals and hoses. Seals and hose inner liners containing "nitrile" are sensitive to these fluids. Excessive seal swelling and resultant friction increase are the common results of this seal non-compatibility. Swelling of the hose inner liner will restrict fluid flow through the hose and result in an increased pressure loss. Fluorinated elastomer compounds may be a better choice for seal material.

Polyglycol Fluids (Polyalkylenglykols)

These water soluble fluids are classified as "HEPG" or "HPG" by ISO. Polyglycol fluids are not compatible with mineral oil. The mixture of these two fluids and water could lead to the formation of a precipitate. It is this precipitate that will plug pressure sensing orifices, lubricating clearances and filters (fig. 10-82). Polyglycol fluids, though exhibiting good lubrication properties, corrosion protection and viscosity similar to mineral oil, are less biodegradable than mineral oil. It is because of these conditions that polyglycol fluids find limited use in mobile equipment.
Mixtures

Because of the potential reaction between the additive packages contained within the fluids, it is recommended that fluids not be mixed. As mentioned earlier, the byproduct of this reaction is the formation of a precipitate. It is this precipitate that plugs filters and orifices. A decrease in filter service life can be between 22-60%, depending upon the fluid. Such decreases in filter life increase the required number of filter changes and associated costs.

Associated Costs

The cost of these bio-fluids can be many times that of standard mineral oil. Therefore, to offset this increased cost the correct bio-fluid must be chosen with care. The fluid must be strong enough to survive extended change intervals, i.e. go longer between oil changes. Also, by down sizing the reservoir, less bio-fluid will be needed. However, with less fluid (smaller reservoir) fluid temperature and system leakage are of greater concern than before. More efficient system design will compensate for some of the temperature increase and correct component selection, along with straight thread or o-ring face seals, which will reduce the potential for leakage (fig. 10-83). In addition, some of the higher cost of the "bio-fluids" will be offset by the lower cost of clean up should a spill occur.

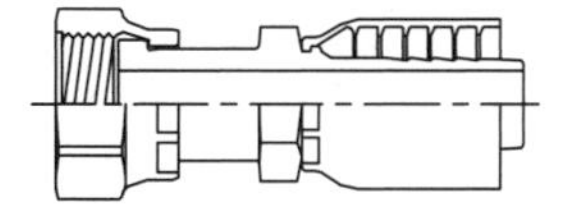
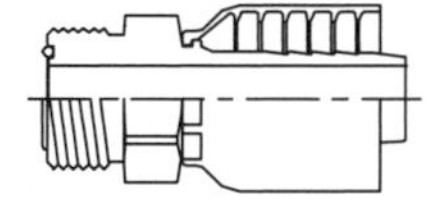

Figure 10-83 O-ring face seal

Extended oil change intervals will mean increased production time, less contaminated fluid and filters to dispose of, less chance of fluid spillage and more time for maintenance personnel to do other jobs.

The following suggestions should help minimize the associated and long term costs of bio-fluid use:

- Pick a top quality bio-fluid, one that meets the equipment and system component manufacturer's performance specifications.
- Design/build in a filtering system capable of removing water and maintaining a contamination level equal to or better than that recommended by the manufacturer. "Off line" or a "kidney loop" filtration system should meet these requirements (fig. 10-84).
- Establish and maintain a regularly scheduled fluid monitoring program. This should include, but not be limited to, frequent sampling of the fluid, analysis of fluid samples and maintaining of individual machine fluid records.

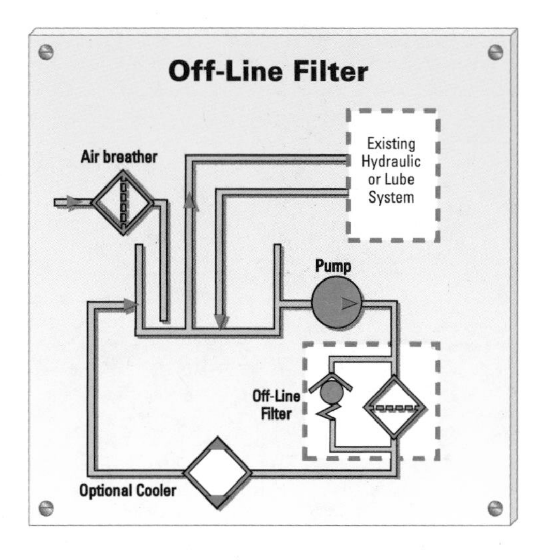

Figure 10-84 Off-line filter

Contact a local oil supplier and a local oil analysis laboratory for test procedures specific to given applications and requirements. The following is a list of commonly used oil analysis test procedures:

Particle Counting - ISO 4406
Wear Metals, Additives, and Contamination
Viscosity - ASTM D-445
Total Solids - ASTM D-91
Total acid number (TAN) ASTM D-664
Water by Karl Fisher ASTM D-1744
Dielectric Strength ASTM D-877

An additional bio-fluid, one that has been around for decades and is available in virtually unlimited supply is water.

Water as a Hydraulic Fluid

There are some compelling reasons to consider this bio-fluid as a system choice. Certainly among them is its availability. Additional reasons are it is "nontoxic" and fire resistance. With these properties fluid leakage becomes less of a concern from the environment standpoint.

However, as with all fluids, there are drawbacks. Water, as a system fluid, has four main limitations. They are:

- Lack of lubrication. Water exhibits good wetting characteristics but poor adherence and "body." (Fig. 10-85)
- Its operating temperature range is limited to between 0°C *(32°F)* and 100°C *(212°F).*
- Water promotes component corrosion.
- Water has a high vapor pressure, that is, it vaporizes easily. This leads to cavitation problems.

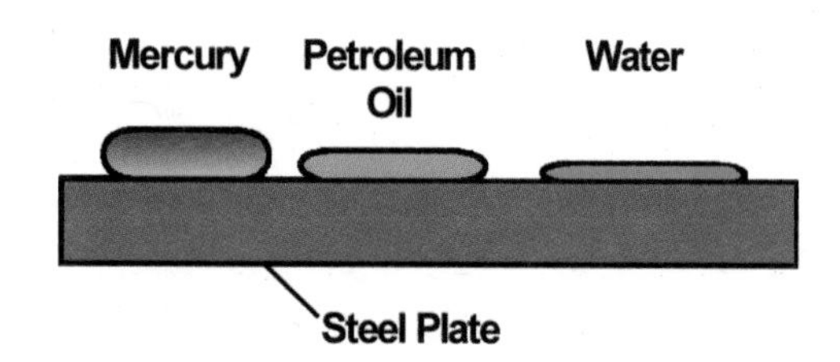

Figure 10-85 As a lubricant, water lacks adherence and body.

ISO has classified water hydraulic fluids by their performance characteristics and make up. The ISO codes for three of these fluid classifications are: HFA, HFB and HFC. These fluids fall under the fire resistant classification.

As with other hydraulic fluids, there are certain characteristics that are of concern:

- dissolved air
- liquid vapor pressure
- viscosity

Dissolved Air

Dissolved air (fig. 10-86) was defined earlier in this text. It is the air that is trapped in the fluid, between the fluid molecules. At sea level and under standard conditions, mineral base hydraulic fluid contains approximately 10% air by volume. Water, on the other hand, contains just over 2% by volume.

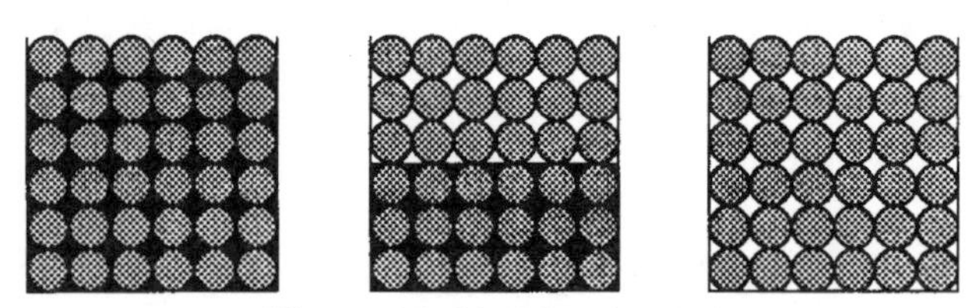

Figure 10-86 Dissolved air

Liquid Vapor Pressure

Of the different fluids used in hydraulic systems,. water is perhaps the one that is the easiest to achieve a change of state, that is, to change from a liquid state to a gaseous state. This ability will quickly lead to cavitation. This phenomena was defined in an earlier chapter. A liquid's vapor pressure is temperature dependent. As a liquid's temperature increases it takes less pressure (vacuum) to cause the change of state to occur. At a liquid temperature of 50°C *(122°F)* the vapor pressure of water (Table 10-4) (vacuum needed to create a change of state) is about 100 mm Hg. The vapor pressure of petroleum oil at the same temperature would be a much stronger 0.001 mm Hg. This is one of the reasons that it is recommended that water hydraulic systems not run at temperatures above 37.8°-49°C *(100°-120°F).*

Vapor Pressure of Water			
Temperature °F / °C		**Vapor Pressure in in./mm Hg Abs. Press.**	
100°F	37.8	2.0"	50.8 mm
110°F	43	2.6"	66.04 mm
120°F	49	3.5"	88.9 mm
130°F	54	4.5"	114.3 mm
140°F	60	5.9"	149.86 mm
150°F	66	7.7"	195.58 mm
212°F	100	29.92"	760 mm

Table 10-4 Vapor Pressure of Water

Viscosity

As defined in earlier chapters, viscosity is an indication of a liquid's ability to pour or move. This characteristic is both temperature and pressure related. If the temperature should decrease or the pressure increase, the viscosity will increase. This change in viscosity will have a profound effect on such component design characteristics as hydrodynamic lubrication and clearance flow. These are in turn affected by component manufacturing tolerances and clearances.

Of the hydraulic fluids discussed in this chapter, water is the least affected by temperature change. While the viscosity-temperature relationship is not linear (not a straight line) the change in viscosity per change in temperature is significantly less for water than for other common hydraulic fluids (fig. 10-87). The viscosity-pressure relationship for mineral oil shows an increase of 1.4-2.0 times depending on the pressure. Such a relationship for water is substantially less.

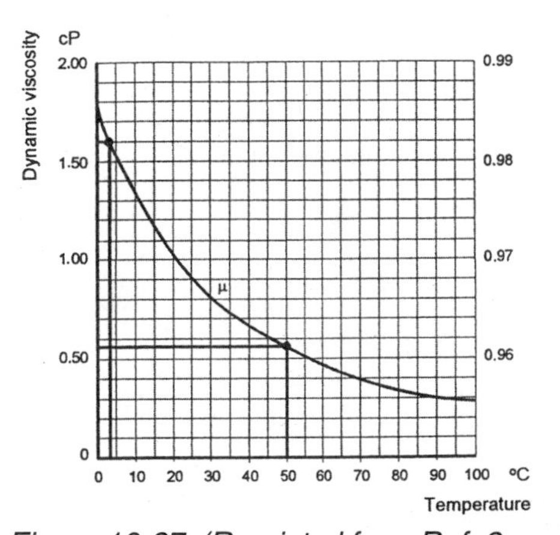

Figure 10-87 (Reprinted from Ref. 2, p. 49, by courtesy of Marcel Dekker, Inc.)

Water Properties

It is **not** recommended to use ordinary "tap" water for a system's hydraulic fluid. Tap water contains chemicals to kill microorganisms and minerals. The amounts of each in the water will vary from location to location. These minerals can block the close clearances that are required due to water's low viscosity.

Because of the potential for corrosion and because water can act as an electrolyte, components are made of materials such as stainless steel, ceramics, brass, bronze, or nickel chrome steels. This results in higher cost components.

Under the right environmental conditions (temperature, etc.) bacterial growth can occur (fig. 10-88). To minimize the potential for this to happen the system and fluid must be as clean as possible at all times. Prevention is cheaper and easier than the removal of bacteria once it appears. However, should it appear, depending upon the amount, the addition of a bactericide

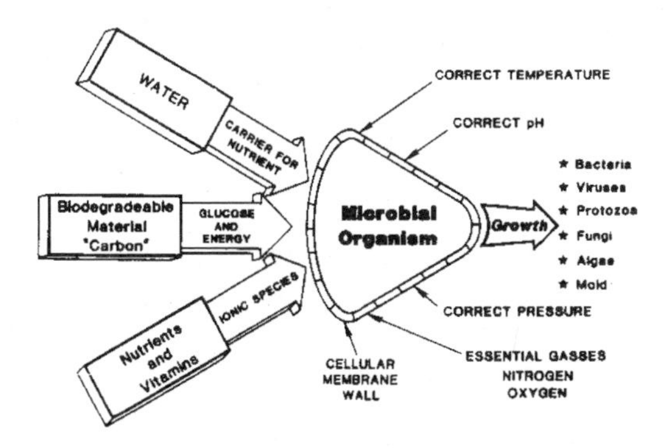

Fig. 10-88 Water and air contribute to microbe growth in fluids. (Courtesy E. C. Fitch, Fluid Contamination Control, FES, Inc.)

may bring it under control. This topic was covered in more detail in a previous chapter.

Conclusion

Before a system is converted from mineral oil to a bio-fluid it is strongly recommended that the equipment manufacturer and the manufacturer(s) of the system components be consulted. They will have experience, recommendations and guidelines for a successful conversion.

The ideal time to consider bio-fluids is during the acquisition phase of a piece of mobile equipment. At this time the equipment manufacturer and the system component manufacturer(s) can supply components and systems designed to work on bio-fluids. In addition, the system will be filled and tested on a bio-fluid and not filled and run on mineral oil and then refilled with a bio-fluid.

References

"Hydraulic Fluids and Alternative Industrial Lubricants" (SP-1384) ISBN 0-7680-0262-1 Copyright © 1998 Society of Automotive Engineers, Inc., 400 Commonwealth Dr., Warrendale, PA 15096-0001

"Water Hydraulics Control Technology" Erik Torstmann, Copyright © 1996 Danfoss A/S, ISBN 0-8247-9680-2, Marcel Dekker, Inc., 270 Madison Ave., New York, NY 10016

"Vegetable Oil Lubricants" Saurabh Lawate, Rick Unger, Chor Huang, Lubrizol Corp., Lubricants World, May, 1999, Pages 43-45

"Tribology Data Handbook" E. Richard Booser (Editor), Copyright © 1997 CRC Press LLC., ISBN 0-8493-3904-9, CRC Press, Boca Raton, FL.

"Canola Oil-based Fluid is Gentle on Environment" R. Adams, P. Kromdyk, T. Noblit, Houghton International Inc., Valley Forge, PA., Hydraulics & Pneumatics, April, 1999, Pages 68, 70, 72

Chapter 10 exercises
Exercise 1

A mobile hydraulic reservoir is filled with Mobil DTE 25 fluid. There is some concern by the maintenance supervisor that the fluid will be too viscous in the winter for his piston pumps. He expects the temperature to drop to 35°F *(1.7°C)* in the machine area during some winter months. Determine how viscous the fluid will become.

Brand Name	Fluid Type	Specific Gravity	Viscosity (SUS)
Mobil DTE 25	PB 876	225 @ *100°F*	49 @ *210°F*

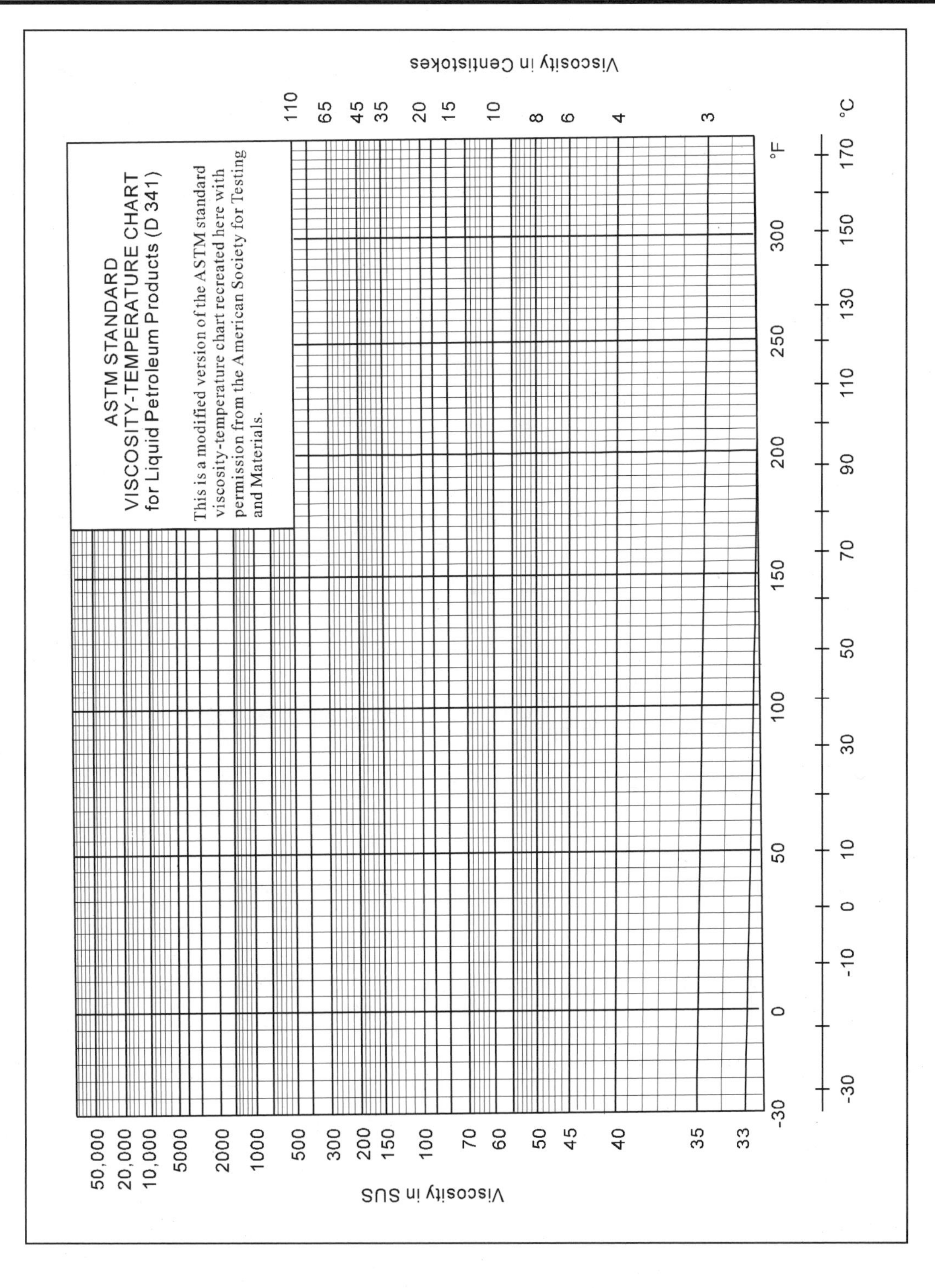

Chapter 10 exercises (cont'd.)
Exercise 2

Observing that a system's fluid had a tendency toward foaming, it was decided to switch to an oil with an anti-foam additive. With the addition of the oil to the system's reservoir, foaming ceased but the oil appeared to oxidize quickly. Offer an explanation for the rapid oxidation of the oil.

Exercise 3

Because of excessive system leakage, a maintenance supervisor decides to switch to a straight mineral oil (an oil without additives) which is relatively inexpensive. The fluid will operate at 152 bar *(2200 psi).*

Describe what probable results can be expected from this action.

Exercise 4

Because of the danger of fire, a mobile hydraulic system operating at 152 bar *(2200 psi)* with petroleum base fluid is required to change to a fire resistant fluid.

Describe what type of fire resistant fluid might be recommended in this case and indicate its affect on existing system seals and protective coatings.

Exercise 5

A certain individual cannot see why fire resistant fluid is needed in a system. "Oil doesn't burn," he says. "Anyone who has worked in a plant has seen someone extinguish a cigarette in a pool of oil."

Explain how hydraulic oil can be a fire hazard.

Chapter 10 exercises (cont'd.)

Exercise 6

On the night shift, a machine's hydraulic system was changed from petroleum oil to an invert emulsion. The day shift maintenance man knows that the viscosity of the petroleum oil was 32cSt *(150 SUS)* @ 37.7°C *(100°F)*. He notices that the invert emulsion has a viscosity of 80.9 cSt *(375 SUS)* @ 37.7°C *(100°F)*. He feels someone has made a mistake.

Has a mistake been made? Explain.

Exercise 7

To be classified "biodegradable" a fluid must meet what two conditions?

Chapter 10 exercises (cont'd.)
Exercise 8

Which of the following is classified as a non-mineral oil? _____

a. vegetable oil
b. synthetic ester
c. polyglycols
d. all of the above
e. a. & c. only

Exercise 9

A maintenance man informs the equipment operator that the ester based hydraulic fluid in the system will "hydrolize" if he is not careful. What does he mean?

Exercise 10

List the four main limitations of water as a system fluid.

1.

2.

3.

4.

Chapter 11

Hydraulic Filters

The components which have been discussed and the circuits which have been illustrated to this point will function as described and perform the job they were intended to do as long as the fluid is clean. The best designed components and the most carefully thought out circuits require clean fluid to achieve optimum performance.

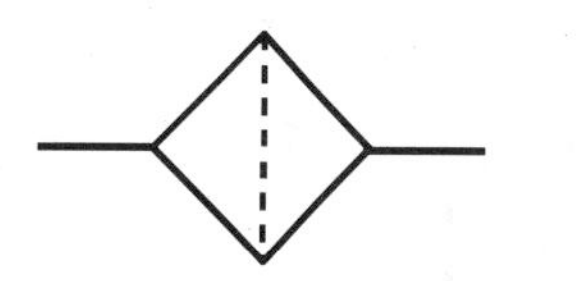

Fig. 11-1 Filter symbol

All hydraulic fluids contain dirt to some degree. But, the need of a filter in a system is many times not recognized. After all, the addition of this particular component does not increase a machine's apparent actions. But, this text would be sorely lacking if it did not clearly point out that dirt in hydraulic fluid is the downfall of even the best designed hydraulic systems. As a matter of fact, experienced maintenance men agree that the great majority of component and system malfunctions is caused by particles of dirt. Dirt particles can bring huge and expensive machinery to its knees.

Dirt interferes with hydraulic fluid

Dirt causes trouble in a hydraulic system because it interferes with the fluid which has four functions:

1. to act as a medium for energy transmission
2. to lubricate internal moving parts of hydraulic components
3. to act as a heat transfer medium
4. to seal clearances between close fitting moving parts

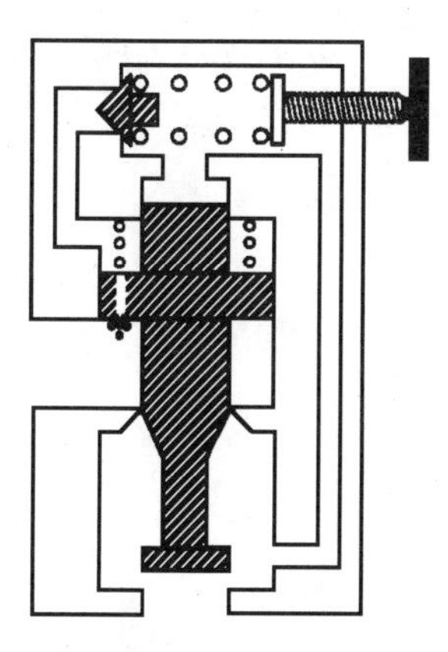

Fig. 11-2 Dirt can plug orifices

Dirt interferes with three of these functions. Dirt interferes with the transmission of energy by plugging small orifices in hydraulic components (fig. 11-2) like pressure valves and flow control valves. In this condition pressure has a difficult time passing to the other side of the spool. The valve's action is not only unpredictable and nonproductive, but unsafe.

Because of viscosity, friction, and changing direction, hydraulic fluid generates heat during system operation. When the liquid returns to the reservoir, it gives the heat up to the reservoir walls (fig. 11-3). Dirt particles interfere with liquid cooling by forming a sludge which makes heat transfer to reservoir walls difficult.

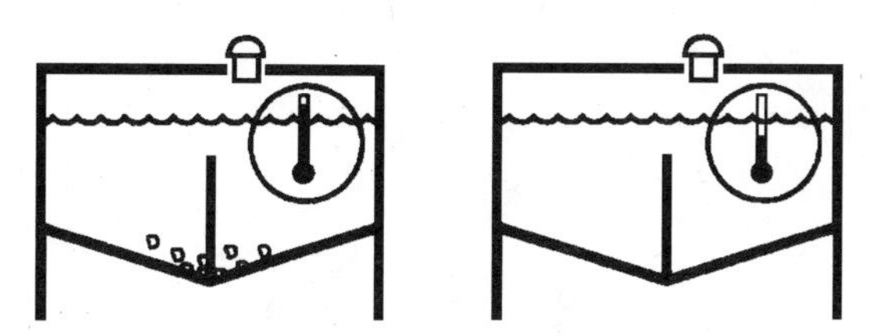

Fig. 11-3 Sludge can form in the reservoir, making heat transfer difficult.

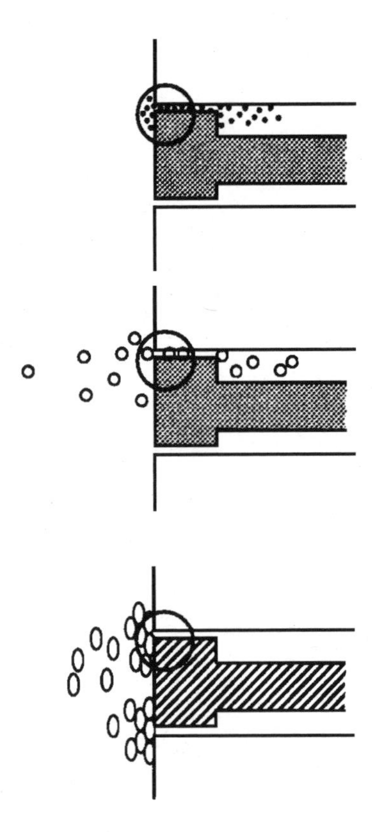

Fig. 11-4 Dirt can cause silting or component wear and block fluid flow.

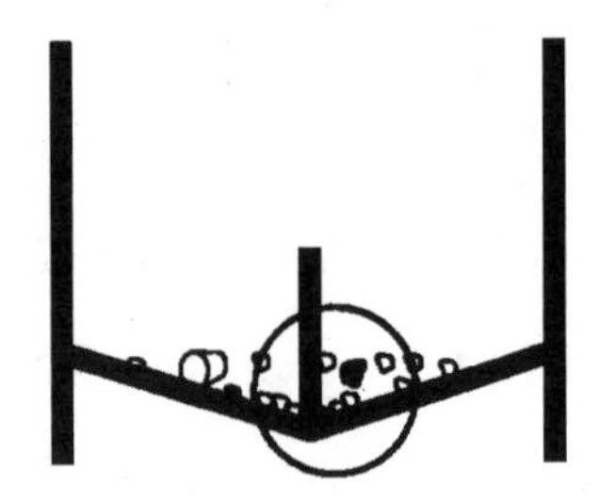

Fig. 11-5 Reservoir contamination

Clean hydraulic systems run cooler than dirty systems. Probably the greatest problem with dirt in a hydraulic system is that it interferes with lubrication.

Dirt can be divided into three sizes with respect to a particular component's clearances; that is, dirt which is smaller than a clearance, dirt which is the same size, and dirt which is larger than a clearance (fig. 11-4).

Extremely fine dirt, which is smaller than a component's clearances, can collect in clearances especially if there are excessive amounts and the valve is not operated frequently. This blocks or obstructs lubricative flow through the passage.

An accumulation of extremely fine dirt particles in a hydraulic system is known as silting.

Dirt which is about the same size as a clearance rubs against moving parts breaking down a fluid's lubricative film.

Large dirt can also interfere with lubrication by collecting at the entrance to a clearance and blocking fluid flow between moving parts.

A lack of lubrication causes excessive wear, slow response, erratic operation, solenoid burn out, and early component failure.

Dirt is pollution

Dirt in a hydraulic system is pollution (fig. 11-5). It is very similar to bottles, cans, paper and old tires floating in your favorite river or stream. The difference is that hydraulic system pollution is measured using a very small scale. The micrometre scale is used to measure dirt in hydraulic systems.

Contamination types and sources

Water contamination

There is more to proper maintenance than just removing particulate matter. Water is virtually a universal contaminant, and just like solid particle contaminants, must be removed from operating fluids.

Water can be either in a dissolved state or in a 'free' state. Free, or emulsified, water is defined as the water above the saturation point of a specific fluid (Table 11-1). At this point, the fluid cannot dissolve or hold any more water.

Free water is generally noticeable as a 'milky' discoloration of the fluid (fig. 11-6).

Damage

- corrosion of metal surfaces (fig. 11-7)
- accelerated abrasive wear
- bearing fatigue (diagram 11-1)
- fluid additive breakdown
- viscosity variance
- increase in electrical conductivity

Anti-wear additives break down in the presence of water and form acids. The combination of water, heat and dissimilar metals encourages galvanic action. Pitted and corroded metal surfaces and finishes result. Further complications occur as temperature drops and the fluid has less ability to hold water. As the freezing point is reached, ice crystals form, adversely affecting total system function. Operating functions may also become slowed or erratic.

Electrical conductivity becomes a problem when water contamination weakens the insulating properties of a fluid, thus decreasing its dielectric kV (kilovolt) strength.

Sources

- worn actuator seals
- reservoir opening leakage
- condensation
- heat exchanger leakage

Filtration facts

- A simple 'crackle test' will tell you if there is free water in your fluid.
  Apply a flame under the container. If bubbles rise and 'crackle' from the point of applied heat, free water is present in the fluid.
- Hydraulic fluids have the ability to 'hold' more water as temperature increases. A cloudy fluid may become clearer as a system heats up.

Table 11-1 Typical saturation points

Fluid type	Parts per million (ppm)	Percent (%)
Hydraulic fluid	300	0.03
Lubrication fluid	400	0.04
Transformer fluid	50	0.005

Fig. 11-7 Typical results of pump wear due to particulate and water contamination.

Fig. 11-6 Fluid contamination can be visual

Filtration facts

- Free water is heavier than oil, thus it will settle to the bottom of the reservoir where much of it can be easily removed by opening the drain valve.
- Absorption filter elements have optimum performance in low flow and low viscosity applications.

Fluids are constantly exposed to water and water vapor while being handled and stored. For instance, outdoor storage of tanks and drums is common.

Water may settle on top of fluid containers and be drawn into the container during temperature changes. Water may also be introduced when opening or filling these containers.

Water can enter a system through worn cylinder and actuator seals or through reservoir openings. Condensation is also a prime water source. As the fluids cool in a reservoir or tank, water vapor will condense on the inside surfaces, causing rust or other corrosion problems.

Prevention

Excessive water can usually be removed from a system. The same preventive measures taken to minimize particulate contamination ingression in a system can be applied to water contamination. However, once excessive water is detected, it can usually be eliminated by one of the following methods.

Absorption

This is accomplished by filter elements that are designed specifically to take out free water. They usually consist of a laminate-type material that

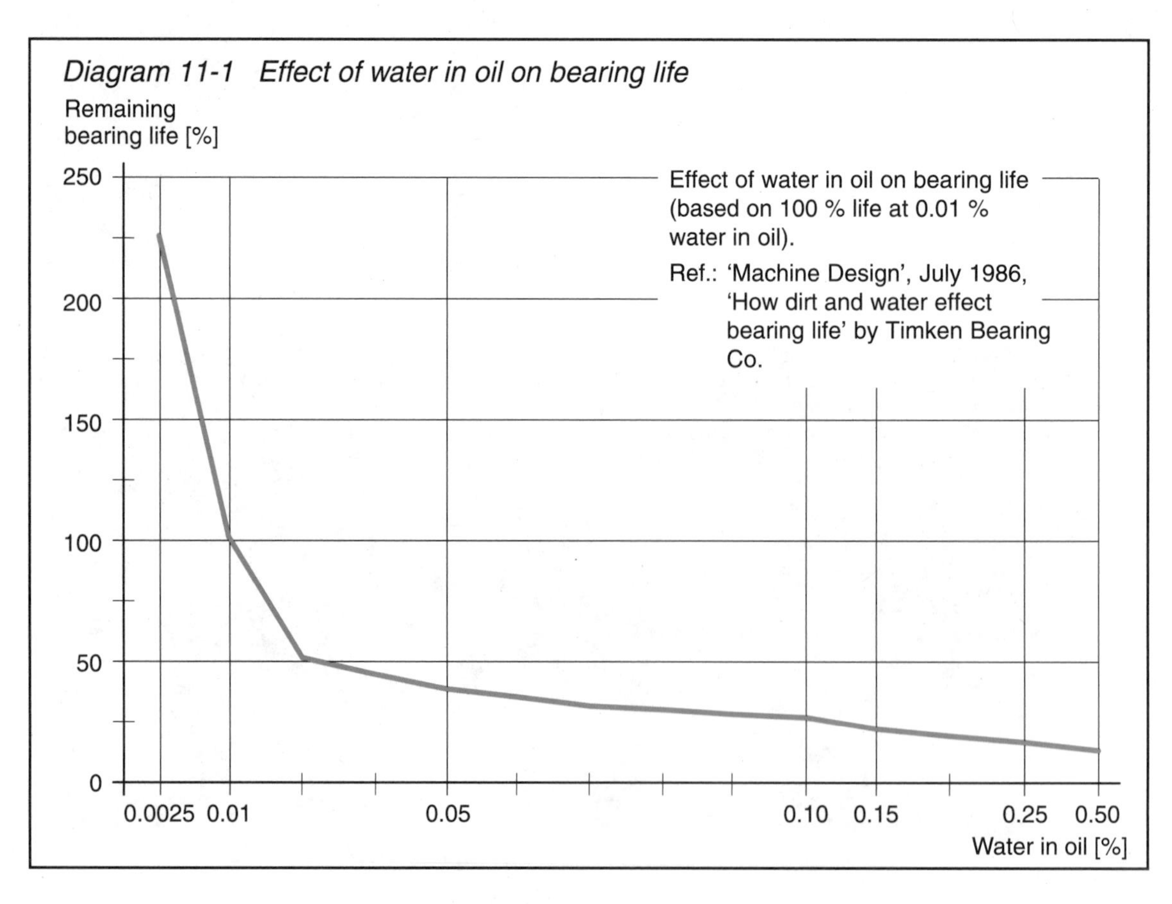

Diagram 11-1 Effect of water in oil on bearing life

transforms free water into a gel that is trapped within the element. These elements fit into standard filter housings and are generally used when small volumes of water are involved.

Centrifugation

Water is separated from oil by a spinning motion. This method is also only effective with free water, but for larger volumes.

Vacuum dehydration

Water is separated from oil through a vacuum and drying process (fig. 11-8). This method is also for larger volumes of water, but is effective with both the free and dissolved water.

Fig. 11-8 Vacuum dehydration system

The micrometre scale

One micrometre (micron) is equal to one millionth of a meter or thirty-nine millionths of an inch. A single micrometre is invisible to the naked eye and is so small that to imagine it is extremely difficult. To bring the size more down to earth, some everyday objects will be measured using the micrometre scale.

An ordinary grain of table salt measures 100 micrometres (µm). The average diameter of human hair measures 70 micrometres (µm) (fig. 11-9). Twenty-five micrometres is approximately one thousandth of an inch.

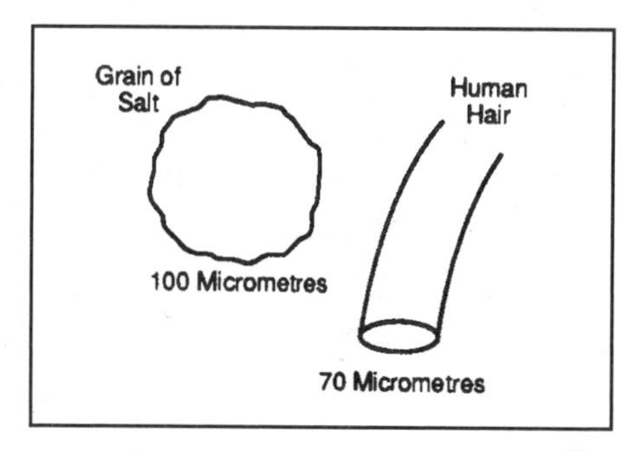

Fig. 11-9

Limit of visibility

The lower limit of visibility for a human eye is 40 micrometres. In other words, the average person can see dirt which measures 40 micrometres and larger. This means that just because a sample of hydraulic fluid looks clean, doesn't necessarily mean that it is clean. Much of the harmful dirt in a hydraulic system is below 40 micrometres.

Determination of fluid cleanliness

Since human vision is not a proper judge, cleanliness of hydraulic fluid is determined by examining a sample of a system's fluid by visual inspection with a microscope or with the use of an automatic particle counter.

In each of these methods, the number of particles in a micrometre size range is the determining factor for cleanliness.

Particle counting

In order to detect or correct problems, a contamination reference scale is used. Particle counting is the most common method to derive cleanliness level standards.

Very sensitive optical instruments are used to count the number of particles in various size ranges. These counts are reported as the number of particles greater than a certain size found in a specified volume of fluid.

The ISO 4406 (International Standards Organization) cleanliness level standard has gained wide acceptance in most industries today. A widely-used modified version of this standard references the number of particles greater than 2, 5, and 15 µm (micrometers) in a known volume, usually 1 ml (milliliter) or 100 ml. **NOTE:** 1 ml (0.001 liter) equals approx. 0.06 in^3.

The number of 2+ µm (greater than 2 µm) and 5+ µm (larger than 5 µm) particles is used as a reference point for 'silt' particles. The 15+ size range indicates the quantity of larger particles present which contribute greatly to possible catastrophic component failure.

Tables 11-2 and 11-3 show a particular ISO code (18/16/13) and the corresponding classification. The ISO 4406 chart (Table 11-4) lists the particle count range for range numbers from 24 to 6.

Component cleanliness level requirements

Many manufacturers of hydraulic and load bearing equipment specify the optimum cleanliness level required for their components. Subjecting components to fluid with higher contamination levels may result in much shorter component life.

In Table 11-5, a few components and their recommended cleanliness levels are shown. It is always best to consult with component manufacturers and obtain their written fluid cleanliness level recommendations. This information is needed in order to select the proper level of filtration.

It may also prove useful for any subsequent warranty claims, as it may draw the line between normal use and excessive or abusive operation.

Filtration facts

- Knowing the cleanliness level of a fluid is a basis for contamination control
- The ISO code index numbers can never increase as the particle sizes increase.

Table 11-2 ISO code (example)

ISO code	**18 / 16 / 13**
Particles • 2 µm	18
Particles • 5 µm	16
Particles • 15 µm	13

Table 11-3 ISO classification of 18/16/13

Range number	Size in µm	Actual particle count range (per ml)
18	• 2	1300 - 2500
16	• 5	320 - 640
13	•15	40 - 80

Table 11-4 ISO 4406 chart

Range number	Number of particles per ml more than	up to and incl.
24	80 000	160 000
23	40 000	80 000
22	20 000	40 000
21	10 000	20 000
20	5 000	10 000
19	2 500	5 000
18	1 300	2 500
17	640	1 300
16	320	640
15	160	320
14	80	160
13	40	80
12	20	40
11	10	20
10	5	10
9	2.5	5
8	1.3	2.5
7	0.64	1.3
6	0.32	0.64

Filter elements

The function of a mechanical filter is to remove dirt from hydraulic fluid. This is done by forcing the fluid stream to pass through a porous filter element which catches the dirt.

Filter elements are divided into depth and surface types.

Depth type elements

Depth type elements force the fluid to pass through an appreciable thickness of many layers of material. The dirt is trapped because of the intertwining path the fluid must take (fig. 11-10).

Treated paper and synthetic materials are commonly used porous media for depth elements.

Pore size in depth type elements

Because of its construction, a depth type filter element has many pores of various size. This fact is shown by the pore size distribution curve.

A point on the curve is the number of pores per unit area of a given size in a typical depth type element (fig. 11-11).

Filtration facts

- Most machine and hydraulic component manufacturers specify a target ISO cleanliness level to equipment in order to achieve optimal performance standards.
- Color is not a good indicator of a fluid's cleanliness level.

Table 11-5 Fluid cleanliness required for typical hydraulic components

Component	ISO code
Servo control valves	16/14/11
Proportional valves	17/15/12
Vane and piston pumps and motors	18/16/13
Directional and pressure control valves	18/16/13
Gear pumps and motors	19/17/14
Flow control valves and cylinders	20/18/15
New, unused fluid	20/18/15

Table 11-6 Cleanliness level correlation table

ISO code	Particles per ml (milliliter) • 2 µm	• 5 µm	• 15 µm	NAS 1638 (1964)	Disavowed SAE level (1963)
23/21/18	80 000	20 000	2 500	12	–
22/20/18	40 000	10 000	2 500	–	–
22/20/17	40 000	10 000	1 300	11	–
22/20/16	40 000	10 000	640	–	–
21/19/16	20 000	5 000	640	10	–
20/18/15	10 000	2 500	320	9	6
19/17/14	5 000	1 300	160	8	5
18/16/13	2 500	640	80	7	4
17/15/12	1 300	320	40	6	3
16/14/12	640	160	40	–	–
16/14/11	640	160	20	5	2
15/13/10	320	80	10	4	1
14/12/9	160	40	5	3	0
13/11/8	80	20	2.5	2	–
12/10/8	40	10	2.5	–	–
12/10/7	40	10	1.3	1	–
12/10/6	40	10	0.64	–	–

NAS: National Aeronautical Standard (US)

SAE: Society of Automotive Engineers (US)

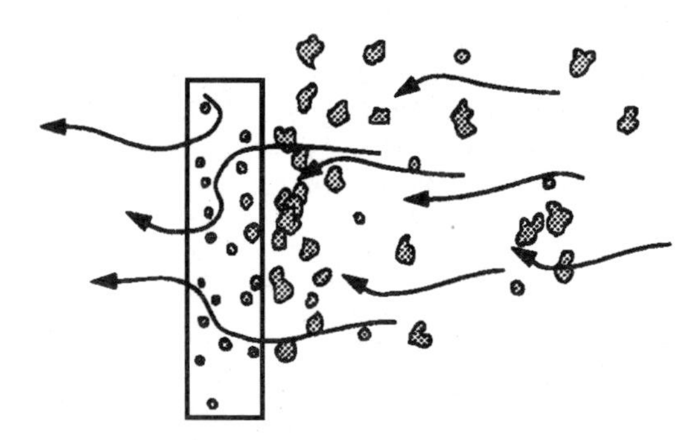

Fig. 11-10 Depth type element

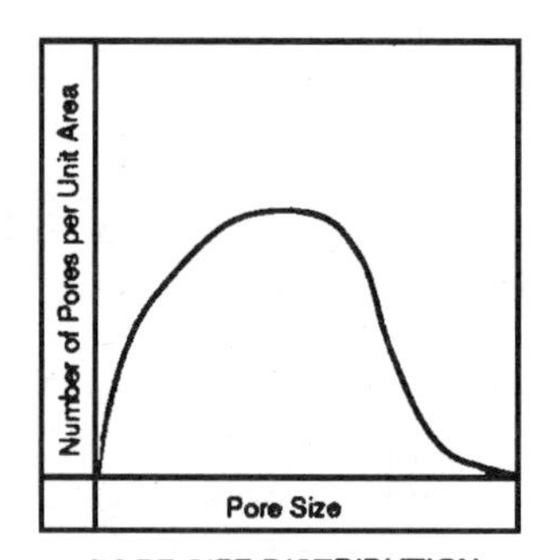

PORE SIZE DISTRIBUTION FOR DEPTH TYPE ELEMENT

Fig. 11-11

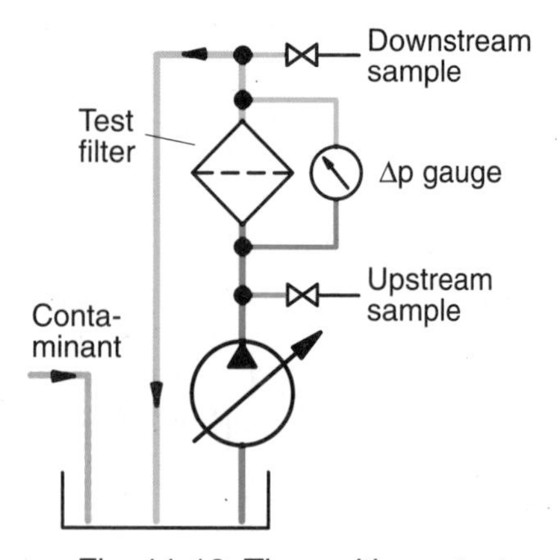

Fig. 11-12 The multipass test

> *Filtration facts*
>
> - Filter media ratings expressed as a 'ß (beta) ratio' indicate a media's particle removal efficiency.
> - Multipass test results are very dependent upon the following variables:
>   - flow rate
>   - terminal pressure differential
>   - contaminant type

The shape of the curve shows that there are a great deal more pores of small size than of relatively large size. This means that a large percentage of flow passes through relatively small holes.

Nominal rating

Since there is no one consistent hole or pore size in a depth type element, it is given a nominal rating which is based on its average pore size.

For example, a depth element with a nominal rating of 40 micrometres means that the element's average pore size is at least 40 micrometres and that initially it will remove dirt of 40 micrometres and smaller and will not remove some dirt which is larger than 40 micrometres.

NOTE: Some manufacturers do not use an element's average pore size as a basis for a nominal rating. In these cases, the nominal rating is usually an arbitrary value which indicates little.

Filter media types and ratings

The multipass test

The filtration industry uses the ISO 4572 'Multipass Test Procedure' to evaluate filter element performance. This procedure is also recognized by ANSI (American National Standards Institute) and NFPA (National Fluid Power Association).

During the multipass test (fig. 11-12), fluid is circulated through the circuit under precisely controlled and monitored conditions. The differential pressure across the test element is continuously recorded, as a constant amount of contaminant is injected upstream of the element. On-line laser particle sensors determine the contaminant levels upstream and downstream of the test element.

This performance attribute, the 'ß (beta) ratio' is determined for several particle sizes.

Three important element performance characteristics are a result of the multipass test:

1. dirt holding capacity
2. pressure differential of the test filter element
3. separation or filtration efficiency, expressed as a 'ß ratio'

ß ratio

The ß ratio (also known as the filtration ratio) is a measure of the particle capture efficiency of a filter element. It is therefore a performance rating (Table 11-7).

In Formula 11-2 is an example of how a ß ratio is derived at from a multipass test. Assume that 50,000 particles, 10 µm and larger, were counted upstream (before) of the test filter, and 10,000 particles at that same size range were counted downstream (after) of the test filter. The corresponding ß ratio would be 5 and the example would read 'ß ten equal to five'.

A ß ratio number alone means very little. It is a preliminary step to find a filter's particle capture efficiency. This efficiency, expressed as a percent, can be found by a simple equation (Formula 11-1). In the example, the particular filter tested was 80 % efficient at removing 10 µm and larger particles. For every 5 particles introduced to the filter at this size range, 4 were trapped in the filter media.

The 'capture efficiency versus ß ratio' (Table 11-8) shows some common ß ratio numbers and their corresponding efficiencies.

Table 11-7 Capture efficiency versus ß ratio

ß ratio (at a given particle size)	Capture efficiency [%] (at same particle size)
1.01	1.0
1.1	9.0
1.5	33.3
2.0	50.0
5.0	80.0
10.0	90.0
20.0	95.0
75.0	98.7
100	99.0
200	99.5
1000	99.9

Table 11-8 Common ß ratio numbers

Upstream particles	Downstream particles	ß ratio $_{(x)}$	$\eta_{(x)}$ [%]
100 000 ≥ (x) µm	50 000	$\frac{100\,000}{50\,000} = 2$	50.0
	5 000	$\frac{100\,000}{5\,000} = 20$	95.0
	1 333	$\frac{100\,000}{1\,333} = 75$	98.7
	1 000	$\frac{100\,000}{1\,000} = 100$	99.0
	500	$\frac{100\,000}{500} = 200$	99.5
	100	$\frac{100\,000}{100} = 1\,000$	99.9

Filter

Formula 11-1 Capture efficiency

$$\eta_x = (1 - \frac{1}{\beta}) \times 100 \text{ [\%]}$$

where: 'η_x' is the capture efficiency in % at removing x µm (or larger) particles.

Example: $\beta_{10} = (1 - \frac{1}{5}) \times 100 = 80$ [%]

Formula 11-2 ß (beta) ratio

$$\beta_X = \frac{\text{Number of particles upstream}}{\text{Number of particles downstream}}$$

where: 'β_X' is the ß ratio at a specific particle size (x).

Example: $\beta_{10} = \frac{50\,000}{10\,000} = 5$

Table 11-9 Filter media selection - lubrication systems

Component type	Suggested cleanliness code	Media efficiency β_x >200	Number of filter placements*	Minimum filter placements
Ball bearings	15/13/11	2	1.5	P or R, and O
		2	1	P or R
Roller bearings	16/14/12	5	2	P or R
		2	0.5	O
Journal bear.; gear boxes	17/15/13	5	1.5	P or R, and O
		10	2.5	P, R and O

* Number of filtration placements in system; more placements are the option of the specifier.

P = Full flow pressure filter (equals one filtration placement)

R = Full flow return filter (equals one filtration placement

O = Off-line (flow rate 10% of reservoir volume equals 0.5 of a filtration placement.

Surface type elements

In a surface type filter element a fluid stream has a straight flow path through one layer of material. The dirt is caught on the surface of the element which faces the fluid flow.

Wire cloth and perforated metal are common types of materials used in surface elements.

Pore size in surface type elements

Since the process used in manufacturing the cloth material and the perforated metal can be very accurately controlled, surface type elements have a consistent pore size (fig. 11-13). Because of this fact, surface type elements are usually identified by their absolute rating.

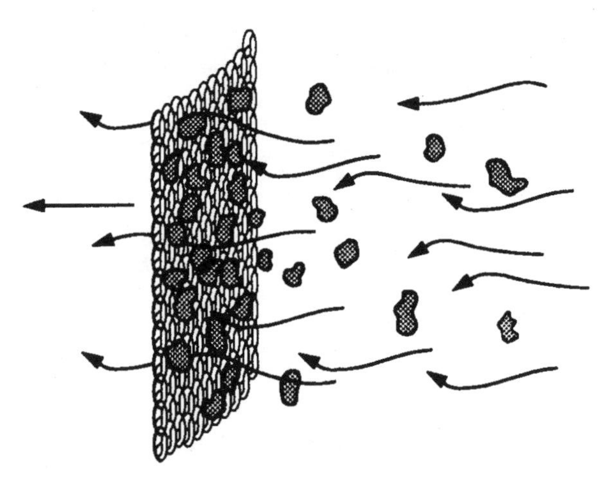

Fig. 11-13 Vacuum dehydration system

Absolute rating

An absolute rating is an indication of the largest opening in a filter element. This rating indicates the diameter of the largest hard spherical particle which can pass through an element.

Since the pore size can be accurately controlled in this type element, basically all the holes in a 200 wire mesh element are 74 micrometres square (fig. 11-14).

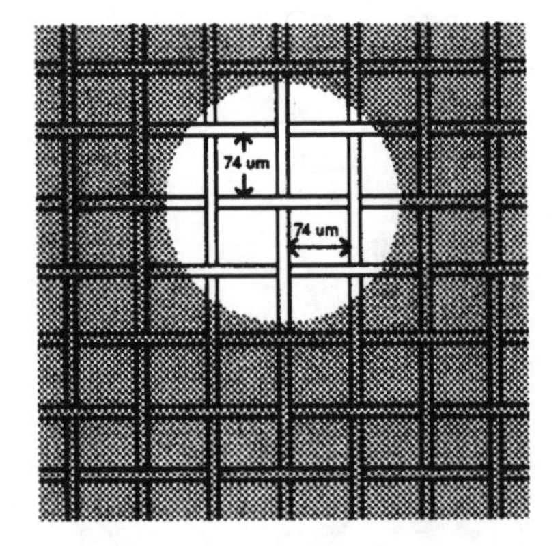

Fig. 11-14 Absolute rating square wire mesh

A 200 mesh wire element with an absolute rating of 74 micrometres means that for every square inch of material there are 200 wires running vertically, 200 wires running horizontally, and the perpendicular distance between the wires is 74 micrometres. Therefore, the largest, hard spherical particle to pass through the element would have a diameter of 74 micrometres.

The absolute rating for a depth type element would be the last point on the pore size axis of the pore size distribution curve. There may only be one hole of that size in the element, but that would still be its absolute rating (fig. 11-15).

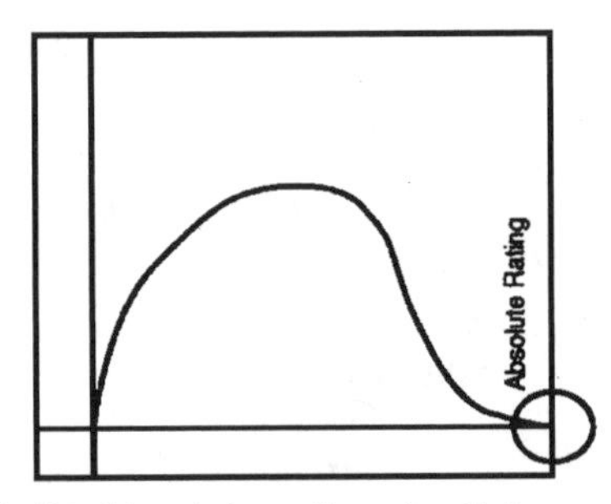

Fig. 11-15 Absolute rating depth type element

Filter ratings in practice

As was pointed out above, an absolute rating indicates the largest hole size in an element. Sometimes, it is deduced that an absolute rating indicates the largest particle which will pass through the element while operating in a system. Since the dirt in the common hydraulic system is not spherical particles, this deduction is erroneous.

Dirt in hydraulics fluids is any insoluble material. It comes in all sizes, shapes, and materials.

An element with an absolute rating of 74 micrometres may have a difficult time catching a long, thin particle. A sliver with an overall length of 150 micrometres and a diameter of 3 micrometres, travels like a rocket in a fluid stream. The sliver penetrates the element with its 3 micrometre dimension and probes its way through the relatively large holes. Consequently, using a 74 micrometre absolute element is not a guarantee that all dirt larger than 74 micrometres will be removed from a fluid stream.

An absolute or nominal rating only indicates a filter element's largest or average hole size. Filter ratings do not guarantee what size particles will be removed or how clean a fluid will become.

To determine how fine and what type of a filter element should be used to protect a specific hydraulic component, consult the component manufacturer or a reputable filter dealer.

Sources of dirt

Dirt was previously defined as any type of insoluble material in a hydraulic fluid. Dirt comes in all sizes, shapes, and materials and has several sources by which it enters a system.

Dirt built into a system

Hydraulic systems which are newly fabricated are often extremely dirty. As a machine is being assembled, the reservoir becomes a collection point for rust, paint chips, dust, cigarette butts, grit and paper cups. Even though the reservoir is "cleaned" before it is filled with fluid, a sample of the fluid taken shortly after start up time will show dirt particles of many foreign materials. Many dirt particles which are harmful to a system are invisible to the naked eye and cannot be removed by wiping with a rag or blowing off with an air hose.

The components which make up a new system may also be sources of system contamination. Because of storage, handling, and installation practices, new directional valves, cylinders, relief valves, and pumps may be equipped with dirt particles which enter a fluid stream after a very short time.

Dirt generated within a system

Another source of dirt is the dirt generated within the system itself. As a machine continues to operate, moving parts naturally begin to wear and generate dirt (fig. 11-16). Every internal moving part in a system can be considered a source of contamination for the entire system.

Fig. 11-16 Moving parts can generate contamination through component wear.

Component housings are continuously flexing from normal stresses and are constantly subjected to hydraulic shock pressures. These actions cause metal and casting sand to break loose from the housing and enter the fluid stream.

Air entering the reservoir during operation contains water vapor (fig. 11-17). When the machine is shut down, the air inside the tank cools and water vapor condenses on the walls. This causes the formation of rust which eventually is washed into the fluid.

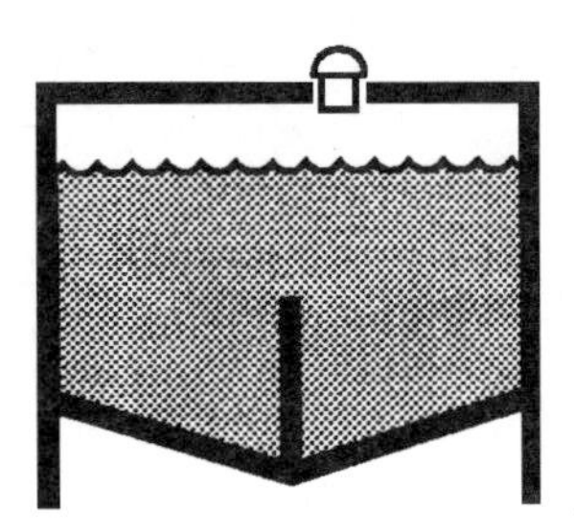

Fig. 11-17 Water vapor can cause rust formation.

Dirt added to a system

An additional source of dirt is the dirt added to a system. This is many times the result of ordinary maintenance practices or a lack of maintenance.

If a pump happens to break down, the maintenance man replaces the component or repairs it right on the spot. In either case he will be working in a dirty environment, which will contaminate the system as soon as a line is cracked (fig. 11-18).

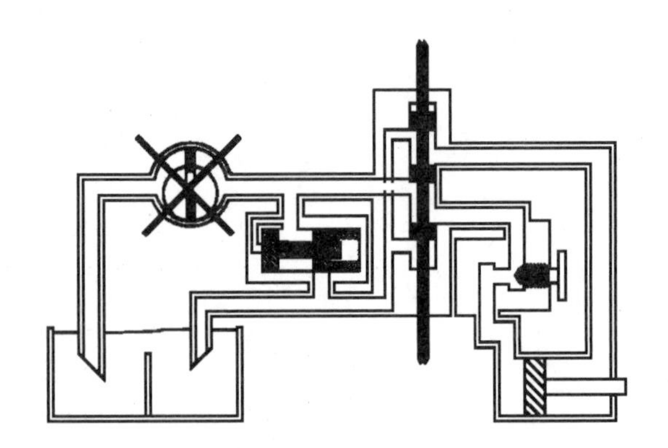

Fig. 11-18 Pump breakdown repairs can intoduce dirt into a system.

If the pump is disassembled, more than likely it will be done on a dirty workbench and the parts will be wiped off with a "not-too-dirty" rag before reassembly. When the pump is replaced, it will spread dirt throughout the system.

The usual first aid measure for a line which has sprung a leak is to put the nearest "clean" bucket under the leak. After it is repaired, the maintenance man must dump the bucket of fluid. It usually happens if no convenient way is seen to dispose of the contaminated fluid, it is poured back into the reservoir (fig. 11-19). It was one bucket short any way.

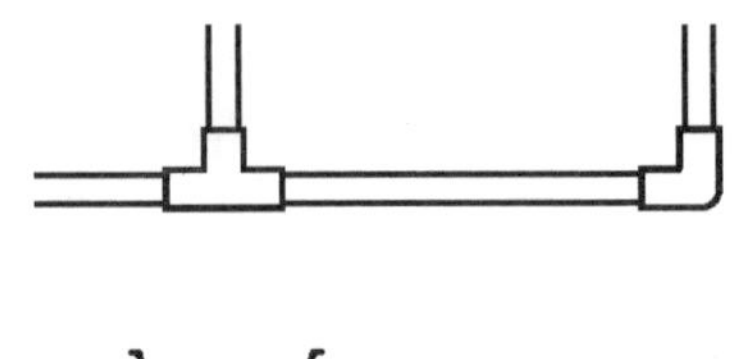

Fig. 11-19 Contamination can enter a system when fluid leaks are poured back into the reservoir.

The atmosphere in which the machine lives contaminates a system's hydraulic fluid. Operation of a hydraulic system depends upon air entering the reservoir through the breather cap and pushing the fluid up to the pump. As actuators in the system are filled and discharged with fluid, the tank actually inhales and exhales dirt-laden air which is filtered by a coarse screen in the breather cap. The dirt which is not filtered out settles into the fluid.

Breather caps are very rarely, if ever, cleaned. Because of this lack of maintenance, the screen becomes plugged. As a result, the breather cap is removed from the reservoir or the atmosphere finds a path through an old, cracked seal or stripped bolt hole. Now a clear path to the fluid is open for almost anything.

Dirt can also be added to a system by means of a cylinder. After a time, a cylinder's rod wiper seal wears. In this condition, dirt is drawn into the system each time the cylinder rod retracts.

Type of filtration by flow

In the early days of hydraulics, filtration was not considered necessary, and this was more or less a correct assumption. Hydraulic components at that time were crude when compared to modern standards. Clearances between moving parts were large, therefore dirty fluid would not affect operation that much. Components were dirt tolerant to a large extent.

As manufacturing processes naturally improved, tolerances were improved. Component operation became much more efficient, but components were less dirt tolerant. The need for filtration was recognized.

Proportional flow filtration

The first measure was to place a filter in a system in such a way that a part of the total volume of a pump was filtered (fig. 11-20). This can be done by placing a filter in a system so that a portion of a pump's flow is bled off through the filter.

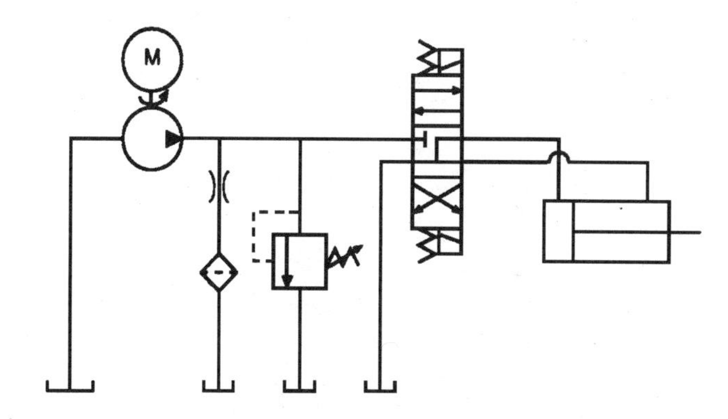

Fig. 11-20 Proportional flow filtration filter placement

Full flow filtration

Proportional flow filtration was found to be inadequate after a time, especially as system components became more and more efficient.

The next step was to place a filter in a system so that all flow from a pump was filtered (fig. 11-21). This full flow filtration is the type of filtration used in most modern hydraulic systems.

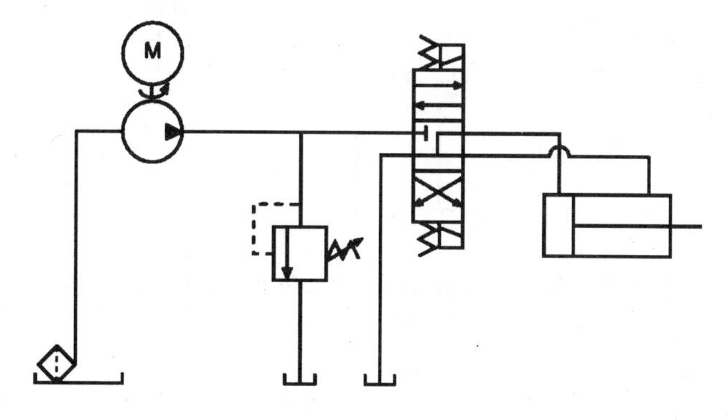

Fig. 11-21 Filter element installed between reservoir and pump

Type of filtration by position in a system

A filter is protection for a hydraulic component. Ideally each system component should be equipped with its own filter, but this is economically impractical in most cases. To obtain best results, the usual practice is to strategically place filters in a system.

In the majority of applications the fluid reservoir is a large source of dirt for a system. Since a pump is the heart of a system, one of the most expensive components in a system, and one of the fastest moving system components, it seems logical that a good place to put a filter is between a reservoir and pump.

Sump strainer

A sump strainer is usually a coarse filter element screwed onto the end of a pump's suction line.

The range of filtration for sump strainers is a perforated metal cylinder with large drilled holes to 74 micrometre wire mesh.

Advantages:

1. Sump strainers protect the pump from dirt in the reservoir.
2. Because they have no filter housing, sump strainers are very inexpensive.

Disadvantages:

1. Being below fluid level, sump strainers are very difficult to service when cleaning is necessary, especially if the fluid is hot.
2. A sump strainer does not have an indicator to tell when it is dirty.
3. A sump strainer may block fluid flow and starve the pump if not sized correctly and properly maintained.
4. A sump strainer does not protect components downstream from the particles generated by the pump.

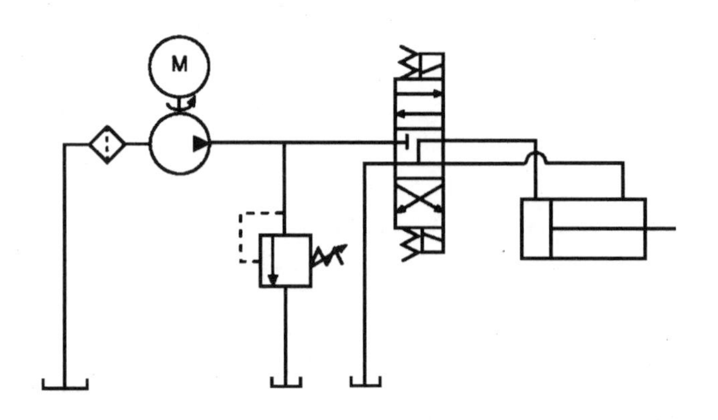

Fig. 11-22 A suction filter is placed outside of the reservoir.

Suction filter

A suction filter is placed in the suction line outside of the reservoir (fig. 11-22). Range of filtration usually found in suction line filters is from 238-25 micrometres.

Advantages:

1. A suction filter protects the pump from dirt in the reservoir.
2. Since the suction filter is outside the reservoir, an indicator telling when the element is dirty can be used.
3. The filter element can be serviced without disassembling the suction line or reservoir.

Disadvantages:

1. A suction filter may starve the pump if not sized or engineered into the system properly.
2. A suction filter does not protect components downstream from the particles generated by the pump.

Pressure filter

A pressure filter is positioned in the circuit between the pump and a system component.

Range of filtration usually found in pressure line filters is from 40-3 micrometres.

A pressure filter can also be positioned between system components (fig. 11-23). If the flow between the components can flow in two directions (as between directional valve and cylinder), the filter must be capable of handling bidirectional flow. Bidirectional pressure filters are used on the downstream side of servo valves and in closed looped hydrostatic transmissions.

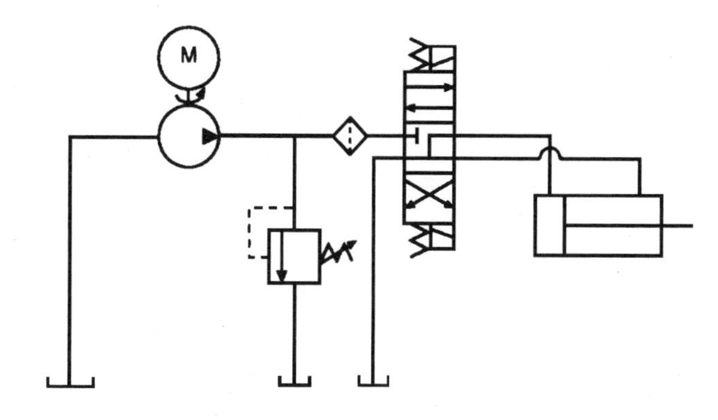

Fig. 11-23 Pressure filter placement

Advantages:

1. A pressure filter can filter very fine since system pressure is available to push the fluid through the element.
2. A pressure filter can protect a specific component from the harm of deteriorating particles generated from an upstream component.

Disadvantages:

1. The housing of a pressure filter must be designed for high pressure because it is operating at full system pressure. This makes the filter expensive.
2. If pressure differential and fluid velocity are high enough, dirt can be pushed through the element or the element may tear or collapse.

Return line filter

A return line filter is positioned in the circuit just before the reservoir. Range of filtration usually found in return line filters is from 40-5 micrometres.

Advantages:

1. A return line filter catches the dirt in the system before it enters the reservoir.
2. The filter housing does not operate under full system pressure and is therefore less expensive than a pressure filter.
3. Fluid can be filtered fine since system pressure is available to push the fluid through the element.

Disadvantages:

1. There is no direct protection for circuit components.
2. In return line full flow filters, flow surges from discharging cylinders, actuators, and accumulators must be considered when sizing.
3. Some system components may be affected by the back pressure generated by a return line filter.

Filter bypass valve

If filter maintenance is not performed, pressure differential across a filter element will increase.

An unlimited increase in pressure differential across a filter on the suction side of a system means that a pump will eventually cavitate.

An unlimited increase in pressure differential across a filter on the pressure side means that the filter element will eventually collapse.

To avoid this situation a simple or direct acting relief valve is used to limit the pressure differential across a full flow filter. This type relief valve is generally called a bypass valve.

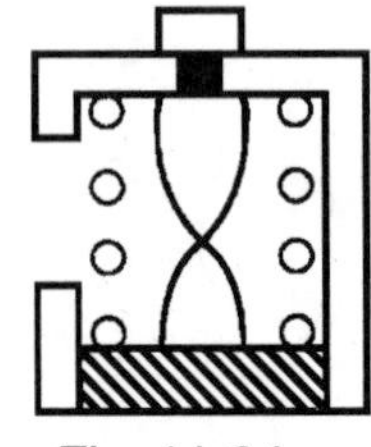

Fig. 11-24

What a bypass valve consists of

A bypass valve basically consists of a movable piston, housing, and a spring which biases the piston (fig. 11-24).

How a bypass valve works

There are several types of bypass valves, but they all operate by sensing the difference in pressure between dirty and clean fluid.

In our illustration, pressure of dirty fluid coming into the filter is sensed at the bottom of the piston. Pressure of the fluid after it has passed through the filter element is sensed at the other side of the piston on which the spring is acting (fig. 11-25).

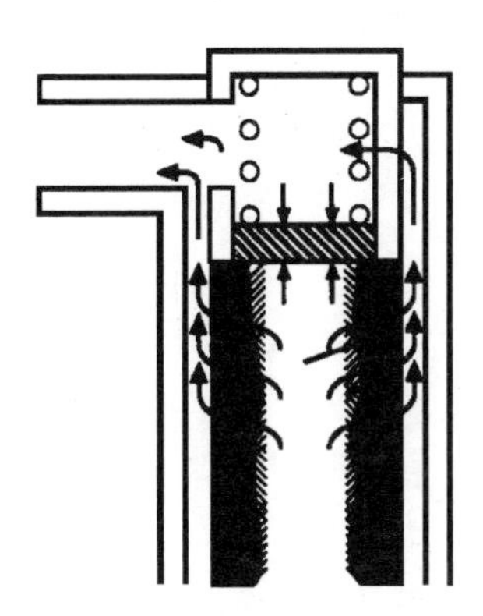

Fig. 11-25

As the filter element collects dirt, the pressure required to push the dirty fluid through the element increases. Fluid pressure after it passes through the element remains the same. When the pressure differential across the filter element, as well as across the piston, is large enough to overcome the force of the spring, the piston will move up and offer the fluid a path around the element (fig. 11-26).

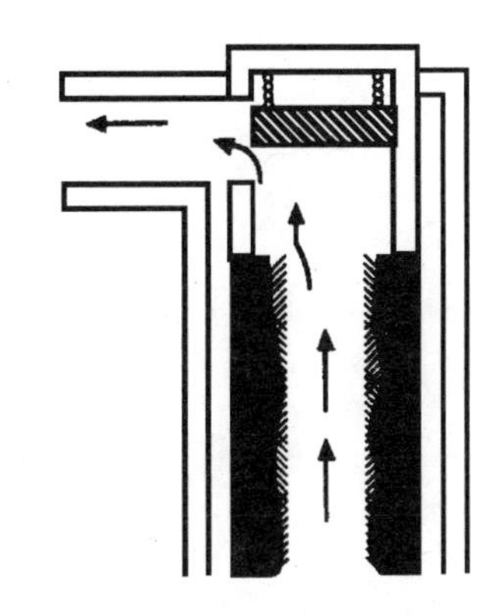

Fig. 11-26 Dirty fluid bypassing

A bypass valve is a fail safe device. In a suction filter, a bypass limits the maximum pressure differential across the filter if it is not cleaned. This protects the pump. If a pressure or return line filter is not cleaned, a bypass will limit the maximum pressure differential so that dirt is not pushed through the element or that the element is not collapsed. In this way, the bypass protects the filter.

The whole key, then, to filter performance centers around cleaning the filter when it needs cleaning. To help in this regard, a filter is equipped with an indicator.

Filter indicator

A filter indicator shows the condition of a filter element. It indicates when the element is clean, needs cleaning, or in the bypassing condition.

What a filter indicator consists of

One common type of filter indicator consists of a helix and a dial indicator which is attached to the helix (fig. 11-27).

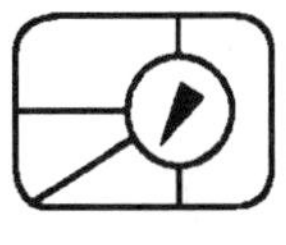

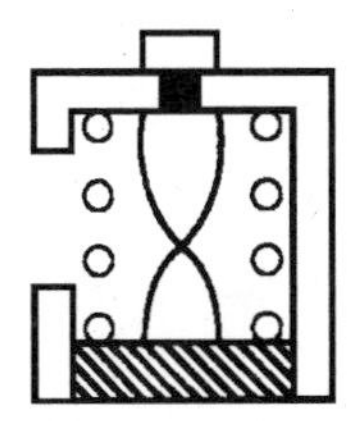

Fig. 11-27 Filter indicator

How a filter indicator works

The operation of a filter indicator is dependent on the movement of the bypass piston. When the element is clean, the bypass piston is fully seated and the indicator shows "clean."

While pressure differential across the piston and element increases to the point where the filter needs cleaning, the piston moves up. During its movement, the piston twists the helix which positions the indicator dial at "needs cleaning."

If the filter element is not cleaned when indicated, pressure differential will continue to increase. The piston will continue to move up and bypass the fluid. At this time the indicator will show a bypassing condition.

Filters must be maintained

A machine may be equipped with the best filters available and they may be positioned in the system where they do the most good; but, if the filters are not taken care of and cleaned when dirty, the money spent for the filters and their installation has been wasted. A filter which gets dirty after one day of service and is cleaned 29 days later gives 29 days of non-filtered fluid. A filter can be no better than the maintenance afforded it.

Chapter 11 exercise
Filters

INSTRUCTIONS: Find a word in column 2 related to a word in column 1. Then, pair up two more words which have the same relationship, taking one from column 3 and one from column 4. For example: bird-air & fish-water. All words are used once.

1	2	3	4
Water in oil	absorption	centrifugation	vacuum dehydration
ISO			
Depth element			
Piston pumps & motors			
Nominal			
ISO Code			
Dirt built-in			
Pressure filter			
Bird	Air	Fish	Water

absorption	suction filter	pump
synthetic	absolute	ß ratio
directional valve	servo valves	reservoir
4406	dirt added	pump
18/16/13	particles 5µm	wire mesh
average	centrifugation	16/14/11
new components	surface element	largest
particles 2µm	multipass test	particles 15µm

Chapter 12

Fluid Conductors

Introduction

This chapter is concerned with the conductors which carry the fluid through a system and with their selection, assembly and installation.

To enable the reader to identify the different fittings used and the proper assembly and installation of these components to achieve a leak-free system.

Leakage must be eliminated in hydraulic systems

Leakage has to be eliminated in the hydraulic system because it increases the cost of operating the system by the loss of fluids and slippery floors are dangerous.

Throughout the year, plants all over the world waste millions of liters *(gallons)* of fluid. In addition to fluid cost, the cost of reworking the system and constant maintenance increase your overall operating cost.

Beyond operating cost, you must consider lost business by opening doors to foreign products that may be more leak free; by opening doors to competing technologies; by missing delivery dates due to downtime caused by leakage, etc.

Reasons for leakage

Leakage in a hydraulic system occurs because of four main reasons:

1. poor system design
2. substandard component quality
3. improper installation
4. abuse

Each one of these areas will be looked at in detail.

System design: Component selection-fitting style

Eight basic fitting connection types (four tube/hose connections and four port connections) can be chosen from for installation in your hydraulic system.

Tube/hose connections:

- 37° flare fitting (SAE J514)
- inch flareless fitting (SAE J514)
- o-ring face seal (O.R.F.S.) (SAE 1453)
- metric flareless fitting (ISO 8434-1).

Port connections:

- pipe thread (NPTF)
- SAE straight-thread (SAE J1926)
- SAE/ISO four bolt-flange (SAE J518)
- metric straight thread (ISO 6149).

37° flare (for metric or inch tubing)

The 37° flare fitting (fig. 12-1) is the most widely used type of fitting. This fitting has been used in industry for quite some time, so many people are familiar with it. This fitting is designed to be used in systems with average operating pressure of 210 bar *(3000 psi).* Also, it is used with tubing which has a low to medium wall thickness. Thick wall tubing is very difficult to flare so it is not recommended with a flare fitting.

The 37° flare fitting is suitable to be used in systems which operate between -55°C and 250°C *(-65°F and 500°F)* with carbon steel tubing. The flare fitting occupies less space than most other fittings and it can easily be adapted to metric tubing.

This fitting is readily available and is one of the lowest in installed cost. The nut draws the sleeve towards the flare and causes a positive seal between the flared tube face and the fitting body.

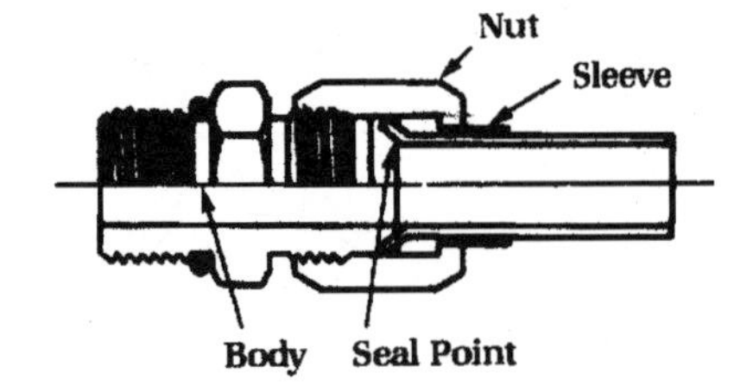

Fig. 12-1 37° flare fitting

Flareless (for inch tubing)

The flareless fitting (fig. 12-2) is very popular in certain markets in the United States. The flareless fitting is suitable for use in both hydraulic and pneumatic systems and has an average working pressure to 210 bar *(3000 psi).* The flareless fitting must be used on either medium or heavy walled tubing because when the ferrule is properly seated, it penetrates into the tube wall.

The flareless fitting requires minimal tube preparation. The nut forces the sleeve against the tapered seat, causing the front edge of sleeve to bite into the tube and creating a positive seal.

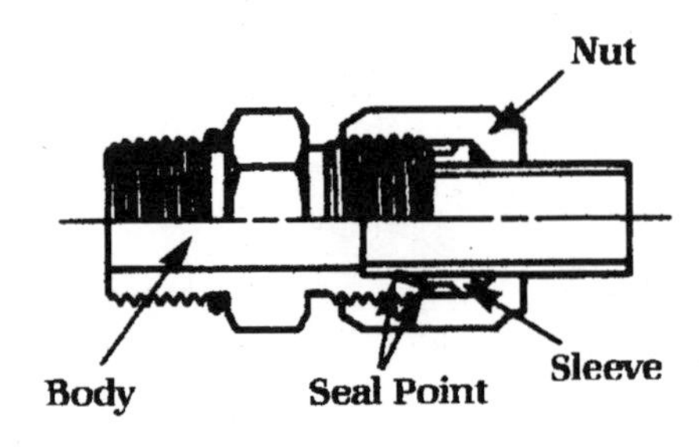

Fig. 12-2 Flareless fitting

Multiple bite flareless (for metric tubing)

The multiple bite fitting has a ferrule with two cutting edges (fig. 12-3). One bite, the leading edge bite, is visible in the front; you can't see the second bite.

This double bite gives the fitting a higher pressure capacity. The operating principle is similar to the fitting with a single bite ferrule.

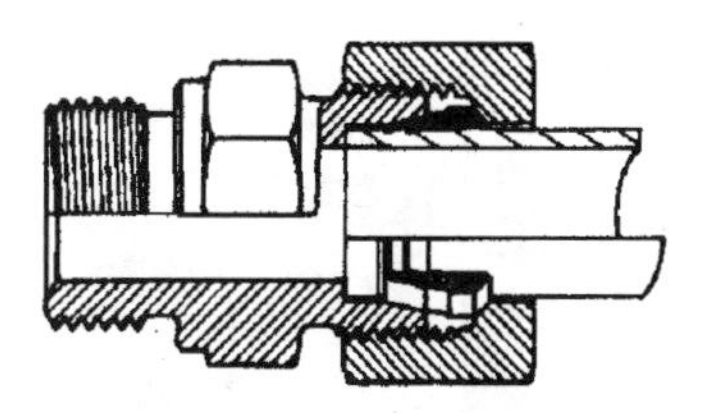

Fig. 12-3 Multiple bite, flareless fitting

Face seal with o-ring (for metric or inch tubing)

The o-ring face seal fitting (fig. 12-4) is best suited for hydraulic and pneumatic systems and it can be used on any wall thickness tubing. Also, it can be used in systems which operate up to 415 bar *(6000 psi),* and in systems which are subjected to high frequency vibrations.

After the sleeve is attached to the tube by brazing or flanging*, the nut is tightened to the body. The o-ring provides a positive seal between the sleeve and the body. The fitting can be repeatedly assembled and disassembled without affecting the seal. This fitting can easily be adapted to metric fastening tubing by using an adapter sleeve.

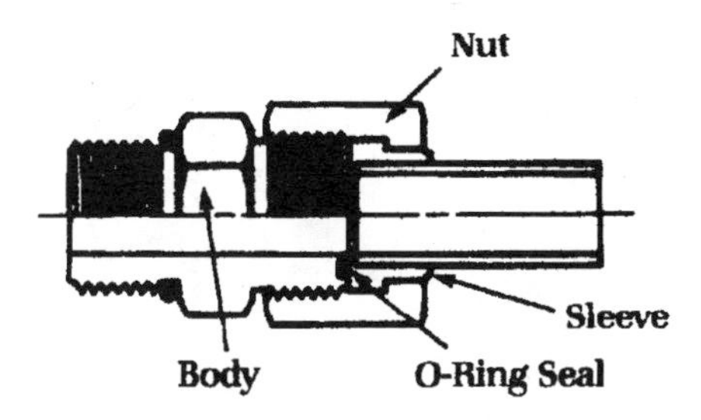

Fig. 12-4 O-ring face seal fitting

**Flanging is a faster, cleaner and less expensive way for attaching a sleeve to the tube.*

Pipe thread port connections

The pipe thread fitting and the mating part (fig. 12-5) have tapered threads which rely on the force generated by the taper to provide a seal. It requires a thread sealer to assure a positive seal. Also, this fitting is prone to loosening and/or cracking when there is vibration and temperature cycling in the system.

NOTE: Thermal expansion and contraction varies with different metals which will cause the threads to loosen.

Because the threads are tapered, repeated assembly and disassembly can cause the threads to distort, leading to leakage. If the fitting is used in a cast iron port, overtightening can cause the port boss to crack.

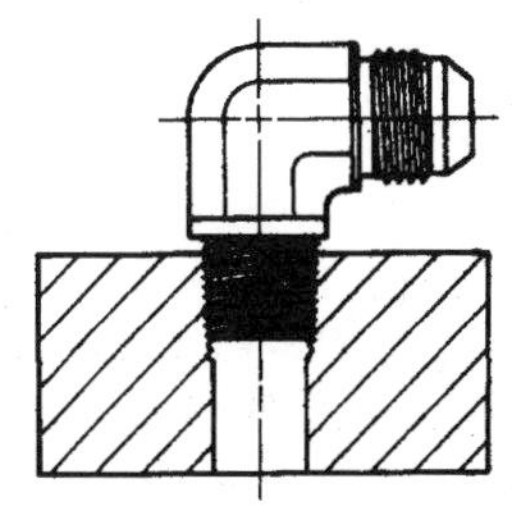

Fig. 12-5 Pipe thread connection

Straight thread with o-ring (inch and metric threads)

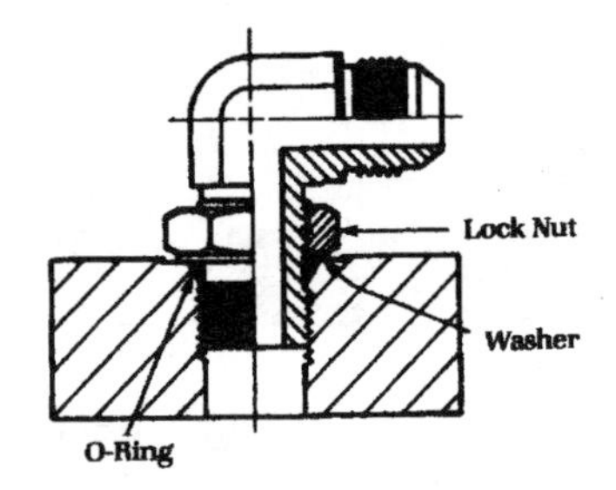

Fig. 12-6 Straight thread o-ring connection

Straight threads are preferred over pipe threads because they rely on an o-ring to provide a positive seal. The o-ring seats on a spot face (fig. 12-6) or in a cavity around the port; a jam nut locks the fitting into position.

This assembly is much less susceptible to vibrational and thermal loosening than pipe threads. Elbows and tees can be positioned easily because the jam nut on an adjustable fitting can be tightened with the fitting in any position. The fitting can be assembled and disassembled repeatedly without causing damage to the seal.

The chance of contamination is minimal with the o-ring seal because there is no need for thread sealing compounds to create the seal, as is the case with pipe threads.

Four-bolt split flange

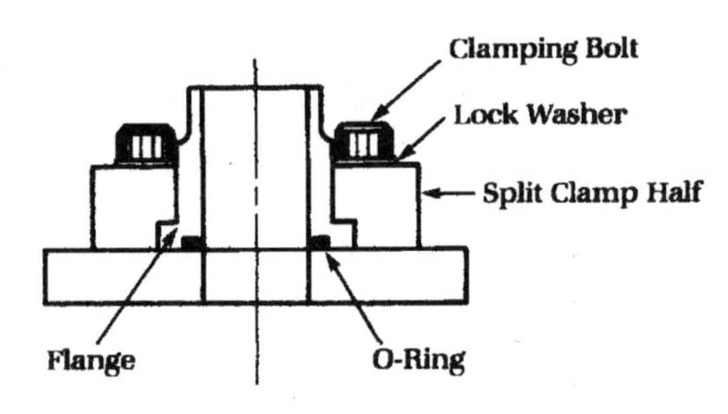

Fig. 12-7 Four-bolt split flange connection

The four-bolt split flange fitting is a face seal type connection (fig. 12-7) especially suited for large port sizes and high pressure systems. It is good for use in hard to reach areas and it has all the same advantages as a straight thread o-ring connection.

Even with these advantages, the four-bolt split flange does take up a lot more space on the component it is mounted onto, installation is more time consuming, and careless torquing of the mounting bolts can cause other problems than just leaks.

Component selection - tube type

Once the type of fittings to be used in your hydraulic system have been selected, you must select the type of tubing to go with the fittings. Not all tubing is suitable for use with all fittings.

There are three major types of tubing available for steel:

1. Welded flash controlled (SAE J356) - It is ***not*** suitable for flared fittings and only marginally suitable for flareless. It is best suited for brazed on face or flanged seal fittings.
2. Welded and drawn (SAE J525) - is suitable for all styles of fittings.
3. Seamless (SAE J524) - works well with flareless and face seal fittings, but is ***not*** as good as welded and drawn (J525) for flared

fittings because of occasional concentricity problems.

Component interchangeability

One of the main criteria for component interchangeability is their design (dimensional) standard. These standards need to be coupled with performance standards to assure trouble-free hydraulic systems.

Such standards as published by ISO, SAE, NFPA, and ASTM are all aimed at providing safe and effective component interchangeability.

System design - routing

When routing piping runs, maintaining joint accessibility is important for proper tightening and servicing of the joints and leakage prevention.

Avoid excessive strain

Although the shortest distance between any two points is a straight line, you may find that because of machine movement, pressure surges, vibrations and temperature changes, a straight piece of tubing can place an excessive amount of strain on the fittings.

By bending the tube on short runs (fig. 12-8), the strain is lessened or even eliminated, thus preventing leaks at the fittings.

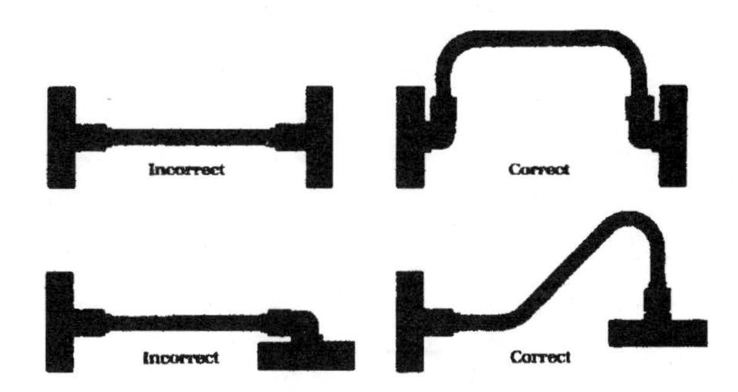

Fig. 12-8 A tube bend lessens strain

Use proper clamping

The use of proper clamping to dampen vibration reduces potential leakage at joints. It is recommended that clamps with resilient inserts be used as illustrated.

When long runs are required, clamp supports should be positioned at various distances along the run according to the tubing outer diameter size (fig. 12-9).

Recommended clamp spacing

Tube OD *(in)*	Spacing
$^1/_4$" – $^1/_2$"	0.9 m *(3 ft)*
$^5/_8$" – $^7/_8$"	1.2 m *(4 ft)*
1"	1.5 m *(5 ft)*
$1^1/_4$" and up	2 m *(7ft)*

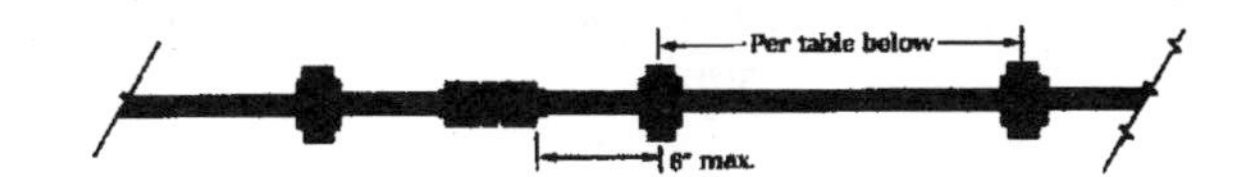

Fig. 12-9 Proper clamping is important

Also, in long runs of tubing you must take into consideration expansion and contraction of the tubing due to pressure surges and temperature variations.

To compensate for this, it is recommended that a 'U' bend or hose be placed in the line (fig. 12-10).

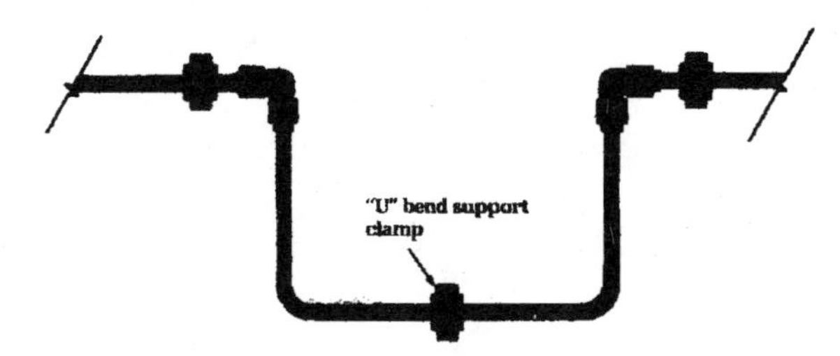

Fig. 12-10 Long lines may require 'U' bends

Allow for movement under load

Another consideration when routing hydraulic systems is to anticipate a certain amount of movement by a system which is under load. For example, a design must acknowledge that cylinders under pressure and temperature changes elongate and contract.

In addition, flexing and rocking makes the mounting method important. A foot-mounted cylinder may triple or quadruple leakage problems because of swinging motion of the heads (fig. 12-11).

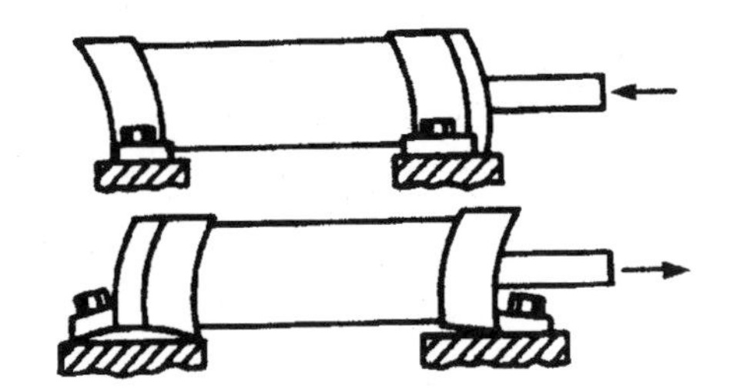

Fig. 12-11 Foot mounted cylinders can develop leakage.

Cylinders with non-centerline-type mountings tend to change length and sway under load and temperature change (fig. 12-11). Any rigid tubing connected to the cylinder heads will be subject to the resulting forces and motion.

Leakage at the port threads (fig. 12-12) is inevitable, regardless of whether they are tapered pipe or straight thread ports. An 'S' bend in the tubing, as previously illustrated in fig. 12-8, is recommended to minimize the strain on fittings.

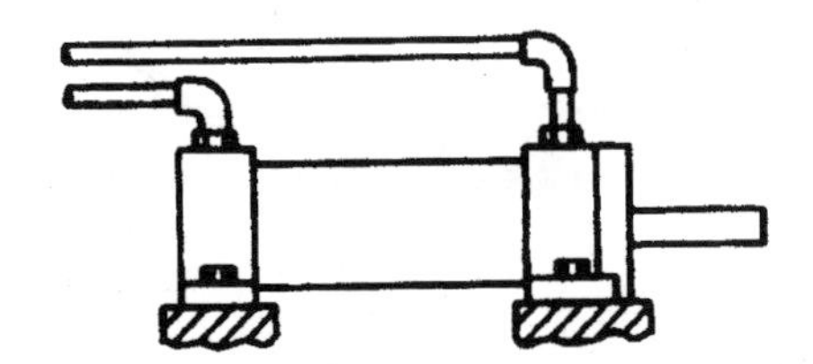

Fig. 12-12 Without countermeasures, port thread leakage can be expected.

Cylinders with non-centerline mountings often require stronger machine members to resist bending, so consider the rigidity of the machine frame. Where one end of a cylinder must be overhung, an additional supporting member should be provided (fig. 12-13).

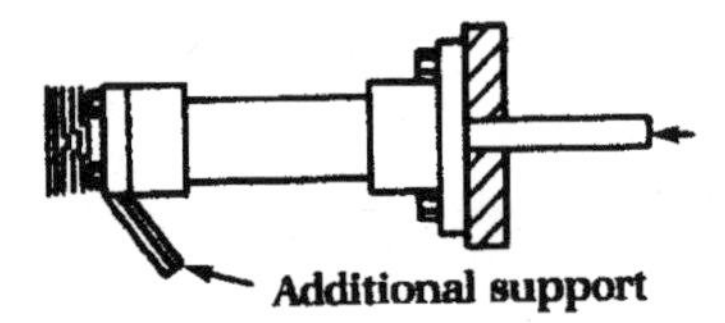

Fig. 12-13 An overhung cylinder may require a support.

Component quality - fitting quality

Check for nicks and burrs on the surfaces and threads as they will cause thread binding and poor sealing.

If you have selected fittings that use o-rings, be certain to check the condition of the o-ring as well as the contact areas of the fitting. Cracks or nicks in the o-ring will decrease the entire fitting's ability to prevent leaks.

Tubing quality

Proper size and roundness along with ductility are critical areas for all types of fittings. Surface quality of the outside dimension is very important for the flareless fittings while interior surface quality and concentricity are important for 37° flare fittings.

Face seal fittings are more tolerant of inside and outside surface imperfections.

Mating ports quality

Inspection of the ports (fig. 12-14) should be done to insure that they are located in such a manner as to provide sufficient strength and surface area for the fitting to seal properly.

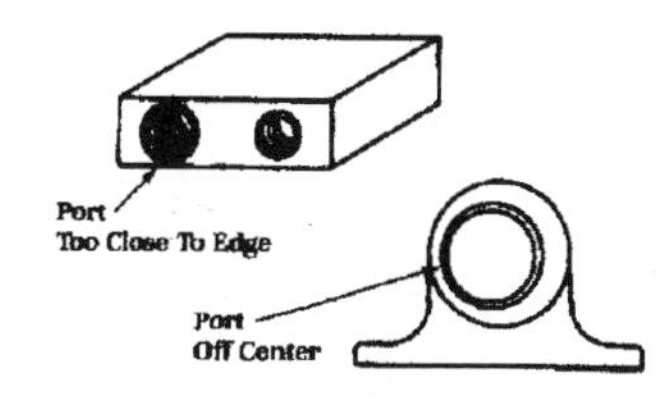

Fig. 12-14 Port threads must be adequately supported.

Installation – tube bends

The assembly of tubing and fittings is critical to leakage elimination within a hydraulic system. This part of the chapter deals with tube preparation for the types of fittings discussed earlier and proper assembly techniques for fittings attached to tubing and ports.

Proper tube bending is essential in optimizing hydraulic system efficiency. Poorly bent tubes (fig. 12-15) which are pried into place cause unnecessary stress on the fittings making them prone to leakage.

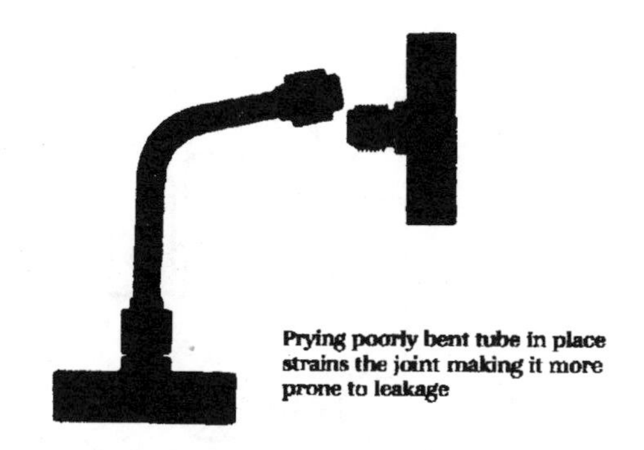

Fig. 12-15 Tubing should fit properly when being assembled.

Tube cutting

One of the first steps in preparing tubing for assembly to the fitting is to cut the tubing end square. To do this, two methods can be used.

The recommended method is to use a power cutoff saw using a toothed cutoff wheel. The other method is use of a hacksaw and saw guide. A properly cut tube is necessary for all fittings. As illustrated in fig. 12-16, the maximum allowable angle for a cut tube end is 2°.

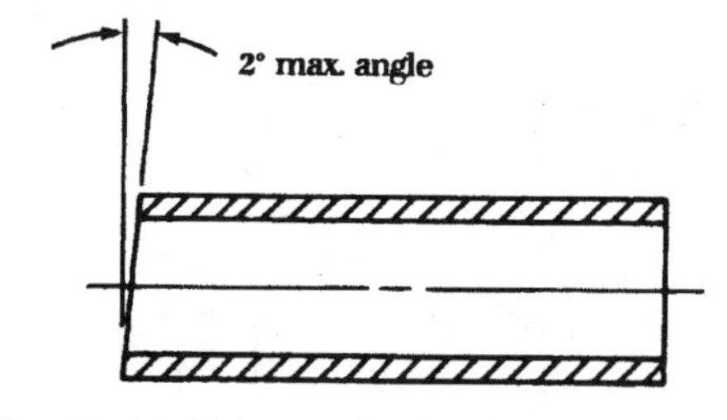

Fig. 12-16 Tube ends should be cut square

Tube flaring

When using 37° flared fittings, the tube end must be flared properly. Flaring can be accomplished by using a suitable flaring tool. The flare should be 37° ±2° as shown in fig. 12-17.

The flared tube should be checked for proper size by sliding the outside sleeve into place. The outside flare diameter should not exceed the outside sleeve diameter; likewise, it should not be less than the inside taper of the sleeve.

The inside surface of the tube flare should also be checked for imperfections such as burrs, nicks, cracks and splits as they will reduce the fittings ability to seal properly.

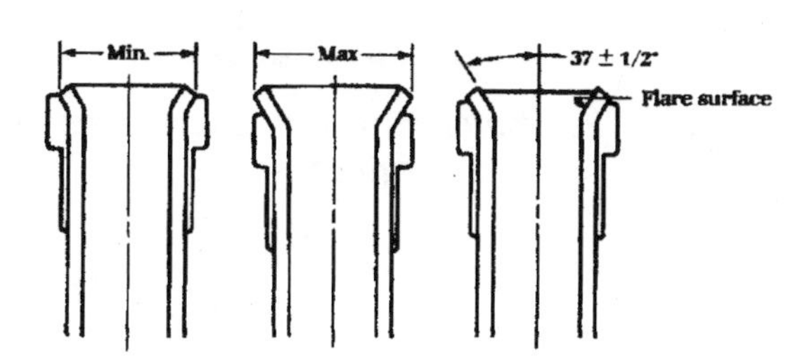

Fig. 12-17 Proper tube flaring

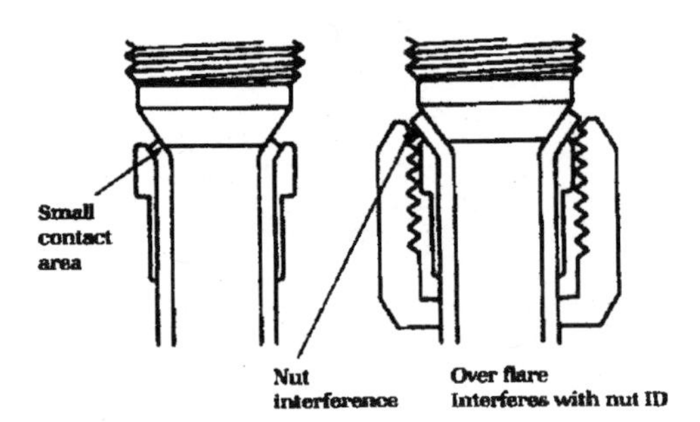

Fig. 12-18 Under- and overflaring

Under and overflaring

Improper flaring can result in a condition of underflaring or overflaring (fig. 12-18). Underflaring will result in insufficient surface area for contact between tube and fitting, while overflaring can interfere with the fitting nut.

Underflaring can result in nose collapse and leakage, and overflaring poses fitting assembly problems.

Flareless preset

The use of flareless fittings does not require flaring of the tube end. To be effective, care must be taken when presetting (attaching) the ferrule to the tube.

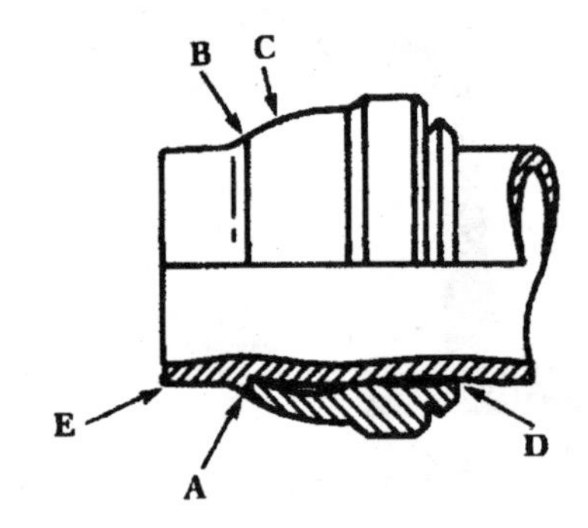

Fig. 12-19 Flareless fitting pre-set

A proper preset should have the following characteristics (refer to fig. 12-19):

A. ridge on the tube raised to at least 50% of the ferrule's front edge thickness
B. the leading edge of the ferrule coined flat
C. a slight bow formed at remaining part of barrel.
D. the back end of ferrule should be snug against tube.
E. there should be a slight indentation 360° around the end of the tube caused by the tube being bottomed out during assembly.

Improper flareless preset

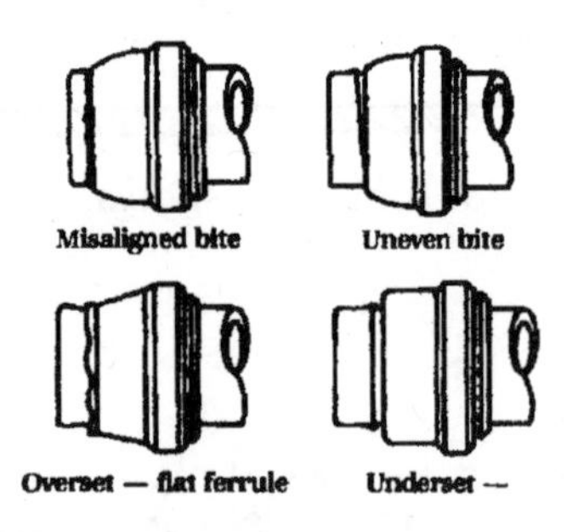

Fig. 12-20 Improperly preset and misaligned ferrules

Fig. 12-20 shows four improperly preset or misaligned ferrules; all of them can result in a leaky joint.

- An **uneven bite** occurs when the ferrule fails to seal along its original cut in the tube.
- A **misaligned bite** occurs when the ferrule is cocked when secured.
- **Overset** or a flat ferrule is caused by overpressuring or overtightening the nut during preset.
- **Underset** is caused by low pressure or undertightening the nut during preset.

Face seal brazing

The face seal with o-ring provides the highest degree of leak resistance when assembled properly. It is necessary for the sleeve of the face seal to be properly brazed to the tube or leakage may occur (fig. 12-21).

When a sleeve is properly brazed to a tube, these characteristics should exist:

- sleeve should be positioned onto the tube with no more than 1.5 mm *(0.060")* gap; preferably zero gap should exist
- sleeve face must be perpendicular to center line of tube (90° ±1°)
- there should be no braze overflow on the face of the sleeve
- the braze fillets should be visible all around the tube.

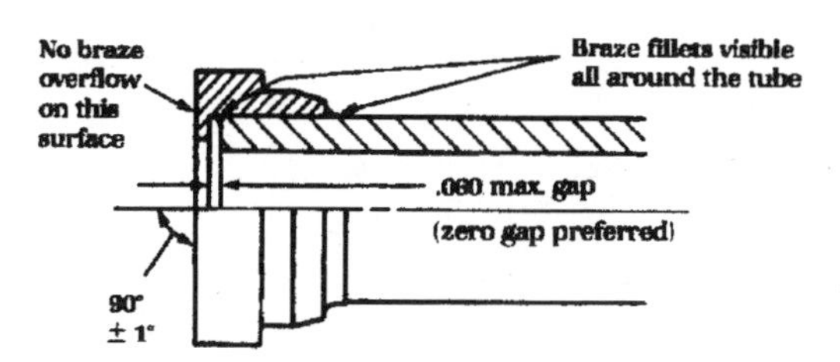

Fig. 12-21 Face seal sleeve brazing

Improper brazing

An improper braze leads to a leaking joint. Should leakage occur, disassemble joint and check for:

- improper placement or misalignment of the sleeve over the tube
- poor braze joint caused by improper cleaning and/or fluxing, uneven or insufficient heat
- braze overflow on the sealing surface.

Assembly of tube to 37° flare fitting – the torque method

This part of the appendix deals with the various methods of assembly for each of the fittings described earlier.

The torque method is used to properly tighten 37° flared fittings. Once the tube and fitting are aligned, tighten the nut to the torque recommended by the manufacturer.

Table 12-1 shows representative torque values for 37° flared fitting for dry torquing. Undertorquing will result in poor contact between the flare and fitting nose and overtorquing can damage the flare or sleeve; either condition may result in a leaky joint.

The size of the fitting is given as a dash size which indicates the tube size in 16ths. For example, size 4 indicates $^{4}/_{16}$" or $^{1}/_{4}$" inch tube OD.

Assembly of mixed platings on 37° flare fittings

The plating combination between the nut and the body changes the friction between the two surfaces. For a given torque, the amount of turns will vary.

As illustrated in fig. 12-22, a $^{3}/_{8}$" flared fitting requires 30 Nm *(275 lbf in)* of dry torque for proper seating. For the prevention of leakage when both

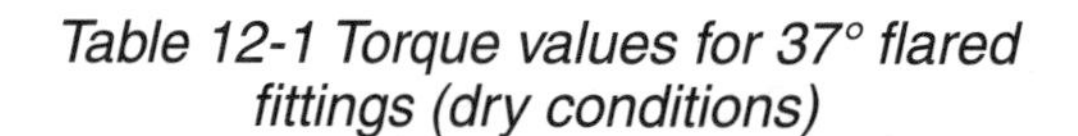

Table 12-1 Torque values for 37° flared fittings (dry conditions)

Size (in 16:ths)	Steel torque *[lbf ft]*	Stainless steel torque *[lbf ft]*
*2	3.3 ± 0.4	4.2 ± 0.4
*3	5.8 ± 0.8	6.7 ± 0.8
4	11.7 ± 0.8	13.3 ± 0.8
*5	15 ± 1.3	19 ± 1.3
6	21 ± 1.3	27 ± 2
8	39 ± 2	52 ± 2
10	54 ± 4	67 ± 4
12	88 ± 4	108 ± 4
*14	104 ± 4	125 ± 4
16	120 ± 4	140 ± 4
20	190 ± 8	210 ± 8
24	250 ± 12	285 ± 16
32	315 ± 16	365 ± 16

* Estimated

NOTE: 1 lbf ft equals approx. 1.36 Nm.

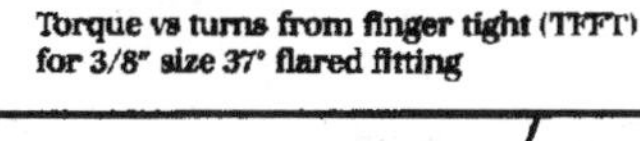

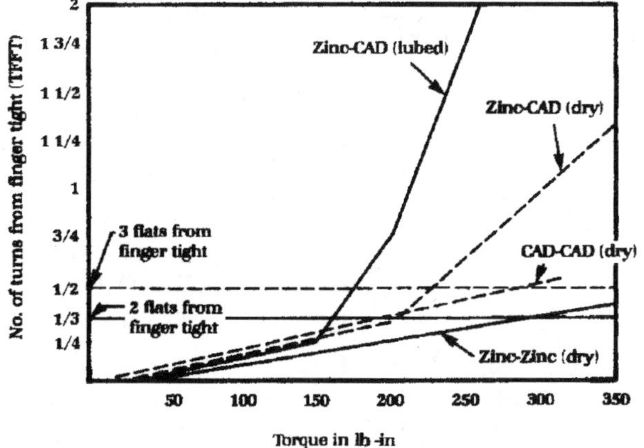

Fig. 12-22 Face seal sleeve brazing

the nut and body are cadmium plated, this torque produces a rotation of 3 flats from finger tight (refer to the 'FFFT' method below). The same size fitting using different platings such as zinc-cad (dry) must be turned more than twice as many flats to achieve the same 30 Nm *(275 lb in)* of torque.

Although both plating combinations are torqued to equal value, the amount of crush created by the zinc-cad combination could damage the fitting leading to leakage. It is, therefore, recommended that plating material be the same between body and nut. Check with the manufacturer for type of plating used.

Assembly of tube to 37° flared fitting - the 'flats from finger tight' method

The following, alternate method of assembly is recommended if the plating combination of 37° flare fitting components is not known. It is called the 'flats from finger tight method' or FFFT. This method requires that the tube and fitting are aligned and tightened by hand until some metal to metal contact causes a resistance to turning.

The nut and body hex should be marked for an initial position (fig. 12-23). Tighten the nut further until it rotates through the number of flats. If desired, you may also mark the final position for later remakes or to provide an indication of proper initial assembly.

Subsequent remakes will require less turns to produce a positive seal.

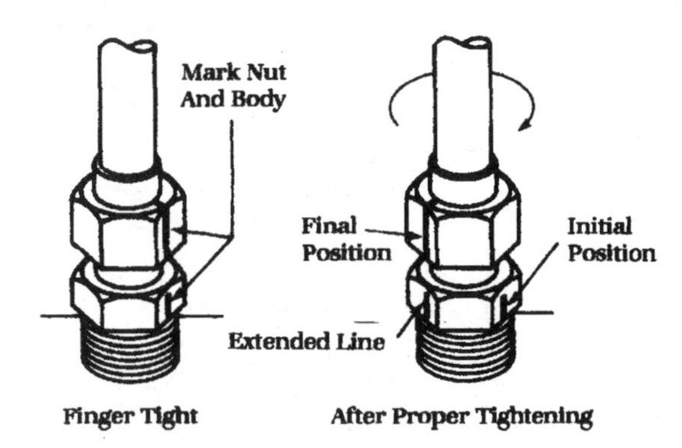

Fig. 12-23 The 'FFFT' method

Table 12-2 'Flats from finger tight' method for 37° flared fittings

Size (in 16:ths)	Number of flats to be turned Steel	Stainless steel
*2	2.25 ± 0.25	2.00 ± 0.25
*3	2.25 ± 0.25	2.00 ± 0.25
4	2.25 ± 0.25	2.00 ± 0.25
*5	2.25 ± 0.25	2.00 ± 0.25
6	2.25 ± 0.25	2.00 ± 0.25
8	2.25 ± 0.25	2.00 ± 0.25
10	2.00 ± 0.25	2.00 ± 0.25
12	2.00 ± 0.25	2.00 ± 0.25
*14	2.00 ± 0.25	2.00 ± 0.25
16	2.00 ± 0.25	2.00 ± 0.25
20	2.00 ± 0.25	2.00 ± 0.25
24	2.00 ± 0.25	2.00 ± 0.25
32	2.00 ± 0.25	2.00 ± 0.25

* Estimated

Flats from finger tight (FFFT) table

The 'flats from finger tight (FFFT) method' requires turning the nut a specific number of flats. Table 12-2 shows the number of flats required at initial assembly. The size of the fitting is found in the code. Examples: A '-4' fitting is a $^1/_4$" fitting, a '-32' is a 2" fitting (sizes are incremented in 16ths).

Torque and FFFT methods compared

A comparison of the torque versus flats method describes the advantages and disadvantages of the two methods.

The torque method is fast, but it is difficult to visually inspect for proper tightening. The FFFT method is slower but provides consistent preload, and visual inspection is easy because of markings on the fitting.

Assembly of a flareless fitting

The flareless fittings require a minimum of instructions. You only need to align the tube and fitting, tighten the nut until finger tight, and turn an additional $^1/_2$ turn. It is important not to overtighten the nut or damage to the ferrule will occur.

On sizes larger than -8 ($^1/_2$" OD tube), it is desirable to preset the bite sleeve on a presetting machine prior to final assembly in the fitting.

Assembly of a face seal fitting

Assembly of the face seal fitting is also easy. Once aligned you need only to tighten the nut to the proper torque listed in the table.

Assembly of a pipe thread fitting to a port

Port fittings come in two basic varieties, pipe thread ports and straight thread. To assemble pipe thread ports (fig. 12-24) first visually inspect both threads so they are free from nicks, burrs, dirt, etc. Apply sealant and tighten the fitting to the number of turns from finger tight as shown in Table 12-2.

A high quality thread sealant will also retain the fitting and is usually more effective than pipe dope. Teflon tape should not be used in hydraulic systems as parts of it can be torn off during assembly and contaminate the system, causing valve spools to be stuck in the housing, and orifices to be closed.

Assembly of a straight thread fitting to a port

A more effective seal can be achieved by using ISO (metric) or SAE straight thread o-ring fittings.

One such type of fitting is the nonadjustable variety. This fitting is quite easy to assemble (fig. 12-25). First inspect the threads, then lubricate the o-ring with a light coating of hydraulic fluid and tighten to the torque level recommended by the manufacturer, or as shown in the table.

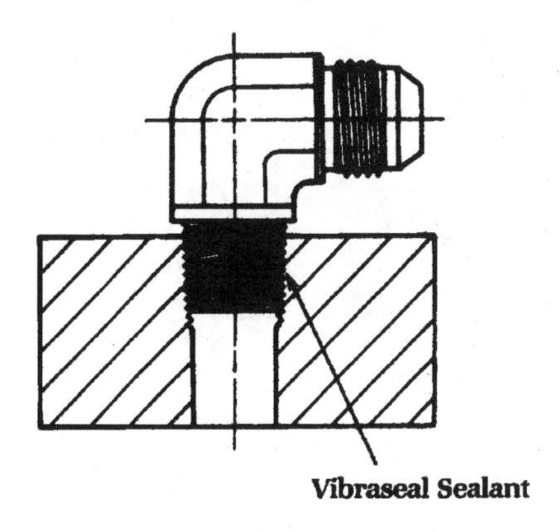

Fig. 12-24 Pipe thread fitting assembly

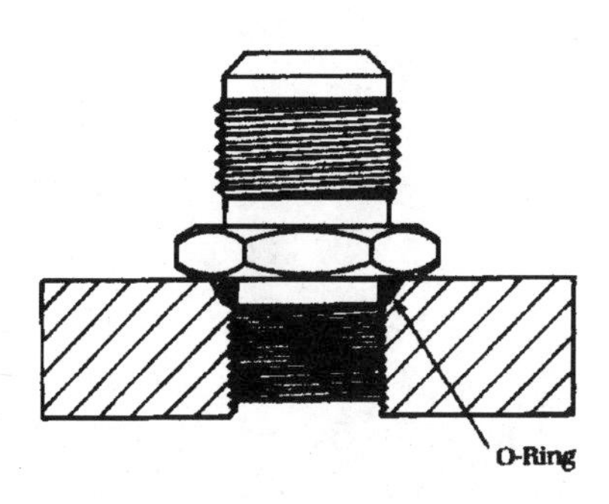

Fig. 12-25 Straight thread fitting assembly

Assembly of an adjustable straight thread fitting

The assembly of an adjustable straight thread o-ring fitting involves slightly more care than the non-adjustable variety. There are six steps that must be followed for proper assembly of adjustable straight thread o-ring fittings:

1. Inspect and correct both mating parts for burrs, nicks, scratches or any foreign particles.
2. Lubricate the o-ring with light coat of oil or compatible fluid.
3. Back off the lock nut as far as possible.
4. Screw the fitting into the port by hand until the backup washer contacts the face of the port.
5. Turn the position of the fitting to the desired position but not more than one full turn.
6. Hold the fitting in the desired position and tighten the lock nut to the torque level listed in table for nonadjustable straight thread o-ring fittings.

Pinched o-rings

Care should be taken to assure that all steps have been properly performed. Improper assembly can pinch the o-ring causing a leaky joint.

Assembly of a four-bolt split flange fitting

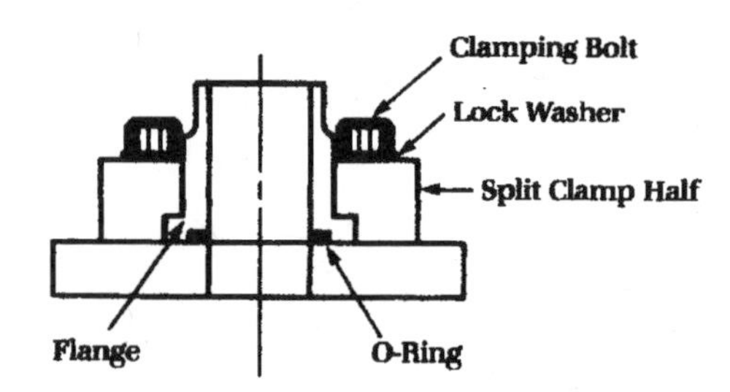

Fig. 12-26 Four-bolt split flange fitting assembly

Fig. 12-26 shows the final assembly of a four-bolt split flange. Note the seal between the port and flange at the o-ring. When assembling four-bolt split flanges, care must be taken to insure that both surfaces are clean and free of burrs or nicks so as not to damage the o-ring during assembly. Position the flanges as shown in the illustration and hand tighten all four bolts.

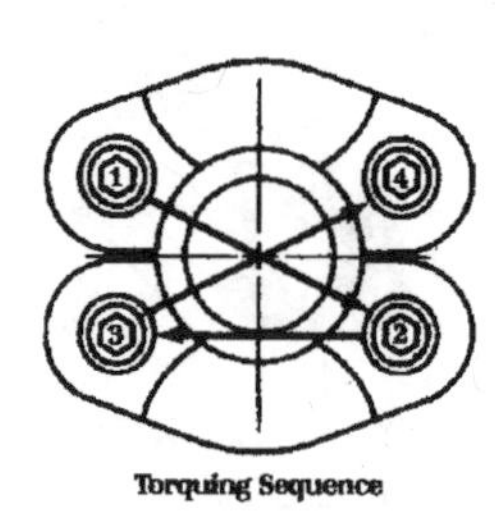

Fig. 12-27 Four-bolt split flange torquing sequence

Tighten the bolts in a diagonal sequence (fig. 12-27) and in small increments to the torque level recommended in Table 12-3.

Four-bolt split flange torque

Table 12-3 shows recommended torque values for four-bolt split flanges. 'SAE Code 61' is used for split flanges in hydraulic systems with working pressures to 210 bar *(3000 psi)*. For smaller sizes Code 61 is valid up to 350 bar *(5000 psi)*.

Table 12-3 Four-bolt split flange torque

Flange size *[in]*	Code 61 Bolt thread	Code 61 Torque *[lbf ft]*	Code 62 Bolt thread	Code 62 Torque *[lbf ft]*
1/2	5/16-18	15 - 19	5/16-18	15 - 19
3/4	3/8-16	21 - 29	3/8-16	25 - 33
1	3/8-16	27 - 35	7/16-14	42 - 50
1 1/4	7/16-14	35 - 46	1/2-13	63 - 75
1 1/2	1/2-13	46 - 58	5/8-11	117 - 133
2	1/2-13	54 - 67	3/4-10	200 - 217

Note: 1 lbf. ft. equals approx. 1.36Nm.

'SAE Code 62' is valid for systems to 415 bar *(6000 psi).*

Troubleshooting fitting failures

The procedure for troubleshooting various types of fitting assembly installations for leakage is:

1. Determine where the leak location is
2. Check the joints for proper tightness
3. If a joint continues to leak, check for correct assembly and possible remake of the joint.

Abuse failures

Leakage in a hydraulic system can often be traced to abuse. Tube and fitting damage can be caused by any of the following:

- Handling and storage. Threads and sealing surfaces can be damaged through rough handling. **DO NOT** remove the protective caps and plugs while putting parts in storage.
- Overtorquing. To think that 'if a fitting leaks, tighten it a little more' is not always correct. Overtightening, more often than not, distorts parts and causes leakage.
- Using tube lines as structural supports, ladder rails, etc., puts excessive strain on joints, causing leakage.

Recommendations for a leak-free system

In order to achieve a leak free hydraulic system follow these recommendations:

1. Think zero leakage from the very start; design for zero leakage.
2. Design in provisions for tube length variations and component movement due to hydraulic loads or changes in temperature.
3. Use proper clamping to dampen vibrations.
4. Use ISO (metric thread M33x2) or SAE straight thread ports for sizes up to 25 mm *(1")* and four-bolt flanges for larger sizes.
5. Specify the most cost effective fitting that meets your requirements best.
6. Specify ISO or SAE standard tubes, tube fittings, hose adapters and hose ends. Add performance requirements where missing
7. Insist on ISO or ANSI standard connections on all equipment, domestic and foreign.

8. Insist on quality components. Buy from reputed suppliers that can provide the necessary technical support including training.
9. Avoid indiscriminate mixing of components from different suppliers.
10. Train assembly personnel on a regular basis.
11. Use cylinder mounts that take thrust on their centerline.
12. Allow for component changes due to external force, pressure and/or temperature changes.
13. Remember that a well designed system put together with quality components by properly trained personnel using proper tools and procedures is the least costly system in the long run.

Quick couplings

Quick disconnect couplings are used for ease of rapidly and repetitively connecting and disconnecting fluid lines without the use of tools. One half is normally attached to a flexible line.

Quick disconnect couplings consist of two mating parts: the female half (sometimes referred to as the "socket" or "coupler body") and the male half (often referred to as the "plug" or "nipple body").

Couplings are locked with detented balls, pins, dogs, pawls, cams or screw threads. The ball type lock is the most widely used.

The three most popular types of quick disconnect couplings are the single shut-off, double shut-off, and straight-thru.

Single shut-off

Single shut-off couplings (fig.12-28) are designed basically for use in pneumatic systems using standard shop air at 21 bar *(300 psig)* or less. The female half includes a shut-off valve, but the male half does not, hence the term "single shut-off."

The female half should be installed on the upstream (supply) end of the line to shut off the air supply when the coupling is disconnected.

There are many types and styles of popular interchangeable designs available.

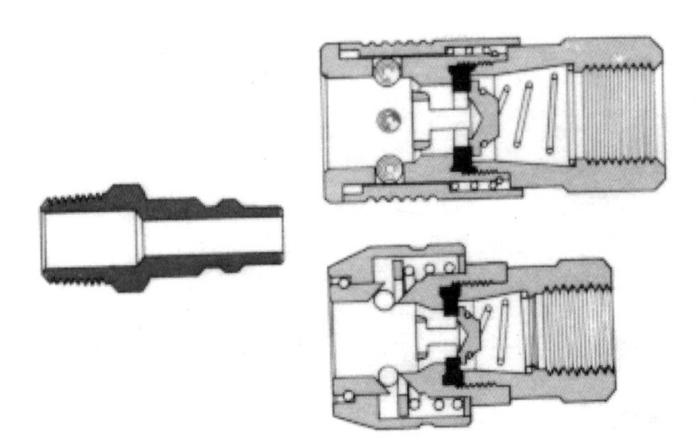

Fig. 12-28 Single shut-off

Double shut-off

Often referred to as hydraulic couplers, the double shut-off (fig.12-29) designs have shut-off valves in both male and female halves.

Since it includes positive shut-off in both halves, the double shut-off type virtually eliminates fluid loss upon disconnecting; thus, this type is ideal for applications on lines carrying liquids.

Quick disconnects using precision steel balls as a valving means are popular on agricultural and mobile equipment for remote hydraulic lines. The steel ball offers long and reliable service for these rugged applications.

Several designs are available for both "open" and "closed" center hydraulic systems, which allow for connecting under pressurized conditions.

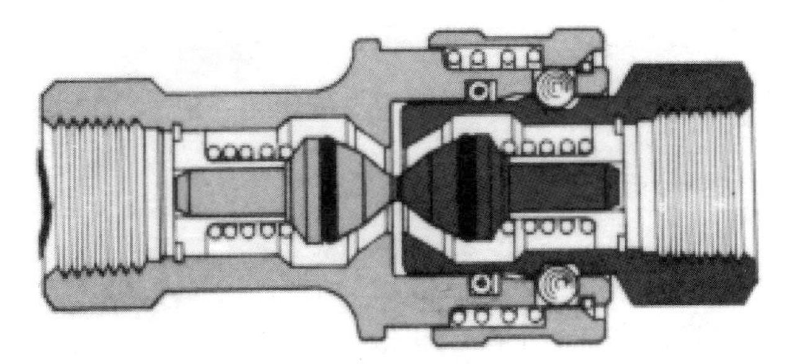

Fig. 12-29 Double shut-off

Straight-thru

Since there is no shut-off valve in either coupling half, the straight-thru coupling (fig. 12-30) provides free, unrestricted flow with very little pressure drop.

As the straight-thru coupling lacks a means for shutting off flow automatically upon disconnect, a shut-off valve should be installed in the line.

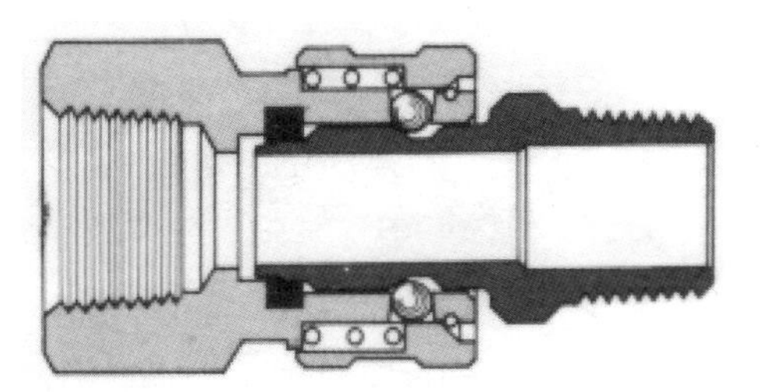

Fig. 12-30 Straight-thru fitting

Coupling materials

Standard materials are brass, aluminum, steel and stainless steel.

Steel, carbon alloy

Applications: General purpose use and high pressures for maximum wear and strength.

Brass

Applications: Non-magnetic, non-sparking, fresh water marine service, high temperature and water systems.

Stainless steel

Applications: Corrosive material handling in chemical, pharmaceutical, beverage and food processing. Excellent for salt water marine applications.

Coupling seal material

Many seal materials are used in quick couplings to help assure "leak free" shutoff. The four most common are: Fluorocarbon Rubber, Nitrile, Ethylene Propylene Rubber and Chloroprene Rubber.

Fluorocarbon rubber (FPM)

FPM has a working temperature range of -29° to +204°C *(-20° to +400°F).*

FPM is recommended for:
- petroleum oils
- di-ester base lubricants
- silicate ester base lubricants
- silicone fluids and greases

Nitrile or Buna N (nbr)

The working temperature range of Nitrile is -54° to +121°C *(-65 to +250°F).* This material (nbr) is recommended for:
- general purpose sealing
- petroleum oils and fluids
- water
- silicone greases and oils
- di-ester base lubricants
- ethylene glycol base fluids

Ethylene propylene rubber (EPM)

EPM is a material with a larger working temperature range, -54° to +149°C *(-65° to +300°F).* EPM material is recommended for applications involving:
- phosphate ester base hydraulic fluids (Skydrol, Cellulube, Phydraul)
- steam (to +204°C/*+400°F*)
- water
- silicone oils and greases
- diluted acids
- diluted alkalies
- ketones
- alcohols
- automotive brake fluids

Chloroprene rubber (CR)

CR, with a working temperature range of -54° to +149°C *(-65° to +300°F)*, is recommended for applications involving:
- refrigerants (Freon, NH^3)
- high aniline point petroleum oils

- mild acid resistance
- silicate ester lubricants

Coupling selection

There are many considerations when selecting quick couplings. Some of these are: system pressure and pressure surges, flow requirements, system media and temperature, environmental and functional considerations.

System pressure and pressure surges

Pressure rating is invariably the most important factor in selecting a quick disconnect coupling, because if the system pressure is higher than the rating of the coupler, it cannot be used. Quick disconnect couplings are pressure-rated according to operating, proof, and burst pressures.

Flow requirements

Pressure drop is usually very important as this factor can drop the overall efficiency of the system, thus increasing heat generation.

System media and temperature

The compatibility of materials and seals with the system media and temperature governs the choice of metallic components and sealing materials.

Environmental and functional considerations

Environmental considerations such as corrosion, maintainability and cleanliness, along with functional characteristics such as forces required for connection and disconnection, fluid loss at disconnect, air inclusion during connection, end terminations, weight and size, vacuum rating, coupling means and type, are all factors affecting the selection of a quick disconnect coupling.

Performance data to assist in the selection of a quick disconnect coupling is available from the manufacturer's specifications.

Hose selection considerations

Choosing the correct hose for a given application requires thorough knowledge of the application. Each application has specific considerations.

However, most applications share several common considerations. These common hose selection considerations include:

- line size
- system pressure
- surge pressure peaks
- temperature range - internal & ambient
- fluid effect on hose material
- service life
- cost
- availability
- installation: Stationary - moving (flexing) - bend radius - routing - amount of slack - use of angle adaptors and end fittings - use of straight thread fittings - use of hose supports - control of twisting - control of abrasion
- ease of maintenance

Additional considerations include the following:

Ambient temperatures

Very high or low ambient (outside of hose) temperatures will affect cover and reinforcement materials, thus influencing the life of the hose.

Bend radius

Recommended minimum bend radius (fig. 12-31) is based upon maximum operating pressures with moderate flexing of the hose. Safe operating pressure decreases when bend radius is reduced below the recommended minimum.

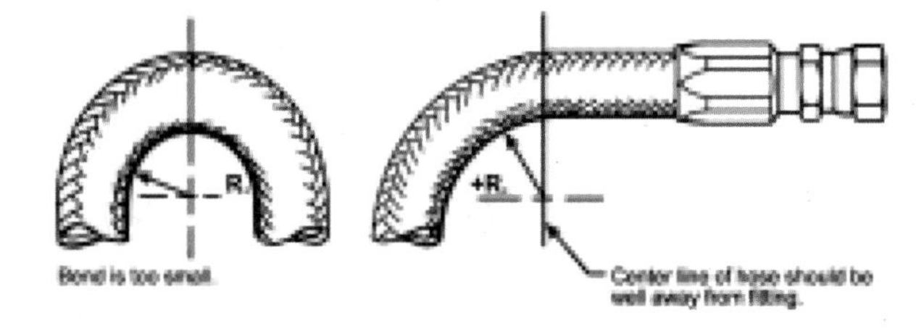

Fig. 12-31 Hose bend radius

Burst pressure

These are test values only and apply to hose assemblies that have not been used and have been assembled for less than one month.

Chemical resistance

Consider chemical resistance of hose cover and tube stock. The flexible materials from which most modern hoses are made are subject to corrosion when the fluid with which they come in contact interact with their particular chemical makeup. The hose liner is the element most likely to be attacked.

The hose cover is also subject to attack from the fluid to which it is exposed. Harmful chemicals may be contained in the environment. If the cover deteriorates, the reinforcement may also be affected. In some cases, the cover serves as protection for the outer surface of the inner tube liner (low pressure hose).

Just as the liner and hose cover must be compatible with the fluid being transmitted and encountered in the environment, so must the fittings.

Charts and general guide lines are available and help to prevent gross misapplication of hose materials. However, there is no substitute for direct contact with the manufacturer to determine if their particular product is right for a given application.

Electrical conductivity

Hose is available for electrically conductive applications. Hose is also available for non-conductive applications such as powerline work. Each must be specified.

Line size

Undersized pressure lines produce excessive pressure drop (ΔP) which results in energy loss and heating. Undersize suction lines can cause cavitation at the pump inlet.

Flow rates desired and operating pressures establish required hose size. Required size of hose depends upon volume and velocity of fluid flow. If fluid velocity is too high, much of the energy is lost in the form of heat. If hose must be a different size from the accessory ports, jump size adapters should be used.

Hose manufacturers identify hose sizes using a dash numbering system. To determine the size of a hose, measure the I.D. (inside diameter) of the hose and convert this measurement to 1/16-inch increments. The dash size coincides with the number of 1/16" increments in the hose I.D. A 4/16" I.D. hose is a -4 size hose (Table 12-4).

Unlike hose, tubing is sized by its O.D. (outside diameter) (fig. 12-32). The number of 1/16" increments in its O.D. represents its dash size . There are exceptions to this, hose built to SAE 100R5 specifications is one.

Operating (working) pressure

Hose lines are rated for continuous operation at the maximum operating pressures specified for

Table 12-4 Dash Sizing Table

I.D.	Dash Sizing	Metric (mm)
1/4	-4	6.3
3/8	-6	10
1/2	-8	12.5
3/4	-12	19
1	-16	25
2	-32	51

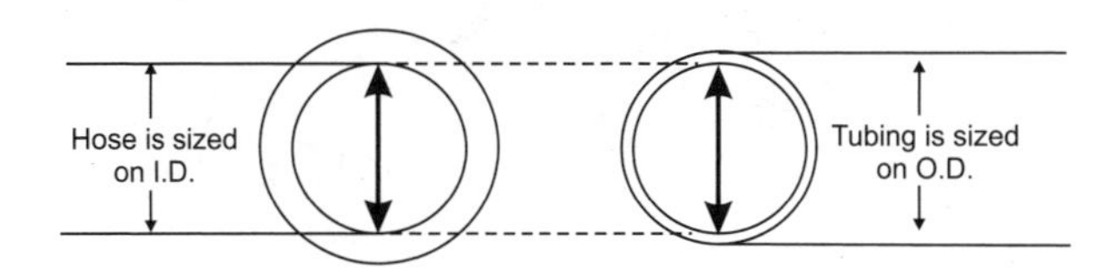

Fig. 12-32 Tube vs. hose sizing

Table 12-5 Hose Assembly Pressure Ratings

Pressure Ratings Hose Assemblies - psi

The maximum dynamic working pressure of the hose assembly is the lesser of the rated working pressure of the hose and the end connections used.

Pressure Rating of Hose End Connections - psi

Hose End Connection Description	Inch Size Fittings									
	-2	-4	-6	-8	-10	-12	-16	-20	-24	-32
Male pipe (NPTF)	12,000	12,000	10,000	10,000		7,500	6,500	5,000	3,000	2,500
Female pipe (NPTF, NPSM)	7,500	7,000	6,000	5,000		4,000	3,000	2,500	2,000	2,000
Pipe (JIS)		5,000	5,000	5,000		4,000	3,000	2,500	1,500	1,500
Seal-Lok		6,000	6,000	6,000	6,000	6,000	6,000	4,000	4,000	
SAE Flanges Code 61				5,000		5,000	5,000	4,000	3,000	3,000

Hose End Connection Description	Metric Size Fittings - mm									
	-6	-8	-10	-12	-15	-18	-22	-28	-35	-42
DIN Metric "L" Series	3,625	3,625	3,625	3,625	3,625	2,320	2,320	1,450	1,450	1,450
DIN Metric "S" Series		5,800	5,800	5,800	5,800	5,800	5,800	5,800	3,625	3,625

Note: 14.5 psi = 1 bar = 10^5 Pa

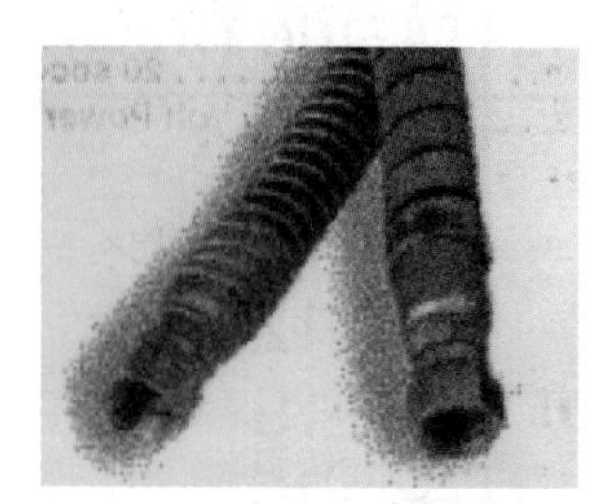

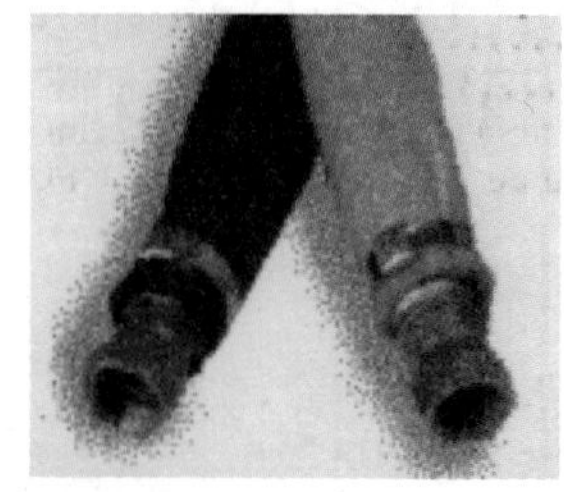

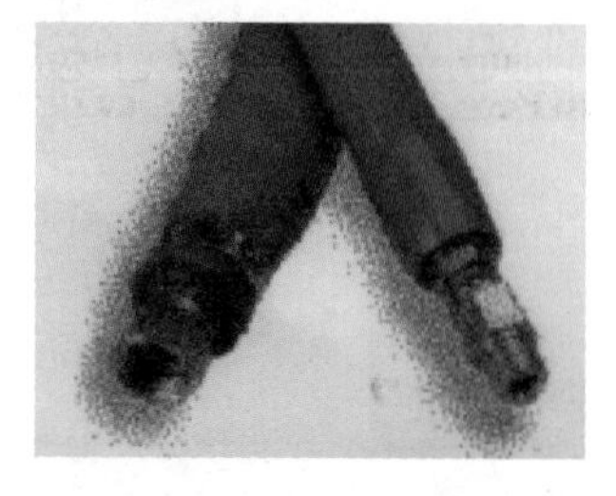

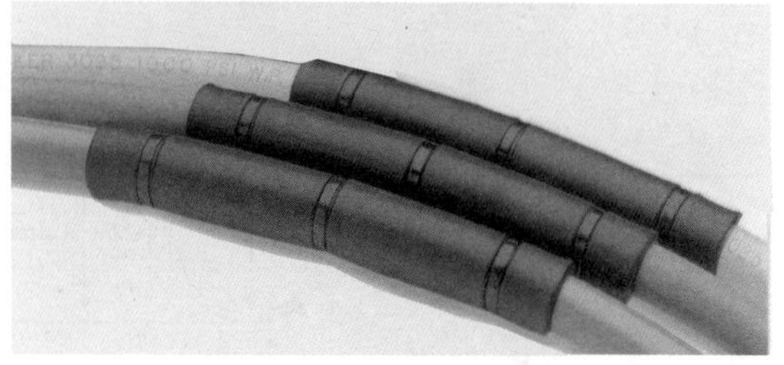

Fig. 12-33 Hose guards

the hose. Never exceed the hose maximum operating pressure.

Hose styles vary in pressure rating according to size and construction. A hose with a small I.D. (inside diameter) will have a higher rating than a hose of similar construction, but larger I.D. (Table 12-5).

Operating temperatures

Operating temperatures specified refer to the maximum temperature of the fluid being conveyed. Continuous operation at or near maximum rated temperatures will materially reduce the service life of the hose. Engineering data concerning critical temperature requirement is available upon request from the manufacturer.

Pressure surges

Many hydraulic systems develop pressure shocks which exceed relief valve settings and affect the service life of all system components. In systems where shocks are severe, select a hose of higher working pressure.

Hose guards

Hose guards (fig. 12-33), such as sleeves, spring guards and shields, prolong the life of hose lines when operating in hostile environments. They distribute bending radii to avoid sharp kinks in hose lines at juncture of couplings and protect the hose from excessive wear and deep cuts. Some hose guards even add a level of fire resistance or protection to the hose.

Hose application considerations

1. Hose should be routed to flex in a single plane. If it must be routed through a compound bend it should be divided into two or more bends by clamping to permit each to flex in a single plane.

2. Clamps should also be used to prevent hose cover abrasion. Abrasion may occur when two hose lines cross, or when a hose line rubs against a fixed point.

3. Sometimes when hose is looped or routed outside the contour of the machine it is used as a convenient hand hold or step which leads to premature hose failure.

4. When a hose is used above its rated working pressure it reduces service life and increases the customer costs by increasing down time and requiring unnecessary replacement. It will also cause accidents.

5. Tension on hose during pressure applications will cause fitting blowoffs. The hose should be the proper length.

6. Protective spring guards should be used whenever a hose is exposed to sharp or abrasive materials. Also, hose lines should be adequately protected from external shock and mechanical or chemical damage. They should also be suitably protected to prevent whiplash action in the event of failure for any reason.

Pressure effects

When flexible hose is subjected to pressure it may change its length as much as +2 to -4%. Installation of lines should compensate for shortening effects by bending the hose or leaving slack in straight runs.

Swivel adapters

- increase hose life
- cut hose costs

Swivel adapters feature 360° swiveling action that especially suits them for use in applications where hose moves, bends or twists (fig. 12-34). Swivel adapters connected to hose assemblies relieve twisting, prevent excessive flexing of hose, and eliminate need for long radius bends.

Results: Longer hose life and reduced equipment downtime.

Hose failure

Twisting is one of the primary causes for hose failure (fig. 12-35). Hose is weakened when installed in twisted position. A 7° twist in a large diameter, flexing hose can reduce service life by as much as 90%. Also, pressure in twisted hose tends to loosen fitting connections.

Vacuum requirements

1. Maximum negative pressures shown for hoses -16 and larger are suitable only for

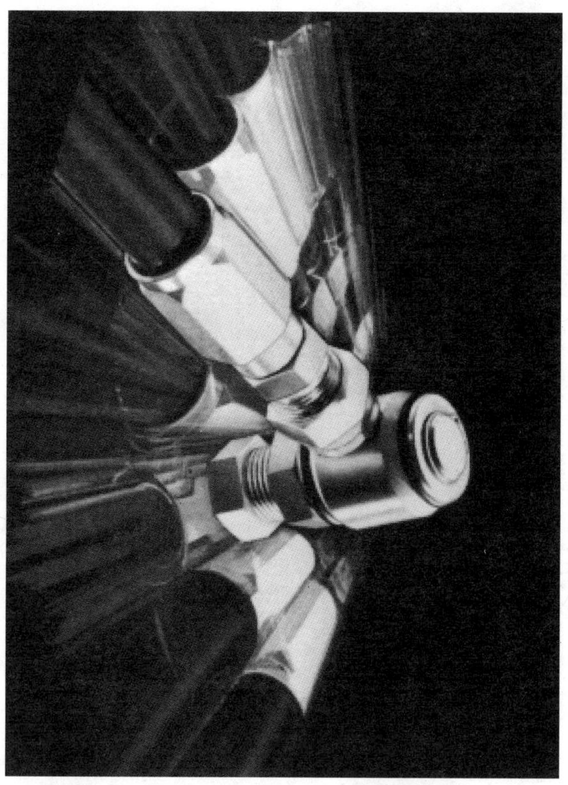

Fig. 12-34 Hose Swivel Adapters

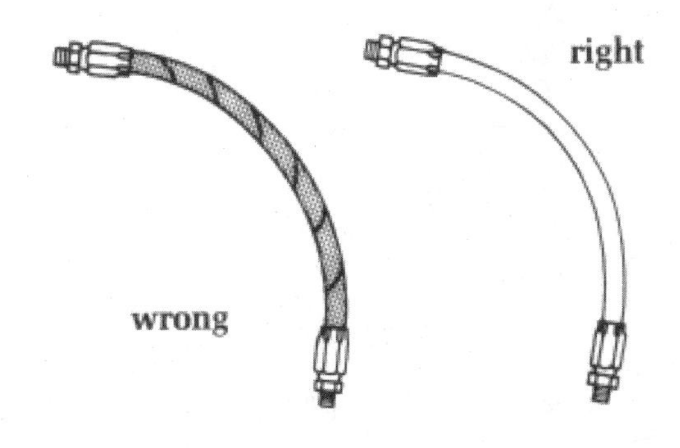

Fig. 12-35 Twisted Hose

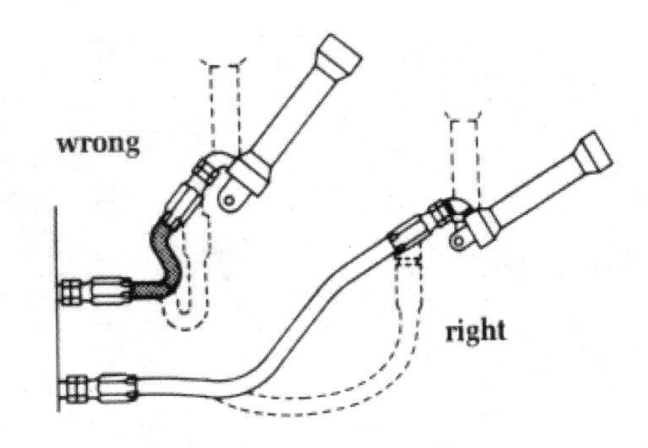

Fig. 12-36 Hose flexing application

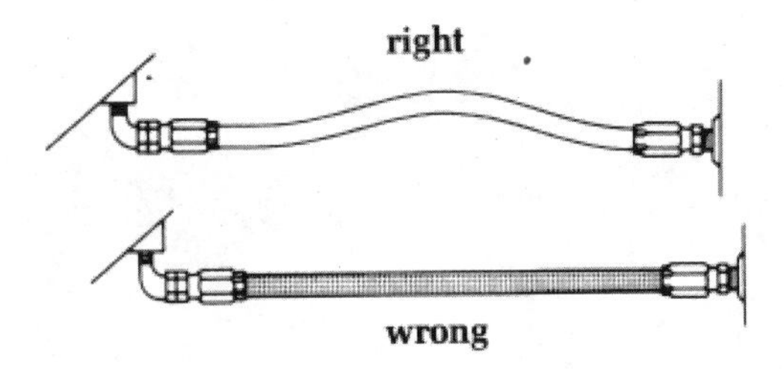

Fig. 12-37 Hose length change with pressure

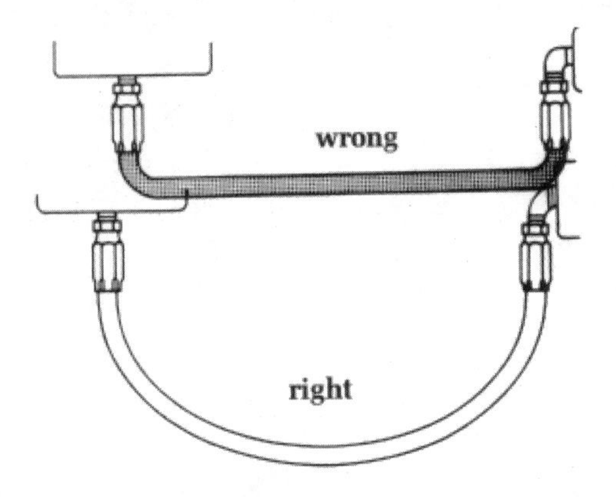

Fig. 12-38 Ample Hose Bend Radius

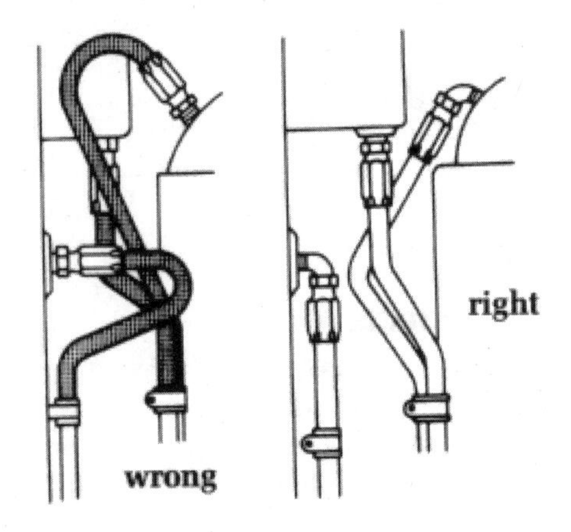

Fig. 12-39 Use of hose elbows and adapters

hose which has suffered no external damage or kinking.

2. If greater negative pressures are required, an internal supporting coil is recommended.

3. No vacuum recommended for multiple wire braid of multiple spiral wrap (4 spiral or more).

Hose installation guidelines

Consideration in routing and installation helps insure efficient system operation and reduces field problems.

When hose assembly is installed in a flexing application, remember that metal hose fittings are not part of the flexible portion. Allow ample free length for flexing (fig. 12-36).

Pressure can change hose length as much as +2% or -4%. Provide slack in line to compensate for hose length changes (fig. 12-37).

Ample bend radius should be provided to avoid collapsing of line and restriction of flow (fig. 12-38). Exceeding minimum bend radius will greatly reduce hose assembly life.

Use elbows or other adapters as necessary to eliminate excess hose length and to insure neater installation for easier maintenance (fig. 12-39).

To avoid weakening the hose due to twisting during installation, swivel fittings (fig. 12-40) should be installed on at least one end of the hose.

Avoid installing hose line close to exhaust manifold or any other hot section (fig. 12-41). If impossible, isolate hose with fire proof boot or other protective means.

Basic hose construction

Hoses used to transmit fluids under pressure are constructed in layers, with each layer filling a specific requirement. Most types of hose consist of three basic elements or parts as illustrated (fig.12-42):

- tube or inner lining
- reinforcement
- cover

1. Inner tube - The tube or liner is the innermost element of the hose. The primary function of the tube or lining is to retain the material — liquids, gases or a combination of these. The type of hose and the service to be encountered determines the materials used and the thickness of the inner tube. This tube must also be compatible with various methods of attaching hose end fittings.

   These materials are compounded to operate in various temperature ranges where elevated temperatures or low temperatures are required, in addition to being resistant to the various materials to be conveyed. The materials most commonly used in inner tube construction are listed in Table 12-6 (page 12-24).

Note: For selection of recommended materials refer to the hose manufacturer's handbook or catalog.

2. Reinforcement - The reinforcement is the fabric, cord, or metal elements built into the body of the hose for purposes of strength, to withstand internal pressure or external forces or a combination of both.

   The materials most commonly used as reinforcement are textile yarns, synthetic yarns, textile fabrics and wire. The yarn may be cotton, rayon, nylon, dacron, glass or similar materials. The wire, usually braided or spirally wound, may be steel, stainless steel, bronze, aluminum or other metal. The type and number of layers or reinforcing materials used depends upon the methods of manufacture and the ser

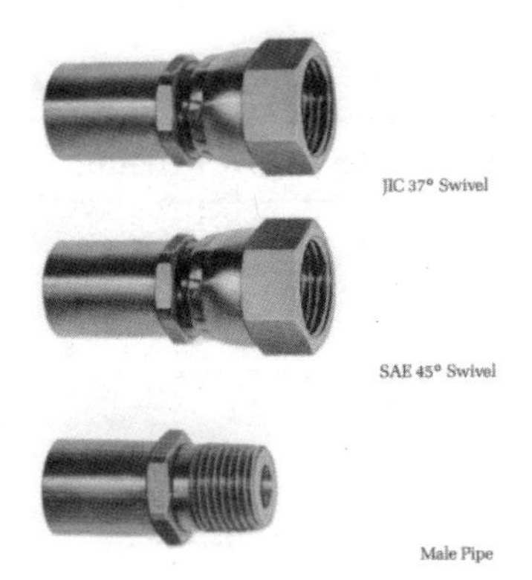

Fig. 12-40 Swivel fittings

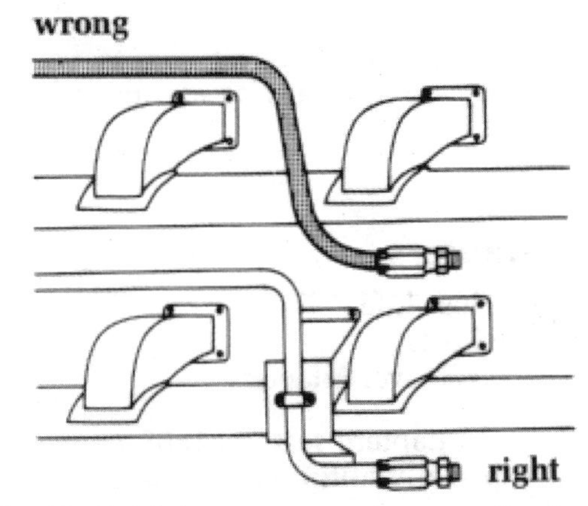

Fig. 12-41 High temperature hose application

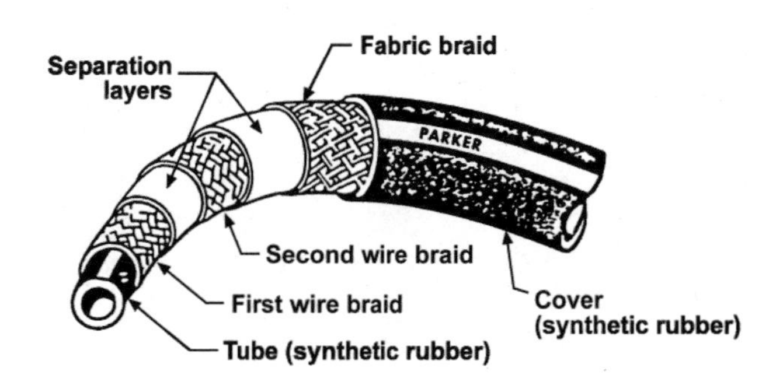

Fig. 12-42 Basic hose construction

Table 12-6 Inner Tube Materials

Common Name	ASTM Designation	Composition	General Properties
Buna-N	N. B. R.	Nitrile-Butadiene	Excellent oil resistance Good resistance to aromatics Good physical properties
Neoprene	CR	Chloroprene	Very good weathering resistance Good oil resistance Good physical properties
ethylene-propylene	EPR, EPDM	Copolymer of ethylene, propylene and (sometimes) another	Excellent weathering resistance Resistance to certain fluids Fair physical properties
Teflon	TFE	Tetrafluoro-ethylene	Excellent high temperature resistance Excellent chemical compatibility Low moisture absorption Excellent weather resistance
PKR	--	Proprietary	Good chemical resistance Outstanding temperature range Excellent weather resistance

vice conditions of the hose. Various materials and compounds are often applied to separate the layers of reinforcement in rubber hose. One of the functions of this separation is to reduce abrasion between reinforcing layers.

3. Cover - The cover is the outermost element of the hose and its primary function is to provide protection against damage to the tube and reinforcement. The cover materials are selected to provide resistance to abrasion, sunlight, hot and cold temperature conditions, and to protect the hose from the various oils, solvents, acids, gasoline and other substances encountered in service.

   Neoprene, synthetic fabrics, nylon, urethane, and other elastomeric materials are commonly used for hose covers.

Hose construction summary

The combination of inner tube, reinforcement, and cover materials leads to hundreds of possible combinations. In practice, the 11 combinations listed in Table 12-7 are the most common with several examples shown in fig. 12-43.

These hose types are listed by a designation or number set by a regulating authority such as ISO and the SAE (Society of Automotive Engineers) in the United States. These regulating bodies (ISO, SAE, etc.) set standards in the hydraulic hose industry. These standards include:

A. Hose and fitting dimension requirements
B. Performance requirements
   1. impulse life
   2. pressure ratings
   3. temperature ratings

Table 12-7 SAE Hose Types

SAE Hose Types			
SAE Number	**Inner Tube**	**Reinforcement**	**Cover**
SAE 100R1	synthetic rubber	1 high tensile steel wire braid	synthetic rubber
SAE 100R2	synthetic rubber	2 wire braids or 2 spiral plies and 1 wire braid	synthetic rubber
SAE 100R3	synthetic rubber	2 textile braids	synthetic
SAE 100R4	synthetic rubber	braided textile fibers/ spiral body wire	synthetic
SAE 100R5	synthetic rubber	1 textile braid and a high tensile steel wire braid	cotton braid
SAE 100R6	synthetic rubber	1 textile braid	synthetic rubber
SAE 100R7	thermoplastic	synthetic fiber	thermoplastic
SAE 100R8	thermoplastic	synthetic fiber	thermoplastic
SAE 100R9	synthetic rubber	4 spiral plies wrapped in alternating directions	synthetic rubber
SAE 100 R10	synthetic rubber	4 spiral plies of heavy wire wrapped in alternating directions	
SAE 100 R11	synthetic rubber	6 spiral plies of heavy wire wrapped in alternating directions	synthetic rubber

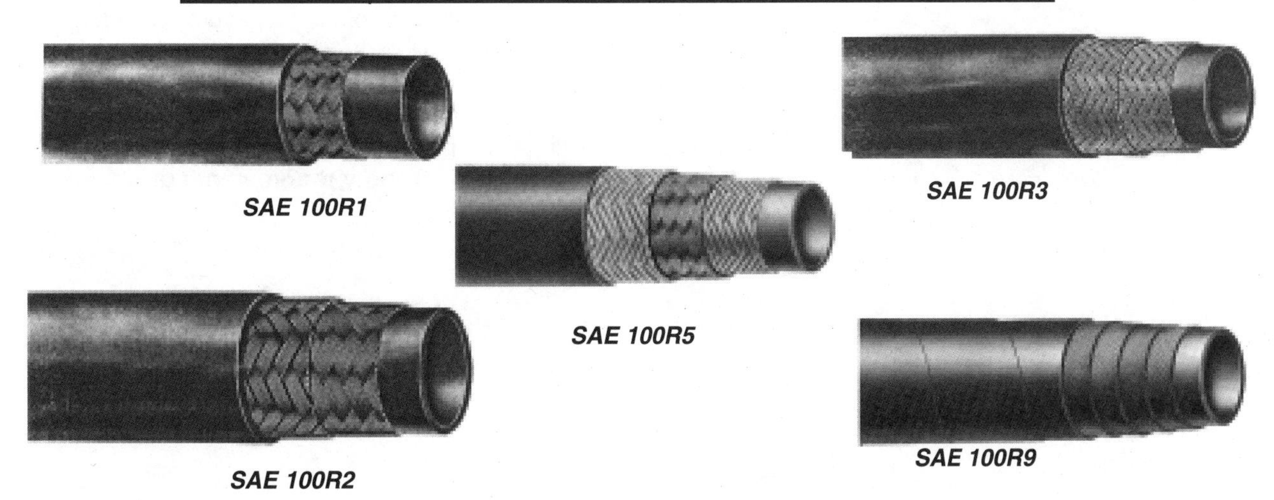

Fig. 12-43 Hose construction types

Reusable, screwed no-skive

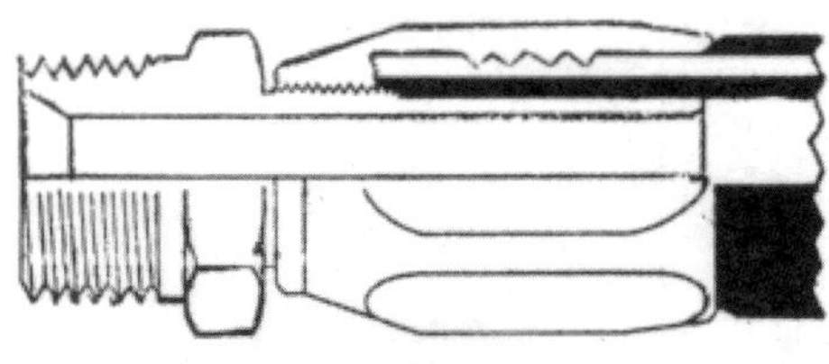

Reusable, screwed, skive

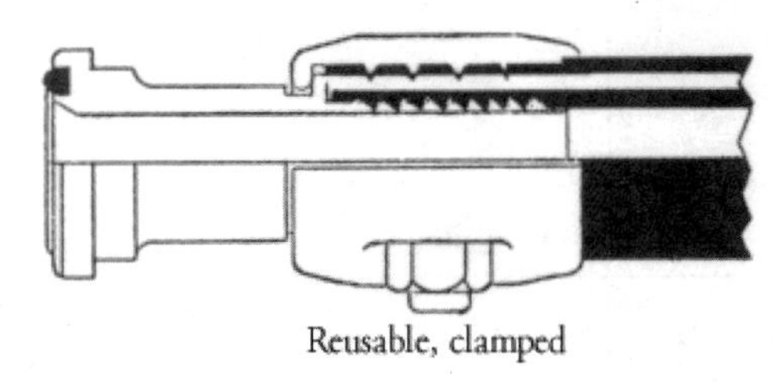

Reusable, clamped

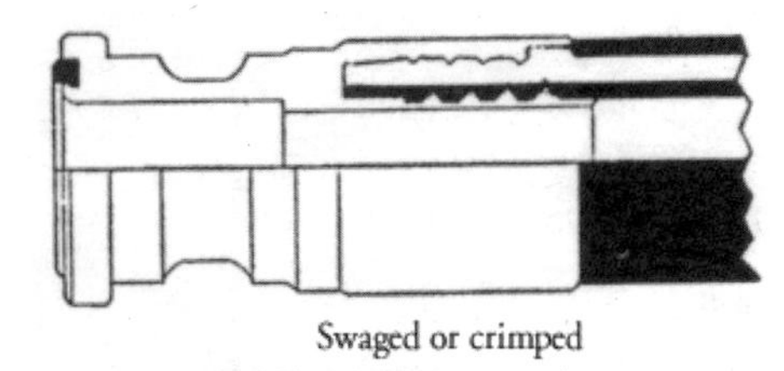

Swaged or crimped

Fig. 12-44 Basic hose fitting types

Hose fitting types

Hose fittings are classified by the method used to attach them to the hose. There are two basic categories, permanently attached and reusable. Both have the same components, a nipple and a socket with the hose clamped between them.

Four commonly used types of hydraulic hose fittings are shown in fig. 12-44.

The **no-skive** type fitting is a screw-on design for use on hose with a thinner outer cover than the skive type, but still thick enough to provide adequate protection for the reinforcement. The hose cover does not require removal and provides a supporting cushion of rubber to reduce stress concentration and fills the voids in the gripper thread of the fitting to protect the hose wire reinforcement from moisture and corrosion.

The **skive** type fitting is a screw-on design for use on hoses with a thick outer cover and this cover must be removed (skived) from the ends of the hose prior to installing this type of fitting.

The **clamp** type fitting is designed with a barbed nipple which is inserted directly into the hose and two clamp halves are then assembled with bolts and nuts on the outside of the hose, and the halves are then bolted together to provide a positive, leakproof grip.

Permanently attached fittings are generally of a one-piece design and the hose is inserted directly into the fitting between the nipple and the socket and the socket is then either swaged or crimped to hold the hose in a viselike grip. The barbs on the o.d. of the nipple provide additional holding power by being forced into the I.D. of the hose inner tube. Because of the possibility of improper assembly of the skive and no-skive fittings, the **permanent** type hose fitting is preferred and **strongly recommended**.

Chapter 12 Exercise
Fluid conductors

Instructions: Complete the sentences below by filling in the blanks with the appropriate word. In each blank, the number of letters for the correct word is shown by dashes. One dash for each word is circled. After all the blanks have been filled in, take all of the circles and form two words which answer the question at the end of the assignment.

1. One reason that leakage occurs in a hydraulic system is ___ ___ ___ ___ ___ ___ (_) ___ installation.
2. A fitting that can be used in systems which operate up to 415 bar (6000 psi), is subjected to high frequency vibrations, and contains an o-ring is known as a (_) ___ ___ ___ ___ ___ ___ ___ .
3. ___ ___ ___ ___ (_) couplings are used for ease of connecting and disconnecting fluid lines without the use of tools
4. The foremost common sealing materials used in couplings to help assure leak-free shutoff are: CR, FPM, nbr, (_) ___ ___ .
5. The safe operating hose pressure decreases when this is reduced below the recommended minimum ___ ___ ___ ___ (_) ___ ___ ___ ___ ___ .
6. Undersizing these fluid conductors will produce excessive pressure drop (ΔP), pressure (_) ___ ___ ___ ___ .
7. Sleeves, springs and shields prolong the life of hose lines and are known as hose ___ ___ (_) ___ ___ ___.
8. The three elements that make up the basic hose construction are: tube or inner lining, reinforcement and the ___ ___ ___ (_) ___ .

Question: Properly chosen, installed and maintained fluid conductors should help achieve what type of system?

Answer: ◯◯◯◯ ◯◯◯◯

Chapter 13

Steering Hydraulic Systems and Accumulators

Many vehicles cannot be steered manually because of their weight and size. Therefore, power steering systems are required to provide the driver of a rubber tired vehicle, power steering control. Fig. 13-1, shows the components found in a typical power steering system. The pump - either gear or piston; a relief valve internal to the pump or external; one or more cylinders; a reservoir; a filtering system; suitable lines and hoses; and a steering valve.

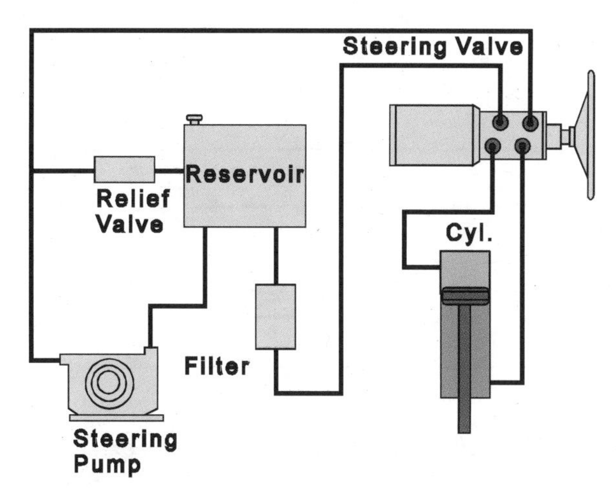

Fig. 13-1 Typical components of a hydraulic steering system

Steering pump - gear

This section will concentrate on the gear type power steering pump as shown in fig. 13-2. The gear pump provides a fixed displacement of fluid flow. The fluid flow is generated by fluid being carried from the inlet to the outlet in pockets or cavities created between the gear teeth and pump housing. As the pump is driven by its prime mover, the pump's maximum output flow will vary dependant upon the rpm of the prime mover.

The standard gear pump consists of a drive gear and a driven gear, pressure plates and seals, and a drive shaft that can be either splined or keyed for connection to the prime mover. This standard type gear pump can be modified to convert it into a power steering pump.

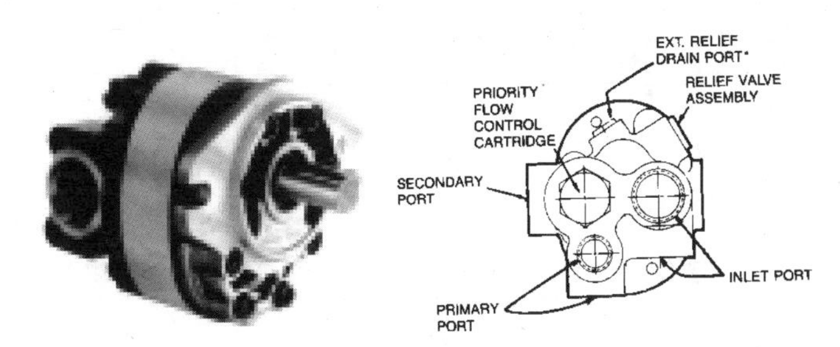

Fig. 13-2 Steering pumps are commonly gear type pumps with built-in flow divider and or relief valve.

Built in relief and flow divider (return to inlet)

In a steering system, the relief valve can be located internal to the pump or external in the system. A steering pump with an internal relief valve, as shown in fig. 13-3, can relieve or limit system pressure in two ways. The schematic shows that when system pressure is reached, the normally non-passing relief valve will open and allow the fluid flow from the pump outlet to return to the inlet side of the pump.

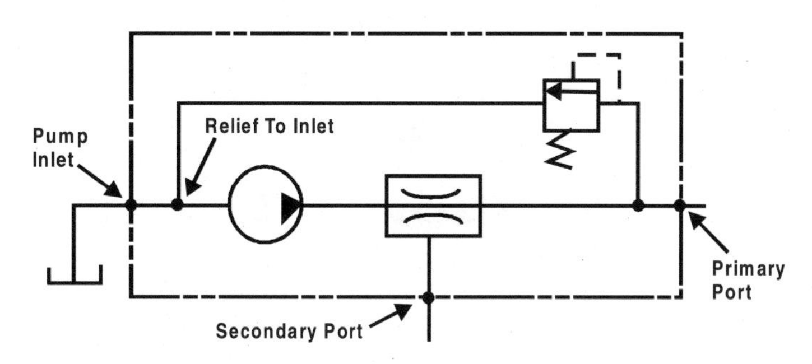

Fig. 13-3 Relief valve flow returns to the pump inlet.

In addition, a flow divider can be installed in the pump. This divider, when employed, will create a secondary flow output from the pump. This secondary flow can be used to supply fluid to another system. To control the maximum pressure in this secondary system, a separate relief valve must be installed.

Built in relief and flow divider (return to tank)

The power steering pump can also be designed to allow the relief valve to direct its flow to tank instead of back to the inlet. Fig. 13-4 shows a schematic representation of this function. When a vehicle, such as a backhoe, uses a power steering pump, it does so sparingly due to its mode of operation. In other words, the backhoe is usually stationary in its work cycles.

The pump output flow is best directed from the relief valve back to tank during this time of steering idleness. The advantage in this type of system is found in the reduction of heat generated. Having the relief valve direct its output flow back to the pump inlet for extended periods of time will tend to heat the hydraulic fluid and thus the entire system. Recirculating the relief valve output flow back to tank, allows for cooler fluid to be drawn in from the reservoir.

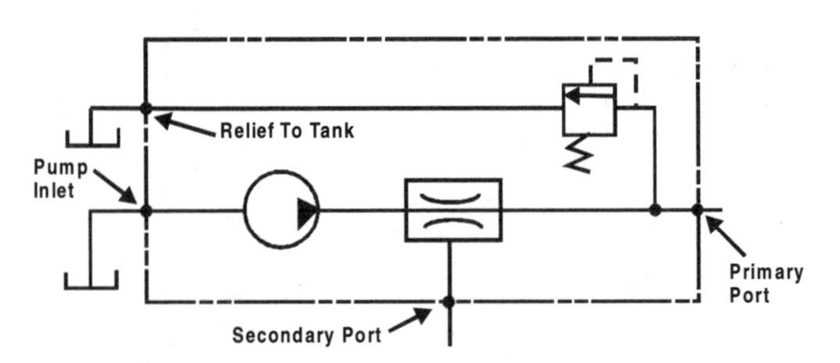

Fig. 13-4 Relief valve flow returns to tank externally.

The flow divider contained in the power steering pump is a fixed flow device. However, during system operation, an increase in flow rate or a decrease in flow rate may occur in the secondary flow system, due to pressure variations in the primary flow system.

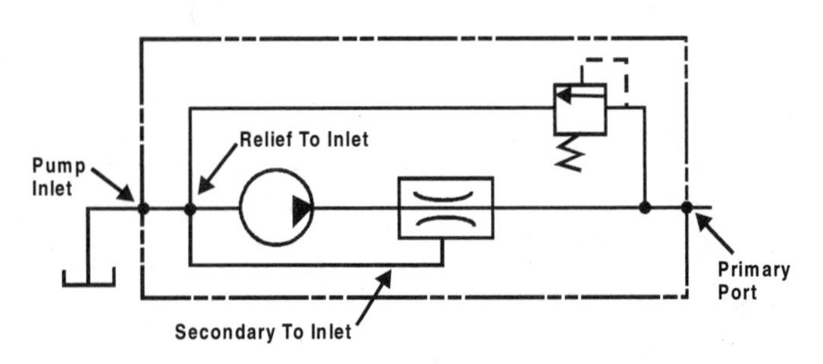

Fig. 13-5 Relief valve flow and secondary flow return to the pump inlet.

Built in relief and flow divider (Return to inlet and secondary flow to inlet)

On vehicles that are designated as "on road" equipment, fig. 13-5 shows a schematic in which both the outlet flow of the relief valve and the secondary flow of the flow divider, return to the pump inlet. This design allows for faster vehicle speeds.

Priority flow divider

The flow divider in fig. 13-6 consists of a fixed bias spring, flow divider spool with a fixed orifice, a priority flow port and a secondary flow port. The bias spring pre-positions the spool to cut off flow to the secondary port. As flow enters the priority port it also flows into the internal flow chamber of the spool and through the fixed orifice.

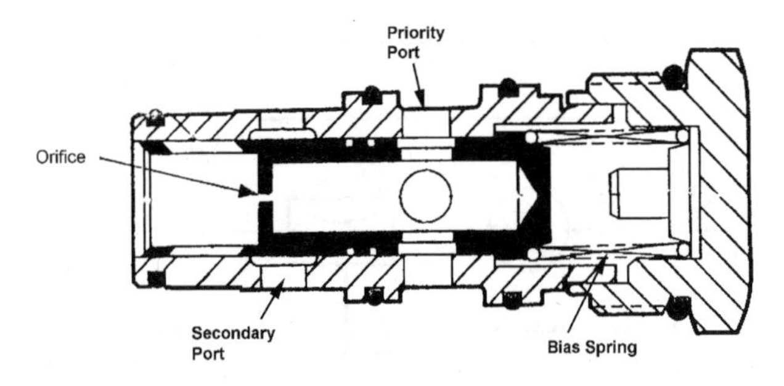

Fig. 13-6 The flow through the priority port is satisfied first, then the spool shifts to let flow exit the secondary port

When the flow requirement through the priority port is satisfied, pressure builds in the left cavity of the spool chamber until the pressure is high enough to overcome the spool bias spring, thus opening the secondary flow port. Flow through the fixed orifice in the spool is now allowed to exit the flow divider.

Flow divider performance

Fig. 13-7 is the performance curve that can be achieved within a power steering pump. Flow (l/min / *gpm*) versus speed (rpm) curves for a typical power steering pump shows a pump with a maximum gpm at 2400 rpm to be 45.5 l/min *(12 gpm).* The primary or priority flow at this rpm is approximately 21 l/min *(5.5 gpm).* While at this same rpm, the secondary flow is the excess flow rate of 25 l/min *(6.5 gpm).*

The flow divider performance curve also shows that as rpm is reduced from 3600 to 1000, the priority flow only slightly reduces and as the rpm is increased, there is a minimal increase in the priority flow. But, at the same rpm, the secondary flow changes significantly due to rpm decrease or increase.

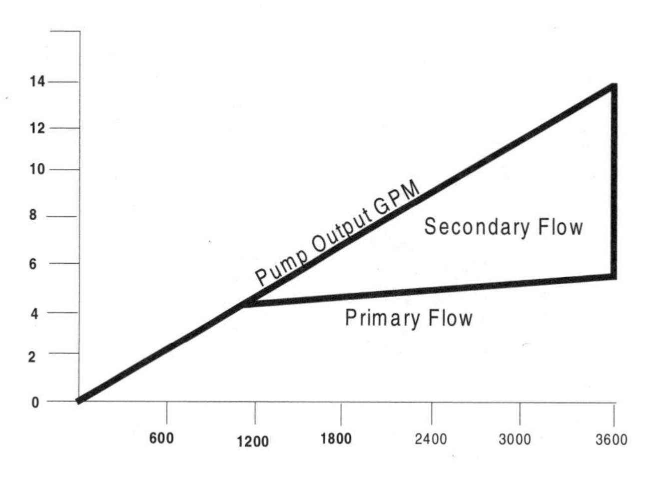

Fig. 13-7 Flow divider performance curve

What a built in relief valve consists of

The function of a pressure relief valve is to limit the maximum allowable pressure in a system. In chapter 6, pressure control valves were discussed. In this section the internal type relief valve for the power steering pump, fig. 13-8 consists of a poppet, a relief bias spring, shims, and both an inlet and outlet port. As a direct acting pressure control valve, system pressure enters the inlet port and acts upon the poppet surface area at the left.

The relief bias spring is preset by the number and thickness of the shims. To increase the system pressure, shims can be added or shims can be removed to lower the maximum pressure.
Note: The power steering pump manufacturer should be consulted for proper pressure adjustment procedure.

Some power steering pump relief valves have an adjustment screw to change the maximum allowable pressure setting.

CAUTION: Regardless of the type of adjustment method used, never operate the pump above the recommended maximum pressure rating.

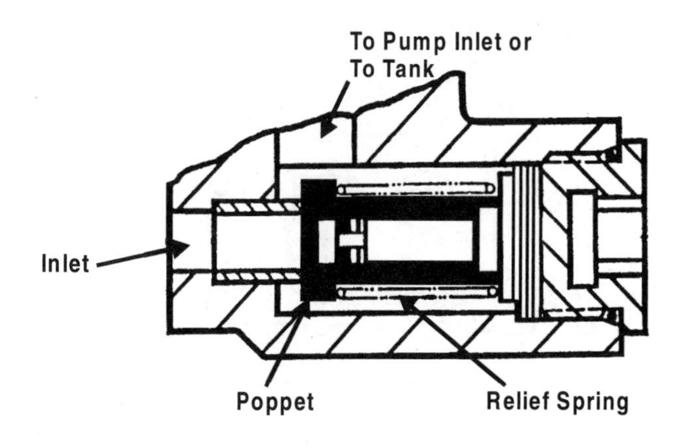

Fig. 13-8 Typical internal relief valve components

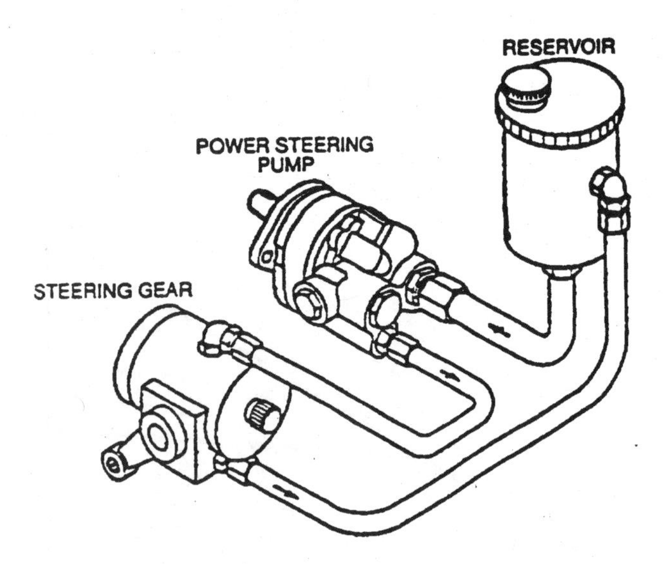

Fig. 13-9 A "two line" power steering circuit is commonly used on "on road" equipment.

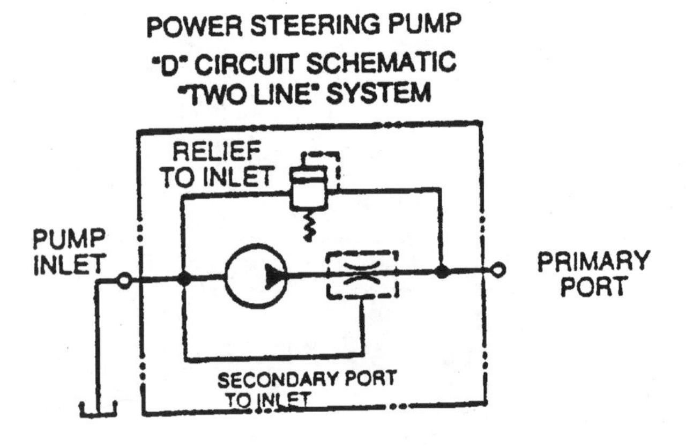

Fig. 13-10 "Two line" power steering circuit schematic

"Two line" power steering pump circuit

A typical power steering pump circuit, commonly referred to as a "two line" system consists of the steering pump, steering gear or valve and a reservoir. As shown in fig. 13-9, the steering pump receives fluid from the reservoir and delivers its output flow to the steering gear or valve. The schematic in fig. 13-10 illustrates the internal flow condition of the pump.

The primary flow exits the pump to the steering gear or valve while the secondary flow returns to the inlet side of the pump. Also if relief valve pressure is reached, the outlet flow of the relief valve is also returned to the inlet of the pump. **CAUTION: excessive heat may be experienced and additional cooling of the fluid may be required.**

In a steering system, the relief valve can be located internal to the pump or external in the system. A steering pump with an internal relief valve, as shown in fig. 13-2, can relieve or limit system pressure in two ways. The schematic in fig. 13-10 shows when system pressure is reached, the normally non-passing relief valve will open and allow the fluid flow from the pump outlet to return to the inlet side of the pump.

This circuit is commonly used in "on road" equipment. With the secondary flow re-entering the pump inlet in the case of over speed, this additional flow will help to prevent cavitation of the pump.

"Three line" power steering pump circuit

In a "three line" power steering pump circuit (fig. 13-11), the secondary flow is directed externally to a second system. A typical application for this type circuit could be a salt spreader highway truck. The secondary flow is used to turn an auger which forces the salt to the back of the truck where a spreader is used.

Again, if the secondary flow is used in another system, there must be a separate relief valve for this system. The internal relief valve of the pump only protects the primary system. If the secondary flow is not to be used, the secondary outlet port must be connected back to the reservoir. **Never plug or block this secondary port.**

Steering circuits

Steering circuits can be classified as:

- open center, non-reversing
- open center, reversing
- open center, power beyond
- open center, demand system
- closed center, non-reversing
- closed center with steering priority valve
- closed center, load sense

Each of these classifications is determined by the type of hydrostatic steering unit being used in the system. The hydrostatic steering unit or valve (fig. 13-12) primarily consists of a linear spool and sleeve, a drive link, a rotor and stator set, a manifold, a commutator ring and commutator, input shaft, torsion bar and housing.

The typical hydrostatic steering valve is divided into two sections, a fluid control valve section and a fluid metering section which are hydraulically and mechanically interconnected. The control valve section contains the mechanically actuated linear spool which is centered by the torsion bar.

The function of the control valve section is to direct the fluid to and from the metering section, to and from the cylinder(s), and to regulate the pressure supplied to the cylinder(s).

The metering section consists of a commutator and bidirectional gerotor element, which contains an orbiting rotor and a fixed stator. The commutator rotates at orbit speed with the rotor and channels the fluid to and from the rotor set and the valve section. The function of the metering section is to meter the fluid to the cylinder(s), maintaining the relationship between the steering handwheel and the steered wheels.

Rotor set operation in the metering element

Each lobe of the rotor has a diametrically opposite lobe; therefore, when one lobe is in a cavity, its opposite lobe is at the crest of the convex form opposite the cavity. As the rotor is rotated, each lobe in sequence is moved out of its cavity to the crest of the stator convex form and this forces each opposite lobe, in sequence, into a cavity. Due to the interaction between the rotor and the stator, there are 42 fluid discharging actions in one revolution of the rotor. (6 lobes times 7 convex forms = 42 fluid discharges).

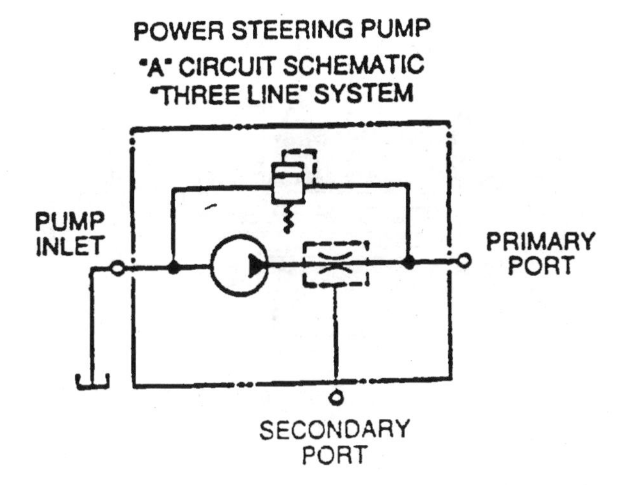

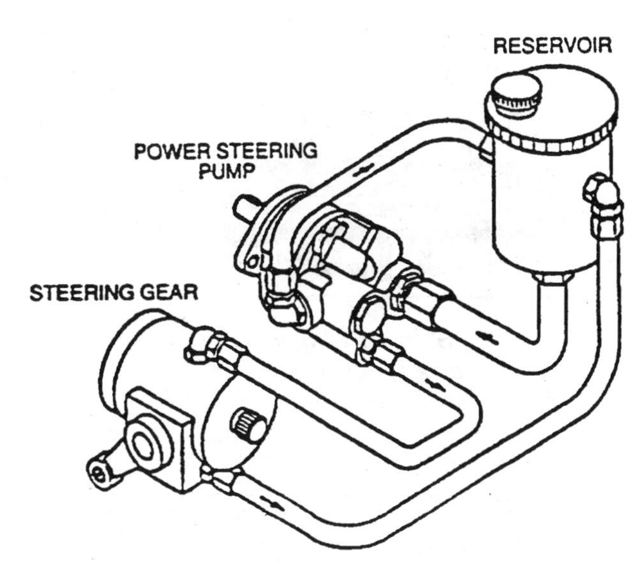

Fig. 13-11 Three Line power steering circuit provides secondary flow to another system

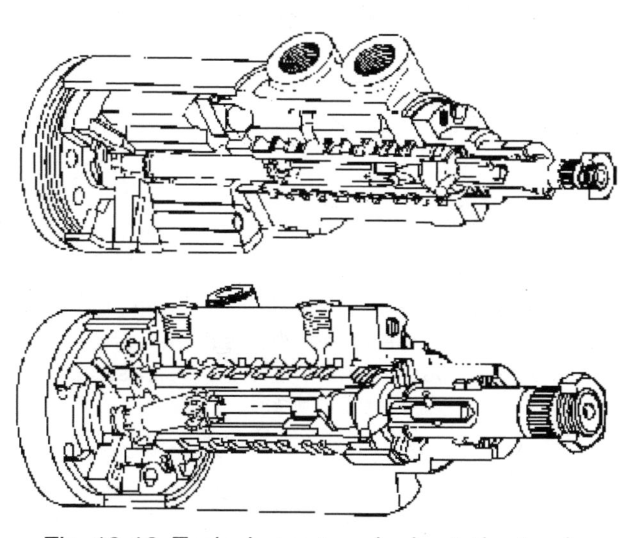

Fig. 13-12 Typical vane type hydrostatic steering valve.

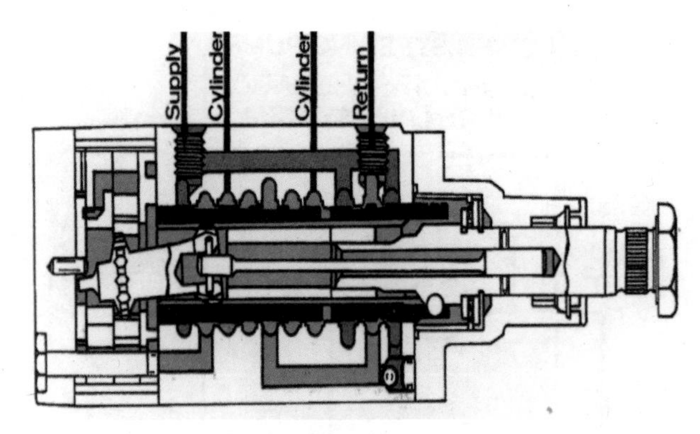

Fig. 13-13 Steering valve in a neutral (straight ahead) mode

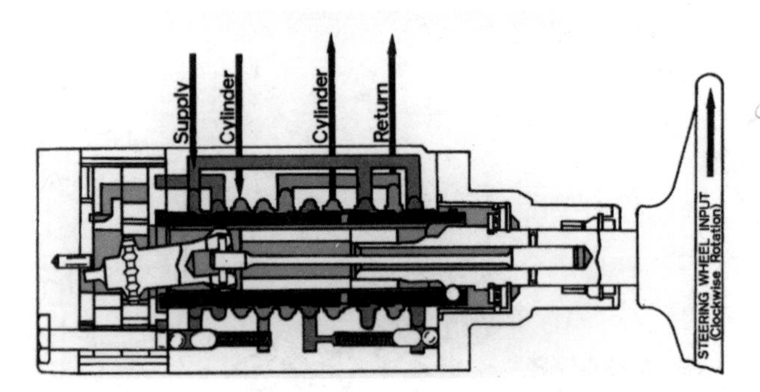

Fig. 13-14 Steering valve in a right turn mode. Valve spool move to the left.

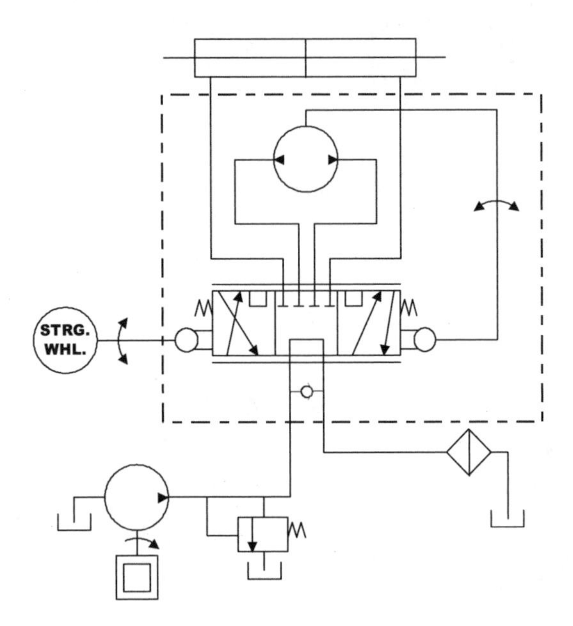

Fig. 13-15 This open center, nonreversing steering valve system will keep the wheels of the vehicle in their steered position when operator releases the steering wheel.

When the rotor is moving, fluid is always flowing out of three of the cavities while fluid is flowing into three other cavities, and one of the cavities is inactive as it changes from one of discharging (output) to one of admitting fluid (intake). The commutator rotates with the rotor and channels the fluid from and to the valve section, and to and from the metering element.

Steering valve operation neutral position

When the spool is in its center or neutral position ,as in fig. 13-13, the hydraulic fluid from the steering pump circulates through the valve section, directly back to the reservoir with sufficient pressure only to overcome friction of the valve channels and lines. There is no circulation of pump flow to or from the cylinder(s). The fluid pressure at the cylinder ports is equal and produces ineffective forces in the cylinder(s).

Steering valve operation right turn

When the operator wants to turn the vehicle, the operator must turn the steering wheel in the direction of the turn. Fig. 13-14 illustrates the spool and sleeve alignment for a right turn.

The initial rotation of the steering wheel rotates the input shaft which tends to rotate the drive link and rotor set through the torsion bar centering spring. Rotation of the rotor set and spool, which are coupled by the drive link, is resisted by the cylinder pressure required to overcome the steering forces. As the input shaft is rotated relative to the spool, the centering spring is torsionally deflected. Axial shift of the spool is induced by the ball which is captive in the spool and engaged in the helical groove provided in the input shaft.

When the spool is axially displaced within the body, fluid channels are selected connecting the power steering pump flow to the intake side of the rotor set via the commutator. The exhaust side of the rotator set is connected through the commutator, to one side of the cylinder(s) while the other side of the cylinder(s) is connected to the reservoir.

Further axial displacement of the spool results in increased system pressure to provide the level of pressure required. A portion or all of the hydraulic fluid at the required pressure from the power steering pump, depending upon the speed of steering, is directed to the cylinder(s) via the metering section, using the cylinder movement to turn the vehicle.

Manual steering operation - no system pressure

If for some reason system pressure is lost, the driver displaces the spool axially by turning the steering wheel. When the spool is displaced in the body, fluid channels are selected connecting the rotor set, which is now acting as a pump, via the commutator to one side of the cylinder(s). The return flow from the other side of the cylinder(s) is channeled through the recirculation valve so that fluid will flow to the intake side of the rotor set via the commutator instead of back to the reservoir.

Open center, non-reversing system

The operating characteristics of a vehicle with an open center, non-reversing hydrostatic steering valve (fig. 13-15) keeps the wheels in the steered position when the operator releases the steering wheel. The cylinder ports are blocked in the neutral valve position. The operator must steer the wheels back to the straight ahead position.

Open center, reversing system

With an open center, reversing hydrostatic steering valve (fig. 13-16) installed in the steering system, the steered wheels are allowed to return to the straight ahead position after the operator releases the steering wheel. This happens only if the steering geometry exerts a centering force on the steering cylinder(s). The cylinder ports are interconnected with the metering section so that the steering wheel follows the wheels back to center position.

Open center, power beyond steering system

Some hydrostatic steering valves have an auxiliary port (fig. 13-17) for power beyond to supply fluid to other functions downstream of the steering valve. This type of system automatically directs priority flow for steering, with the remainder available for auxiliary functions. When not steering, all flow is available to auxiliary functions. This system eliminates a flow divider, such as the one which can be found in the power steering

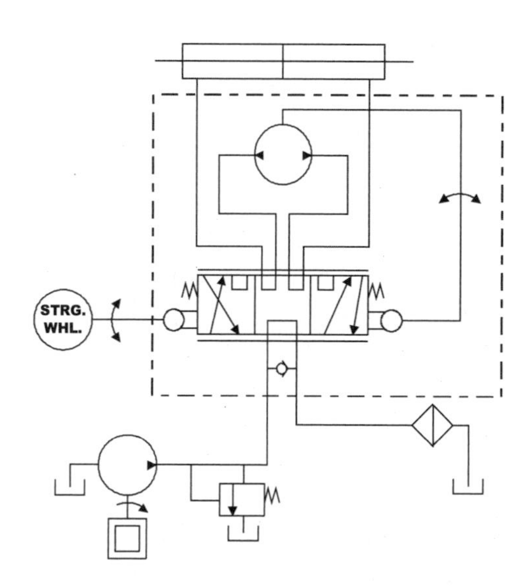

Fig. 13-16 The open center, reversing steering valve system allows the vehicle wheels to return to the straight ahead position when steering wheel is released.

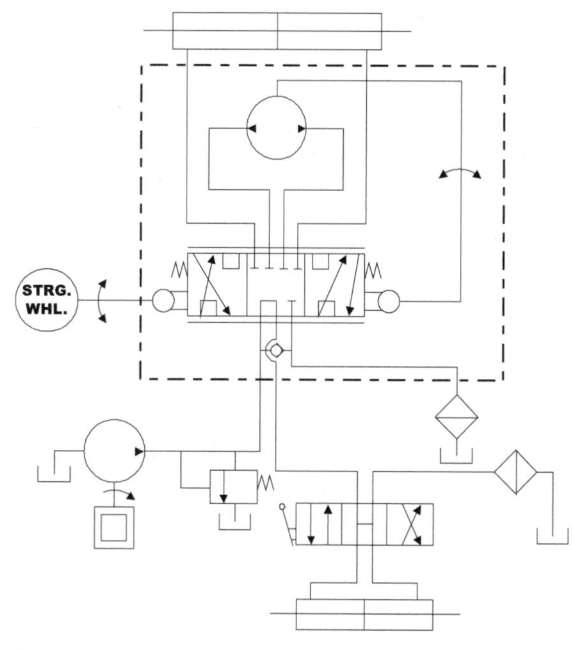

Fig. 13-17 "Open center, power beyond" provides supply flow downstream of the steering valve system.

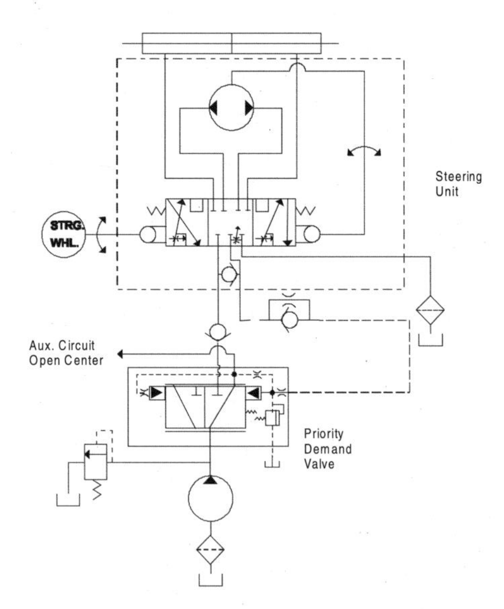

Fig. 13-18 When steering is required the priopity demand valve diverts pump flow to the steering system.

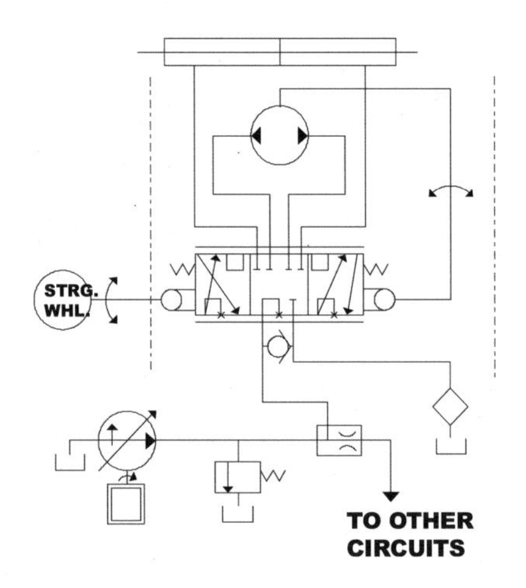

Fig. 13-19 A closed center, non-reversing steering system.

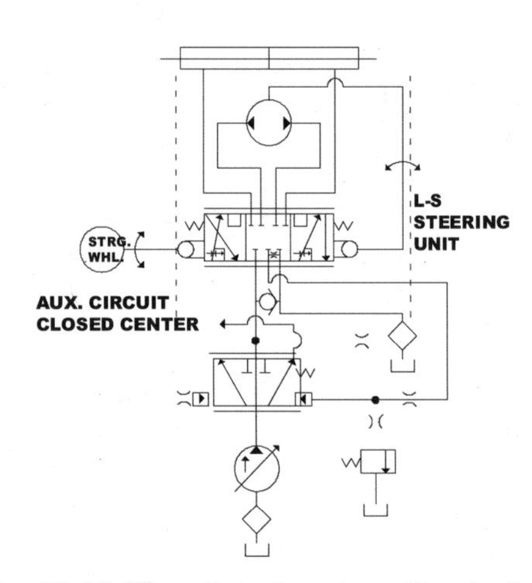

Fig. 13-20 The closed center with priority valve insure adequate supply flow to the steering system.

pump. **Note:** The auxiliary circuit function(s) requires a separate relief valve.

Open center, demand system

The schematic in fig. 13-18, shows a system that uses a fixed displacement pump, a priority demand valve to guarantee an adequate amount of flow to the steering valve, a closed center load sensing steering valve and an open center auxiliary circuit valve(s).

With the priority demand valve centered and the steering valve in its neutral block center position, all of the pump flow is directed to the auxiliary circuit. When the operator steers the vehicle, the steering valve will communicate pressure through the load sense passage and external line to the priority demand valve. Pressure at the right pilot, of the priority demand valve increases, forcing the valve to shift diverting the pump flow to the steering system.

Closed center, non-reversing steering system

A closed center steering system (fig. 13-19) uses a variable displacement pump to provide variable flow to the steering circuit. All ports of the steering valve are blocked when the vehicle is not being steered. The amount of flow through the steering circuit depends upon steering speed and displacement of the steering valve.

Closed center system with steering priority valve

Fig. 13-20 shows a system that uses a variable volume, pressure compensated pump, a steering priority demand valve, a closed center load sense steering valve and closed center auxiliary valve(s). With the exception of the pump and the closed center auxiliary valve(s), the circuit operation is the same as previously discussed in the section.

Open center, demand system

Note: If auxiliary circuit(s) requires a large demand from the pump, such that an inadequate amount of pump flow is available for steering, then a flow limiting control valve such as a needle or flow control valve should be applied to the auxiliary circuit. This is required in order for the steering circuit to operate properly under all conditions.

Closed center, load sense steering system

In fig. 13-21 the load sense circuit uses a load sense steering valve with a sense line for actuating the priority valve. The function of the priority valve is to ensure a supply of pressurized fluid to the steering valve regardless of the downstream demand of the auxiliary valve(s).

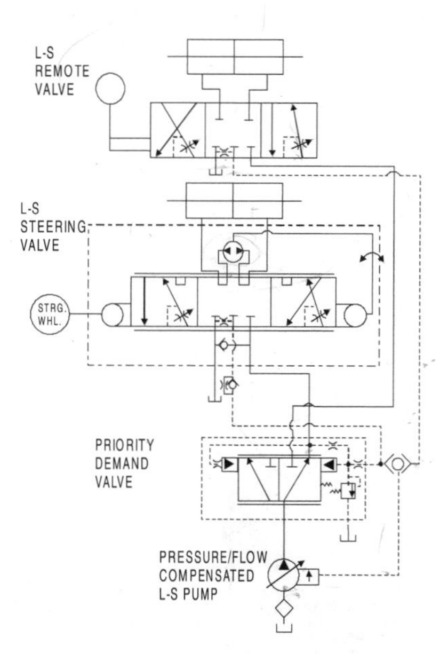

Fig. 13-21 Load sensing is accomplished with a load sense steering valve and remote sense pilot line to the priority valve.

Accumulator in the steering system

A hydro-pneumatic accumulator such as a piston type or bladder type is commonly used in mobile hydraulic steering systems. They can be used for shock absorption and/or for backup fluid flow in emergency situations. These type accumulators apply a force to a liquid by using an inert compressed gas (nitrogen).

NOTE: In most cases of hydro-pneumatic accumulators applied to mobile systems, an inert gas like dry nitrogen is used. ***Compressed air and especially oxygen should never be used because of the danger of exploding an air-oil vapor.***

Hydro-pneumatic accumulators are divided into piston, diaphragm and bladder types. The name of each type indicates the device separating the gas from the liquid.

Piston type accumulator

A piston type accumulator (fig. 13-22) consists of a cylinder body and moveable piston with resilient seals. Gas occupies the volume above the piston and is compressed as the cylinder body is charged with fluid. As fluid discharges from the accumulator, gas pressure pushes the piston down when all liquid has been discharged, the piston should have reached the end of its stroke.

Bladder type accumulator

A bladder type accumulator (fig. 13-22) consists of a synthetic rubber bladder inside a metal shell; the bladder contains the gas. As fluid enters the shell, gas in the bladder is compressed. Gas pressure decreases as fluid flows from the shell. When all liquid has been discharged, gas pressure attempts to push the bladder through the outlet. But, as the bladder contacts the poppet valve at the outlet, flow from the accumulator is automatically shut off.

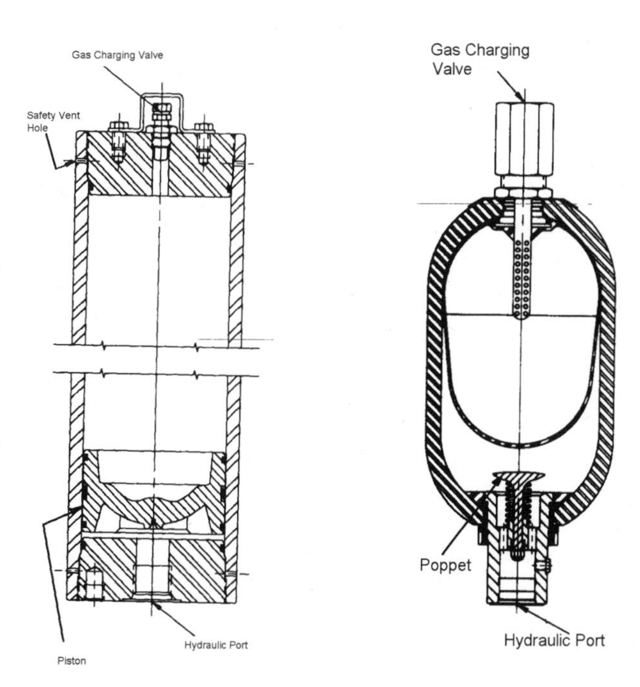

Fig. 13-22 (a) Piston type accumulator; note the position of the U-shape piston. (b) Bladder type accumulator

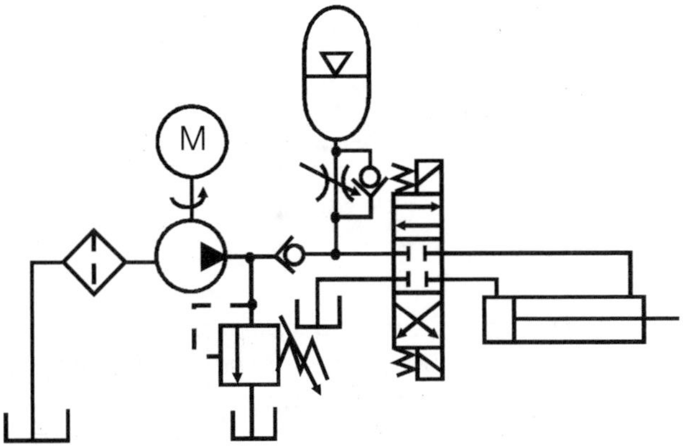

Fig. 13-23 Hydro-pneumatic accumulator used in a system for developing flow to a cylinder

Accumulators in a circuit

Accumulators can perform a variety of functions in a hydraulic system. Two of these in a mobile hydraulic system are developing system flow and absorbing system shock.

Developing flow

Developing liquid flow is one accumulator application (fig. 13-23). Since charged accumulators are a source of hydraulic potential energy, stored energy of an accumulator can be used to develop system flow when the main pressure source is lost, as in the case of an emergency steering situation. For instance, if a machine is designed to steer infrequently, a small displacement pump can be used to fill an accumulator over a period of time. When the moment arrives for the machine to be steered in an emergency situation, the steering wheel is rotated and the accumulator delivers the required pressurized flow to the steering valve and the cylinder(s). Using an accumulator in combination with a small pump in this manner conserves peak horsepower.

Precharge pressure

The gas pressure present in a hydro-pneumatic accumulator, when it is drained of hydraulic fluid, is the accumulator precharge pressure. This pressure significantly affects a hydro-pneumatic accumulator's usable volume and operation as a shock absorber.

Precharge affects shock absorber operation

Precharge of a hydro-pneumatic accumulator affects its operation as a shock absorber. Shock generation in a hydraulic system is the result of fast pressure rises due to an external mechanical force acting on a cylinder or hydraulic motor, or the result of liquid crashing into a component as a valve is suddenly closed. An accumulator acts to reduce the shock effect by limiting pressure rise.

In a hydraulic system as shock pressures develop, these high pressures attempt to displace or push the fluid to another part of the system. But since liquid is relatively incompressible, it won't move or compress. Without an accumulator in the line, shock pressures can rise to a high value.

Above a certain system pressure, as shock pressure begins to build, an accumulator absorbs the volume of liquid the shock attempts to compress or displace. The line in which the accumulator is located becomes compressible above a certain point.

Gas precharge for a hydro-pneumatic accumulator used as a shock absorber is generally set 60% to 65% of the maximum working pressure of the line in which it is located. If the maximum pressure happens to be determined by the relief valve setting, gas precharge can be set at 60% to 65% of the relief valve setting.

As an accumulator operates in a system as a shock absorber, it is generally required to get rid of the fluid it has accumulated in a controlled fashion (fig. 13-24). Commonly, accumulators in these applications are once again equipped with a restriction and bypass check valve. With this arrangement, an accumulator can accept its required fluid, yet any fluid accumulation can bleed off through the restriction.

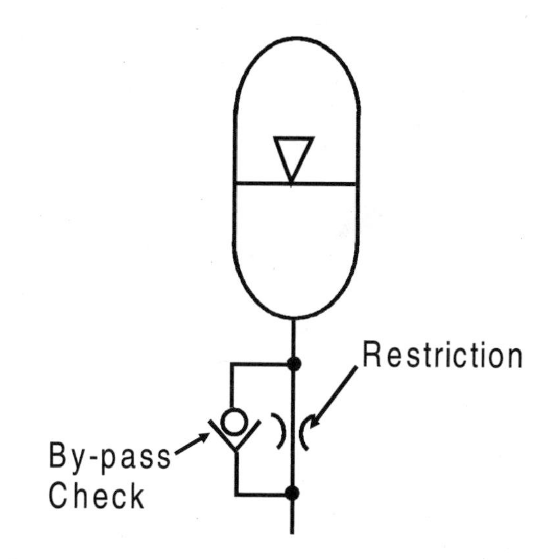

Fig. 13-24 A fixed or an adjustable restriction is used to smooth out the flow from the accumulator. The bypass check allows for free flow into the accumulator.

Since proper gas precharge is such an important factor in hydro-pneumatic accumulator operation, it is discussed in the following section how an accumulator can lose its precharge and how precharge pressure can be checked.

Losing gas precharge pressure

Just because a hydro-pneumatic accumulator is charged once to the proper gas precharge, it does not mean it will remain charged to that pressure indefinitely. As accumulators operate, gas pressure can seep out through the gas valve. This can be due to a faulty or deteriorated seal in the valve, or an improperly seating poppet in the valve core.

Hydro-pneumatic accumulators also lose gas precharge when discharging fluid. With bladder and diaphragm type accumulators, this usually occurs in a catastrophic manner as a result of a rupture in the synthetic rubber separator. When a piston type accumulator discharges, gas pressure can escape across the piston due to a worn seal. Piston type accumulators give an indication of wear as gas precharge gradually dissipates.

Checking gas precharge

Since proper gas precharge is an important consideration in hydro-pneumatic accumulator per-

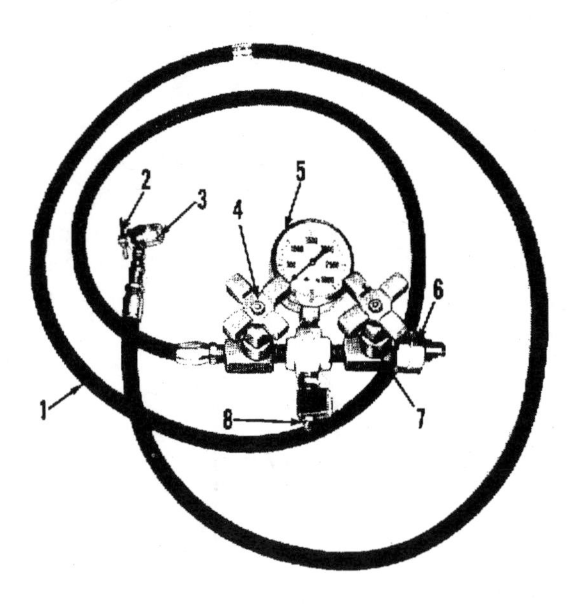

Fig. 13-25 Accumulator charging assembly

formance, precharge should be checked periodically. A necessary piece of equipment for checking gas pressure is a precharging and gaging assembly (fig. 13-25). This assembly is primarily made up of:

1 - line
2 - valve chuck
3 - swivel nut
4 - accumulator valve
5 - pressure gauge
6 - gland nut
7 - tank valve
8 - bleed valve

To check precharge, discharge the accumulator of liquid and remove the protective cap which frequently is found on the gas valve in the accumulator top. Screw the valve chuck (2) handle of the assembly all the way out; check to see that the bleeder valve (8) is closed. Attach the assembly to the accumulator gas valve at the valve chuck (2). Using a wrench, tighten valve chuck swivel nut (3) securely onto the accumulator charging gas valve. Turn the valve chuck (2) stem in; this depresses the core in the accumulator charging gas valve registering a gage pressure. This is the accumulator precharge pressure.

If the accumulator is properly charged, back the valve chuck (2) handle out and open the bleeder valve (8) venting the assembly. Loosen the valve chuck swivel nut (3) and remove the assembly. Replace the accumulator charging gas valve protective cover. If it is found that the accumulator is overcharged, excess pressure can be bled off through the bleeder valve (8).

If it is necessary to increase the gas pressure, back the valve chuck (2) handle out. Open the bleeder valve (8) venting the assembly; then reclose the bleeder valve (8). At this point, the charging assembly will have to be connected to a nitrogen gas bottle.

With the nitrogen bottle gas valve closed, connect the accumulator charging assembly by screwing down on the gland nut (6). Slowly open the tank valve (7); the pressure on the gauge (5) is the tank pressure. Open accumulator valve (4) slowly and charge accumulator to the desired precharge pressure.

To check accumulator charge, close tank valve (7) and relieve the trapped pressure between the pressure gauge (5) and tank valve (7) by momentarily opening the bleed valve (8). The gauge (5) pressure will lower to the correct pressure reading of the accumulator charge.

Once the required precharge pressure has been reached, close the accumulator valve (4) and the tank valve (7). Open the bleed valve (8) to vent any trapped pressure. Remove the gland nut (6) from the nitrogen bottle.

At the accumulator, rotate the valve chuck (2) handle counterclockwise until it stops. Make sure the accumulator's gas charging valve lock nut is tight and secure. Then loosen the swivel nut (3) and remove the accumulator charging assembly.

Note: Check the accumulator charging valve for leakage using soapy water.

Chapter 13 exercise
Steering systems

Instructions: In this assignment the answers are already given - you write the questions.

Answer 1 Open center, non-reversing, reversing, power beyond, demand system; closed center, non-reversing, steering priority vale and load sense

Answer 2 Linear spool and sleeve, drive link, rotor, and stator set, manifold, commutator ring, commutator, input shaft, torsion bar and housing

Answer 3 Fluid control valve section and a fluid metering section

Answer 4 Keeps wheels in the steered position when steering wheel is released

Answer 5 Used to absorb shock and/or for backup fluid in an emergency

Answer 6 Dry nitrogen

Answer 7 Built in relief and flow divider

Answer 8 Two- and three-line circuits

Answer 9 Uses a variable displacement pump

Answer 10 Steered wheels are allowed to return to the straight ahead position after the steering wheel is released

Chapter 14

Reservoirs and coolers

Hydraulic reservoirs

Hydraulic reservoir symbol

The obvious function of a hydraulic reservoir is to contain or store a system's hydraulic fluid.

What a hydraulic reservoir consists of

Hydraulic reservoirs can be found in many shapes and sizes. They can be a separate device attached to the mobile machine or can be an integral part of the machine such as the transmission or final drive case. Where space and weight are not a problem consideration can be given to good design, i.e. hydraulic reservoirs should consists of four walls (usually steel); a dished bottom; a flat top with mounting plate; four legs, suction, return, and drain lines; drain plug; oil level gage; filler/breather cap; cleanout covers; and baffle plate (fig. 14-1). But, due to the weight and space penalty incurred on mobile equipment the typical stand alone mobile reservoir consists of: four walls, a flat bottom, flat top, suction and return lines, filler/breather cap and a baffle plate.

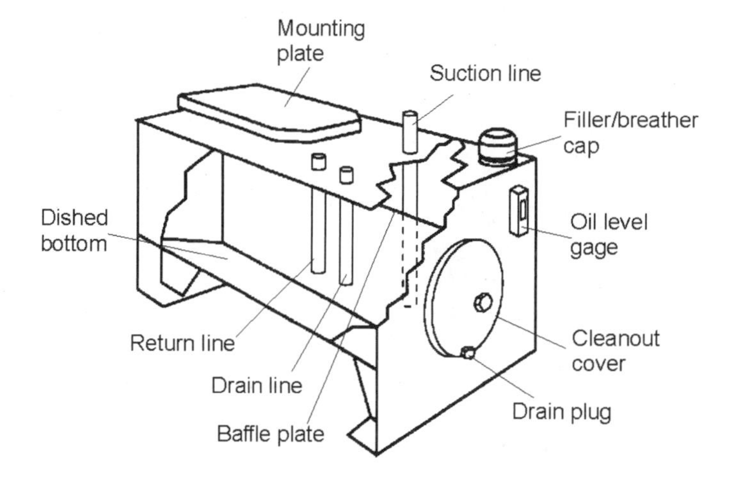

Fig. 14-1 Hydraulic reservoir

How a reservoir works

Besides acting as a fluid container, a reservoir serves also to cool the fluid, to allow contamination to settle out, to allow entrained air to escape, and a convenient location to install environmental controls such as heaters and/or coolers.

With fluid returning to a reservoir, a baffle plate blocks the returning fluid from going directly to the suction line (fig. 14-2). This creates a quiet zone which allows large dirt to settle out, air to rise to the fluid surface, and gives a chance for the heat in the fluid to be dissipated to the reservoir walls.

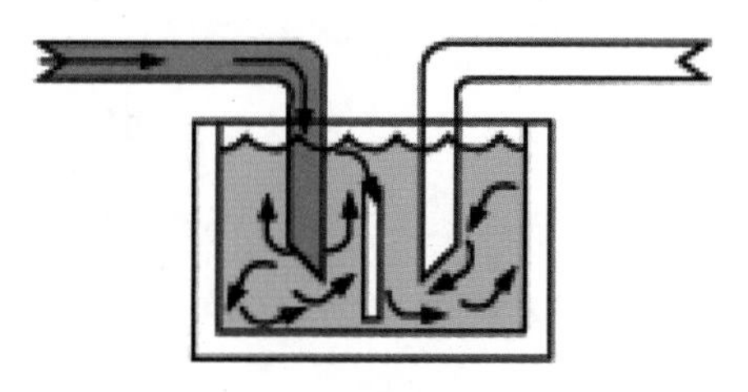

Fig. 14-2 Baffle plate in reservoir

Fluid baffling is a very important part of proper reservoir operation. For this reason, all lines which return fluid to the reservoir should be located below fluid level and at the baffle side opposite the suction line. In addition, because of the movement of the oil within the reservoir, due to the movement of the machine, additional baffles are a necessity.

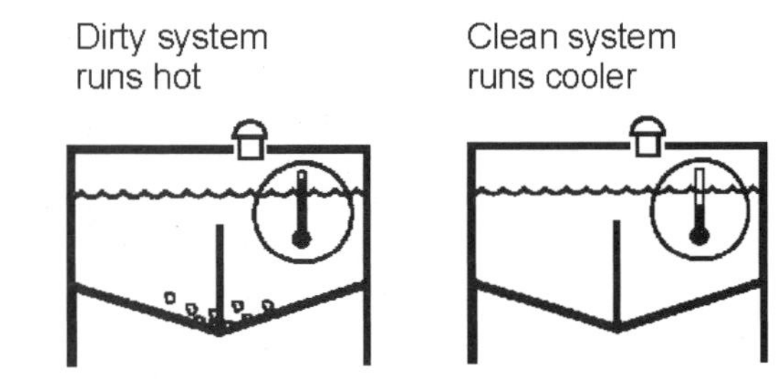

Fig. 14-3 Effects on heat transfer in a reservoir

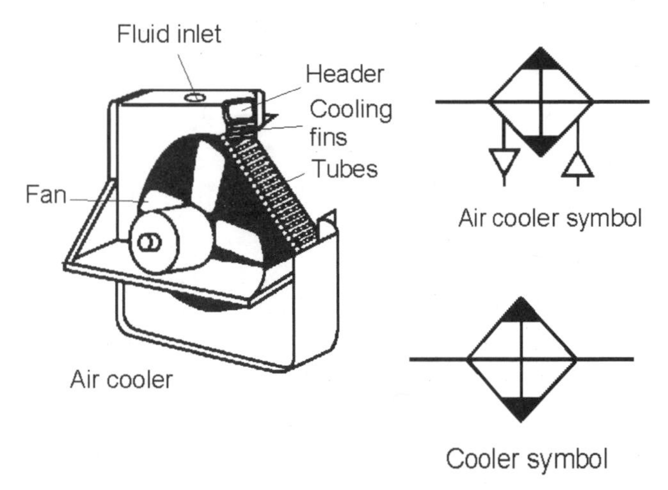

Fig. 14-4 Air coolers

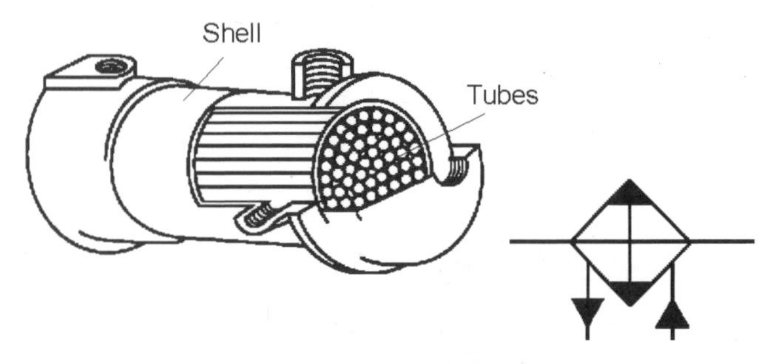

Fig. 14-5 Water cooler

Fig. 14-6 Dredge

Because of viscosity, friction, and changing direction, hydraulic fluid generates heat during system operation. When the liquid returns to the reservoir, it gives the heat up to the reservoir walls. Dirt particles interfere with liquid cooling by forming a sludge which makes heat transfer to reservoir walls difficult (fig. 14-3).

Clean hydraulic systems run cooler than dirty systems.

Coolers

Inefficiency in the form of heat can be expected in all hydraulic systems. Even well designed hydraulic systems can be expected to turn some portion of its input horsepower into heat. It is not uncommon for the fluid in a mobile hydraulic system to operate at temperatures in excess of 225°F. Hydraulic reservoirs are sometimes incapable of dissipating all of this heat. In these cases a cooler is required.

Coolers are divided into air coolers, most common in mobile applications, and water coolers, seen on marine or skid mounted systems. The latter could be considered a "semi-mobile" system.

Air cooler

In an air cooler (fig. 14-4), fluid is pumped through tubes to which fins are attached. To dissipate heat, air is blown over the tubes and fins by a fan. The operation is exactly like an automobile radiator.

Air coolers are generally used where water is not readily available or too expensive.

Water cooler

A water cooler (fig. 14-5) basically consists of a bundle of tubes encased in a metal shell. In this cooler, a system's hydraulic fluid is usually pumped through the shell and over the tubes which are circulated with cooling water. A variation of this would be the "Keel Cooler" onboard a marine vessel (fig. 14-6).

This cooler is also known as a "shell-and-tube" type heater exchanger. It is a true heat exchanger since hydraulic fluid can also be heated with this device by simply running hot water through the tubes.

Coolers in a circuit

Coolers are usually rated at a relatively low operating pressure (10 bar/*150 psi*). This requires that they be positioned in a low pressure part of a system (fig. 14-7). If this is not possible, the cooler may be installed in its own separate circulating system.

To insure that a pressure surge in a line does not damage a "shell-and-tube" type cooler, they are generally piped into a system in parallel with a 4 bar *(65 psi)* check valve.

Coolers can be located in a system's return line, after a relief valve, or in a case drain line of a variable volume, pressure compensated pump.

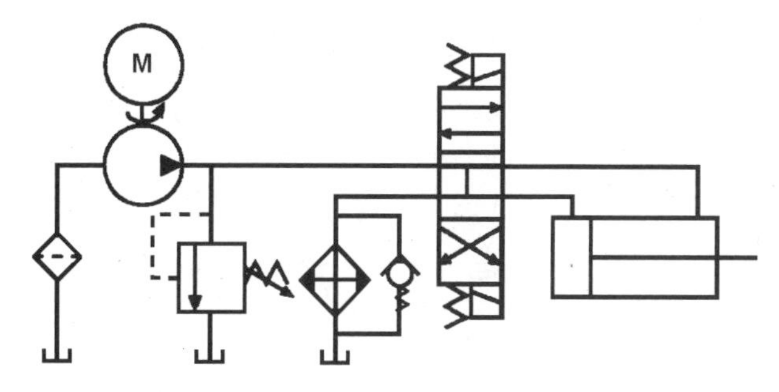

Fig. 14-7 Coolers in a circuit

Mechanical forces on the reservoir

Reservoirs are continuously flexing from stresses due to the normal movement and vibration of the equipment (fig. 14-8). These forces cause metal to distort resulting in welding slag breaking loose along with settled contamination being re-entrained in the fluid and reentering the fluid stream.

Fig. 14-8 Environmental stresses

In addition to withstanding these structural forces, the reservoir must be able to withstand impact blows from debris and other equipment. This consideration can be minimized by inclosing it with a protective structure or by concealing it within the machine. However, this makes routine maintenance much more difficult and may lead to lack of maintenance. Reservoirs need to be strong, protected and yet accessible.

Filler/breather cap

The atmosphere in which the machine lives contaminates a system's hydraulic fluid. Operation of a hydraulic system depends upon air entering the reservoir through the breather cap and pushing the fluid up to the pump. As actuators in the system are filled and discharged with fluid, the tank actually inhales and exhales dirt-laden air which is filtered by a coarse screen in the filler/breather cap (fig. 14-9). The dirt which is not filtered out settles into the fluid.

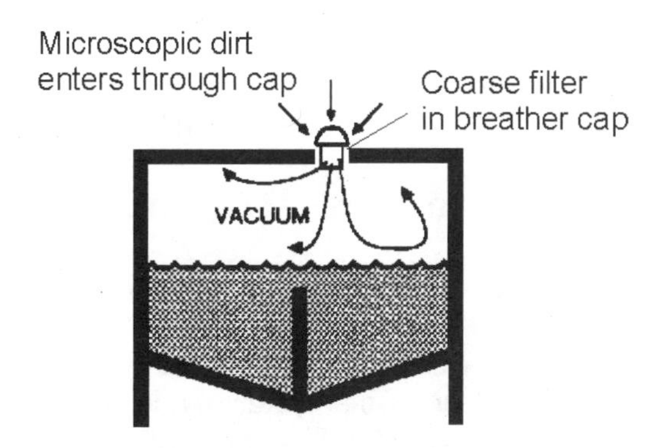

Fig. 14-9 Filler/breather cap

Breather caps are very rarely, if ever, cleaned. Because of this lack of maintenance, the screen becomes plugged. As a result, the breather cap is removed from the reservoir or the atmosphere finds a path through an old, cracked seal or stripped bolt hole. Now a clear path to the fluid is open for almost anything. Such a condition is not acceptable, but, particulate contamination is not the only potential reservoir consideration.

Air entering the reservoir during operation also contains water vapor (fig. 14-10). When the machine is shut down, the air inside the tank cools and water vapor condenses on the walls. This causes the formation of rust which eventually finds it way into the fluid.

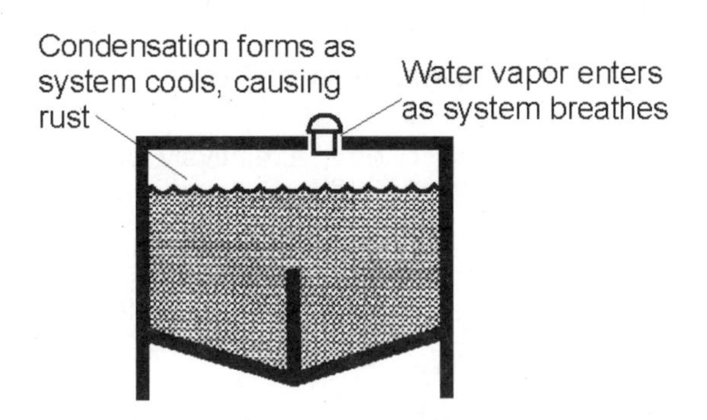

Fig. 14-10 Water vapor entering the reservoir

When operating in heavily contaminated environments it is advantageous to replace the standard course filler/breather cap with a high capacity "spin-on" style air filter (fig. 14-11). These air breather units are available with levels of filtration as low as 5 micron. For specific applications, such air breather elements can be equipped with a pressurization valve (fig. 14-11). However, the reservoir must be designed to withstand the pressure.

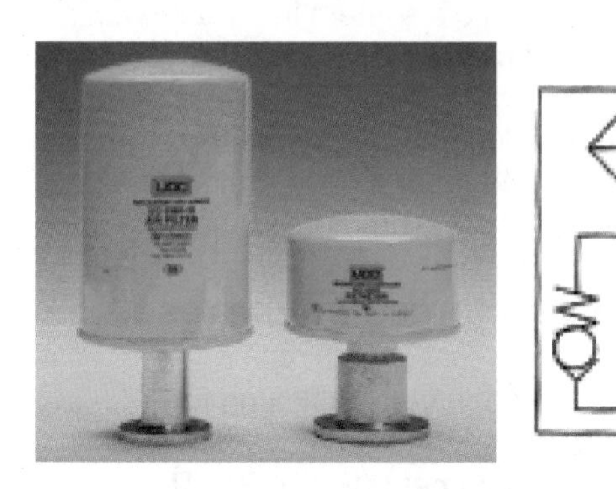

Fig. 14-11 Spin-on air breathers and pressurization valve circuit

If an additional level of protection is required or desired, then a filler/breather element containing particulate filter(s), a chemical dryer to remove water vapor and activated carbon to remove oil (exhaust) vapor and associated odors (fig. 14-12) is called for.

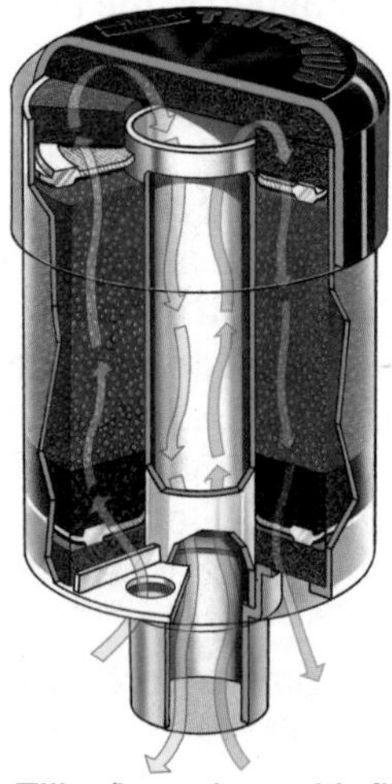

Fig. 14-12 Filler/breather with filter, chemical dryer and activated carbon

To minimize spillage an "anti-splash" type filler/ breather cap (fig. 14-13) should be employed on the reservoir. This device prevents oil splash during vehicle movement. If necessary, such devices are available with locks (fig. 14-13), to add a level of security.

If, as a result of environmental legislation, it is required that there be no oil spill, even when the reservoir (equipment) is turned over, then a sealed or closed reservoir must be considered.

Additional reservoir accessories

Fluid level temperature gages (fig. 14-14) are not a convenience in a mobile hydraulic system, but are a necessity. Due to the smaller reservoir size (compared to industrial systems) and therefore, smaller volume of fluid; its temperature and level are critical and must be monitored. The fluid level temperature gage gives the operator and/or maintenance personnel a convenient, easy and non-invasive way to monitor both.

Fig. 14-13 Anti-splash and filler/breather caps with locks

Because of reduced reservoir and conductor size in mobile applications, the oil returning to the reservoir will do so at an elevated flow velocity. Velocities in excess of the accepted guideline of 10-12 f.p.s. (metric number) are typical. To reduce the return flow turbulence in the reservoir, a "diffuser" should be considered (fig. 14-15). This simple device can achieve flow velocity reductions as high as 19:1 with very little pressure drop.

Fig. 14-14 Oil level and temperature gages

Integral reservoirs

Integral reservoirs are those that are not only the hydraulic system reservoir but also the reservoir (sump) for such components as the vehicles transmission or final drive case (fig. 14-16).

Such sharing can create problems. The first is the added heat load the oil must carry due to the other functions, transmission and/or final drive operation.

Fig. 14-15 Diffuser

Another problem, especially in the case of the final drive section, is the increased fluid aeration caused by the gears partially operating in the fluid. Such a condition could lead to erratic system operation and premature pump failure. In addition to this, the wear debris created by the gears increases the contamination level of the system hastening system failure because of it. Only a properly designed system with the correct filtration chosen for that application and properly maintained can stave off the catastrophic system failure that would otherwise result.

Fig. 14-16 Intregral reservoir

Chapter 14 exercise
Reservoirs and coolers

INSTRUCTIONS: In the maze of letters are hidden words which are the answers to the questions below. Circle the words.

M	E	G	A	I	P	L	T	P	P	O	S	R	C	A	I	R	T	J
O	S	V	E	G	I	S	R	O	R	D	T	P	A	B	I	C	E	W
T	H	I	R	K	S	M	L	X	I	M	P	G	V	M	F	L	M	L
H	T	N	A	S	T	B	F	N	W	Y	S	P	I	N	O	N	P	S
V	S	W	T	C	O	C	A	M	A	J	N	T	T	L	E	T	E	P
P	C	E	I	F	C	H	E	A	T	X	T	O	R	C	E	S	R	A
L	V	J	O	K	C	E	O	S	E	D	K	T	T	E	W	P	A	N
Z	O	A	N	U	F	X	S	V	R	C	I	O	I	R	S	A	T	R
T	H	W	R	K	E	G	A	S	K	A	E	F	O	I	E	S	U	I
R	B	L	P	I	D	N	J	G	I	A	R	X	F	P	E	Q	R	C
J	W	F	X	E	A	Y	C	E	W	B	W	T	T	U	L	K	E	F
B	B	O	P	E	N	B	B	A	F	F	L	E	S	E	S	T	M	H
S	B	W	Y	U	S	R	L	E	V	E	L	E	C	T	N	E	E	R
M	T	V	A	P	O	R	E	E	B	N	C	F	U	F	G	R	R	I

1. The device used to reduce the velocity of the return fluid in the reservoir is called a(n) ____________ .

2. Integral reservoirs can create two main problems. The first is an added ____________, and the second is increased ____________ .

3. Due to smaller reservoirs in mobile hydraulic systems, a fluid ____________ and ____________ gauge is critical.

4. In heavily contaminated environments, a standard filler/breather cap should be replaced with a ____________ style air filter.

5. As air in the reservoir cools, ____________ condenses on the reservoir walls.

6. Reservoirs on mobile equipment must be strong, protected, but yet ____________.

7. There are two types of oil coolers used in hydraulic systems. They are the water and the ____________ cooler.

8. Oil coolers are usually positioned in ____________ pressure lines.

9. Due to the movement of the mobile machine, additional ____________ are a necessity in the mobile hydraulic reservoir.

10. Reservoirs on mobile equipment are subjected ____________ due to the normal movement and vibration of the equipment.

Appendix A - Unit Conversions

1.1 Metric and SI units

Name of unit	Value	Symbol
Length		
Meter	Base unit	m
Centimeter	0.01 m	cm
Millimeter	0.001 m	mm
Micrometer	0.000 001 m	µm
Kilometer	1 000 m	km
International nautical mile (for navigation)	1 852 m	n mile
Mass		
Kilogram	Base unit	kg
Milligram	0.000 001 kg	mg
Gram	0.001 kg	g
Tonne	1 000 kg	t
Time		
Second	Base unit	s
Minute	60 s	min
Hour	60 min	h
Day	24 h	d
Area		
Square meter	SI unit	m^2
Square millimeter	0.000 001 m^2	mm^2
Square centimeter	0.000 1 m^2	cm^2
Hectare	10 000 m^2	ha
Square kilometer	1 000 000 m^2	km^2
Volume		
Cubic meter	SI unit	m^3
Cubic centimeter	0.000 001 m^3	cm^3

Name of unit	Value	Symbol
***Volume** (fluids)*		
Liter	0.001 m^3	l
Milliliter	0.001 l	ml
Kiloliter	1 000 l (1 m^3)	kl
Velocity		
Metre per second	SI unit	m/s
Kilometre per sec	0.27 m/s	km/h
Knot	1 naut. mile/h or 0.514 m/s	kn
Force		
Newton	SI unit	N
Kilonewton	1 000 N	kN
Meganewton	1 000 000 N	MN
Energy		
Joule	SI unit	J
Kilojoule	1 000 J	kJ
Megajoule	1 000 000 J	MJ
Power		
Watt	SI unit	W
Kilowatt	1 000 W	kW
Megawatt	1 000 000 W	MW
Density		
Kilogram per cubic meter	SI unit	kg/m^3
Tonne per cubic meter	1 000 kg/m^3	t/m^3
Gram per cubic meter	0.001 kg/m^3	g/m^3

1.1 Metric and SI units (continued)

Name of unit	Value	Symbol
***Density** (fluids)*		
Kilogram per litre	1000 kg/m³	kg/l
Pressure		
Pascal	SI unit (N/m²)	Pa
Kilopascal	1000 Pa	kPa
Megapascal	1000 000 Pa	MPa
***Pressure** (meteorology)*		
Millibar	100 Pa	mb
Electric current		
Ampère	unit	A
Milliampère	0.001 A	mA
Potential difference		
Volt	SI unit	V
Microvolt	0.000001 V	μV
Millivolt	0.001 V	mV
Kilovolt	1000 V	kV
Megavolt	1,000,000 V	MV
Electrical resistance		
Ohm	SI unit	
Microhm	0.000001	
Megohm	1,000,000	
Frequency		
Hertz	SI unit	Hz
Kilohertz	1,000 Hz	kHz
Megahertz	1,000,000 Hz	MHz
Gigahertz	1,000,000,000	GHz
Temperature		
°Kelvin	SI unit	K
°Celsius	-273.15 °K	C

1.2 Metric units of length

Unit	Equals
1 meter	39.4 inches
	3.28 feet
	1.094 yards
	1000 mm
	100 cm
	10 dm
	0.001 kilometers
1 centimeter	0.394 inch
	0.032 8 foot
	10 mm
	0.01 meters
1 millimeter	39.4 mils
	0.0394 inch
	0.001 meter
1 Kilometer	3 280 feet
	1094 yards
	0.621 mile
	1000 meters

1.3 Imperial units of length

Unit	Equals
1 inch	1 000 mils
	0.0833 foot
	0.0278 yard
	25.4 mm
	2.54 cm
1 foot	12 inches
	0.333 yard
	0.000189 miles
	0.305 meter
	30.5 cm
1 yard	36 inches
	3 feet
	0.000568 mile
	0.914 meter
1 mile	63,360 inches
	5280 feet
	1760 yards
	320 rods
	8 furlongs
	1 609 meters
	1.609 kilometers

1.4 Metric units of volume

Unit	Equals
1 cubic meter	61,023 cubic inch 35.31 cubic feet 1.308 cubic yard 1000 liters
1 cubic decimeter	61.02 cubic inch 0.0353 cubic foot 1000 cubic centimeters 0.001 liter
1 cubic centimeter	0.000035 cubic foot 0.061023 cubic inch 1000 cubic millimetres 0.001 liter
1 cubic millimeter	0.000061 cubic inch 0.000000035 cubic foot 0.001 cubic centimeter
1 liter	1 cubic decimeter 61.02 cubic inches 0.353 cubic foot 1000 cubic centimeters 0.001 cubic meter 2.202 lbs of water

1.5 Imperial units of volume

Unit	Equals
1 cubic yard	46,656 cubic inches 27 cubic feet 0.764 cubic meter
1 cubic foot	1278 cubic inches 0.037 cubic yard 28.32 cubic decimeters 0.028317 cubic meter 7.481 gallons
1 cubic inch	16.387 cubic centimeters
1 gallon (British)	4.544 liters
1 gallon (U.S.)	3.785 liters

1.6 Metric units of weight

Unit	Equals
1 gram	15.43 grains 0.022046 lb 0.353 oz
1 kilogram	1000 grams 2.205 lb 35.27 oz
1 metric ton	2204 pounds 0.984 206 ton of 2 240 pounds 22.046 cwt 1.102 ton of 2000 pounds 1000 kilograms

1.7 Imperial units of weight

Unit	Equals
1 ounce	437.5 grains 0.0625 pounds 28.35 grams
1 pound	7000 grains 16 ounces 454 grams 0.454 kilograms
1 ton (2 240 pounds)	1.016 metric tons 1016 kilograms

1.8 Metric prefixes

Prefix	Symbol
Giga	G One thousand million 1,000,000,000
Mega	M One million 1,000,000
Kilo	k One thousand 1 000
Milli	m One thousandth 0.001
Micro	µ One millionth 0.000001
Nano	n One thousand millionth 0.000000001

2.1 Metric-to-imperial conversions

Name of unit	Value
Length	
1 cm	0.394 in
1 m	3.28 ft
1 m	1.09 yd
1 km	0.621 mile
Mass	
1 g	0.0353 oz
1 kg	2.20 lb
1 tonne	0.984 ton
Area	
1 cm^2	0.155 in^2
1 m^2	10.8 ft^2
1 m^2	1.20 yd^2
1 ha	2.47 ac
1 km^2	247 ac
Volume	
1 cm^3	0.061 in^3
1 m^3	35.3 ft^3
1 m^3	1.31 yd^3
1 m^3	27.5 bushels
Volume *(fluids)*	
1 ml	0.0352 fl oz
1 liter	1.76 pint
1 m^3	220 gallons
Force	
1N (newton)	0.225 lbf
Pressure	
1 kPa (kilopascal)	0.145 psi
Velocity	
1 km/h	0.621 mph
Temperature	
°F	$°C \times \frac{9}{5} + 32$
Energy	
1 kJ (kilojoule)	0.948 Btu (British thermal unit)
Power	
1 kW =	1.34 hp
Fuel consumption	
1 liter/1 km	2.35 mpg (miles per gallon)

2.2 Imperial-to-metric conversions

Name of unit	Value
Length	
1 in	25.4 mm
1 ft	30.5 cm
1 yd	0.914 m
1 mile	1.61 km
Mass	
1 oz	28.3 g
1 lb	454 g
1 ton	1.02 tonne
Area	
1 in^2	6.45 cm^2
1 ft^2	929 cm^2
1 yd^2	0.836 m^2
1 ac^2	0.405 ha
1 $mile^2$	259 ha
Volume	
1 in^3	16.4 cm^3
1 ft^3	0.02383 m^3
1 yd^3	0.765 m^3
1 bushel	0.0364 m^3
Volume *(fluids)*	
1 fl oz	28.4 ml
1 pint	568 ml
1 gallon	4.55 litre
Force	
1 lbf (pound force)	4.45 N
Pressure	
1 psi (lb/in^2)	6.89 kPa 0.069 bar
Velocity	
1 mph	1.61 km/h
Temperature	
°C	$(°F - 32) \times \frac{5}{9}$
Energy	
1 Btu (British thermal unit)	1.06 kJ
Power	
1 hp	0.746 kW
Fuel consumption	
1 mpg	2.35 liters/1km

2.3 Fraction-to-decimal conversion table

Fractions of inches to decimal values of mm and in.

Fraction	mm	in	Fraction	mm	in
1/64	0.397	0.016	33/64	13.097	0.516
1/32	0.794	0.031	17/32	13.494	0.531
3/64	1.191	0.047	35/64	13.891	0.547
1/16	1.588	0.063	9/16	14.288	0.563
5/64	1.985	0.078	37/64	14.684	0.578
3/32	2.381	0.094	19/32	15.081	0.594
7/64	2.778	0.109	39/64	15.478	0.609
1/8	3.175	0.125	5/8	15.875	0.625
9/64	3.572	0.141	41/64	16.272	0.641
5/32	3.969	0.156	21/32	16.669	0.656
11/64	4.366	0.172	43/64	17.066	0.672
3/16	4.763	0.188	11/16	17.463	0.688
13/64	5.160	0.203	45/64	17.859	0.703
7/32	5.556	0.219	23/32	18.256	0.719
15/64	5.953	0.234	47/64	18.653	0.734
1/4	6.350	0.250	3/4	19.050	0.750
17/64	6.747	0.266	49/64	19.447	0.766
9/32	7.144	0.281	25/32	19.844	0.781
19/64	7.541	0.297	51/64	20.241	0.797
5/16	7.938	0.313	13/16	20.638	0.813
21/64	8.334	0.328	53/64	21.034	0.828
11/32	8.731	0.344	27/32	21.431	0.844
23/64	9.128	0.359	55/64	21.8281	0.859
3/8	9.525	0.375	7/8	22.225	0.875
25/64	9.922	0.391	57/64	22.622	0.891
13/32	10.319	0.406	29/32	23.019	0.906
27/64	10.716	0.422	59/64	23.416	0.922
7/16	11.112	0.438	15/16	23.813	0.938
29/64	11.509	0.431	61/64	24.209	0.953
15/32	11.906	0.469	31/32	24.606	0.969
31/64	12.303	0.484	63/64	25.003	0.984
1/2	12.700	0.500	1	25.400	1.000

2.4 °C-to-°F conversion table

°C	°F	°C	°F	°C	°F
-9	15.8	21	69.8	51	123.8
-8	17.6	22	71.6	52	125.6
-7	19.4	23	73.4	53	127.4
-6	21.2	24	75.2	54	129.2
-5	23	25	77	55	131
-4	24.8	26	78.8	56	132.8
-3	26.6	27	80.6	57	134.6
-2	28.4	28	82.4	58	136.4
-1	30.2	29	84.2	59	138.2
0	32	30	86	60	140
1	33.8	31	87.8	61	141.8
2	35.6	32	89.6	62	143.6
3	37.4	33	91.4	63	145.4
4	39.2	34	93.2	64	147.2
5	41	35	95	65	149
6	42.8	36	96.8	66	150.8
7	44.6	37	98.6	67	152.6
8	46.4	38	100.4	68	154.4
9	48.2	39	102.2	69	156.2
10	50	40	104	70	158
11	51.8	41	105.8	71	159.8
12	53.6	42	107.6	72	161.6
13	55.4	43	109.4	73	163.4
14	57.2	44	111.2	74	165.2
15	59	45	113	75	167
16	60.8	46	114.8	76	168.8
17	62.6	47	116.6	77	170.6
18	64.4	48	118.4	78	172.4
19	66.2	49	120.2	79	174.2
20	68	50	122	80	176

2.5 °F-to-°C conversion table

°F	°C	°F	°C	°F	°C
0	-17.8	60	15.6	120	48.9
2	-16.7	62	16.7	122	50
4	-15.6	64	17.8	124	51.1
6	-14.4	66	18.9	126	52.2
8	-13.3	68	20	128	53.3
10	-12.2	70	21.1	130	54.4
12	-11.1	72	22.2	132	55.6
14	-10	74	23.3	134	56.7
16	-8.9	76	24.4	136	57.8
18	-7.8	78	25.6	138	58.9
20	-6.7	80	26.7	140	60
22	-5.6	82	27.8	142	61.1
24	-4.4	84	28.9	144	62.2
26	-3.3	86	30	146	63.3
28	-2.2	88	31.1	148	64.4
30	-1.1	90	32.2	150	65.6
32	0	92	33.3	152	66.7
34	1.1	94	34.4	154	67.8
36	2.2	96	35.6	156	68.9
38	3.3	98	36.7	158	70
40	4.4	100	37.8	160	71.1
42	5.6	102	38.9	162	72.2
44	6.7	104	40	164	73.3
46	7.8	106	41.1	166	74.4
48	8.9	108	42.2	168	75.6
50	10	110	43.3	170	76.7
52	11.1	112	44.4	172	77.8
54	12.2	114	45.6	174	78.9
56	13.3	116	46.7	176	80
58	14.4	118	47.8	178	81.1

Appendix B - Applications

Application - Excavator

The hydraulic excavator is one of the most common mobile machines today. It is used for a wide variety of jobs, from the smallest mini-excavator being used e.g. in gardening to the biggest crawler excavator utilized as a highly productive machine in the mining industry. Today, around 150 000 machines are produced every year.

Since the early days of the first cable control excavators, a lot of things have happened on the technical side. When hydraulics were brought onboard in the 1950's, the excavator became much more flexible and easier to control.

From that time, excavators have developed either to be even more flexible and easy to control, always with high capacity in the background, or to be mainly productive and efficient but still easy to control. Demands may differ somewhat but the application is always to be regarded as "heavy duty with high precision".

The major excavator market consist of so called "single bucket" excavators. They can be divided into three major groups:

- mini-excavators of less than 6 tons
- wheeled excavators of more than 6 tons
- crawler excavators of more than 6 tons

The machine weight may range from 300 kg *(650 lb)* to more than 500 tons *(1,100,000 lb)*. A typical wheeled excavator weighs no more than 25 tons *(55,000 lb)*. Irrespective of weight, the motion control is hydraulic, today almost without exception.

Certain things characterize an excavator. The diesel engine is normally underpowered which puts special demands on the hydraulic control system and, especially so, on the pump control and the directional control valve.

The excavator is very much a heavy duty application and demands are consequently high on each component. This is especially true for components used in the swing (slew) function which may be operating 75% of the working time.

Function description

The main functions are the same for all excavators:

- boom
- dipper-stick
- bucket
- swing
- travel (wheels or crawler tracks)

Two remote control levers, "joysticks", are attached to the front of the armrests of the driver's seat. With two foot pedals, and the two joysticks, the operator can control all main excavator functions.

Ergonomic aspects are very important for the driver in this type of machine since he, or she, may be working all day with levers and pedals in the cabin. This is also the reason why hydraulic remote control has become almost universal on excavators, except on the smallest ones, as both high speed movements and low speed precision maneuvers are possible.

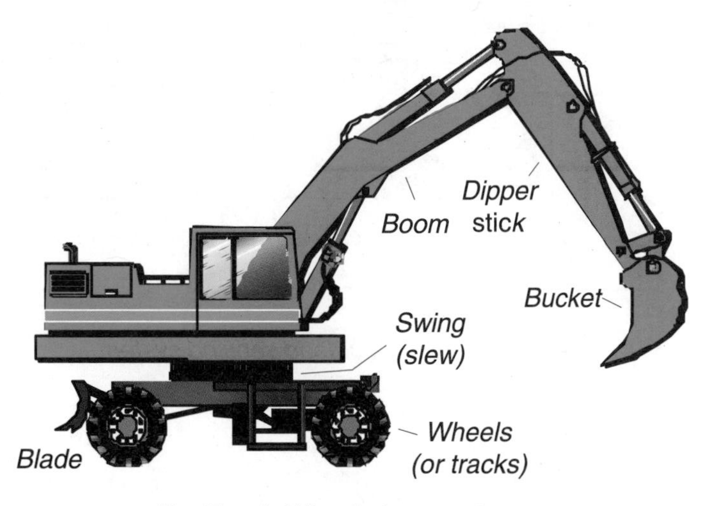

Fig. Exc-1 Wheeled excavator

In recent years, however, levers for electrohydraulic control are becoming more and more common.

Much effort has been put on controllability in order to make the machine easier to operate. The machine should behave consistently under all working conditions. The operator is able to control several functions at the same time, e.g. boom, bucket and stick at the same time as the swing. This is the reason why load sensing systems are becoming more common nowadays, especially in multipurpose machines.

Fig. Exc-2 illustrates a modern load sensing system. All functions are load sensing except for the swing function which is open centre. This gives the swing function priority over other functions. At the same time, these other functions do not interfere with each other.

Many excavator functions can be exposed to cavitation, either in one or in both directions. For example, the swing and the track drives need to be protected in both directions; boom, stick and bucket mainly in only one. To avoid cavitation, the flow from the fixed pump is utilized in connection with a counterpressure valve. If, for example, the boom is lowered by its own weight, the pressure on the piston rod side of the corresponding cylinder decreases. The counterpressure valve increases the pump pressure and part of the pump flow is directed over an anti-cavitation check valve in the directional control valve to the piston rod side, which helps prevent cavitation.

Especially critical to cavitation are functions such as the dipper stick where the load is on the piston rod side. In this case additional makeup flow is required to the piston side as the flow exhausted from the rod side is insufficient.

The solution, once again, is to use pump flow from the open center valve, pressurized by a counterpressure valve in the return line.

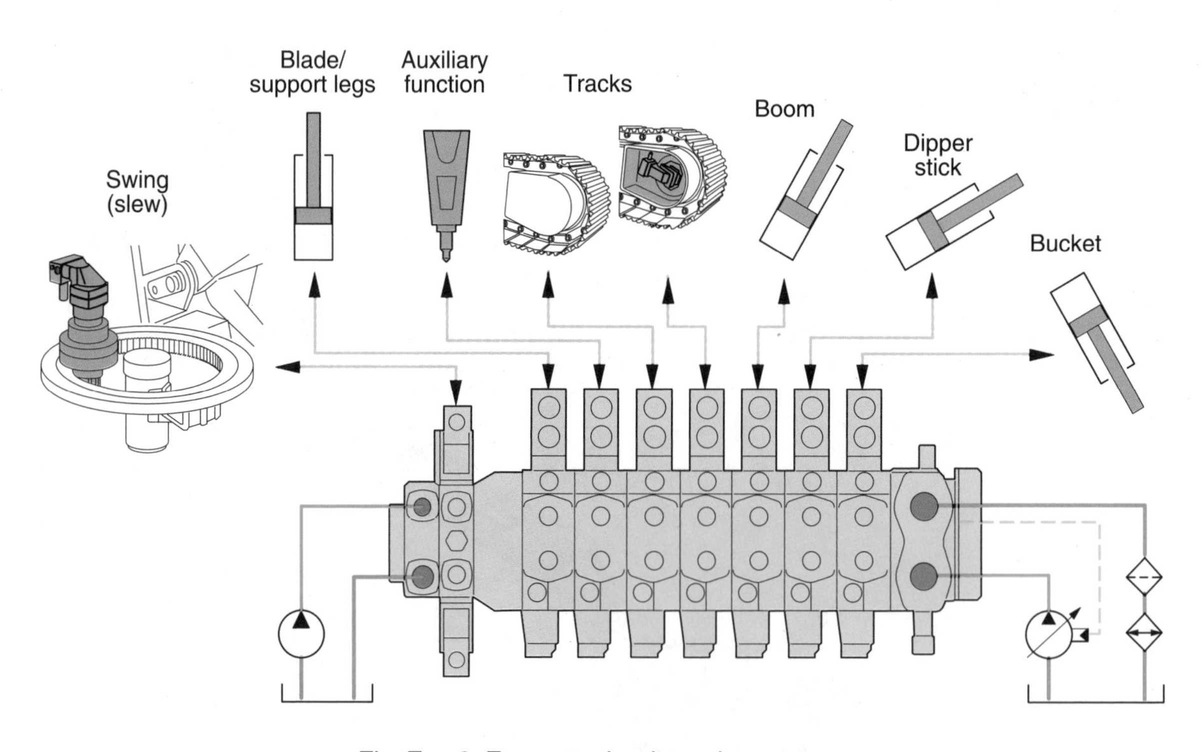

Fig. Exc-2 Excavator load sensing system

Application - Forestry

Years ago, utilizing horses, donkeys, elephants or other animals used to be the main way of transporting logs out of a forest. Within a relatively short period of time, the harvesting of wood has changed from being a fully manual to a fully industrialized process. Most of the development has taken place over the past 25 years.

Due to varying climates in different parts of the world, there are different ways of taking care of the forest. The time it takes from planting the tree until it is fully grown and ready to be cut down varies from 6 to more than 100 years.

In slow growing forests, specifically in cold climate regions, the thinning operation takes place before the forest is cut down. Cutting the forest has more and more become a very advanced process and is now fully mechanized.

Today mechanized harvesting falls into two methods:
- full-tree harvesting
- cut-to-length harvesting

Full-tree harvesting

The harvesting process starts with felling and bunching the trees. A tracked "feller-buncher" built on an excavator chassis (fig. For-1) is probably the most advanced and most productive machine for this purpose.

The hydraulic system of a feller-buncher contains variable displacement motors with integrated planetary gear-boxes for propelling the tracks.

The upper structure has a fixed displacement motor with integrated planetary gearbox for the swing *(slew)* function and hydraulic cylinders for the hoist and stick boom as well as the tilt and wrist functions.

Fig. For-1 Tracked feller-buncher with rotating saw head

The saw head (fig. For-2) has a continuously rotating, 50 mm *(2")* thick disc with a diameter of up to one meter *(40")*, which is equipped with cutting teeth. At maximum speed, typically 1000 to 1300 rpm, the inertia of the disc can momentarily transmit a power of up to 600 kW *(800 hp)* to the cutting teeth. The variable displacement motor restores the energy in the disc after the tree has been cut.

The saw head contains clamping arms for grabbing the tree after the cut and "accumulator arms" that gather several small diameter trees.

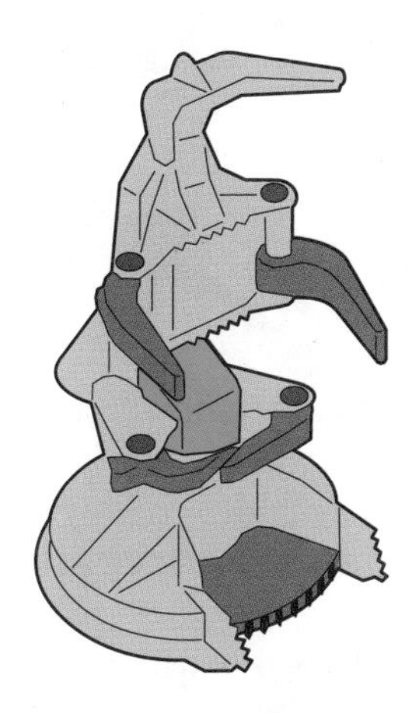

Fig. For-2 Rotary saw head

The schematic of a typical system (fig. For-3) shows a dedicated pump for the saw motor driving the saw disc and a fixed displacement pump for the clamp and wrist cylinder functions.

The main, variable displacement pump supplies the track drives and the swing and boom functions.

Two-position, on-off control valves in a separate manifold activate e.g. low speed for the track drives when the motors are locked in max. displacement. The swing brake is also released with a valve function in this manifold.

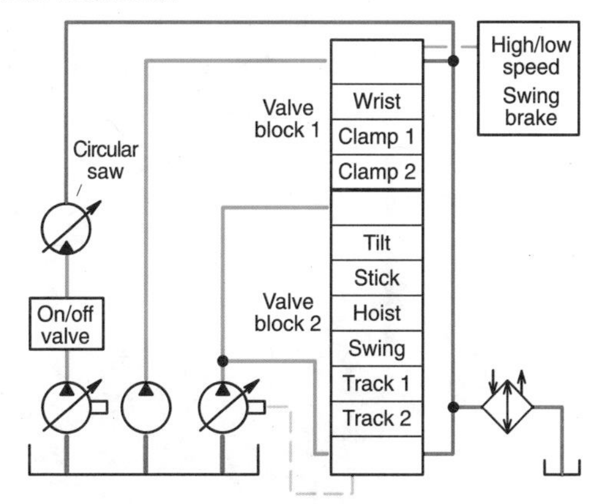

Fig. For-3 Simplified schematic for a feller-buncher with a circular saw head

A “skidder” is used for transporting the trees to the roadside where a “log loader” with a de-limbing unit cuts off the branches. When the tree is intended for pulp, the log loader (provided with a saw head) cuts it into 4 meter *(13 ft)* logs.

In addition, the log loader is utilized for loading logs (fig. For-4) onto a truck which takes them to a plant for further processing.

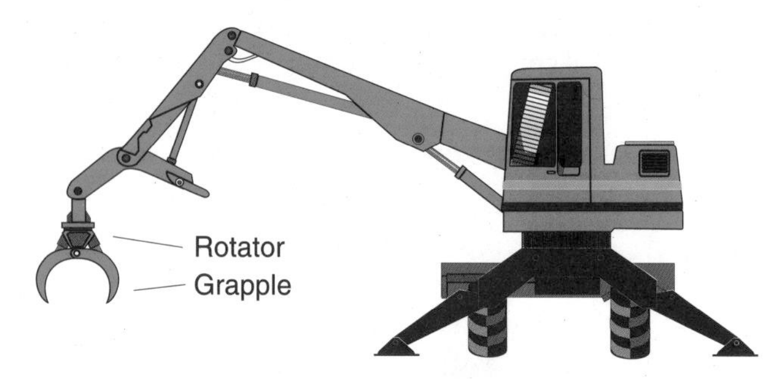

Fig. For-4 Load loader (shown with a grapple)

Cut-to-length harvesting

The harvesting process starts with a ‘harvester’ which not only cuts the tree at the root but also de-limbs and cuts it into logs of suitable lengths.

The machine carries a harvesting head (fig. For-5) which performs several processing steps. The hydraulic functions of the head are:

- arms holding the tree while being cut at the root
- cylinders pulling the drive rollers against the tree
- drive rollers pulling the tree through the head
- de-limbing knifes cutting off the branches
- chain saw driven by a high-speed hydraulic motor.

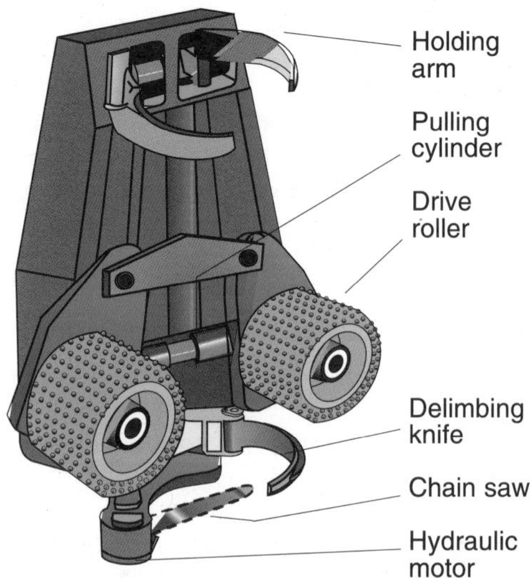

Fig. For-5 Harvesting head with chain saw

When the tree is being pulled through the head, sensors measure the diameter and the length of the tree. The information is fed to an on-board computer which determines the cuts so that the treee is utilized in an optimized way.

The harvesting head is carried either by a dedicated rubber-tired harvester with a crane or attached to the stick boom of an excavator; refer to fig. For-1.

When the tree has been cut into logs, a forwarder (fig. For-6) hauls them to the roadside. A forwarder is typically propelled by a hydrostatic transmission (fig. For-7), permitting the operator to control traction and speed independently, while maintaining full control of the vehicle when negotiating ground obstacles. With two joysticks, each having three proportional functions, the operator can control all 6 crane functions.

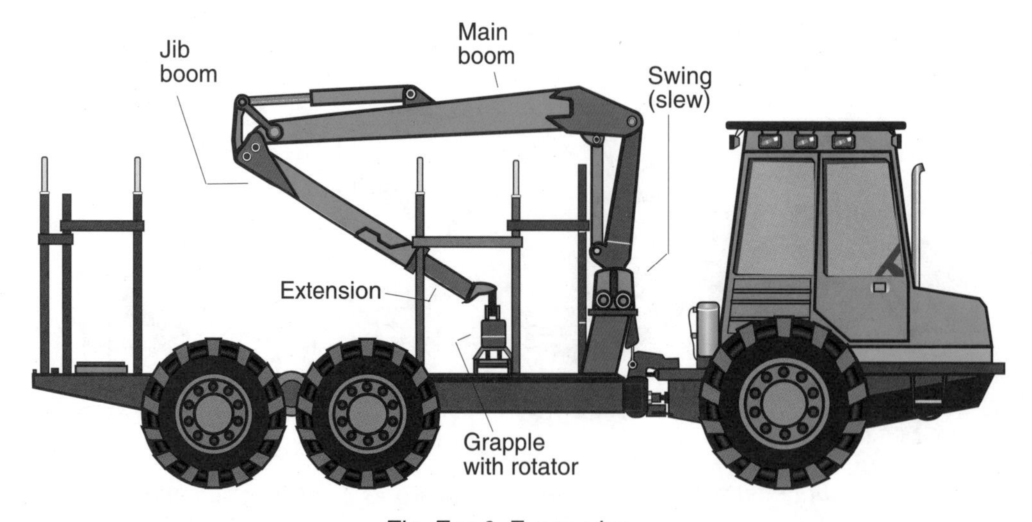

Fig. For-6 Forwarder

Other hydraulic functions are:

- high or low gear
- articulated steering
- differential lock
- AWD (all-wheel drive) control
- dozer blade
- "ladder" (securing the load)

The loading capacity of a typical forwarder is 8 to 15 tons.

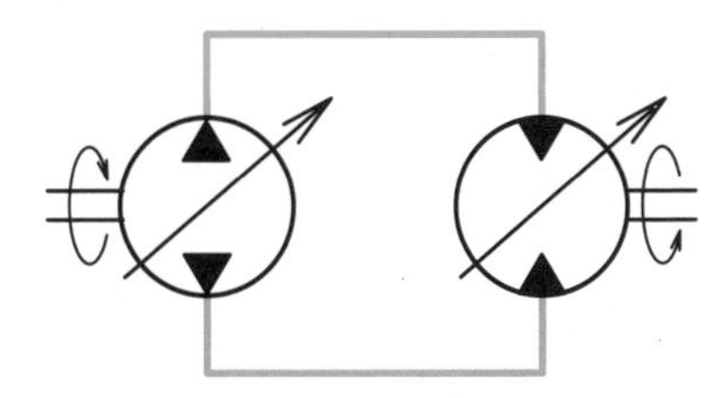

Fig. For-7 Simplified ydrostatic propel transmission

A common hydraulic system includes a dedicated hydrostatic transmission (fig. For-7) with a pump and a motor. The main load-sensing pump supplies a load-sensing valve with 7 sections (fig. For-8). It also supplies a manifold with control valves for auxiliary functions.

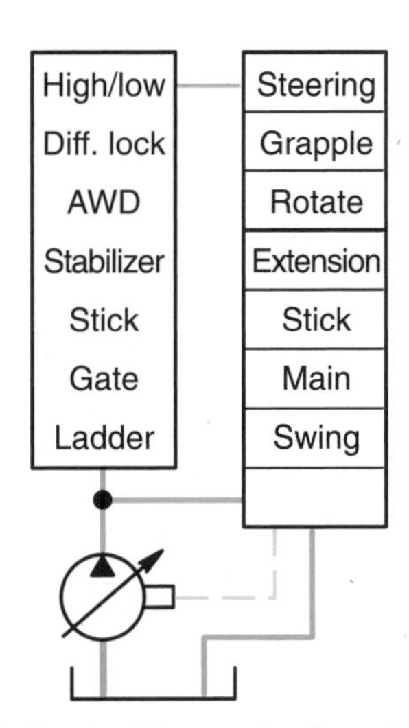

Fig. For-8 Typical forwarder function schematic

The forwarder dumps the logs at the roadside. Later, a truck equipped with a timber crane (fig. For-9) picks up the logs and transports them to the processing plant. The truck normally has a fixed displacement, bent-axis pump installed on the gearbox PTO (power-take-off).

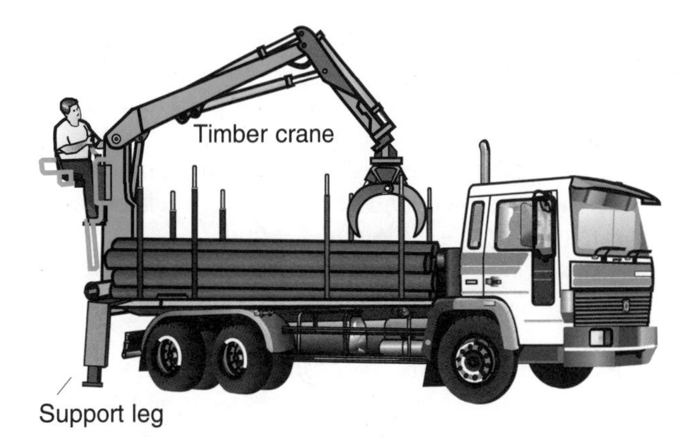

Fig. For-9 Truck with timber crane

A 6-spool, open-center valve controls the following functions on the articulated timber crane:

- swing (slew)
- main boom
- jib boom
- extension
- grapple
- grapple rotator

Hydraulic support legs provide increased stability when te truck is loading or unloading.

Application - Mining

Over the past 50 years, drilling has improved drastically due to mechanization. As an example, drilling capacity in hard rock has increased from 10 m/hour *(30 ft/hr)* and operator to over 300 m/hr *(1000 ft/hr)*.

The general term "drill rig" covers several machine types used in the construction industry, utilized for:

- mining above and under ground
- quarrying
- tunnelling

A particular task determines the drilling method and type of machine choosen. For this reason, there are many ways of categorizing drill rigs according to:

- Task
- Drilling method used
- Cutting removal method used.

Fig. Min-1 shows the main components of a typical drill rig (here utilizing an auger tool).

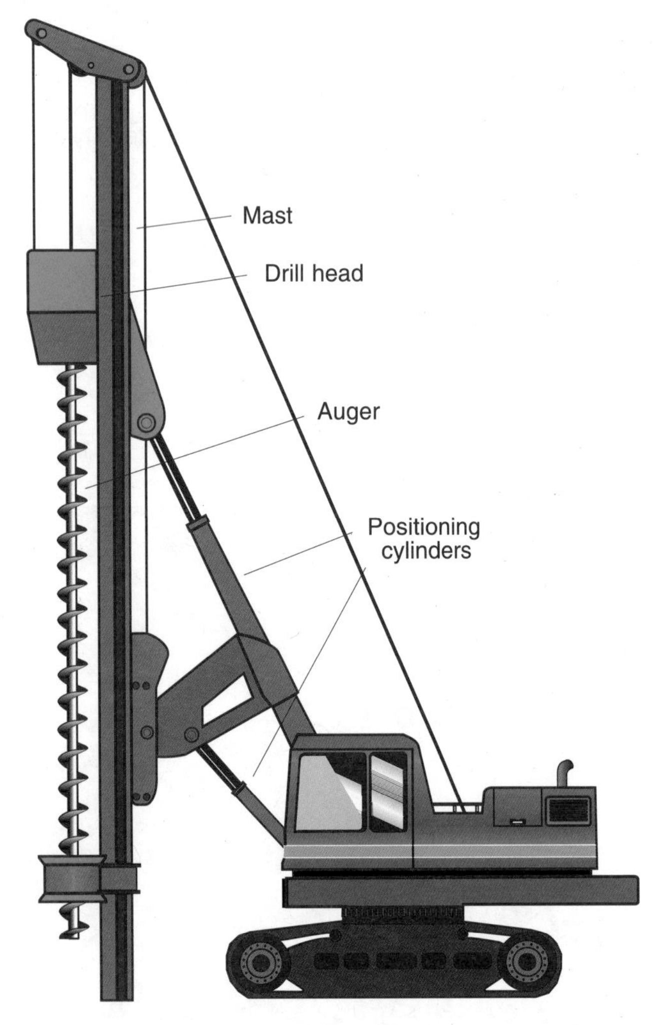

Fig. Min-1 Auger drill rig

Construction work

Basically, drill rigs can be divided into the following four groups:

1. Percussive drilling - Percussive drilling is a method where the hole is formed by striking the drilling tool. The ground or rock at the end of the tool is then being crushed; cuttings are removed periodically.

2. Rotary percussive drilling - Rotary percussive drilling is performed by a piston which strikes directly on the drill bit ("down the hole" hammer drill) or by percussive energy transmitted via a drill rod to the drill bit. Either hydraulic or compressed air powers the piston.

During the drilling process, the drill rod is rotated either continuously or intermittently. The cuttings are continuously removed by a flushing medium, carried to the drill bit inside the drill rod.

3. Rotary drilling - Rotary drilling is a method where torque is transmitted to the drill bit by means of a drill rod where, at the same time, a feed force is applied through the drill rod feeding system. The ground or rock at the bottom of the hole is crushed (with a rolling action) or cut (with a shearing action).

The cuttings are removed either periodically or continuously. If the ground is not self-supporting, a casing or a drilling fluid can be used to stabilize the surrounding wall.

4. Pile drilling - Pile drilling is normally a rotary drilling method for producing a hole in the ground which is either filled with concrete or used for installation of a prefabricated pile.

The pile drilling tool may collect the cuttings within its body. From time to time, the tool is lifted out of the bore with the feed mechanism and the cuttings discharged beside the bore hole by turning the rig sideways with the swing (slew) function.

Instead of the above method, a "continuous flight" auger (fig. Min-1) or a pure rotary drilling method can be utilized.

Most machines today are very sophisticated and use a lot of hydraulic power for various machine functions.

Main functions

The main drill rig functions are:

- drill bit rotation
- percussion (if applicable)
- feed

The performance of these functions determine the drilling capacity. Fast and accurate positioning of the drill bit is another important function that helps increase the effectiveness of the machine.

The block diagram in figure Min-2 shows an example of the hydraulic functions of a large drill rig for earth drilling. The rig can be used for foundation work, drilling water wells or for environmental purpose.

Control valves for the individual drill rig functions are grouped into three valve blocks mainly because:

- certain functions are far apart
- it is an advantage to have the valve block as close
- it is difficult to put several sections together due to design limitations of the basic valve

Function description

Individual valves in the valve blocks can be remotely controlled from the cabin of the machine or from a work station, either by means of hydraulic or electro-hydraulic pilot control. Today, more and more drill rigs are marketed with sophisticated electronic controls for automatic operation, particularly in the mining and tunneling industry.

Valve sections connected to various functions control flow and pressure, many times with a load sensing system. Tracks move the drill rig from one place to another. Tracks and winches require a lot of power but are not used simultaneously; these functions are, therefore, usually located on the same valve block.

Main and auxiliary winches are used mainly for rod and tool handling.

Tool *rotation* and *feed* are the most important functions of the drill rig. Rotational torque on a large drill rig (fig. Min-1) may be 250 000 Nm *(185,000 lbf ft)* or more. The operating speed of the drill tool in an earth drill is between 5 and 10 rpm when drilling with an auger (as shown in the illustration).

When the auger is fully "loaded", it has to be emptied. The auger is then lifted out of the hole by means of the feeding system and turned sideways with the swing (slew) function. In order to dispose of the accumulated cuttings, the rotation is reversed and the speed increased. The entire procedure is repeated until the required depth of the hole has been reached.

Tool *feed force* is provided by pressurizing a feed cylinder or feed motor, whichever is utilized. The set feed force may have to be reduced during the drilling operation to compensate for the increasing weight of the "kelly" bar, drill rod or drill pipe.

For drilling operations like geological drillings, the weight of a very long drill rod may require a cylinder feed force in the opposite, pulling direction in order to prevent excessive drilling pressure.

Positioning cylinders support the drill mast. For the positioning and auxiliary functions, a low-flow valve block is choosen as the drilling and positioning functions are not engaged at the same time.

The *swing* function is required to be smooth and accurate. A special valve block installed directly on the motor ports provides gradual and shock-free acceleration and retardation.

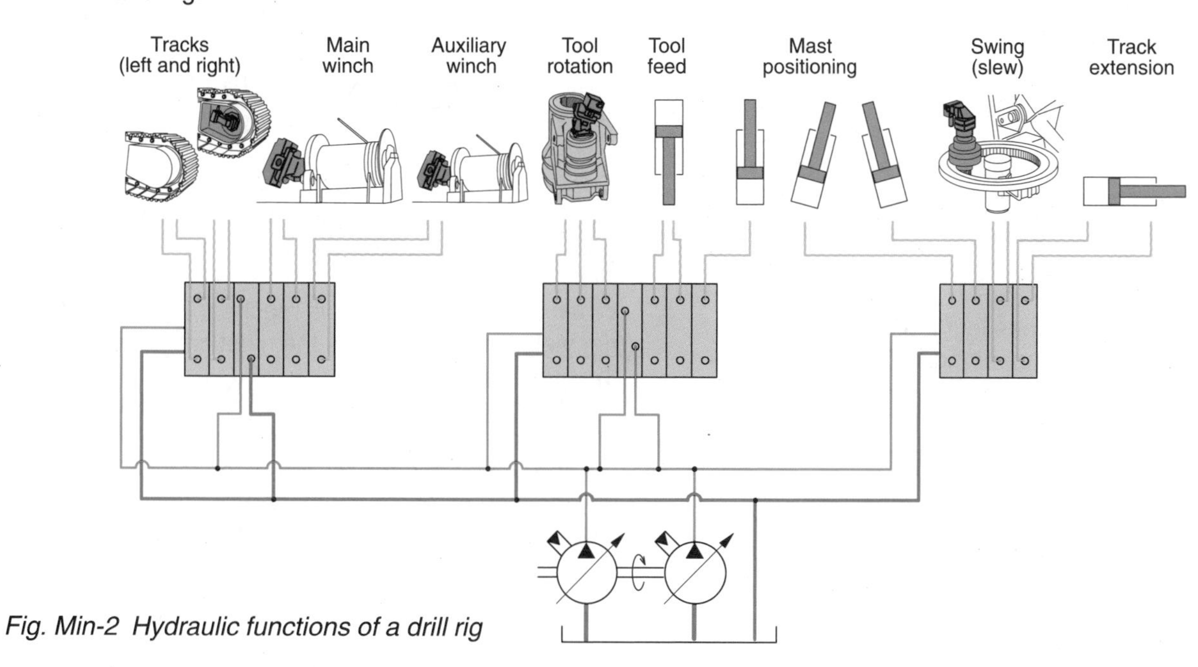

Fig. Min-2 Hydraulic functions of a drill rig

Application - Refuse collecting vehicle Rear end loader

The rear end loader (REL) is typically used to pick up domestic refuse. A crew of two to four men is normally engaged in the loading.

An increasing number of machines are equipped with bin lifts (fig. RCV-1).

Fig. RCV-1 Rear end loader

Nowadays, machines are operated in more than one working shift per day, thereby putting higher demands on durability and economy.

Automatic transmissions are now the industry standard for propelling RELs. This puts demand on the hydraulic system to handle the 'torque conflict' when selecting a gear at low engine speed and simultaneously starting the packing cycle.With very few exceptions, these machines are all built on commercial chassis.

The REL is classified of European Community (CE) as a "dangerous machine" and is (in Europe) exposed to safety approval.

An open-back REL has four functions:

- sweeping the hopper (fig. RCV-2)
- packing
- ejecting the refuse
- opening and closing the hopper door

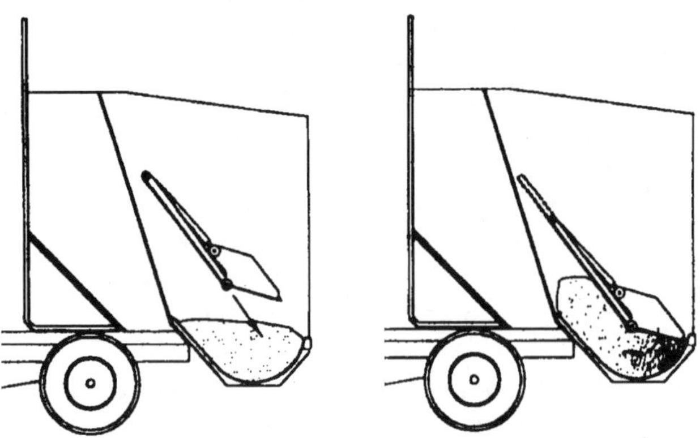

Fig. RCV-2 Packing blade ready to sweep (left); sweeping the hopper (right)

Compaction

The refuse is dumped into the rear compartment, normally known as the hopper. When full, the packing sequence is started by the operator activating a switch or a handle (fig. RCV-3).

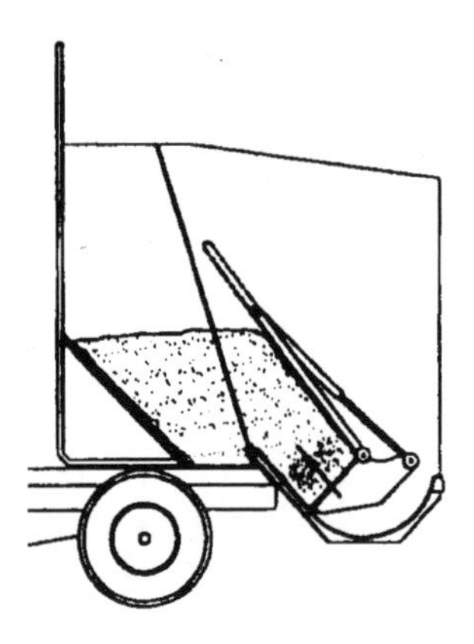

Fig. RCV-3 Packing

(If the REL is of the open back type, i.e. without bin hoist, the switch must be "hold to run" until the sweep blade has passed the hopper edge).

Once the sweep has completed its stroke, the packing function takes over, starting to compress the refuse towards the ejector-barrier.

When sufficient pressure is reached, the ejector-barrier yields forward. This is the most critical func-tion of the process as:

- the yielding takes place towards a telescopic cylinder with varying area and angle
- the weight and the friction of the refuse varies
- the type of refuse varies and, in addition to this, the rear axle is already heavily loaded

A typical compaction ratio is between 1:5 and 1:7.

Discharging

When full, the REL is moving to an appropriate dump-site where the refuse is dumped.

The entire rear end is then lifted and the ejector barrier moves rearward, thereby pushing the refuse out of the container.

Hydraulic system

A single flow or twin-flow, fixed displacement pump can be utilized, driven from the PTO (power-take-off) or the engine crankshaft (fig. RCV-4).

When a bin lift is provided, it has its own dedicated pump. Alternatively, depending on the size of the lift, a shared pump would be used with full priority to the lift function.

The hydraulic valve functions could be located either in one valve bank or be split into two banks. The first bank is then mounted at the front of the container, including the ejector barrier and hopper lift functions.

The second bank is then mounted at the rear end, on top of the hopper, comprising the sweep and packing functions. The packing circuit has normally an integrated, regenerative function to lower the cycle time.

Constantly increasing environmental demands will, eventually, call for variable displacement pumps (fig. RCV-5) in order to lower the noise level, fuel consumption and engine RPM. This will also improve the 'driveability'of the vehicle, in turn improving productivity. The same environmental demands also calls for oil spill protection and biodegradeable hydraulic fluids.

Control system

The REL is subject to a vast number of safety regulations but it also to high productivity demands.

In addition, the cycle is automated to a high degree. This makes the vehicle very suitable for an electronic control system (fig. RCV-5) which also helps the fleet operator monitor the "LCC" (life cycle cost).

As these machines are highly productive, maintenance, monitoring and fast troubleshooting must be addressed seriously.

Hydraulic functions

In the following, the main features and hydraulic components of the various RCV functions are highlighted; refer to the corresponding hydraulic schematics.

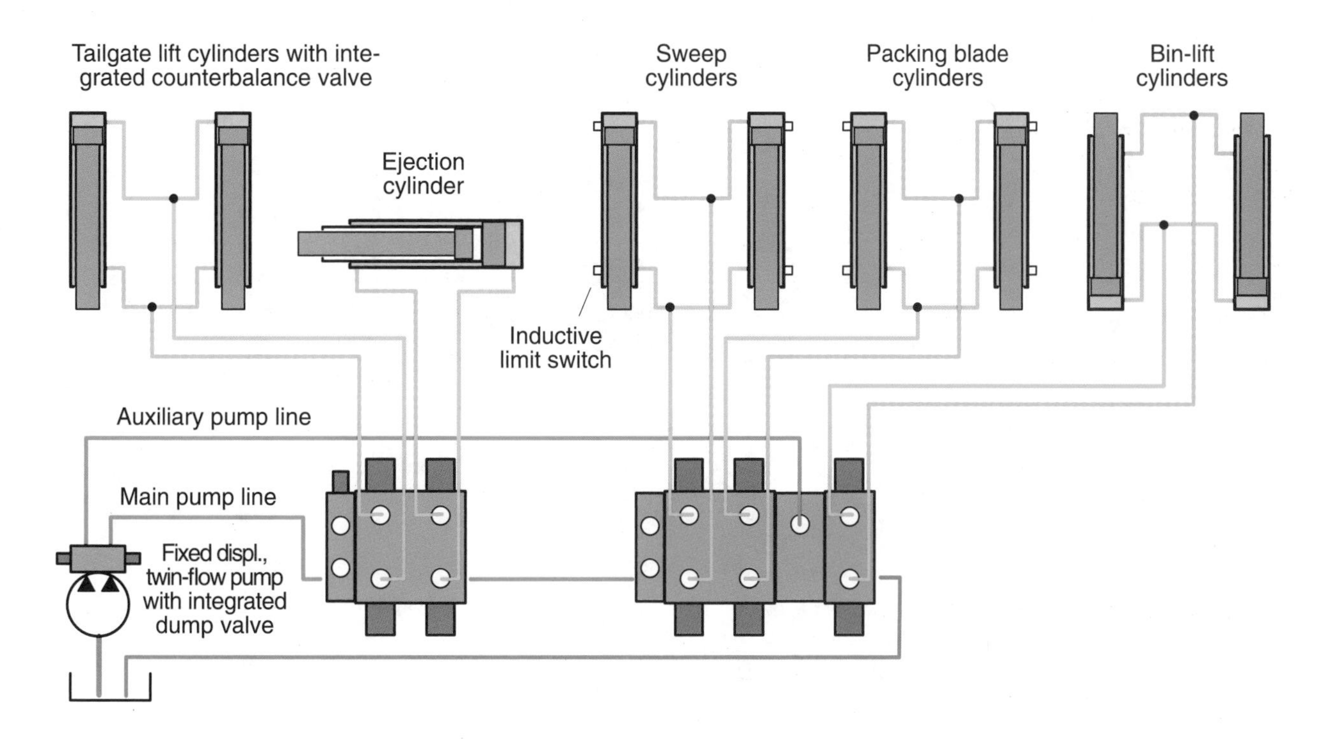

Fig. RCV-4 REL circuit with twin-flow pump and open center valves

Hopper door function

- directional control valve with a special spool adapted to work with external counterbalance valves
- controlled closing of the hopper door by a built-in flow control function in the spool (EC regulation) (fig. RCV-6)
- softer opening of the door (less mechanical stress)
- max pressure reduced on “closing” to avoid rubber seal damage
- Spool leakage compensated to avoid “ghost movement” of the hopper door when operating the packing mechanism

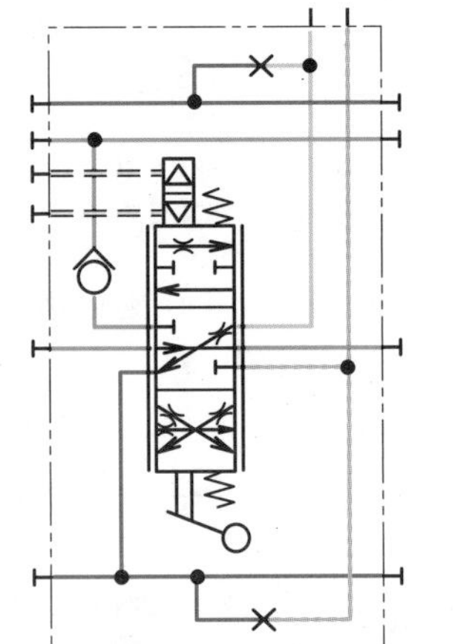

Fig. RCV-6 Hopper door circuit

Ejecting function

- Controlled movement of the ejector barrier by a proportional, integrated sequence valve
- easily accessible and sealed for service and maintenance
- service line relief valve with ‘flat’ characteristic to avoid buckling of the telescopic cylinder on ejection
- service line relief valve with extremely well defined opening and closing characteristics to avoid creep movements during the first packing phase (when the small area of the ejector cylinder is active).

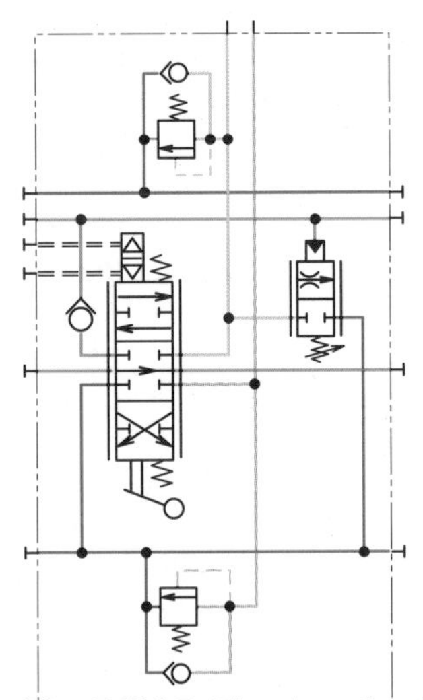

Fig. RCV-7 Ejection circuit

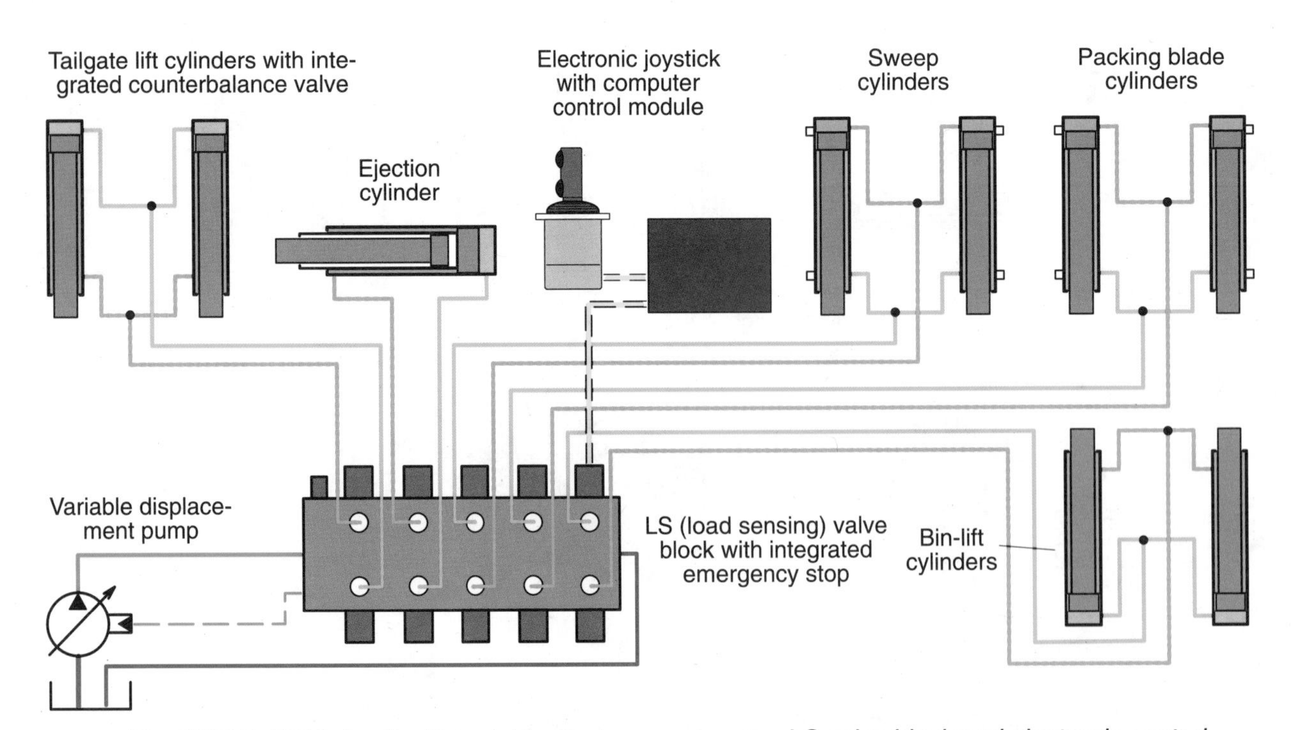

Fig. RCV-5 RCV circuit with variable displacement pump, LS valve block and electronic control

Sweeping function

- Double acting spool, optimized to handle the large return flow during the return stroke
- service line relief valve with extremely well defined opening and closing characteristics to allow a controlled rollback
- built-in and preset pressure switch in the "sweeping plus" port (fig. RCV-9)

Packing function

- Regenerative spool operating the packer plate for high-speed compaction
- built-in and preset pressure switch in the "packing minus" port
- vented neutral position of the piston rod side to eliminate parasite signals to the pressure switch
- service line relief valve allowing the packing cylinders to retract if the sweep cylinders hits a too big an obstacle during the first part of the sweep stroke.

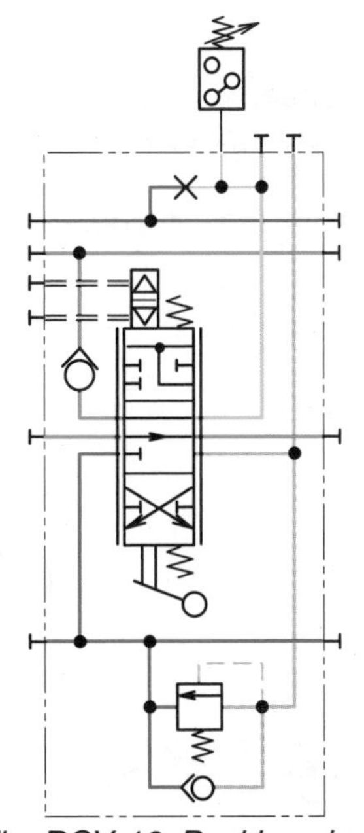

Fig. RCV-10 Packing circuit

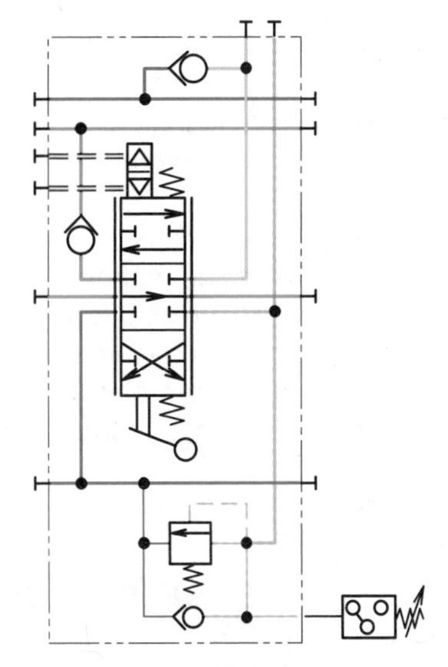

Fig. RCV-9 Sweeping circuit

Application - Truck Crane

Selfloading cranes are used to load and unload own cargo. They are normally mounted on commercially produced chassis.

Small size cranes are controlled manually by hand from either side of the truck (lorry). Bigger cranes use, alternatively, radio controlled remote control systems. All cranes of 4 tm (ton-meters) and up are equipped with overload system and emergency stop.

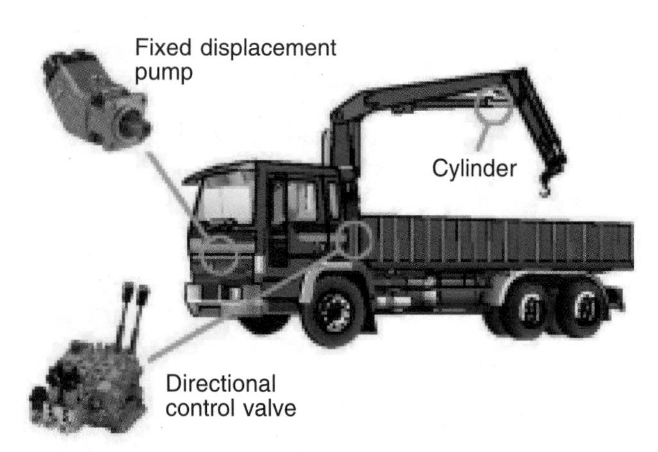

Fig. Crane-1 Truck with cargo crane and main hydraulic components

Functions

A CFO (constant flow, open circuit) system is normally used but many manufacturers are more and more changing to an LS (load sensing) system and radio control. The flow rate is 30–200 l/min *(8–50 gpm)* and the operating pressure between 210 and 350 bar *(3000–5000 psi).*

Crane

The crane has normally between 4 and 9 functions depending on type. The valves have to be configured so that the spool functions meet legislation requirements.

Swing

Small cranes usually use a cylinder type swing (slew); bigger cranes use hydraulic motors.

Due to the varying torque requirement as well as the big differences between starting and operating torque, jerkiness on the swing function is felt by the operator and leads to very uncomfortable operating conditions. Consequently, it is important with a smoothly functioning pressure control (so called 'force feedback'). Suitable pressure compensation is also important in order to avoid interference from other functions.

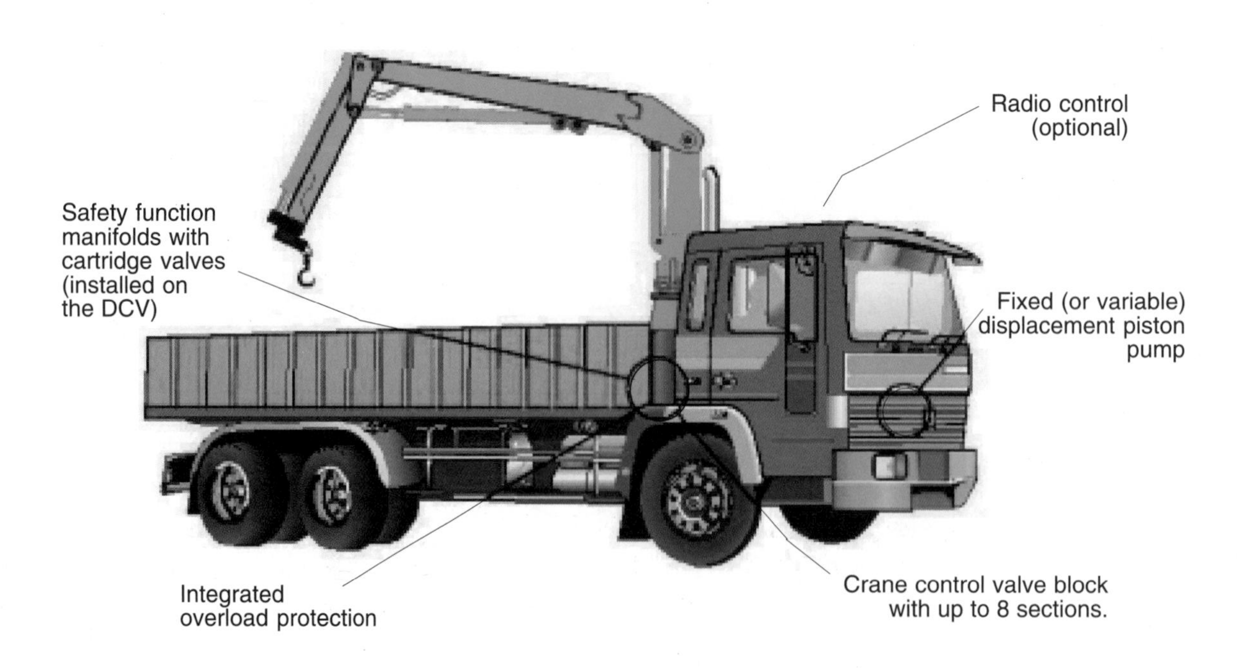

Fig. Crane-2 Cargo crane with selected control equipment

Main boom

The main boom needs good pressure control. (When using load sensing, force feedback is only required on the lift side).

Some crane manufacturers have equipped the main boom with a "lowering brake", which secures the same lowering speed as the lift speed, independent of load pressure.

Second boom

The second boom is sometimes called "knuckle arm". Normally, it has the same control requirements as the main boom but without force feedback.

In some installations, due to legislation, the spool of the control valve must operate in the opposite direction as compared to the main boom.

Telescope

The telescope sometimes utilizes a regenerative control function in order to reduce the pump flow requirement. A prerequisite is, however, that the telescope operates at a reduced pressure.

Winch

Winch motors normally need a pressure controlled start-up flow due to high internal leakage.

Support legs

The support leg functions are not as critical as the previous functions.

Special functions
Emergency stop

In Europe, the crane must have emergency stop equipment in accordance with the EN 418 regulation. When the emergency stop is activated, the pump supply must be cut off and all dangerous movements stop without creating any additional hazards.

Counterbalance valve

There are certain things to consider when selecting a DCV spool in connection with a counterbalance (overcenter) valve.

On the market today are various types of counterbalance valves. It must be determined if the valve is dependent of backpressure and if it has a separate relief valve.

A separate relief valve is always preferred because it has a better pressure-versus-flow characteristic. Due to the high hysteresis of the counterbalance valve, the internal relief function must be set 30% above max allowed load.

When the counterbalance valve has an internal relief and is dependent of back pressure, a type 'D' spool with large leakage grooves or an 'M' spool must be chosen in order to prevent damaging pressure spikes. The best function is obtained by choosing as low an area ratio as possible for the counterbalance valve.

Leakage grooves

There are various size leakage grooves on a spool of a directional control valve. Comparatively large grooves are used when the DCV valve is installed in a CFO system.

In LS and CFC systems where the spool controls flow instead of pressure, small leakage grooves can stabilize a function against oscillations.

Speed reduction

The internal speed limiting function, found in some valve designs, limits the flow to a specific consumer to a selected, max value.

The limiting function does not influence the incoming pump flow (when other consumers are operating); it just reduces the outgoing flow to the actuator.

Overload protection

According to EC regulations, all cranes of four metric tons or more must be equipped with an overload protection system.

In the past, this system has been fully hydraulic where a sequence valve was connected to the main boom, measuring the load pressure (fig. Crane-3). When pressure reached the selected level, the sequence valve sent a signal to the POL control (or controls) of the DCV.

The control then pushed the spool back to the center (neutral) position, preventing movements that could otherwise increase the load moment of the crane to unsafe levels.

Today more and more electronic control equipment is used on cranes, including the overload device (fig. Crane-3).

The sequence valve is replaced by a pressure gauge which, with a spool position indicator and a small computer, detects an overload situation. This, in turn, activates the emergency stop function (or functions).

A disadvantage with the system may be that all functions stop when an overload situation has been detected.

Some valves use the internal logic system. The load pressure signal goes via separate cartridge solenoid valves to those functions that can cause overload. When any of the solenoid valves is de-energized, the load signal is cut off.

Chock valve

It is important to remember, that separate chock valves should be installed when the counterpressure valve is backpressure dependent.

Directional control valve selection and installation

- Select a DCV that offers the the best compromise between maneuverability and energy efficiency.
- Determine where the valve will be installed on the machine and how the operator wants the control levers organized.
- With a counterpressure valve in the circuit, specific DCV requirements must be met.
- Find out if the DCV will be controlled from one or two operator positions.

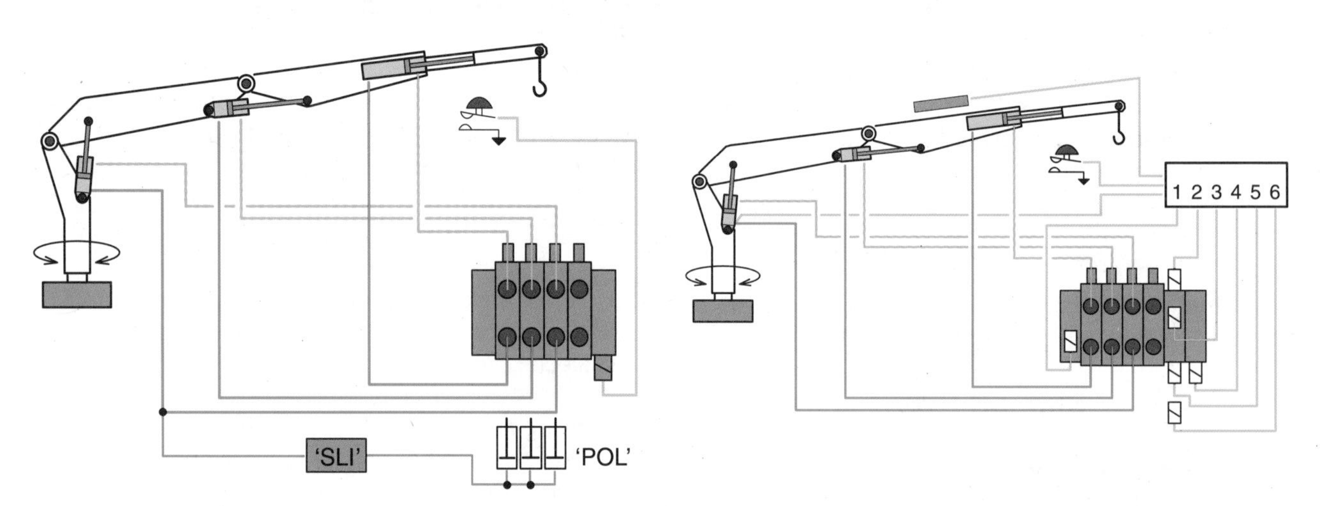

Fig. Crane-3 All hydraulic overload protection system (left); Electronic overload protection system (right)

Application Wheel loader

The use of wheel loaders is well spread around the world. The total number of machines is probably well over 700,000. Wheel loaders can be divided into three groups:

- compact (less than 50 kW/*65 hp*)
- midsize (50 to 110 kW/*65 to150 hp*)
- large (more than 110 kW/*150 hp*)

Compact loaders

Basically, compact loaders (fig. Load-1) are machines used for utility purposes. They are very common in e.g. the middle and southern parts of Europe. The German market is the single largest consumer of this size machine, and the volume is growing.

Small loaders up to 50 kW *(65 hp)* are found in rental fleets, communities and agriculture use. Price and reliability are more essential than efficiency, fuel consumption and ergonomics.

Flexibility is an important feature of these multipurpose machines, available with a variety of attachments like forks, etc.

The compact loader is mostly equipped with hydrostatic transmissions, built on a rigid or articulated frame. Most control valves are manually operated through linkages. The flow from the fixed displacement pump is often shared with steering, which has full priority (fig. Load-2). The hydrostatic function uses a simple mechanical or hydraulic automotive control.

Midsize loaders

The midsize loader has an engine power between 50 and 110 kW *(65 and 150 hp)*. It is normally used as a multipurpose machine, available with a variety of multiple attachments. The midsize loader is often utilized as a service machine and a tool carrier. It has a good capacity as a production loader with a short cycle time.

Fig. Load-1 Compact loader

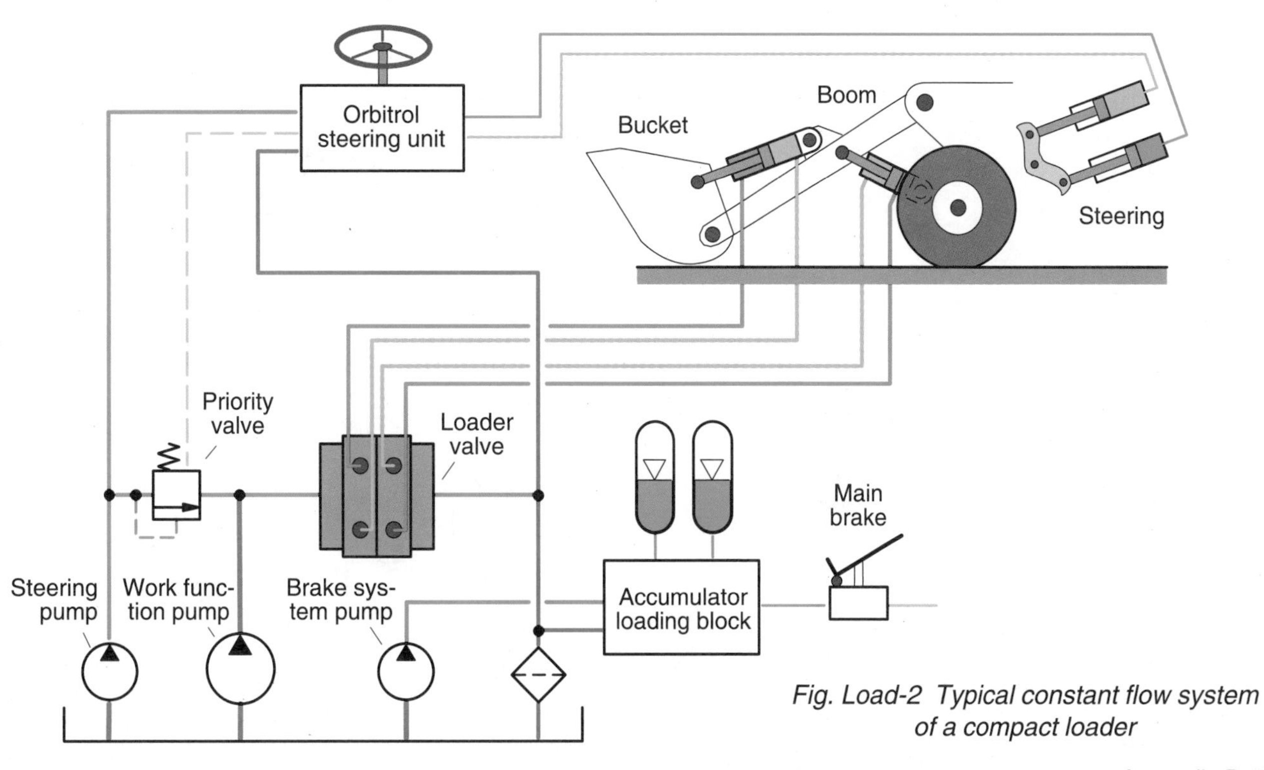

Fig. Load-2 Typical constant flow system of a compact loader

Half of the machines are equipped with a hydrodynamic, the other half with a hydrostatic transmission. The control system is rather sophisticated compared to that of the smaller machine and the cab is ergonomically designed for high productivity and comfort.

A variable displacement pump supplies a load sensing (LS) system (fig. Load-3) for implements and steering. Most machines have LS steering in order to reduce cab noise, lower power losses and allow better operating control.

In most cases, the valves use hydraulic pilot control with solenoid detents for semiautomatic functions. Linear remote control valves are as common in the cab as joystick controls.

Large loaders

Loaders above 110 kW *(150 hp)* are used as a professional loader. They are mainly equipped with a hydrodynamic transmission installed in an articulated frame and control valves are hydraulic-pilot assisted. The full-size loader (fig. Load-4) is mostly used as a production machine in quarries, open-pit mining, construction sites and various industrial applications.

Fuel economy, ergonomics and, of course, reliability are essential features of a production machine. A low operating cost and high productivity is more important than what is valid for small and midsize loaders.

Boom function

The boom is normally equipped with two double-acting cylinders with piston-side lifting function (plus movement). The boom doesn't require sophisticated metering qualities as lifting speed often is controlled by the speed of the engine. Handling and stopping heavy loads when lowering are, however, very essential. A float position on the boom-down function is very common and is often arranged as a forth position in the directional control valve.

Fig. Load-4 Large loader

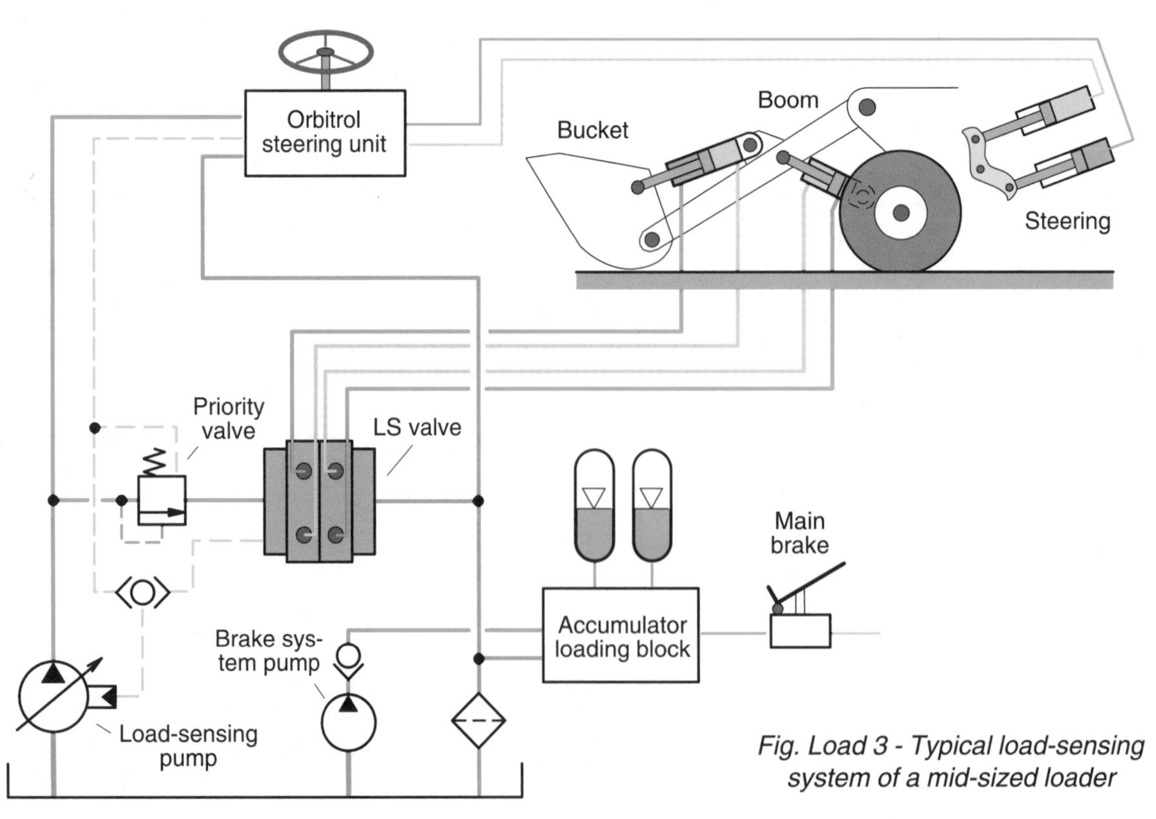

Fig. Load 3 - Typical load-sensing system of a mid-sized loader

Boom suspension

This function is common on large loaders. It increases operator comfort and operating maneuverability and reduces machine fatigue. An accumulator is fitted to the piston side of the lift cylinders; the opposite, rod, side is vented to tank.

Bucket

The bucket tilting function uses one or two cylinders connected to a linkage system. It permits "parallel lifting" of the bucket.

Different philosophies are followed when it comes to linkage and geometry, but all are compromises in trying to achieve the best possible function. The most common is the "Z" linkage; "Tp" and "parallel linkage" are others.

Auxiliary functions

Wheel loaders are mostly equipped with a third valve block for auxiliary tools, bucket locks and other functions. Attachments like a rotating brush and a snow blower are common.

Valve control system

The most common directional control valve is the constant flow, open-center (CFO) valve supplied from a fixed displacement pump. Gear pumps are used on small loaders up to 60 kW *(80 hp)*; vane pumps on large.

Large machines are equipped with pilot operated DCVs, actuated by hydraulic joysticks or linear levers. Hydraulic pilot control is reliable, requires only low operating forces, and can easily be equipped with solenoid detents for semiautomatic operation. Linear levers and joysticks cover about 50% each of the market.

Smaller loaders have direct operating, manual levers, mostly of the joystick type which are installed directly on the valve.

Steering

All articulated wheel loaders are equipped with a steering wheel and, mostly, hydrostatic steering of the orbitrol type.

On large machines with high flow demands, the steering system is also equipped with a flow amplifier. The system is supplied by a separate pump or has first priority from a shared pump. The largest machines above 60 kW *(80 hp)* have variable displacement pumps and load-sensing steering units.

Hydrodynamic transmission

Large loaders above approx. 75 kW *(100 hp)* are to a large extent equipped with hydrodynamic transmissions and automatic gearboxes where low cost, simplicity and high functional reliability are important features. The transmission also allows very good traction control, and energy efficiency at high speeds is also good.

Hydrostatic transmission

Most small and midsized loaders are equipped with a hydrostatic transmission and a hydraulic, automotive type, drive control system. The hydrostatic transmission has several advantages such as "spin-steer" function and an energy efficient anti-stall control. In addition, it eliminates the need for a clutch/brake system.

Appendix C - Hydraulic fluid filter selection

Lubrication and wear

The pressures produced in hydraulic systems can be very high. In some gas compressor lube systems, the pressure can rise as high as 4000 bar (*60,000 psi*). In mobile hydraulic systems, an operating pressure of 350 bar (*5000 psi*) is common.

Such pressures require precise machining to produce small mechanical clearances between moving parts of system components.

For example (fig. C-1), one type of lubricant metering valve uses a piston and bore matched and fitted within a mechanical tolerance of ± 0.005 mm (5 µm), which corresponds to ± 0.0002" (two ten thousandths of an inch). In electrohydraulic devices, clearances can be less than three µm, or 117 millionths of an inch (table C-1).

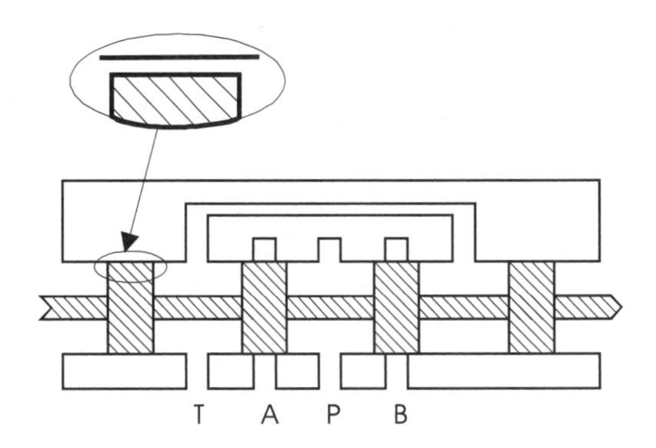

Fig. C-1 Hydraulic components have small clearances.

Ideally, a lubricating film is thick enough to completely fill the clearance between moving parts and always keep them separated. This condition is known as 'hydrodynamic' or 'full film' lubrication and results in very little wear. The thickness of the film depends on fluid viscosity, applied load, and relative speed of the two surfaces.

Table C-1 Typical component clearances

Component	µm
Slide bearings (vane pump)	0.5
Tip of vane (control valve)	0.5
Roller element bearings	0.1 - 1
Hydrostatic bearings	1 - 25
Gears	0.1 - 1
Gear pump (tooth tip to case)	0.5 - 5
Vane pump (vane tip)	0.5 - 1
Piston pump (piston to bore)	5 - 40
Servo valves (spool sleeve)	1 - 4

In many hydraulic and lubricating systems, mechanical loads (pressures) are so high that they squeeze the lubricant into a very thin film, less than one µm thick. This is elastohydrodynamic (EHD), or thin film lubrication. If loads become high enough, the film will be punctured by the asperities of the two moving parts (fig. C-2). This is called 'boundary' lubrication.

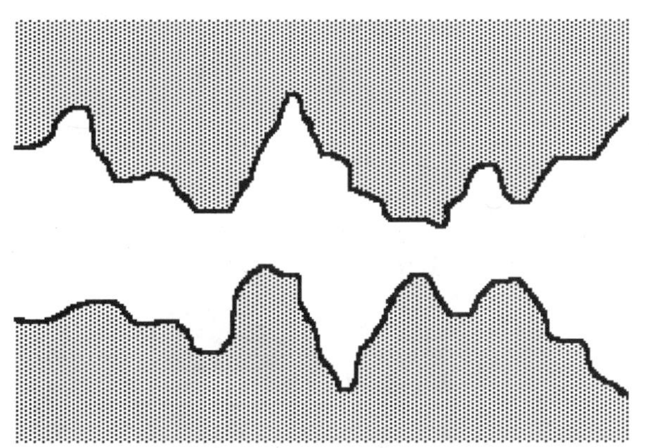

Fig. C-2 If loads become high enough, the film will be punctured by the asperities of the two moving parts.

Many hydraulic systems operate at least part of the time with boundary lubrication. It is during boundary lubrication that asperities of moving surfaces contact each other, and may be torn away from the parent material. These particles spread throughout the fluid system if not removed by filtration.

Effects of particle generated wear and interference
As pointed out earlier in this text, fluid particles can lead to additional wear. Also, they enter the fluid from a variety of sources (fig. C-3). These particles can interact with moving parts as two-body or three-body wear mechanisms.

For example, if a particle is about the size of the space between parts, it can enter and interact mechanically with surface asperities.

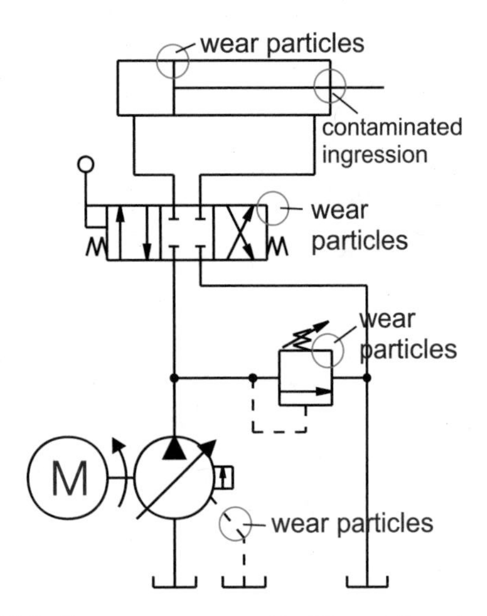

Fig. C-3 Some sources of unfiltered flow in a hydraulic circuit

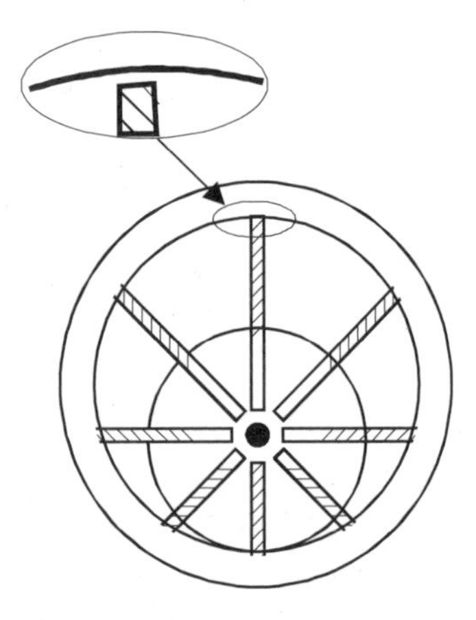

Fig. C-4 Clearances will increase with time due to component wear.

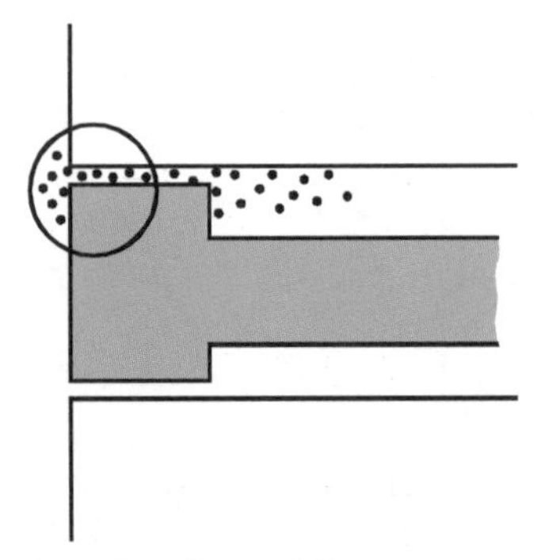

Fig. C-5 Small soft particles can cause silting interference; this contaminant could cause a valve to stick in place.

This three-body mechanism will accelerate wear under boundary lubrication conditions by promoting the tearing away of asperities.

In pumps (fig. C-4), wear may be detected first as reduced flow rate. This is due to increased clearance dimensions resulting from erosive wear. This condition is sometimes called increased slippage, and means that the pump is less efficient than it once was. When pump flow rate decreases, the fluid system may become sluggish.

Hydraulic cylinders move slower, and metering devices deliver less flow. Pressure at some locations in the system also may decrease. At some point, there may be a sudden catastrophic failure of the pump. In extreme cases, this can occur within a few minutes after initial start-up of the system

A machine tool spindle is another type of rotating device that is very sensitive to the amount and cleanliness of lubricant. If excessive wear begins, precisely balanced spindle bearings can be destroyed in a matter of minutes.

In valves, wear may cause increased leakage. The effect this has on the system depends on the type of valve involved. For example, in valves used to control flow, increased leakage usually means increased flow. In valves designed to control pressure, increased leakage may reduce the circuit pressure set by the valve.

Not all particles are a result of wear. Many enter from outside a system that has worn seals or lacks a proper filter/breather on the reservoir. With higher concentrations of airborne contaminants, more particles are likely to be drawn into a system. These particles can be either hard or soft depending on their origin.

Without adequate filtration, a large number of soft particles may build up in the fluid. Although they may not contribute to wear, a condition called silting can occur (fig. C-5). This happens when the particles are small and fill the space between parts. This can cause valves and variable flow pump parts to become sticky and operate erratically.

A particularly dangerous situation occurs when valves stick in place, because this can result in loss of control of the hydraulic system.

Economic consequences of downtime

Most hydraulic systems are installed on heavy duty production equipment. Because of the equipment's high cost, it is expected to maintain high output with minimum down time. Downtime can easily cost the equipment owner thousands of dollars per hour. Therefore, incurring filtration costs to increase the time between failures is a good investment. Still, this benefit has to be properly balanced against the cost of replacement filter elements, and downtime required to service filtration equipment.

Careful filtration system design and component selection (fig. C-6) will help to minimize maintenance downtime and filter service costs. This objective will be achieved through one or more of these intermediate goals:

- Meet or exceed minimum standards for fluid contamination (cleanliness)
- Reduce fluid system/component maintenance
- Improve the performance of the system and/or its fluid
- Improve the quality of the final product by avoiding faulty machine operation
- Enhance safety/reduce risk of injury to operating personnel (for example, by eliminating the need for maintenance on or around operating equipment).

Filter selection considerations

Step 1. Determine fluid cleanliness required in critical circuit branches

Focus attention on circuit branches where critical components are located. Those components are the ones most sensitive to contaminants, and typically have the highest replacement cost. The particle concentration in those branches should be at a level the user and/or system manufacturer believes will give acceptable component life.

The concentration should be stated for a location immediately downstream of the filter. It is expressed as the number of particles larger than a specified size per volume of fluid, or as an ISO code number.

Table C-2, 'ISO cleanliness levels' shows particle counts at two, five, and 15 µm diameters for various ISO code numbers.

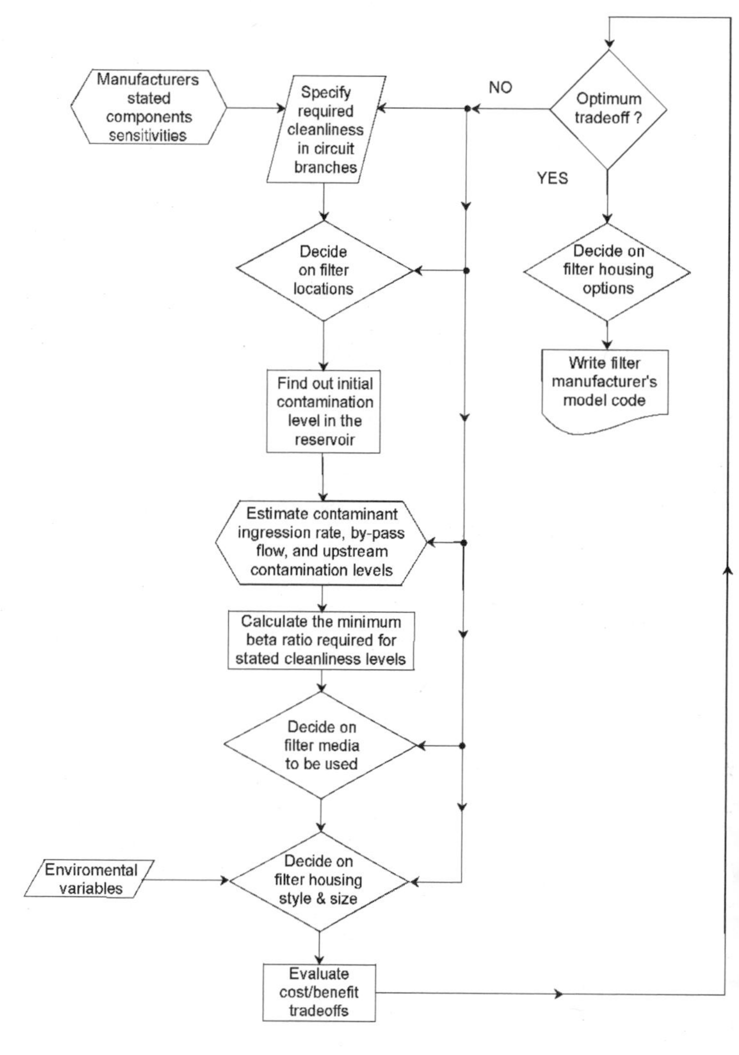

Fig. C-6 Filter selection flow chart

Table C-2 ISO cleanliness levels

ISO Code	No. of particles per ml (millilitre) • 2 µm	• 5 µm	• 15 µm
23/21/18	80 000	20 000	2 500
22/20/18	40 000	10 000	2 500
22/20/17	40 000	10 000	1 300
22/20/16	40 000	10 000	640
20/18/15	10 000	2 500	320
19/17/14	5 000	1 300	160
18/16/13	2 500	640	80
17/15/12	1 300	320	40
16/14/12	640	160	40
16/14/11	640	160	20
15/13/10	320	80	10
14/12/9	160	40	5
13/11/8	80	20	2.5
12/10/8	40	10	2.5
12/10/7	40	10	1.3
12/10/6	40	10	0.64

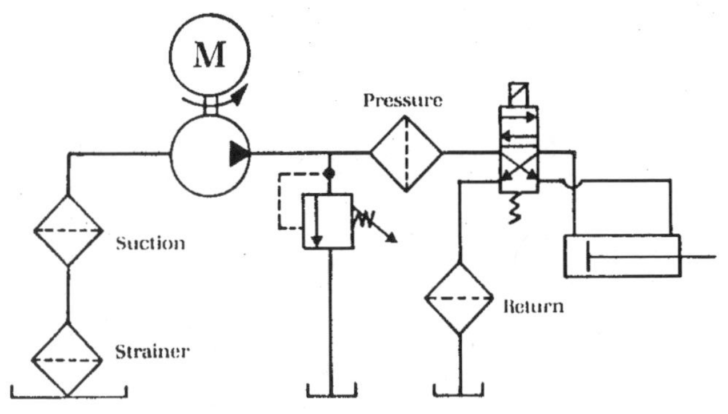

Fig. C-7 Potential filter locations

Table C-3 Filter locations

Location	Advantages	Disadvantages
Suction (externally mounted)	Last chance protection for the pump	Must use relatively coarse media, or large housing size to keep pressure drop low. Cost is relatively high
	Much easier to service than a sump strainer	Does not protect downstream components from pump wear debris
Pressure	Protects downstream components from pump wear	Housing is relatively expensive beacause it must handle full system pressure
	Provides last protection of specific components	Does not catch wear debris from downstream working components
	Can use high efficiency filter media with minimum consideration for pressure drop	Does not catch particles entering the system through worn cylinder rod seals and other working components
Return	Catches wear debris from components and dirt entering through worn cylinder rod seals, before it enters the reservoir	Does not protect specific components
	Lower pressure ratings result in lower costs	No protection from pump generated contamination
	May be in-line or in-tank for easier installation	Return line flow surges may reduce filter performance
Off-line	Servicing is possible without loss of production	No direct component protection
	Fixed flow rate eliminates surges, allowing for optimal element life and performance	Relative initial cost is high
	Fluid cooling may be easily incorporated	Requires more space than a single filter

Step 2. Make a tentative decision regarding filter locations (fig. C-7 and table C-3)

The advantages and disadvantages for individual filter locations have been discussed earlier in this chapter. Other points are presented below. These points use the component sensitivity analysis carried out in Step 1.

Pump protection

Frequently, the pump is one of the most expensive components in a hydraulic system. At the same time, the pump is very sensitive to contaminants. Therefore, it is important to maintain particle concentration at a level recommended by the pump manufacturer.

The pump should also be protected from large debris which can enter the reservoir during system construction and maintenance operations.

In many hydraulic systems, these requirements can be satisfied with a suction strainer and a return line filter. The suction strainer takes care of the large debris, and the return line filter removes ingress particles before they can enter the reservoir.

High pressure components

The outlet side of the pump, and many components downstream of the outlet, experience full system pressure. Generally speaking, higher pressure equates to higher mechanical loading and accelerated wear, particularly when hard particles are present in the fluid.

If a component was manufactured with precision tolerances and a fine surface finish, it will be even more sensitive to particle contamination. A good example of this type of component is an electrohydraulic valve.

Even bearings and gears in mechanical power transmission systems can have precision surface finishes and tight manufacturing tolerances.

These precision components should be protected with a pressure line filter upstream of their location. The number and locations of such components will dictate where pressure filters are placed in the circuit.

If the sensitive components are located in a single branch of the circuit, a 'last chance' filter (fig. C-8) can be placed downstream of the first one to protect that branch. This location has the added advantage of requiring a filter sized for only the flow in that branch.

On the other hand, the precision components may be scattered throughout various branches of the system. This may require a pressure filter located near the pump outlet to filter the total system flow before it splits into the various branch flows.

Other options

In some cases either just a pressure filter, or just a return filter could be used without sacrificing oil quality. In other cases, the contaminant ingression rate may be so high that even with both pressure and return line filters in the circuit, the fluid will not be clean enough.

In that case, a separate off-line filter loop is needed for additional particle removal.

Step 3. Determine initial contamination level

This step is useful for two reasons:

1. It makes the user aware of possible damage from contamination before the system is first started. This contamination can be the result of poor or nonexistent system flushing, or because relatively contaminated fluid was added to the reservoir.
2. The initial contamination level is one of the factors that determines how long it takes for the fluid to reach an equilibrium contamination level.

It is good practice to do a particle count on fresh fluid to be added to the reservoir. This fluid should be prefiltered if necessary (fig. C-9) to reduce the contamination to the target level set for the system.

Likewise, a reservoir sample from a newly purchased system should be analyzed before starting the equipment. If necessary, the reservoir fluid can be cleaned up with a portable filtration unit operating as an off-line system.

Step 4. Estimate contaminant ingression rate

For large hydraulic systems, the contaminant ingression rate can be quite high (table C-4).

It may be possible to create a more accurate estimate of the contaminant ingression rate by using a filtration model. The ingression rate (can be modeled by using a reference system similar in

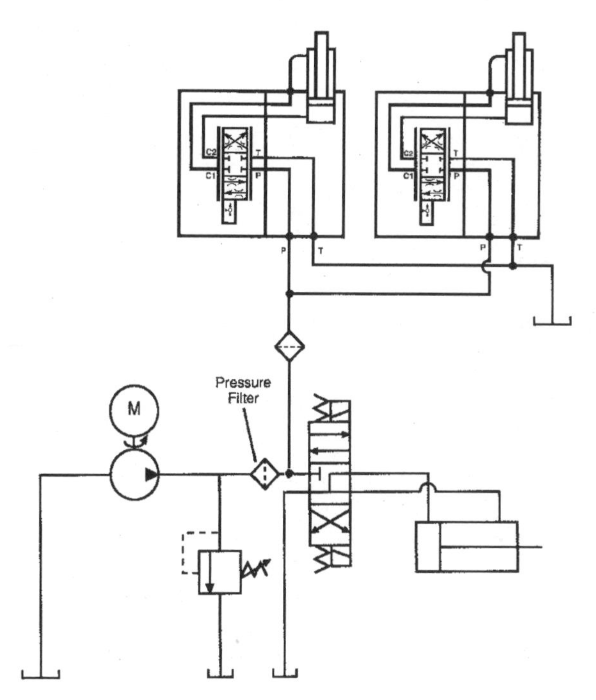

Fig. C-8 A 'last chance' filter is placed downstream of the pressure filter in the supply lines to the servo valves.

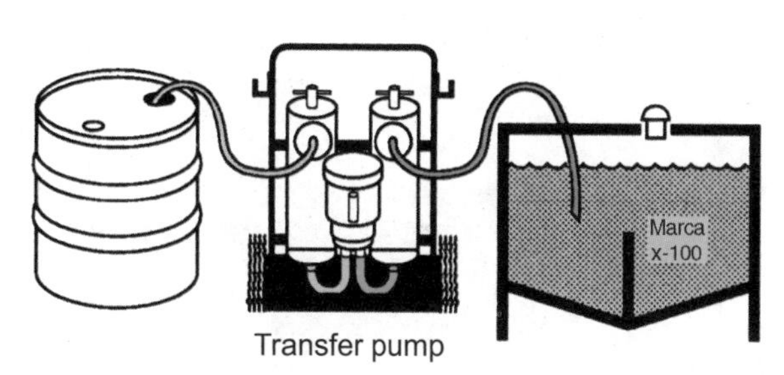

Fig. C-9 Prefilter system fluid as the reservoir is filled.

Table C-4 Ingression rates for typical mobile systems

System	Ingression rate*
Earthmoving and off-highway (extreme conditions)	$10^9 - 10^{10}$
Farm and other mobile equipment	$10^8 - 10^9$

* Number of particles greater than 10 µm per minute (ingressed into the system from all sources).

Table C-5 Fluid cleanliness required for typical hydraulic components

Component	ISO classification
Servo control valve	16/14/11
Vane and piston pump/motor	18/16/13
Directional and pressure control valve	18/16/13
Gear pump/motor	19/17/14
Flow control valve; cylinder	20/18/15

Table C-6 How bypass degrades element performance

Amount of bypass	Efficiency multiplier
0 %	1.00
1 %	0.99
10 %	0.90
50 %	0.50
90 %	0.10

size, components, and operating conditions, to the one being designed.

A particle count obtained from a fluid sample from the reference system is combined with other known filter performance data to calculate the ingression rate. This rate is then used as the estimate for the new system design.

Some approximate rates for various operating environments are listed in table C-5.

Step 5. Calculate unfiltered flow percentage

Flow which bypasses the filter(s) reduces particle removal efficiency (table C-6). In other words, contaminants not exposed to filter media can not be removed.

There are four common reasons for bypass flow:

1. Circuit design, either inadvertent or intentional, allows certain branch flows to circumvent the filter(s).
2. Pumps and other components may have drain ports, called case drains, that return fluid to the reservoir. This flow is usually unfiltered and can have a high concentration of metallic wear debris. A typical pump has a case drain flow rate less than 5% of its maximum output.
3. A filter equipped with a bypass valve is allowed to operate with the valve open (due to a clogged element) for a significant part of the system duty cycle.
4. Imperfect seals around the filter element and or bypass valve allow some of the flow to bypass the filter media. This may not be much of a problem when the filter is new, but can increase with age. Ageing of elastomeric seals is accelerated as operating temperature increases.

Each of these potential sources of bypass should be estimated or measured. To measure leakage through the bypass valve or around element seals, the filter can be tested with a solid element shaped cylinder. This cylinder replaces the standard element and blocks flow through the filter. Any flow that takes place must be around the bypass valve or element.

Typically, leakage is measured at a range of pressure drops up to the bypass setting. Then average leakage flow can be calculated as a percentage of total flow.

Leakage flow from pump or valve drain lines can be measured by directly measuring flow through those lines.

Step 6. Estimate contaminant level upstream of the filters

It is necessary to estimate or measure upstream concentration in order to find the 'ß ratio' required to achieve the desired downstream concentration. Upstream contaminant levels can be estimated either manually or through computer software.

Manual methods of estimating upstream particle concentration require the use of ballpark ingression rates for the environment such as provided previously in table C-4.

Then the individual component ingression rates due to wear are added to the environmental ingression rates (table C-7).

Computer software can do this for all branches of the circuit, taking into account any particles removed by filters. Such a process is not practical for manual calculation. (The details of the computer software process are beyond the scope of this course; contact your filter supplier about current software.)

Step 7. Choose the filter types and media

The goal now is to find media with the 'ß ratio' required to remove enough of the upstream contamination (predicted in the previous step) in order to achieve the acceptable level of downstream contaminant.

In addition to the locations chosen for the filters, some other things that must be considered are:

- Fluid compatibility of media, seals, and hardware
- Suitability of the media and element construction for system operating conditions
- Filter housing pressure ratings.

Fluid compatibility

Media, seals, housing, and hardware must be compatible with the base fluid and its additives. These materials must be compatible over the entire operating range of temperatures and pressures. Check with both fluid and filter suppliers to determine which material will work well together.

Most hydraulic applications will allow use of media made from cellulose, 'Fiberglass', or polyester fi-

Table C-7 Contaminant generation due to wear

Component	10 μm particles/min at 4 l/min (*1 gpm*)	10 μm particles/min at 200 l/min* (*50 gpm*)*
Directional valve	3.5	180
Piston pump	3 100	1 600 000
Gear pump	3 400	170 000
Vane pump	12 000	11 000*
Cylinder	Stroked rod area	

* 125 mm (*5"*) cylinder, 64 mm (*2 1/2"*) rod, 300 mm (*12"*) stroke, 0.2 m/s (*8"/s*) velocity, equal extend and retract times, 58 % duty cycle

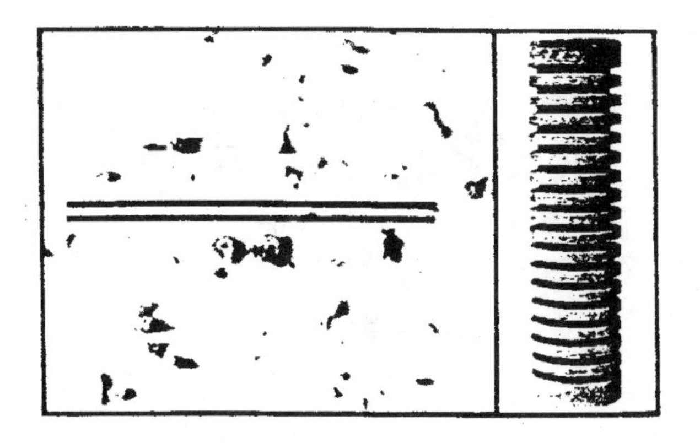

Fig. C-10 Possible media migration must be considered when selecting a filter.

bers. If minimum media migration is desired, 'Fiberglass' is usually chosen. (Media migration, fig. C-10, refers to small pieces of media entering the flow stream. These are subsequently recaptured along with other particles in the fluid, but may cause problems if a large number remain in solution.)

Certain hardware surface treatments may not be compatible with some exotic fluids. If the fluid is an unusual type, check hardware compatibility with the fluid and filter suppliers.

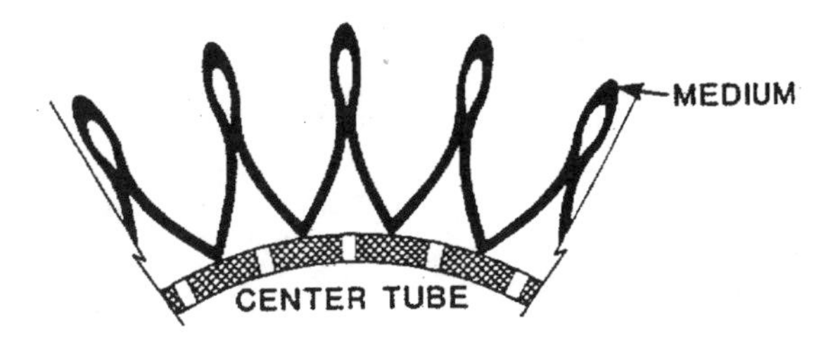

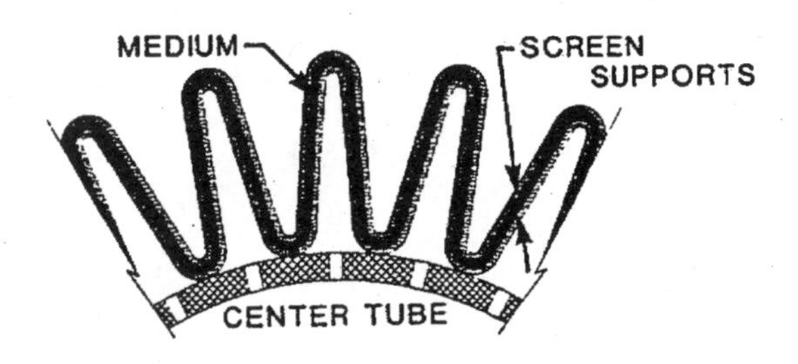

Fig. C-11 Pleat pinching and how to avoid it
(Courtesy E. C. Fitch,
Fluid Contamination Control, FES, Inc.)

Media and element construction

Normally, media will be pleated in cartridge filters to maximize surface area and dirt holding capacity. Still, at high flow rates and when there is surge or cyclic flow, pleats can be forced together (fig. C-11). This cuts down on the effective surface area and service life of the element.

For these types of applications, it may be necessary to select an element which has its media sandwiched with wire mesh. The mesh prevents the pleats from touching each other, even under severe flow conditions.

The offset is that not as many pleats can be fitted into the same element diameter. The wire mesh backing takes up some of the available space.

Pressure rating

If the filter is a non-bypassing type, then the element must be constructed to withstand full system pressure. In such a filter, more and more of the system pressure builds up across the element as it is clogged with contaminant.

This goes on until the element support structure collapses, the media ruptures, or contaminant is forced through the media by the pressure. Even if the filter has a bypass valve, the element collapse rating must be higher than the pressure at which the bypass valve starts to open. Otherwise the element faces the same structural problems as in non-bypassing filters.

For non-bypassing filters, elements must have added features which allow them to withstand high differential pressure. These features may include one or more of the following:

- Heavier support material (typically constructed of heavy gauge steel) located on the down stream side of the media (fig. C-12)

Fig. C-12 Support cylinders increase an element's differential pressure rating.

- A wrapping of relatively fine secondary filter media around the support cylinder to stop any particles being forced through the primary media
- Relatively fine wire mesh backing on both sides of the primary filter media
- Multiple layers of primary filter media, commonly with a slightly coarser grade on the upstream side.

Filter assembly selection procedures

Thus far, filter locations, media, and preliminary selection of acceptable filter types have been made. The remaining four steps in the filter and media selection process result in filter housing and element specifications.

The manufacturer's specifications should cover all the items listed in table C-8.

Frequently, the final selection is a compromise, even when a filter design is modular and flexible. It simply isn't cost effective for a manufacturer to offer every possible combination of features. Also, the initial choice of features may result in a price which is too high.

Therefore, the steps described below may have to be repeated a few times until the best combination of features and price is found.

Step 8. Select the filter housing and element

The selection can be broken down into three parts:

a) Selection of a housing pressure rating
b) Sizing the housing
c) Selecting other physical features required.

The objective is to get only what the fluid system needs.

a) Select a housing style that meets the pressure rating requirements of the application. Housing pressure ratings for hydraulic filters are covered by various standards. Two pressure ratings should be considered. They are the housing's static and fatigue ratings. These ratings can be found in the manufacturers' catalogs or by contacting them at a local distributor's.
b) Select a housing size for the circuit flow rate. After finding housings to meet pressure requirements, the next step is to identify those that will handle the maximum flow in that circuit. Analyzing the flow/pressure drop curves of the proposed housings and elements does this.

Table C-8 Essential filter assembly specifications

- Housing/bowl/element design features, including options
- Materials of construction
- Weight and how measured (wet, dry, or shipping weight)
- Principle dimensions, particularly those needed for installation, and clearance needed to change the element
- Flow/pressure-drop characteristics at the operating viscosity of the fluid for the housing, element size and media selected
- Element dirt holding capacity or cost per gram of contaminant captured prior to element replacement
- Conditions under which the filter media ß ratios were specified or tested
- Operating and environmental limits.

Table C-9 *Important physical characteristics of filters*

- Flow/pressure drop characteristics
- Outside dimensions/element removal clearance
- Bowl and housing style
- Port sizes and types
- Materials of construction
- Mounting methods

Table C-10 Annualized cost of filtration

- Annual depreciation of original filter cost
- Replacement element cost multiplied by the number of times it has to be replaced annually
- Annual labor cost associated with element replacement (usually a fully burdened hourly rate times the total element service hours for the year)
- Other routine maintenance costs associated with the filter, such as inspections, etc.

Fig. C-13 Hydraulic and lubrication filters

c) Select a housing that meets mounting, dimensional, and other physical requirements

In most applications, flow/pressure drop characteristics outweigh other physical requirements of a filter. Therefore, sizing the housing was done before selecting other features. If another characteristic is more important than housing size, then that characteristic should be used first to narrow the choice of acceptable filters.

The physical features of greatest importance are listed in table C-9. Additional considerations would include:

1. Materials of construction
2. Outside dimension
3. Port sizes and types
4. Bowl and housing style
5. Mounting methods.

Step 9. Examine cost trade-offs.

Look at the cost benefit trade-offs for various types of filter, media, sizes, and elements. Be sure to consider internal maintenance cost associated with element service, as well as the price of the element and the cost per gram of dirt removed; refer to table C-10, 'Annualized cost of filtration'.

Step 10. Repeat steps 2 through 9

Repeat the selection process and cost/benefit analysis for other combinations of filters, locations, media, sizes, contaminant concentrations, etc. The idea is to find the best cost/benefit trade-off among the list of acceptable filters and media.

Table C-11 shows advantages and disadvantages for some filter configurations.

Step 11. Specify optional items.

For the type(s) of filter(s) selected, specify options such as indicator, drain ports, magnets, air bleed, accessory kits, etc. These options add convenience and cut down on maintenance time. Still, this step was saved until last because other selection criteria were considered to be more important.

Some of the more common accessories, options, and their benefits include the following:

- **Element condition indicator**
  The indicator senses differential pressure across the element to let the user know when service is needed.

- **Differential pressure switch**
  The switch senses differential pressure across the element and closes switch contacts when service is needed; it allows transmission of this information to a remote location.
- **Internal magnet**
  A magnet collects ferrous particles and extends the life of the element.
- **Mounting flange and bracket kits**
  These parts allow filter installation without resorting to weldments; filter may be removed as necessary for machine maintenance, overhaul, etc.
- **Drain port**
  The drain port allows filter to be drained on bowl-up designs; a more convenient and more complete removal of trapped contaminant is possible.
- **Air bleed port**
  A bleed port allows removal of air from filter housing after servicing; it helps prevent air entrainment in the fluid.

Chapter Summary

This chapter presented a detailed explanation of each step required for hydraulic and lube oil filter selection. The selection process must also include tentative decisions about filter locations and media. Then, actual filter and element assemblies should be selected based upon physical characteristics required by the application. It is recommended that the filtration system designer calculate costs associated with the filters and elements selected. Then the entire process should be repeated for other filters and elements to find the combination having the lowest cost/benefit ratio.

On the next page, "Table C-12 Filter media selection - hydraulic systems" provides additional guidelines for the cleanliness level, media efficiency and placement of system filters for various component types.

Table C-11 Filter configurations - Advantages and disadvantages

Configuration	Advantages	Disadvantages
Bowl-up top cover removal for service	Easier to remove a cover than entire bowl particularly on a heavy pressure filter	Some contaminant may remain in bowl, particularly if elements has outside-to-inside flow
Bowl-down removal for service	Any contaminant in bowl can be easily flushed out	Bowl removal and replacement can be difficult on heavy pressure filters
Reservoir mounted suction and return line filters, usually with top access cover	Saves space, easy to install and maintain, low cost,other advantages of top access filters	Same as for other top access filters ; will require shutoff valve if installed below fluid surface.
Line mounted suction filter	More convenient than reservoir mounted for installation and service in flooded suction applications	Takes up space outside reservoir; may have smaller housing and element than reservoir mouted filter; shorter service life
Spin-on style return line filter	Low cost	Limited pressure rating and convenience features
Duplex filter	Allows element servicing while machine is running	Expensive; can cost twice as much as a comparable single filter
Outside-to-inside element flow path	Less prone to contaminant re-entrainment for a given element size and pressure rating; uses a smaller diameter; less costly media support cylinder than opposite flow direction element	Contaminat is on the outside of the element where it can be knocked off and back into a bowl; may require perforated outer cylinder or wire mesh media backing to prevent pleat collapse
Inside-to-outside element flow path	Contaminant is on the inside of the element and less likely to get knocked off and back into the filter	Media support cylinder is larger and more costly for a given element size and pressure rating.

Table C-12 *Filter media selection - hydraulic systems*

Component type	System pressure bar *(psi)*	Suggested cleanliness code	Media efficiency $\beta_x > 200$	Number of filter placements	Minimum filter placements
Servo valves	< 70 *(1000)*	16/14/12	2	1	P
	“ “	“ “ “	5	2	P and R
	70-200 *(1000-3000)*	15/13/11	2	1.5	P and O
	> 200 *(3000)*	15/12/10	2	2	P and R
Proportional valves	< 70 *(1000)*	17/15/13	2	1	P
	“ “	“ “ “	5	1.5	P and O
	“ “	“ “ “	10	2.5	P, R and O
	70-200 *(1000-3000)*	17/14/12	2	1	P
	“ “	“ “ “	5	2	P and R
	> 200 *(3000)*	16/14/11	2	1.5	P and O
	“ “	“ “ “	5	2.5	P, R and O
Variable displacement pumps	< 70 *(1000)*	18/16/14	5	1	P or R
	“ “	“ “ “	10	2	P and R
	70-200 *(1000-3000)*	17/16/14	2	0.5	O
	“ “	“ “ “	5	1.5	P or R, and O
	“ “	“ “ “	10	2.5	P, R, and O
	> 200 *(3000)*	17/15/13	2	1	P or R
	“ “	“ “ “	5	2	P and R
Vane and fixed displ. pumps; cartridge valves	< 70 *(1000)*	19/17/15	5	0.5	O
	“ “	“ “ “	10	1.5	P or R, and O
	70-200 *(1000-3000)*	18/17/14	5	1	P or R
	“ “	“ “ “	10	2	P and R
	> 200 *(3000)*	18/16/13	5	1.5	P or R, and O
	“ “	“ “ “	10	2.5	P, R and O
Gear pumps, flow controls; cylinders	< 70 *(1000)*	20/18/16	10	1	P, or R
	“ “	“ “ “	20	2.5	P, R and O
	70-200 *(1000-3000)*	19/17/15	10	1.5	P or R, and O
	> 200 *(3000)*	19/17/14	5	0.5	O
	“ “	“ “ “	10	1.5	P or R, and O

Index

Symbols

A

D

E

F

G

H

I

N

O

P

Q

R

S

T

U

V

W

Z